T0221013

Causality in a Social World

Causality in a Social World

Moderation, Meditation and Spill-over

Guanglei Hong

Department of Comparative Human Development,
University of Chicago, USA

This edition first published 2015
© 2015 John Wiley & Sons, Ltd

Registered Office
John Wiley & Sons, Ltd, The Atrium, Southern Gate, Chichester, West Sussex, PO19 8SQ, United Kingdom

For details of our global editorial offices, for customer services and for information about how to apply for permission to reuse the copyright material in this book please see our website at www.wiley.com.

Library of Congress Cataloging-in-Publication Data applied for.

ISBN: 9781118332566

A catalogue record for this book is available from the British Library.

Set in 10/12pt Times by SPi Global, Pondicherry, India

1   2015

Contents

Part III   Mediation 211

9   Concepts of mediated treatment effects and experimental designs for investigating causal mechanisms 213

Preface

A scientific mind in training seeks comprehensive descriptions of every interesting phenomenon. In such a mind, data are to be pieced together for comparisons, contrasts, and eventually inferences that are required for understanding causes and effects that may have led to the phenomenon of interest. Yet without disciplined reasoning, a causal inference would slip into the rut of a "casual inference" as easily as making a typographical error.

As a doctoral student thirsting for rigorous methodological training at the University of Michigan, I was very fortunate to be among the first group of students on the campus attending a causal inference seminar in 2000 jointly taught by Yu Xie from the Department of Sociology, Susan Murphy from the Department of Statistics, and Stephen Raudenbush from the School of Education. The course was unique in both content and pedagogy. The instructors put together a bibliography drawn from the past and current causal inference literature originated from multiple disciplines. In the spirit of "reciprocal teaching," the students were deliberately organized into small groups, each representing a mix of disciplinary backgrounds, and took turns to teach the weekly readings under the guidance of a faculty instructor. This invaluable experience revealed to me that the pursuit of causal inference is a multidisciplinary endeavor. In the rest of my doctoral study, I continued to be immersed in an extraordinary interdisciplinary community in the form of the Quantitative Methodology Program (QMP) seminars led by the above three instructors who were then joined by Ben Hansen (Statistics), Richard Gonzalez (Psychology), Roderick Little (Biostatistics), Jake Bowers (Political Science), and many others. I have tried, whenever possible, to carry the same spirit into my own teaching of causal inference courses at the University of Toronto and now at the University of Chicago. In these courses, communicating understandings of causal problems across disciplinary boundaries has been particularly challenging but also stimulating and gratifying for myself and for students from various departments and schools.

Under the supervision of Stephen Raudenbush, I took a first stab at the conceptual framework of causal inference in my doctoral dissertation by considering peer spill-over in school settings. From that point on, I have organized my methodological research to tackle some of the major obstacles to causal inferences in policy and program evaluations. My work addresses issues including (i) how to conceptualize and evaluate the causal effects of educational treatments when students' responses to alternative treatments depend on various features of the organizational settings including peer composition, (ii) how to adjust for selection bias in evaluating the effects of concurrent and consecutive multivalued treatments, and (iii) how to conceptualize and analyze causal mediation mechanisms. My methodological research has been driven by the substantive interest in understanding the role of social institutions—schools in particular—in shaping human development especially during the ages of fast growth (e.g., childhood, adolescence, and early adulthood). I have shared methodological advances with a broad audience in social sciences through applying the causal inference methods to prominent substantive issues such as grade retention, within-class grouping, services for English language learners, and welfare-to-work strategies. These illustrations are used extensively in this book.

Among social scientists, awareness of methodological challenges in causal investigations is higher than ever before. In response to this rising demand, causal inference is becoming one of the most productive scholarly fields in the past decade. I have had the benefit of following the work and exchanging ideas with some fellow methodologists who are constantly contributing new thoughts and making breakthroughs. These include Daniel Almirall, Howard Bloom, Tom Cook, Michael Elliot, Ken Frank, Ben Hansen, James Heckman, Jennifer Hill, Martin Huber, Kosuke Imai, Booil Jo, John R. Lockwood, David MacKinnon, Daniel McCaffrey, Luke Miratrix, Richard Murnane, Derek Neal, Lindsey Page, Judea Pearl, Stephen Raudenbush, Sean Reardon, Michael Seltzer, Youngyun Shin, Betsy Sinclair, Michael Sobel, Peter Steiner, Elizabeth Stuart, Eric Thetchen Tchetchen, Tyler VanderWeele, and Kazuo Yamaguchi. I am grateful to Larry Hedges who offered me the opportunity of guest editing the *Journal of Research in Educational Effectiveness* special issue on the statistical approaches to studying mediator effects in education research in 2012. Many of the colleagues whom I mentioned above contributed to the open debate in the special issue by either proposing or critically examining a number of innovative methods for studying causal mechanisms. I anticipate that many of them are or will soon be writing their own research monographs on causal inference if they have not already done so. Therefore, readers of this book are strongly recommended to browse past and future titles by these and other authors for alternative perspectives and approaches. My own understanding, of course, will continue to evolve as well in the coming years.

One has to be ambitious to stay upfront in substantive areas as well as in methodology. The best strategy apparently is to learn from substantive experts. During different phases of my training and professional career, I have had the opportunities to work with David K. Cohen, Brian Rowan, Deborah Ball, Carl Corter, Janette Pelletier, Takako Nomi, Esther Geva, David Francis, Stephanie Jones, Joshua Brown, and Heather D. Hill. Many of my substantive insights were derived from conversations with these mentors and colleagues. I am especially grateful to Bob Granger, former president of the William T. Grant Foundation, who reassured me that "By staying a bit broad you will learn a lot and many fields will benefit from your work."

Like many other single-authored books, this one is built on the generous contributions of students and colleagues who deserve special acknowledgement. Excellent research assistance from Jonah Deutsch, Joshua Gagne, Rachel Garrett, Yihua Hong, Xu Qin, Cheng Yang, and Bing Yu was instrumental in the development and implementation of the new methods presented in this book. Richard Congdon, a renowned statistical software programmer, brought revolutionary ideas to interface designs in software development with the aim of facilitating users' decision-making in a series of causal analyses. Richard Murnane, Stephen Raudenbush, and Michael Sobel carefully reviewed and critiqued multiple chapters of the manuscript draft. Additional comments and suggestions on earlier drafts came from Howard Bloom, Ken Frank, Ben Hansen, Jennifer Hill, Booil Jo, Ben Kelcey, David MacKinnon, and Fan Yang. Terese Schwartzman provided valuable assistance in manuscript preparation. The writing of this book was supported by a Scholars Award from the William T. Grant Foundation and an

Institute of Education Sciences (IES) Statistical and Research Methodology in Education Grant from the U.S. Department of Education. All the errors in the published form are of course the sole responsibility of the author.

Project Editors Richard Davis, Prachi Sinha Sahay, and Liz Wingett and Assistant Editor Heather Kay at Wiley, and project Managers P. Jayapriya and R. Jayavel at SPi Global have been wonderful to work with throughout the manuscript preparation and final production of the book. Debbie Jupe, Commissioning Editor at Wiley, was remarkably effective in initiating the contact with me and in organizing anonymous reviews of the book proposal and of the manuscript. These constructive reviews have shaped the book in important ways. I cannot agree more with one of the reviewers that "causality cannot be established on a pure statistical ground." The book therefore highlights the importance of substantive knowledge and research designs and places a great emphasis on clarifying and evaluating assumptions required for identifying causal effects in the context of each application. Much caution is raised against possible misusage of statistical techniques for analyzing causal relationships especially when data are inadequate. Yet I maintain that improved statistical procedures along with improved research designs would greatly enhance our ability in attempt to empirically examine well-articulated causal theories.

Part I
OVERVIEW

1

Introduction

According to an ancient Chinese fable, a farmer who was eager to help his crops grow went into his field and pulled each seedling upward. After exhausting himself with the work, he announced to his family that they were going to have a good harvest, only to find the next morning that the plants had wilted and died. Readers with minimal agricultural knowledge may immediately point out the following: the farmer's intervention theory was based on a correct observation that crops that grow taller tend to produce more yield. Yet his hypothesis reflects a false understanding of the cause and the effect—that seedlings pulled to be taller would yield as much as seedlings thriving on their own.

In their classic *Design and Analysis of Experiments*, Hinkelmann and Kempthorne (1994; updated version of Kempthorne, 1952) discussed two types of science: descriptive science and the development of theory. These two types of science are interrelated in the following sense: observations of an event and other related events, often selected and classified for description by scientists, naturally lead to one or more explanations that we call "theoretical hypotheses," which are then screened and falsified by means of further observations, experimentation, and analyses (Popper, 1963). The experiment of pulling seedlings to be taller was costly, but did serve the purpose of advancing this farmer's knowledge of "what does not work." To develop a successful intervention, in this case, would require a series of empirical tests of explicit theories identifying potential contributors to crop growth. This iterative process gradually deepens our knowledge of the relationships between supposed causes and effects—that is, causality—and may eventually increase the success of agricultural, medical, and social interventions.

1.1 Concepts of moderation, mediation, and spill-over

Although the story of the ancient farmer is fictitious, numerous examples can be found in the real world in which well-intended interventions fail to produce the intended benefits or, in many cases, even lead to unintended consequences. "Interventions" and "treatments," used interchangeably in this book, broadly refer to actions taken by agents or circumstances experienced by an individual or groups of individuals. Interventions are regularly seen in education, physical and mental health, social services, business, politics, and law enforcement. In an

Causality in a Social World: Moderation, Meditation and Spill-over, First Edition. Guanglei Hong.
© 2015 John Wiley & Sons, Ltd. Published 2015 by John Wiley & Sons, Ltd.

education intervention, for example, teachers are typically the agents who deliver a treatment to students, while the impact of the treatment on student outcomes is of ultimate causal interest. Some educational practices such as "teaching to the test" have been criticized to be nearly as counterproductive as the attempt of helping seedlings grow by pulling them upward. "Interventions" and "treatments" under consideration do not exclude undesired experiences such as exposure to poverty, abuse, crime, or bereavement. A treatment, planned or unplanned, becomes a focus of research if there are theoretical reasons to anticipate its impact, positive or negative, on the well-being of individuals who are embedded in social settings including families, classrooms, schools, neighborhoods, and workplaces.

In social science research in general and in policy and program evaluations in particular, questions concerning whether an intervention works and, if so, which version of the intervention works, for whom, under what conditions, and why are key to the advancement of scientific and practical knowledge. Although most empirical investigations in the social sciences concentrate on the average effect of a treatment for a specific population as opposed to the absence of such a treatment (i.e., the control condition), in-depth theoretical reasoning with regard to how the causal effect is generated, substantiated by compelling empirical evidence, is crucial for advancing scientific understanding.

First, when there are multiple versions or different dosages of the treatment or when there are multiple versions of the control condition, a binary divide between "the treatment" and "the control" may not be as informative as fine-grained comparisons across, for example, "treatment version A," "treatment version B," "control version A," and "control version B." For example, expanding the federally funded Head Start program to poor children is expected to generate a greater benefit when few early childhood education alternatives are available (call it "control version A") than when there is an abundance of alternatives including state-sponsored preschool programs (call it "control version B").

Second, the effect of an intervention will likely vary among individuals or across social settings. A famous example comes from medical research: the well-publicized cardiovascular benefits of initiating estrogen therapy during menopause were contradicted later by experimental findings that the same therapy increased postmenopausal women's risk for heart attacks. The effect of an intervention may also depend on the provision of some other concurrent or subsequent interventions. Such heterogeneous effects are often characterized as *moderated* effects in the literature.

Third, alternative theories may provide competing explanations for the causal mechanisms, that is, the processes through which the intervention produces its effect. A theoretical construct characterizing the hypothesized intermediate process is called a *mediator* of the intervention effect. The fictitious farmer never developed an elaborate theory as to what caused some seedlings to surpass others in growth. Once scientists revealed the causal relationship between access to chemical nutrients in soil and plant growth, wide applications of chemically synthesized fertilizers finally led to a major increase in crop production.

Finally, it is well known in agricultural experiments that a new type of fertilizer applied to one plot may spill-over to the next plot. Because social connections among individuals are prevalent within organizations or through networks, an individual's response to the treatment may similarly depend on the treatment for other individuals in the same social setting, which may lead to possible *spill-overs* of intervention effects among individual human beings.

Answering questions with regard to moderation, mediation, and spill-over poses major conceptual and analytic challenges. To date, psychological research often presents well-articulated theories of causal mechanisms relating stimuli to responses. Yet, researchers often lack rigorous analytic strategies for empirically screening competing theories explaining the observed effect. Sociologists have keen interest in the spill-over of treatment effects transmitted through social interactions yet have produced limited evidence quantifying such effects. As many have pointed out, in general, terminological ambiguity and conceptual confusion have been prevalent in the published applied research (Holmbeck, 1997; Kraemer *et al.*, 2002, 2008).

A new framework for conceptualizing moderated, mediated, and spill-over effects has emerged relatively recently in the statistics and econometrics literature on causal inference (e.g., Abbring and Heckman, 2007; Frangakis and Rubin, 2002; Heckman and Vytlacil, 2007a, b; Holland, 1988; Hong and Raudenbush, 2006; Hudgens and Halloran, 2008; Jo, 2008; Pearl, 2001; Robins and Greenland, 1992; Sobel, 2008). The potential for further conceptual and methodological development and for broad applications in the field of behavioral and social sciences promises to greatly advance the empirical basis of our knowledge about causality in the social world.

This book clarifies theoretical concepts and introduces innovative statistical strategies for investigating the average effects of multivalued treatments, moderated treatment effects, mediated treatment effects, and spill-over effects in experimental or quasiexperimental data. Defining individual-specific and population average treatment effects in terms of potential outcomes, the book relates the mathematical forms to the substantive meanings of moderated, mediated, and spill-over effects in the context of application examples. It also explicates and evaluates identification assumptions and contrasts innovative statistical strategies with conventional analytic methods.

1.1.1   Moderated treatment effects

It is hard to accept the assumption that a treatment would produce the same impact for every individual in every possible circumstance. Understanding the heterogeneity of treatment effects therefore is key to the development of causal theories. For example, some studies reported that estrogen therapy improved cardiovascular health among women who initiated its use during menopause. According to a series of other studies, however, the use of estrogen therapy increased postmenopausal women's risk for heart attacks (Grodstein *et al.*, 1996, 2000; Writing Group for the Women's Health Initiative Investigators, 2002). The sharp contrast of findings from these studies led to the hypothesis that age of initiation moderates the effect of estrogen therapy on women's health (Manson and Bassuk, 2007; Rossouw *et al.*, 2007). Revelations of the moderated causal relationship greatly enrich theoretical understanding and, in this case, directly inform clinical practice.

In another example, dividing students by ability into small groups for reading instruction has been controversial for decades. Many believed that the practice benefits high-ability students at the expense of their low-ability peers and exacerbates educational inequality (Grant and Rothenberg, 1986; Rowan and Miracle, 1983; Trimble and Sinclair, 1987). Recent evidence has shown that, first of all, the effect of ability grouping depends on a number of conditions including whether enough time is allocated to reading instruction and how well the class can be managed. Hence, instructional time and class management are among the additional moderators to consider for determining the merits of ability grouping (Hong and Hong, 2009; Hong *et al.*, 2012a, b). Moreover, further investigations have generated evidence that contradicts the long-held belief that grouping undermines the interests of low-ability students. Quite the contrary, researchers have found that grouping is beneficial for students with low or medium prior ability—if adequate time is provided for instruction and if the class is well managed. On the other hand, ability grouping appears to have a minimal impact on high-ability students' literacy. These results were derived from kindergarten data and may not hold for math instruction and for higher grade levels. Replications with different subpopulations and in different contexts enable researchers to assess the generalizability of a theory.

In a third example, one may hypothesize that a student is unlikely to learn algebra well in ninth grade without a solid foundation in eighth-grade prealgebra. One may also argue that the progress a student made in learning eighth-grade prealgebra may not be sustained without the further enhancement of taking algebra in ninth grade. In other words, the earlier treatment (prealgebra in eighth grade) and the later treatment (algebra in ninth grade) may reinforce one other; each moderates the effect of the other treatment on the final outcome (math achievement at the end of ninth grade). As Hong and Raudenbush (2008) argued, the cumulative effect of two treatments in a well-aligned sequence may exceed the sum of the benefits of two single-year treatments. Similarly, experiencing two consecutive years of inferior treatments such as encountering an incompetent math teacher who dampens a student's self-efficacy in math learning in eighth grade and then again in ninth grade could do more damage than the sum of the effect of encountering an incompetent teacher only in eighth grade and the effect of having such a teacher only in ninth grade.

There has been a great amount of conceptual confusion regarding how a moderator relates to the treatment and the outcome. On one hand, many researchers have used the terms "moderators" and "mediators" interchangeably without understanding the crucial distinction between the two. An overcorrective attempt, on the other hand, has led to the arbitrary recommendation that a moderator must occur prior to the treatment and be minimally associated with the treatment (James and Brett, 1984; Kraemer et al., 2001, 2008). In Chapter 6, we clarify that, in essence, the causal effect of the treatment on the outcome may depend on the moderator value. A moderator can be a subpopulation identifier, a contextual characteristic, a concurrent treatment, or a preceding or succeeding treatment. In the earlier examples, age of estrogen therapy initiation and a student's prior ability are subpopulation identifiers, the manageability of a class which reflects both teacher skills and peer behaviors characterizes the context, literacy instruction time is a concurrent treatment, while prealgebra is a preceding treatment for algebra. A moderator does not have to occur prior to the treatment and does not have to be independent of the treatment.

Once a moderated causal relationship has been defined in terms of potential outcomes, the researcher then chooses an appropriate experimental design for testing the moderation theory. The assumptions required for identifying the moderated causal relationships differ across different designs and have implications for analyzing experimental and quasiexperimental data. Randomized block designs are suitable for examining individual or contextual characteristics as potential moderators; factorial designs enable one to determine the joint effects of two or more concurrent treatments; and sequential randomized designs are ideal for assessing the cumulative effects of consecutive treatments. Multisite randomized trials constitute an additional type of designs in which the experimental sites are often deliberately sampled to represent a population of geographical locations or social organizations. Site memberships can be viewed as implicit moderators that summarize a host of features of a local environment. Replications of a treatment over multiple sites allow one to quantify the heterogeneity of treatment effects across the sites. Chapters 7 and 8 are focused on statistical methods for moderation analyses with quasiexperimental data. In particular, these chapters demonstrate, through a number of application examples, how a nonparametric marginal mean weighting through stratification (MMWS) method can overcome some important limitations of other existing methods.

Moderation questions are not restricted to treatment–outcome relationships. Rather, they are prevalent in investigations of mediation and spill-over. This is because heterogeneity of treatment effects may be explained by the variation in causal mediation mechanisms across subpopulations and contexts. It is also because the treatment effect for a focal individual may depend on whether there is spill-over from other individuals through social interactions. Yet similar to the confusion around the concept of moderation among applied researchers, there has been a great amount of misunderstanding with regard to mediation.

1.1.2   Mediated treatment effects

Questions about mediation are at the core of nearly every scientific theory. In its simplest form, a theory explains why treatment Z causes outcome Y by hypothesizing an intermediate process involving at least one mediator M that could have been changed by the treatment and could subsequently have an impact on the outcome. A causal mediation analysis then decomposes the total effect of the treatment on the outcome into two parts: the effect transmitted through the hypothesized mediator, called the indirect effect, and the difference between the total effect and the indirect effect, called the direct effect. The latter represents the treatment effect channeled through other unspecified pathways (Alwin and Hauser, 1975; Baron and Kenny, 1986; Duncan, 1966; Shadish and Sweeney, 1991). Most researchers in social sciences have followed the convention of illustrating the direct effect and the indirect effect with path diagrams and then specifying linear regression models postulated to represent the structural relationships between the treatment, the mediator, and the outcome.

In Chapter 9, we point out that the approach to defining the indirect effect and the direct effect in terms of path coefficients is often misled by oversimplistic assumptions about the structural relationships. In particular, this approach typically overlooks the fact that a treatment may generate an impact on the outcome not only by changing the mediator value but also by changing the mediator–outcome relationship (Judd and Kenny, 1981). An example in Chapter 9 illustrates such a case. An experiment randomizes students to either an experimental condition which provides them with study materials and encourages them to study for a test or to a control condition which provides neither study materials nor encouragement (Holland, 1988; Powers and Swinton, 1984). One might hypothesize that encouraging experimental group members to study will increase the time they spend studying, a focal mediator in this example, which in turn will increase their average test scores. One might further hypothesize that even without a change in study time, providing study materials to experimental group members will enable them to study more effectively than they otherwise would. Consequently, the effect of time spent studying might be greater under the experimental condition than under the control condition. Omitting the treatment-by-mediator interaction effect will lead to bias in the estimation of the indirect effect and the direct effect.

Chapter 11 describes in great detail an evaluation study contrasting a new welfare-to-work program emphasizing active participation in the labor force with a traditional program that guaranteed cash assistance without a mandate to seek employment (Hong, Deutsch, and Hill, 2011). Focusing on the psychological well-being of welfare applicants who were single mothers with preschool-aged children, the researchers hypothesized that the treatment would increase employment rate among the welfare recipients and that the treatment-induced increase in employment would likely reduce depressive symptoms under the new program.

They also hypothesized that, in contrast, the same increase in employment was unlikely to have a psychological impact under the traditional program. Hence, the treatment-by-mediator interaction effect on the outcome was an essential component of the intermediate process.

We will show that by defining the indirect effect and the direct effect in terms of potential outcomes, one can avoid invoking unwarranted assumptions about the unknown structural relationships (Holland, 1988; Pearl, 2001; Robins and Greenland, 1992). The potential outcomes framework has further clarified that in a causal mediation theory, the treatment and the mediator are both conceivably manipulable. In the aforementioned example, a welfare applicant could be assigned at random to either the new program or the traditional one. Once having been assigned to one of the programs, the individual might or might not be employed due to various structural constraints, market fluctuations, or other random events typically beyond her control.

Because the indirect effect and the direct effect are defined in terms of potential outcomes rather than on the basis of any specific regression models, it becomes possible to distinguish the *definitions* of these causal effects from their *identification* and *estimation*. The identification step relates a causal parameter to observable population data. For example, for individuals assigned at random to an experimental condition, their average counterfactual outcome associated with the control condition is expected to be equal to the average observable outcome of those assigned to the control condition. Therefore, the population average causal effect can be easily identified in a randomized experiment. The estimation step then relates the sample data to the observable population quantities involved in identification while taking into account the degree of sampling variability.

A major challenge in identification is that the mediator–outcome relationship tends to be confounded by selection even if the treatment is randomized. Chapter 9 reviews various experimental designs that have been proposed by past researchers for studying causal mediation mechanisms. Chapter 10 compares the identification assumptions and the analytic procedures across a wide range of analytic methods. These discussions are followed by an introduction of the ratio-of-mediator-probability weighting (RMPW) strategy in Chapter 11. In comparison with most existing strategies for causal mediation analysis, RMPW relies on relatively fewer identification assumptions and model-based assumptions. Chapters 12 and 13 will show that this new strategy can be applied broadly with extensions to multilevel experimental designs and to studies of complex mediation mechanisms involving multiple mediators.

1.1.3   Spill-over effects of a treatment

It is well known in vaccination research that an unvaccinated person can benefit when most other people in the local community are vaccinated. This is viewed as a spill-over of the treatment effect from the vaccinated people to an unvaccinated person. Similarly, due to social interactions among individuals or groups of individuals, an intervention received by some may generate a spill-over impact on others affiliated with the same organization or connected through the same network. For example, effective policing in one neighborhood may drive offenders to operate in other neighborhoods. In evaluating an innovative community policing program, researchers found that a neighborhood not assigned to community policing tended to suffer if the surrounding neighborhoods were assigned to community policing. This is an instance in which a well-intended treatment generates an unintended negative spill-over effect. However, being assigned to community policy was found particularly beneficial when the surrounding neighborhoods also received the intervention (Verbitsky-Savitz and Raudenbush,

2012). In another example, the impact of a school-based mentoring program targeted at students displaying delinquent behaviors may be enhanced for a focal student if his or her at-risk peers are assigned to the mentoring program at the same time.

The previous examples challenge the "stable unit treatment value assumption" (SUTVA) that has been invoked in most causal inference studies. This assumption states that an individual's potential response to a treatment depends neither on how the treatment is assigned nor on the treatment assignments of other individuals (Rubin, 1980, 1986). Rubin and others believe that, without resorting to SUTVA, the causal effect of a treatment becomes hard to define and that causal inference may become intractable. However, as Sobel (2006) has illustrated with the Moving to Opportunity (MTO) experiment that offered housing vouchers enabling low-income families to move to low-poverty neighborhoods, there are possible consequences of violating SUTVA in estimating the treatment effects on outcomes such as safety and self-sufficiency. Because social interactions among individuals may affect whether one volunteers to participate in the study, whether one moves to a low-poverty neighborhood after receiving the housing voucher, as well as housing project residents' subjective perceptions of their neighborhoods, the program may have a nonzero impact on the potential outcome of the untreated. As a result, despite the randomization of treatment assignment, the mean difference in the observed outcome between the treated units and the untreated units may be biased for the average treatment effect. Rather, the observable quantity is the difference between the effect of treating some rather than treating none for the treated and the effect of treating some rather than treating none for the untreated, the latter being the pure spill-over effect for the untreated.

In making attempts to relax SUTVA, researchers have proposed several alternative frameworks that incorporate possible spill-over effects (see Hong and Raudenbush, 2013, for a review). Hong (2004) presented a model that involves treatments and treatment settings. A treatment setting for an individual is a local environment constituted by a set of agents and participants along with their treatment assignments. An individual's potential outcome value under a given treatment is assumed stable when the treatment setting is fixed; the potential outcome may take different values when the treatment setting shifts. One may investigate whether the treatment effect depends on the treatment setting. Applying this framework, Hong and Raudenbush (2006) examined the effect on a child's academic growth of retaining the child in kindergarten rather than promoting the child to first grade when a relatively small proportion of low-achieving peers in the same school are retained as opposed to when a relatively large proportion of the peers are retained. Hudgens and Halloran (2008) presented a related framework in which the effect on an individual of the treatment received by this individual is distinguished from the effect on the individual of the treatment received by others in the same local community. These effects can be identified if communities are assigned at random to different treatment assignment strategies (such as retaining a large proportion of low-achieving students as opposed to retaining a small proportion of such students) and, subsequently, individuals within a community are randomized for treatment. Chapter 14 reviews these frameworks and discusses identification and estimation strategies.

Social contagion may also serve as an important channel through which the effect of an intervention is transmitted. For example, a school-wide intervention may reduce aggressive behaviors and thereby improve students' psychological well-being by improving the quality of interpersonal relationships in other classes as well as in one's own class. This is because children interact not simply with their classmates but also with those from other classes in the hallways or on the playground (VanderWeele et al., 2013). In this case, the spill-over between classes becomes a part of the mediation mechanism. In a study of student mentoring, whether an individual actually participated in the mentoring program despite the initial treatment assignment is typically viewed as a primary mediator. The proportion of one's peers that participated in the program may act as a second mediator. A treatment may also exert its impact partly through regrouping individuals and thereby changing peer composition (Hong and Nomi, 2012). Chapter 15 discusses analytic strategies for detecting spill-over as a part of the causal mediation mechanism.

1.2   Weighting methods for causal inference

In most behavioral and social science applications, major methodological challenges arise due to the selection of treatments in a quasiexperimental study and the selection of mediator values in experimental and quasiexperimental studies. Statistical methods most familiar to researchers in social sciences are often inadequate for causal inferences with regard to multivalued treatments, moderation, mediation, and spill-over. The book offers a major revision to understanding of causality in the social world by introducing two complementary weighting strategies, both featuring nonparametric approaches to estimating these causal effects. The new weighting methods greatly simplify model specifications while enhancing the robustness of results by minimizing the reliance on some key assumptions.

The propensity score-based *MMWS* method removes selection bias associated with a large number of covariates by equating the pretreatment composition between treatment groups (Hong, 2010a, 2012; Huang *et al.*, 2005). Unlike propensity score matching and stratification that are mostly restricted to evaluations of binary treatments, the MMWS method is flexible for evaluating binary and multivalued treatments by approximating a completely randomized experiment. In evaluating whether the treatment effects differ across subpopulations defined by individual characteristics or treatment settings, researchers may assign weights within each subpopulation in order to approximate a randomized block design. To investigate whether one treatment moderates the effect of another concurrent treatment, researchers may assign weights to the data to approximate a factorial randomized design. The method can also be used to assess whether the effect of an initial treatment is amplified or weakened by a subsequent treatment or to identify an optimal treatment sequence through approximating a sequential randomized experiment. Even though such analyses can similarly be conducted through inverse-probability-of-treatment weighting (IPTW) that has been increasingly employed in epidemiological research (Hernán, Brumback, and Robins, 2000; Robins, Hernán, and Brumback, 2000), IPTW is known for bias and imprecision in estimation especially when the propensity score models are misspecified in their functional forms (Hong, 2010a; Kang and Schafer, 2007; Schafer and Kang, 2008; Waernbaum, 2012). In contrast, the nonparametric MMWS method displays a relatively high level of robustness despite such misspecifications and also gains efficiency, as indicated by simulation results (Hong, 2010a).

To study causal mediation mechanisms, the *RMPW* method decomposes the total effect of a treatment into an "indirect effect" transmitted through a specific mediator and a "direct effect" representing unspecified mechanisms. In contrast with most existing methods for mediation analysis, the RMPW-adjusted outcome model is extremely simple and is nonparametric in nature. It generates estimates of the causal effects along with their sampling errors while adjusting for pretreatment covariates that confound the mediator–outcome relationships through weighting. The method applies regardless of the distribution of the outcome, the distribution of the mediator, or the functional relationship between the outcome and the mediator.

Moreover, the RMPW method can easily accommodate data in which the mediator effect on the outcome may depend on the treatment assignment (Hong, 2010b). This is the case when a treatment produces its effects not only through changing the mediator value but also in part by altering the mediational process that normally produces the outcome (Judd and Kenny, 1981). One may use RMPW to further investigate whether the mediation mechanism varies across subpopulations and to disentangle complex mediation mechanisms involving multiple con-current or consecutive mediators. A combination of RMPW with MMWS enables researchers to conduct mediation analysis when the treatment is not randomized. The RMPW approach to mediation analysis will be shown to have broad applicability to single-level and multilevel data (Hong & Nomi, 2012; Hong, Deutsch, and Hill, in press) when compared with path analysis, structural equation modeling (SEM), the instrumental variable method, and their recent extensions.

1.3   Objectives and organization of the book

The book consists of four major units: an overview followed by three units focusing on mod-eration, mediation, and spill-over, respectively. The first unit provides an overview of the con-cepts of causal effects, moderation, mediation, and spill-over. After reviewing the existing research designs and analytic methods for causal effect estimation, it introduces the basic rationale of weighted estimation of population parameters in survey sampling and explains the extension of the weighting approach to causal inference. Each subsequent unit addresses research questions of increasing complexity around one of the three focal topics. Part II con-siders treatment effects moderated by individual or contextual characteristics, by a concurrent treatment, or by a prior or subsequent treatment. Part III shows how to investigate the case of a single mediator, of multiple mediators, and of moderated mediation effects. Part IV discusses the spill-over of treatment effects and the spill-over of mediated effects.

Throughout the book, the mathematics is kept to a minimum to ease reading. Derivations and proofs are left to appendices for readers with technical interest. Assuming that some read-ers may have had little prior training in basic probability theory, a topic not always covered in a systematic way in an introductory-level applied statistics course, Chapter 2 provides a glos-sary of the basic notation employed in causal inference. Readers who have already taken a course in causal inference or its equivalent may easily skip Chapters 2 and 3. Chapters 4–11 are suitable for use in an undergraduate- or graduate-level applied statistics course dealing with moderation and mediation analyses. Those who are working on cutting-edge problems related to causal mediation and spill-over may find the last four chapters of the book particularly engaging.

This book aims to make the new analytic methods readily accessible to a wide audience in the social and behavioral sciences. Readers with sufficient understanding of multiple regression and analysis of variance (ANOVA) will quickly grasp the logic of the weighting methods and will find them easy to implement with existing statistical software. For example, after applying MMWS to a sample, researchers can estimate the effects of multivalued treatments or a mod-erated treatment effect simply within the ANOVA framework. For mediational studies, the RMPW method generates parameter estimates corresponding to the direct effect and the indi-rect effect along with their estimated sampling variability and therefore greatly simplifies hypothesis testing. In addition to the templates, along with data examples, for implementing MMWS and RMPW in SAS, Stata, and R, the book is accompanied by stand-alone

MMWS and RMPW software programs. The program interfaces are designed not only to greatly ease computation but also to assist the applied user with analytic decision-making. All these materials are available online free of charge at the publisher's website: http://www.wiley.com/go/social_world.

Numerous examples from education, human development, psychology, sociology, and public policy motivate the development of innovative methods and help illustrate the concepts in the book. Analytic procedures are demonstrated through a series of case studies such as those discussed earlier. The examples are chosen to represent the causal questions often raised in behavioral and social science research and hence serve as prototypes for applying the analytic methods. In each case, statistical solutions are offered in sufficient detail to allow for replications.

1.4   How is this book situated among other publications on related topics?

Many other authors have contributed to research on moderation and mediation analyses and on causal inference in general. Such work has been systematically presented in four categories, in my view. The first category is textbooks on path analysis and SEM. The most popular ones include Kenneth A. Bollen's (1989) *Structural Equations with Latent Variables*, David P. MacKinnon's (2008) *Introduction to Statistical Mediation Analysis*, and Andrew F. Hayes' (2013) *Introduction to Mediation, Moderation, and Conditional Process Analysis: A Regression-Based Approach*. Bollen's book provides a lucid and precise introduction to SEM focusing on combining structural models with measurement models. MacKinnon's book summarizes later developments extending SEM to multilevel data, longitudinal data, and categorical data and relaxing distributional assumptions by employing computer-intensive estimation methods. Hayes's book offers a gentle introduction to the most basic mediation and moderation analyses with the aim of integrating the two within the regression framework. These books, however, do not place a major emphasis on clarifying the identification assumptions crucial for distinguishing between association and causation.

Books in the second category provide a healthy antidote to the standard methods that have been employed by social scientists in addressing causal questions. Among the most well known are Steven D. Levitt and Stephen J. Dubner's (2005) *Freakonomics: A Rogue Economist Explores the Hidden Side of Everything* and Charles Manski's (1995) *Identification Problems in the Social Sciences*. These books were aimed at raising critical awareness among social scientists and the general public with regard to the potentials and limitations of the existing methods. They were excellent in revealing analytic difficulties and uncertainties in social science research.

The third category includes books offering a general overview of research designs and analytic methods for causal inference, represented by William R. Shadish, Thomas D. Cook, and Donald T. Campbell's (2002) *Experimental and Quasi-Experimental Designs for Generalized Causal Inference*, Stephen L. Morgan and Christopher Winship's (2015) *Counterfactuals and Causal Inference: Methods and Principles for Social Research*, and Richard J. Murnane and John B. Willett's (2011) *Methods Matter: Improving Causal Inference in Educational and Social Science Research*. Each book gives a comprehensive survey of a range of design options and analytic options for students in social sciences, yet none is focused on methods for investigating moderation, mediation, and spill-over.

The fourth category is research monographs each presenting a distinct cutting-edge approach to causal analysis. In this category, one may find Paul R. Rosenbaum's (2002) *Observational Studies* and his (2009) *Design of Observational Studies*, Donald B. Rubin's (2006) *Matched Sampling for Causal Effects*, and Judea Pearl's (2009) *Causality: Models, Reasoning, and Inference*. The books by Rosenbaum or Rubin provide detailed discussions of how to use propensity score matching to overcome overt biases associated with the observables and how to use sensitivity analysis to account for hidden biases associated with the unobservables. The authors restricted the discussions primarily to evaluating the average effect of a binary treatment. Pearl's book is unique in defining direct and indirect effects as mathematical functions of counterfactual outcomes. The definitions reflect an integrated understanding of probabilistic relationships and structural relationships. The book is also unique in its extensive use of graphical models and causal diagrams for determining identification. Yet the book may not satisfy practitioners searching for step-by-step analytic solutions directly applicable to their data.

The current book is built on an accumulation of past research represented in the aforementioned publications and in numerous journal articles. However, unlike the publications in the first category, this one is exclusively based on the potential outcomes causal framework with a clear definition for each causal effect. The nonparametric weighting methods presented in this book provide useful alternatives to SEM by relaxing key assumptions and simplifying modeling specification and estimation. Unlike the books in the second category, this one not only criticizes naïve approaches to causal reasoning but also demonstrates analytic strategies enabling researchers to conduct in-depth investigations of real-world problems and arrive at meaningful substantive conclusions. Unlike those in the third category, this book is exclusively focused on moderation, mediation, and spill-over effects and makes original contributions by introducing innovative analytic methods designed for addressing these fundamental issues. This book also distinguishes itself from those in the fourth category by teaching innovative weighting methods that can be readily implemented with most existing statistical software. Finally, most of the books mentioned previously have not given particular considerations of spill-over effects. Discussions related to social interference and spill-over have been ongoing in the fields of econometrics (Durlauf and Young, 2001; Manski, 1993, 2000) and political and social network analysis (Scott and Carrington, 2011; Sinclair, 2012; Wasserman and Faust, 1994) and have increasingly drawn attention from statistics. The current book contributes to the discourse by illuminating a central role of spill-over in social intervention theories from a causal inference perspective.

In summary, this book complements yet cannot replace many other important publications mentioned earlier. Readers who need a broad overview of a wide range of research designs and causal inference methods will likely benefit from reading the third category of publications; those who seek in-depth information of a particular causal inference perspective or strategy other than the weighting approaches presented here are referred to the fourth category of publications.

References

Abbring, J.H. and Heckman, J.J. (2007). Econometric evaluation of social programs. Part III: Distributional treatment effects, dynamic treatment effects, dynamic discrete choice, and general equilibrium policy evaluation, in *Handbook of Econometrics*, vol. 6B (eds J. Heckman and E. Leamer), North-Holland, Amsterdam, pp. 5145–5303.

Alwin, D.F. and Hauser, R.M. (1975). The decomposition of effects in path analysis. *American Socio-logical Review, 40*, 37–47.

Baron, R.M. and Kenny, D.A. (1986). The moderator-mediator variable distinction in social psycholog-ical research: conceptual, strategic, and statistical considerations. *Journal of Personality and Social Psychology, 51*, 1173–1182.

Bollen, K.A. (1989). *Structural Equations with Latent Variables*, Wiley, New York.

Duncan, O.D. (1966). Path analysis: sociological examples. *The American Journal of Sociology, 72*(1), 1–16.

Durlauf, S.N. and Young, H.P. (2001). *Social Dynamics*, MIT Press, Cambridge, MA.

Frangakis, C.E. and Rubin, D.B. (2002). Principal stratification in causal inference. *Biometrics, 58*, 21–29.

Grant, L. and Rothenberg, J. (1986). The social enhancement of ability differences: teacher student interactions in first- and second-grade reading groups. *The Elementary School Journal, 87*(1), 29–49.

Grodstein, F., Stampfer, M., Manson, J. *et al.* (1996). Postmenopausal estrogen and progestin use and the risk of cardiovascular disease. *New England Journal of Medicine, 335*(7), 453–461.

Grodstein, F., Manson, J.E., Colditz, G.A. *et al.* (2000). A prospective, observational study of postme-nopausal hormone therapy and primary prevention of cardiovascular disease. *Annals of Internal Medicine, 133*(12), 933–941.

Hayes, A.F. (2013). *Introduction to Mediation, Moderation, and Conditional Process Analysis: A Regression-Based Approach*, The Guilford Press, New York.

Heckman, J.J. and Vytlacil, E.J. (2007a). Econometric evaluation of social programs. Part I: Causal models, structural models and econometric policy evaluation, in *Handbook of Econometrics*, vol. *6B* (eds J. Heckman and E. Leamer), North-Holland, Amsterdam.

Heckman, J.J. and Vytlacil, E.J. (2007b). Econometric evaluation of social programs. Part II: Using the marginal treatment effect to organize alternative econometric estimators to evaluate social programs, and to forecast their effects in new environments, in *Handbook of Econometrics*, vol. 6B (eds J. Heckman and E. Leamer), North-Holland, Amsterdam, pp. 4875–5143.

Hernán, M.A., Brumback, B.B. and Robins, J.M. (2000). Marginal structural models to estimate the causal effect of zidovudine on the survival of HIV-positive men. *Epidemiology, 11*(5), 561–570.

Hinkelmann, K. and Kempthorne, O. (1994). *Design and Analysis of Experiments, Volume I: Introduc-tion to Experimental Design*, Wiley, New York.

Holland, P.W. (1988). Causal interference, path analysis, and recursive structural equations models. *Sociological Methodology, 18*, 449–484.

Holmbeck, G.N. (1997). Toward terminological, conceptual, and statistical clarity in the study of med-iators and moderators: examples from the child-clinical and pediatric psychology literatures. *Journal of Consulting and Clinical Psychology, 65*(4), 599–610.

Hong, G. (2004). *Causal inference for multi-level observational data with application to kindergarten retention*. Ph.D. dissertation. Department of Educational Studies, University of Michigan, Ann Arbor, MI.

Hong, G. (2010a). Marginal mean weighting through stratification: adjustment for selection bias in multilevel data. *Journal of Educational and Behavioral Statistics, 35*(5), 499–531.

Hong, G. (2010b). *Ratio of Mediator Probability Weighting for Estimating Natural Direct and Indirect Effects*. JSM Proceedings, Biometrics Section, pp. 2401–2415. American Statistical Association, Alexandria, VA.

Hong, G. (2012). Marginal mean weighting through stratification: a generalized method for evaluating multi-valued and multiple treatments with non-experimental data. *Psychological Methods, 17*(1), 44–60.

Hong, G. and Hong, Y. (2009). Reading instruction time and homogeneous grouping in kindergarten: an application of marginal mean weighting through stratification. *Educational Evaluation and Policy Analysis, 31*(1), 54–81.

Hong, G. and Nomi, T. (2012). Weighting methods for assessing policy effects mediated by peer change. *Journal of Research on Educational Effectiveness*, special issue on the statistical approaches to studying mediator effects in education research, *5*(3), 261–289.

Hong, G. and Raudenbush, S.W. (2006). Evaluating kindergarten retention policy: a case study of causal inference for multi-level observational data. *Journal of the American Statistical Association, 101*, 901–910.

Hong, G. and Raudenbush, S.W. (2008). Causal inference for time-varying instructional treatments. *Journal of Educational and Behavioral Statistics, 33*(3), 333–362.

Hong, G. and Raudenbush, S.W. (2013). Heterogeneous agents, social interactions, and causal inference, in *The Handbook of Causal Analysis for Social Research* (ed S.L. Morgan), Springer, New York, pp. 331–352.

Hong, G., Deutsch, J., & Hill, H. D. (in press). Ratio-of-mediator-probability weighting for causal mediation analysis in the presence of treatment-by-mediator interaction. *Journal of Educational and Behavioral Statistics*.

Hong, G., Corter, C., Hong, Y. and Pelletier, J. (2012a). Differential effects of literacy instruction time and homogeneous grouping in kindergarten: who will benefit? Who will suffer? *Educational Evaluation and Policy Analysis, 34*(1), 69–88.

Hong, G., Pelletier, J., Hong, Y., and Corter, C. (2012b). Does literacy instruction affect kindergartners' externalizing problem behaviors as well as their literacy learning? Taking class manageability into account. The University of Chicago and the Ontario Institute for Studies in Education of the University of Toronto.

Huang, I.C., Diette, G.B., Dominici, F. *et al.* (2005). Variations of physician group profiling indicators for asthma care. *American Journal of Managed Care, 11*(1), 38–44.

Hudgens, M.G. and Halloran, M.E. (2008). Toward causal inference with interference. *Journal of the American Statistical Association, 103*, 832–842.

James, L.R. and Brett, J.M. (1984). Mediators, moderators, and test for mediation. *Journal of Applied Psychology, 69*, 307–321.

Jo, B. (2008). Causal inference in randomized experiments with mediational processes. *Psychological Methods, 13*(4), 314–336.

Judd, C.M. and Kenny, D.A. (1981). Process analysis: estimating mediation in treatment evaluation. *Evaluation Review, 5*, 602–619.

Kang, J.D. and Schafer, J.L. (2007). Demystifying double robustness: a comparison of alternative strategies for estimating a population mean from incomplete data. *Statistical Science, 22*(4), 523–539.

Kraemer, H.C., Stice, E., Kazdin, A. *et al.* (2001). How do risk factors work together? Mediators, moderators, and independent, overlapping, and proxy risk factors. *American Journal of Psychiatry, 158*, 848–856.

Kraemer, H.C., Wilson, G.T., Fairburn, C.G. and Agras, W.S. (2002). Mediators and moderators of treatment effects in randomized clinical trials. *Archives of General Psychiatry, 59*(10), 877–883.

Kraemer, H., Kiernan, M., Essex, M. and Kupfer, D. (2008). How and why criteria defining moderators and mediators differ between the Baron & Kenny and MacArthur approaches. *Health Psychology, 27*, 1–14.

Levitt, S.D. and Dubner, S.J. (2005). *Freakonomics: A Rogue Economist Explores the Hidden Side of Everything*, Harper Collins Publishers, New York.

MacKinnon, D.P. (2008). *Introduction to Statistical Mediation Analysis*, Lawrence Erlbaum Associates, New York.

Manski, C. (1993). Identification of endogenous social effects: the reflection problem. *Review of Economic Studies*, *60*, 531–542.

Manski, C. (1995). *Identification Problems in the Social Sciences*, Harvard University Press, Boston, MA.

Manski, C. (2000). Economic analysis of social interactions. *Journal of Economic Perspectives*, *14*, 114–136.

Manson, J.E. and Bassuk, S.S. (2007). Invited commentary: hormone therapy and risk of coronary heart disease why renew the focus on the early years of menopause? *American Journal of Epidemiology*, *166*(5), 511–517.

Morgan, S.L. and Winship, C. (2015). *Counterfactuals and Causal Inference: Methods and Principles for Social Research*, 2nd ed. Cambridge University Press, New York.

Murnane, R.J. and Willett, J.B. (2011). *Methods Matter: Improving Causal Inference in Educational and Social Science Research*, Oxford University Press, New York.

Pearl, J. (2001). *Direct and Indirect Effects*. Proceedings of the 17th Conference on Uncertainty in Artificial Intelligence, Morgan Kaufmann, San Francisco, CA, pp. 1572–1581.

Pearl, J. (2009). *Causality: Models, Reasoning, and Inference*, 2nd edn, Cambridge University Press, New York.

Popper, K. (2014). *Conjectures and Refutations: The Growth of Scientific Knowledge*, Routledge, New York.

Powers, D.E. and Swinton, S.S. (1984). Effects of self-study for coachable test item types. *Journal of Educational Psychology*, *76*, 266–278.

Robins, J.M. and Greenland, S. (1992). Identifiability and exchangeability for direct and indirect effects. *Epidemiology*, *3*, 143–155.

Robins, J.M., Hernán, M.A. and Brumback, B.B. (2000). Marginal structural models and causal inference in epidemiology. *Epidemiology*, *11*(5), 550–560.

Rosenbaum, P.R. (2002). *Observational Studies*, 2nd edn, Springer, New York.

Rosenbaum, P.R. (2009). *Design of Observational Studies*, Springer, New York.

Rossouw, J.E., Prentice, R.L., Manson, J.E. *et al.* (2007). Postmenopausal hormone therapy and risk of cardiovascular disease by age and years since menopause. *JAMA*, *297*(13), 1465–1477.

Rowan, B. and Miracle, A.W. (1983). Systems of ability grouping and the stratification of achievement in elementary schools. *Sociology of Education*, *56*(3), 133–144.

Rubin, D.B. (1980). Discussion of "randomization analysis of experimental data in the Fisher randomization test" by Basu. *Journal of the American Statistical Association*, *75*, 591–593.

Rubin, D.B. (1986). Statistics and causal inference: comment: which ifs have causal answers. *Journal of the American Statistical Association*, *81*(396), 961–962.

Rubin, D.B. (2006). *Matched Sampling for Causal Effects*, Cambridge University Press, New York.

Schafer, J.L. and Kang, J. (2008). Average causal effects from nonrandomized studies: a practical guide and simulated example. *Psychological Methods*, *13*(4), 279–313.

Scott, J. and Carrington, P.J. (2011). *The SAGE Handbook of Social Network Analysis*, SAGE Publications, Thousand Oaks, CA.

Shadish, W.R. and Sweeney, R.B. (1991). Mediators and moderators in meta-analysis: there's a reason we don't let dodo birds tell us which psychotherapies should have prizes. *Journal of Consulting and Clinical Psychology*, *59*, 883–893.

Shadish, W.R., Cook, T.D. and Campbell, D.T. (2002). *Experimental and Quasi-Experimental Designs for Generalized Causal inference*, Houghton Mifflin Company, Boston, MA.

Sinclair, B. (2012). *The Social Citizen: Peer Networks and Political Behavior*, University of Chicago Press, Chicago, IL.

Sobel, M.E. (2006). What do randomized studies of housing mobility demonstrate? Causal inference in the face of interference. *Journal of the American Statistical Association*, *101*, 1398–1407.

Sobel, M.E. (2008). Identification of causal parameters in randomized studies with mediating variables. *Journal of Educational and Behavioral Statistics*, *33*(2), 230–251.

Trimble, K.D. and Sinclair, R.L. (1987). On the wrong track: ability grouping and the threat to equity. *Equity and Excellence*, *23*, 15–21.

VanderWeele, T., Hong, G., Jones, S. and Brown, J. (2013). Mediation and spillover effects in group-randomized trials: a case study of the 4R's educational intervention. *Journal of the American Statistical Association*, *108*(502), 469–482.

Verbitsky-Savitz, N. and Raudenbush, S.W. (2012). Causal inference under interference in spatial settings: a case study evaluating community policing program in Chicago. *Epidemiologic Methods*, *1*(1), 107–130.

Waernbaum, I. (2012). Model misspecification and robustness in causal inference: comparing matching with doubly robust estimation. *Statistics in Medicine*, *31*, 1572–1581.

Wasserman, S. and Faust, K. (1994). *Social Network Analysis: Methods and Applications* (Vol. 8). Cambridge, UK: Cambridge University Press.

Writing Group for the Women's Health Initiative Investigators (2002). Risks and benefits of estrogen plus progestin in healthy postmenopausal women: principal results from the Women's Health Initiative randomized controlled trial. *JAMA*, *288*(3), 321–333.

2

Review of causal inference concepts and methods

2.1 Causal inference theory

Most human beings are engaged in analytic reasoning on a daily basis. Understanding causality in the natural world, typically through observation and experimentation, has enabled our ancestors to survive and thrive in competition with other species in an environment full of risks. Yet human language does not always adequately clarify what constitutes causal understanding and what does not. Many descriptive and predictive conclusions are often mistaken for causal conclusions; and such misunderstandings can be consequential.

2.1.1 Attributes versus causes

A statement such as "this student is high achieving *because* he is smart" views intelligence as a cause of achievement. The statement *describes* a visible pattern in the student population in which those who possess a higher level of intelligence tend to score higher on tests. This same statement may also be useful for *predicting* a student's future academic performance given prior knowledge of his intelligence level. Scientific rationales can be provided for explaining the relationship between intelligence and achievement. Although such information is useful in many ways, the description and prediction do not indicate how any given student could become a high achiever.

An alternative statement such as "this student is high achieving *because* he works hard" provides a behavioral explanation for achievement in which the behavior could arguably be changed. This second statement attributes the student's achievement to his earlier action, suggesting that a conceivably different action taken by the same student (should he have not studied hard) might have led to a different outcome (he would have scored lower on the test). The counterfactual action, even though not directly observed, is conceivable because the student could have received a discouragement to study or could even have been prevented from studying by the experimenter or by a natural event.

Causality in a Social World: Moderation, Meditation and Spill-over, First Edition. Guanglei Hong.
© 2015 John Wiley & Sons, Ltd. Published 2015 by John Wiley & Sons, Ltd.

Holland (1986) made a sharp distinction between attributes and causes and argued that an attribute such as gender, race, and intelligence cannot be a cause unless it can be experimentally manipulated. By his criterion, whether a student studies hard is at least partly a voluntary act that may reflect the student's attribute (e.g., being persistent). He expressed doubt with regard to whether a student could be experimentally exposed to hard work although he acknowledged that one could be prevented from studying (Holland, 1988). The bottom line is whether a student could possibly have studied a different number of hours. The effect of a cause is undefined, according to Holland (1986), without a contrast with a possibly counterfactual cause such as spending either more time, less time, or no time studying. This guideline is summarized in the motto: no causation without manipulation.

Statisticians who have made major contributions to the causal inference literature are generally in agreement with Holland's motto except for pointing out that attributes such as gender, race, and intelligence could conceivably take a counterfactual form by creative experimental manipulation (Pearl, 2009; Rubin, 1986; VanderWeele and Hernán, 2012). For example, on a form, an African American may be identified as white for experimental purposes. In this regard, even though the question of racial disparity in employment may be considered descriptive rather than causal, the question of whether being perceived as a member of a racial minority reduces or increases one's employment opportunity can arguably be addressed as a causal question when the experimenter manipulates an applicant's racial identity on job applications submitted to potential employers (Bertrand and Mullainathan, 2004).

2.1.2   Potential outcomes and individual-specific causal effects

The previous discussion highlights the role of a contrary-to-the-fact event for a given individual in defining a causal effect. For example, rather than being granted time to study before a test, a student might be involved in athletic activities and therefore might find no time to study for the test. In theory, the causal effect of studying versus not studying before a test on the student's test score is the difference between the student's potential test score after studying and the same student's potential test score without studying. One would conclude that studying before the test has no effect if the student would score the same with or without studying.

To formalize these concepts, we need some notation. Throughout the entire book, we use a capitalized italic letter to represent a random variable and an uncapitalized italic letter for a particular value of the random variable. Here, we use Z to denote a random variable indicating whether a student studies before a test. This random variable may take two possible values: $z = 1$ represents studying before a test and $z = 0$ for not studying before a test. For individual student i, $Y_i(z=1)$ denotes the student's potential test score if studying before a test, and $Y_i(z=0)$ denotes the same student's potential test score if not studying before a test. Hence, the causal effect of studying versus not studying on student i's test score denoted by Δ_i is the difference between these two potential outcomes:

$$\Delta_i = Y_i(z=1) - Y_i(z=0). \tag{2.1}$$

The alternative events ($z = 1$ vs. $z = 0$) are usually called treatments or exposures. In the following, we use $Y_i(1)$ and $Y_i(0)$ as a shorthand for $Y_i(z=1)$ and $Y_i(z=0)$, respectively, and write the individual-specific causal effect as $Y_i(1) - Y_i(0)$. Table 2.1 provides a glossary of notation for all of the basic concepts in causal inference introduced in this chapter.

Table 2.1   Glossary of basic concepts in causal inference

Notation	Definition
Z	Random variable for treatment assignment
$z = 1$	Assignment to the experimental condition
$z = 0$	Assignment to the control condition
Y	Random variable for the observed outcome
$Y(z)$	Potential outcome associated with treatment z
$E[Y(z)]$	Population average potential outcome associated with treatment z
$\Delta = Y(1) - Y(0)$	Individual-specific treatment effect
$\delta = E(\Delta) = E[Y(1) - Y(0)]$	Population average treatment effect
$E[Y|Z = 1]$	Average observed outcome of the treated units in the population
$E[Y|Z = 0]$	Average observed outcome of the control units in the population
$\delta_{PF} = E[Y|Z = 1] - E[Y|Z = 0]$	Prima facie causal effect
$\delta_{PF} - \delta$	Selection bias in identifying population average treatment effect
$E[Y(1)|Z = 1]$	Average potential outcome associated with the experimental condition for the treated units in the population
$E[Y(0)|Z = 1]$	Average counterfactual outcome associated with the control condition for the treated units in the population
$E[Y(0)|Z = 0]$	Average potential outcome associated with the control condition for the control units in the population
$E[Y(1)|Z = 0]$	Average counterfactual outcome associated with the experimental condition for the control units in the population
$E[\Delta|Z = 1]$	Average treatment effect on the treated
$E[\Delta|Z = 0]$	Average treatment effect on the untreated
$pr(Z = 1)$	Probability of treatment assignment, that is, proportion of the population treated
$pr(Z = 0) = 1 - pr(Z = 1)$	Proportion of the population untreated
$E[Y(z)|Z = z] = E[Y(z)]$	Independence assumption, also called the ignorability assumption
X	Pretreatment covariate
$E[Y(z)|X = x]$	Average potential outcome associated with treatment z in the subpopulation with pretreatment characteristic x
$E[\Delta|X = x]$	Average treatment effect in the subpopulation with pretreatment characteristic x
$E[Y|Z = z, X = x]$	Average observed outcome of individuals assigned to treatment z in the subpopulation with pretreatment characteristic x
$E[Y(z)|Z = z, X = x]$	Average potential outcome associated with treatment z of individuals assigned to treatment z in the subpopulation with pretreatment characteristic x

Table 2.1   (*Continued*)

Notation	Definition
$E[Y(z)\|Z = z, X = x] = E[Y(z)\|$ $X = x]$	Conditional independence assumption, also called the strong ignorability assumption
$\text{pr}(X{=}x)$	Proportion of individuals with pretreatment characteristic x in the population
$\text{pr}(Z = 1\|X = x)$	Probability of treatment assignment in the subpopulation with pretreatment characteristic x
$\bar{Y}^{(E)}$	Sample estimate of the population average observed outcome of the treated
$\bar{Y}^{(C)}$	Sample estimate of the population average observed outcome of the untreated
$\hat{\delta}_{\text{PF}} = \bar{Y}^{(E)} - \bar{Y}^{(C)}$	Sample estimate of the prima facie causal effect
$E(\hat{\delta}_{\text{PF}} - \delta_{\text{PF}})$	Sampling bias in estimating the prima facie causal effect
$\text{Var}(\hat{\delta}_{\text{PF}})$	Sampling variability in estimating the prima facie causal effect
$\sqrt{\text{Var}(\hat{\delta}_{\text{PF}})}$	Standard error of the sample estimate of the prima facie causal effect
$\sigma^2_{Y\|Z}$	Within-group variance in the observed outcome assumed to be constant across the treatment groups
$\bar{Y}^{(E)}_x$	Sample estimate of the average observed outcome of the treated in the subpopulation with pretreatment characteristic x
$\bar{Y}^{(C)}_x$	Sample estimate of the average observed outcome of the untreated in the subpopulation with pretreatment characteristic x
$\bar{Y}^{(E)}_x - \bar{Y}^{(C)}_x$	Sample estimate of the average treatment effect in the subpopulation with pretreatment characteristic x
$\sigma^2_{Y\|Z,X}$	In the subpopulation with pretreatment characteristic x, within-group variance in the observed outcome assumed to be constant across the treatment groups and across levels of $X{=}x$

This framework of defining causal effects in terms of potential outcomes, initially proposed by Neyman (1923) in the context of agricultural research and serving as a conceptual foundation for randomization (Fisher, 1935; Neyman, Iwaszkiewicz, and Kolodziejczyk, 1935; Rubin, 1978), is often cited as the "Neyman–Rubin causal model" and is now widely adopted in causal inference research (also see Cox, 1958; Holland, 1986; Weisberg, 1979). The interest in modeling causality has had a long tradition in the econometrics literature as well. For example, Haavelmo (1943) made a distinction between economic models for prediction and those for representing causation. Roy (1951) discussed economic principles of occupation selection on the basis of anticipated prices, which directly influenced the development of selection models in labor economics. The concept of potential outcomes was explicitly employed in empirical causal analyses to account for selection bias in this stream of literature in the late 1970s (Heckman, 1976, 1979; Willis and Rosen, 1979).

The individual-specific treatment effect as defined in (2.1) implies a number of conditions:

a. Most notably, when contrasting a treatment ($z = 1$) with an alternative treatment ($z = 0$), the causal effect is defined at the level of an individual unit who could possibly experience either of the two treatments. We have used a random variable Z_i to denote student i's treatment assignment; Z_i can possibly take values 0 or 1 in the current binary contrast of studying versus not studying. The causal effect is undefined for a student who would choose not to study even if granted study time before the test since Z_i becomes a constant equal to 0. For such a student, the potential outcome associated with studying before a test $Y_i(z = 1)$ is inconceivable.

b. When using (2.1) to define a causal effect, one must assume that there is a single potential outcome associated with each treatment for each individual unit. In other words, potential outcome $Y_i(z)$ is strictly a function of and is therefore determined by the corresponding treatment z for individual i. This is the "stable unit treatment value assumption" (SUTVA) that Rubin (1980, 1986) has repeatedly emphasized. For example, suppose that John would score 80 if studying before the test and would score 70 if not studying before the test. SUTVA stipulates that the value of $Y_i(z)$ for individual i when exposed to treatment z will stay fixed (i.e., 80 if $Z_i = 1$, and 70 if $Z_i = 0$ for John) no matter what mechanism is used to assign the treatment to the individual and no matter what treatments the other individuals in the population are exposed to. In the current example, SUTVA implies that there is a single version of each treatment and that there is no contextual influence such as parental input or peer interference. SUTVA would be violated, for example, (i) if John would score higher when volunteering to study than when urged to study by his parents; (ii) if he would score higher in anticipation of a reward for his hard work than without an anticipated reward; (iii) if he would score higher when he and his best friend were both assigned to study by the experimenter than when he was assigned to study, while his best friend was assigned not to study; and (iv) if he would score higher when studying with high-performing peers than with low-performing peers. In the first two examples, volunteering versus being urged to study may be viewed as two different versions of the treatment; with or without an anticipated reward when John studies are two other versions of the treatment. In the last two examples, John's test score may depend on the peers' treatment assignments.

For simplicity, in the "Overview" section and the section on "Moderation," most of the causal effects will be defined, identified, and estimated under SUTVA. However, in the section on "Mediation," we will allow an individual's potential mediator under a given treatment to be a random variable that could possibly take different values. The section on "Spillover" is devoted to the development of concepts and methods when SUTVA is clearly inconsistent with social science theories with regard to the importance of social interactions. We will assess the consequences for causal inference when SUTVA is invoked yet does not hold and will modify the causal framework to allow for peer interference.

2.1.3   Inference about population average causal effects

Even though (2.1) has provided a conceptual foundation for defining causal effects of interest, causal inference is hindered by a fundamental problem (Holland, 1986), that is, one would observe $Y_i(1)$ but not $Y_i(0)$ if student i studied and would observe $Y_i(0)$ but not $Y_i(1)$ if the

Table 2.2   Potential outcomes and population average causal effect

| Unit | Treatment Z | Potential outcomes | | Causal effect |
		$Y(1)$	$Y(0)$	$\Delta = Y(1)-Y(0)$
1	1	80	70	10
2	1	70	50	20
3	1	50	40	10
4	1	100	95	5
5	1	60	45	15
6	0	65	40	25
7	0	40	15	25
8	0	15	5	10
Population average		$E[Y(1)] = 60$	$E[Y(0)] = 45$	$E[\Delta] = \delta = 15$
Observed average		$E[Y(1)|Z = 1] = 72$	$E[Y(0)|Z = 0] = 20$	

student did not study. Clearly, the individual-specific causal effect cannot be calculated because either the student has studied or not studied, but it cannot be both within a defined time frame. Holland (1986) viewed this as a missing data problem. Nonetheless, he argued that a statistical solution is available for making causal inference about the population average treatment effect under an important assumption to be explained shortly.

Table 2.2 describes a hypothetical population of eight students. The individual-specific causal effect of studying versus not studying as shown in the last column ranges from 5 to 25 in test score with an average of 15. This average of the individual-specific causal effect over all the individuals in the population is the population average treatment effect. In general, in a finite population of N individuals, the population average causal effect is defined as

$$\delta = E(\Delta) = \frac{\sum_{i=1}^{N} [Y_i(1) - Y_i(0)]}{N}. \tag{2.2}$$

Here, $E(\Delta)$ is simply the expected or average value of the random variable Δ in the population where Δ denotes the individual-specific causal effect.

More generally, let $Y(1)$ and $Y(0)$ be each a random variable. We have that $\Delta = Y(1) - Y(0)$. The population average causal effect is represented mathematically as

$$\delta = E(\Delta) = E[Y(1) - Y(0)] = E[Y(1)] - E[Y(0)]. \tag{2.3}$$

The last equation in (2.3) holds because the mean of a difference is always equal to the difference between the means. In the hypothetical example in Table 2.2, $E[Y(1)] = 60$ represents the population average potential outcome should all the individuals in this hypothetical population study before the test, and $E[Y(0)] = 45$ represents the population average potential outcome should none of the individuals study before the test. Hence, the difference between these two population average potential outcomes is equal to the population average causal effect: $60 - 45 = 15$.

2.1.3.1   Prima facie effect

In this same example, suppose that the first five students actually studied and the last three did not study before the test, that is, $Z = 1$ for the first five students and $Z = 0$ for the last three students. When a student actually studied, under SUTVA, the observed test score Y would equal the student's potential outcome associated with studying before a test $Y(1)$; similarly, the observed test score Y of a student who did not study would equal the student's potential outcome associated with not studying $Y(0)$. The last row of Table 2.2 summarizes the average observed score of each of these two groups of students. The average observed score of the five students who studied is 72 and is denoted by $E[Y|Z = 1]$, read as the conditional expectation of Y for the $Z = 1$ group. Here, the condition statement "$Z = 1$" to the right of the vertical line "|" defines the subpopulation of individual units for whom we take the average value of the random variable Y. When a student actually studied, under SUTVA, the observed outcome Y is equivalent to the potential outcome $Y(1)$. Therefore, we have that $E[Y|Z = 1] = E[Y(1)|Z = 1]$. Similarly, the average observed score of the three students who did not study is 20 and is denoted by $E[Y|Z = 0]$, which equals their average potential outcome $E[Y(0)|Z = 0]$. The mean difference in the observed score between those who studied and those who did not study is 52. As discussed earlier, the population average causal effect of studying versus not studying is 15. If one mistakenly takes the mean difference between the two groups for the population average causal effect, the bias is as large as 37, an enormous overestimation of the benefit of studying before the test!

Holland (1986) called the mean difference in the observed outcome between an experimental group and a control group "the prima facie causal effect":

$$\delta_{PF} = E[Y|Z = 1] - E[Y|Z = 0]. \tag{2.4}$$

In many cases, δ_{PF} is unequal to the population average causal effect δ. In the current hypothetical example, two selection factors contribute to the bias. The first is that in comparison with the group of students who did not study, the group that actually studied would have *scored higher* on average even if they were prevented from studying, that is, $E[Y(0)|Z = 1] = 60$ is unequal to and in fact higher than $E(Y|Z = 0)$, In general, δ_{PF} would contain a positive bias if the experimental group would enjoy an advantage over the control group even in the absence of the treatment. However, this first bias is hidden because the potential outcome associated with not studying cannot be observed for the individuals who actually studied, and hence, $E[Y(0)|Z = 1]$ is unknown to the researcher. The second factor is that the group of students who did not study, should they have studied instead, would have *benefitted more* than the group that actually studied. This is because the causal effect for the group who did not study is $E[Y(1)|Z = 0] - E[Y(0)|Z = 0] = 40 - 20 = 20$, while the causal effect for the group who actually studied is $E[Y(1)|Z = 1] - E[Y(0)|Z = 1] = 72 - 60 = 12$. In general, if two groups would not gain equally from the treatment, δ_{PF} would contain a negative bias if the group that would benefit more from the treatment makes less use of it; δ_{PF} would contain a positive bias if, on the contrary, the treatment is delivered primarily to the individuals for whom the benefit would be maximized, should rational choices be made. This second bias is also hidden exactly because $E[Y(1)|Z = 0]$ and $E[Y(0)|Z = 1]$ are unknown quantities to the researcher.

In summary, a prima facie effect may not be equal to a population average causal effect because the former may contain two potential sources of bias: *Bias1* comes from selection based on the potential outcome in the absence of the treatment, that is, the mean difference between the experimental group and the control group in $Y(0)$; *Bias2* comes from selection on the gain, that is, the mean difference in the treatment effect Δ between the experimental group and the control group. The sum of these two potential sources of bias is

$$\delta_{PF} - \delta = Bias1 + Bias2$$

where

$$Bias1 = E[Y(0)|Z=1] - E[Y(0)|Z=0] \text{ and } Bias2 = \{E[\Delta|Z=1] - E[\Delta|Z=0]\} \times pr(Z=0).$$

$$(2.5)$$

Here, $pr(Z=0)$ denotes the proportion of the population that has been assigned to the control condition. When applied to the current example, $E[\Delta|Z=1]$ is the average treatment effect for those who actually studied, and $E[\Delta|Z=0]$ is the average treatment effect for those who did not study. The importance of the second source of bias increases as the proportion of individuals in the population who did not study increases. Appendix 2.1 shows the derivation. This appendix also reveals that only *Bias1* is of concern if the researcher is primarily interested in the average treatment effect on the treated defined as $E[\Delta|Z=1]$.

2.1.3.2   Ignorability assumption

The prima facie effect becomes unbiased for the population average treatment effect when a crucial assumption is satisfied. This assumption states that the treatment assignment Z is *independent* of the potential outcomes $Y(1)$ and $Y(0)$, in which case the treatment assignment is considered to be *ignorable* in the Bayesian sense (Rubin, 1978) or "worry-free" in intuitive terms. The treatment could be assigned by unknown forces in nature, by the experimenter, by an agency that provides the treatment, or self-assigned by the individual participant. In the current example, Z is *independent* of $Y(1)$ if the students who actually studied and those who did not study before the test would have scored the same on average had they all studied, that is, if

$$E[Y(1)|Z=1] = E[Y(1)|Z=0] = E[Y(1)],$$

in which case the mean observed outcome of the group of students who actually studied, denoted by $E[Y|Z=1]$, would be unbiased for $E[Y(1)]$, the population average potential outcome associated with studying before the test. In parallel, Z is *independent* of $Y(0)$ if the students who actually studied and those who did not study before the test would have scored the same on average had none of them studied, that is, if

$$E[Y(0)|Z=1] = E[Y(0)|Z=0] = E[Y(0)].$$

The mean observed outcome of those who did not study, denoted by $E[Y|Z=0]$, would then be unbiased for $E[Y(0)]$, the population average potential outcome associated with not studying before the test.

When the ignorability assumption holds, the two potential sources of bias are eliminated. Interestingly, if Z is independent of $Y(0)$ but not independent of $Y(1)$, the first source of bias will disappear, while the second source will remain. In everyday language, evaluation results often invite criticisms when an experimental group and a control group are found "incompa-rable." The ignorability assumption clarifies the elements that need to be comparable between the treatment groups. To be explicit, causal inference requires that the experimental group and the control group have comparable distributions of $Y(1)$ and comparable distributions of $Y(0)$. Of course, the ignorability assumption can never be empirically verified because $Y(0)$ is unobserved when $Z=1$ and because $Y(1)$ is unobserved when $Z=0$.

In practice, two different approaches have been taken in attempts to satisfy the ignorability assumption. One is through research design and the other through extra data collection and statistical adjustment. The former has been categorized as the experimental approach, and the latter the quasiexperimental or observational approach. In an experimental study, indivi-duals are assigned at random to either the experimental condition or the control condition dis-regarding how they would respond under each treatment condition. Hence, the randomization procedure guarantees the ignorability of treatment assignment.

In quasiexperimental studies, acknowledging possible selection of individual participants into a certain treatment condition, researchers often seek to satisfy the ignorability assumption within subpopulations defined by measured pretreatment characteristics. We use X to denote a pretreatment measure. For example, if the pretest scores of all students are available and given that pretest scores are usually the best predictor of posttest scores, one would hope that among those who scored the same on the pretest, those who studied and those who did not study before the posttest would have scored the same again on average on the posttest had they all studied or had they all not studied. Let $X=x$ define a subpopulation of students whose pret-est scores share the same value x. A relatively strong version of the ignorability assumption states that conditioning on $X=x$, that is, within a subpopulation defined by $X=x$, treatment assignment becomes independent of the potential outcomes:

$$E[Y(1)|Z=1,\ X=x]=E[Y(1)|Z=0,\ X=x]=E[Y(1)|X=x];$$
$$E[Y(0)|Z=1,\ X=x]=E[Y(0)|Z=0,\ X=x]=E[Y(0)|X=x].$$

One may obtain the mean difference in the observed outcome between the experimental group and the control group within the subpopulation, $E[Y|Z=1,\ X=x]-E[Y|Z=0,\ X=x]$. When the strong ignorability assumption holds, this is unbiased for the subpopulation-specific treatment effect $E[\Delta|X=x]=E[Y(1)-Y(0)|X=x]$. One then evaluates the population average treatment effect δ by pooling the subpopulation-specific treatment effect that is, taking an average of the subpopulation-specific treatment effect over all the subpopulations:

$$\delta=E\{E[\Delta|X=x]\}. \tag{2.6}$$

For example, if pretest score X takes possible integer values between 0 and 100, in eval-uating the population average treatment effect, each subpopulation-specific treatment effect is to be weighted by the proportion of individuals in that subpopulation denoted by $\mathrm{pr}(X=x)$ for $x=0,\ldots,100$:

$$\delta=\sum_{x=0}^{100}E[\Delta|X=x]\times\mathrm{pr}(X=x).$$

The strong ignorability assumption invoked in a quasiexperimental study is considerably stronger than the ignorability assumption warranted by an experimental design. In practice, there are numerous cases in which the strong ignorability assumption is misused so that empirical evidence generated from quasiexperimental data can become seemingly perplexing and misleading. In the following are two classical examples, one described by Frederic M. Lord to exemplify Lord's paradox in real-world evaluation studies (Lord, 1967, 1975) and the other an illustration of Simpson's paradox. We will show that the potential outcomes causal framework sheds important light on each of these paradoxes.

2.2   Applications to Lord's paradox and Simpson's paradox

2.2.1   Lord's paradox

Lord's paradox (Lord, 1967, 1975) refers to a common flaw in observational research that was seen, for example, in the widely publicized Westinghouse report (Westinghouse Learning Corporation and Ohio University, 1969). This report was the first major evaluation of the federally funded Head Start programs targeted at disadvantaged preschool-aged children in the United States. In a simple setup described by Lord (1975), a group of disadvantaged children are pretested at the beginning of the year, are then enrolled in the Head Start program, and receive a posttest at the end of the year; a control group of children receive similar pretest and posttest but are not enrolled in the Head Start program. The Head Start group has a concentration of the most disadvantaged children and therefore has lower average scores in both pretest and posttest than does the control group. For simplicity, suppose that the test scores have been normalized such that a score represents an individual's relative standing in the population of same-age children at a given time point. We use X to denote the pretest score and Y for the posttest score. Let $Z = 1$ if a child was enrolled in a Head Start program and 0 otherwise.

In Lord's example, the average pretest score is equal to the average posttest score in the Head Start group; the average pretest and posttest scores are also equal in the control group. Lord showed that two fictitious statisticians analyzing the same data would draw contradictory conclusions. Because the average gain scores are equal between the Head Start group and the control group, the first statistician concludes that there is no evidence of an effect of the Head Start program on student test score. This is illustrated in Figure 2.1. The two ellipses represent the joint distribution of pretest and posttest scores for the Head Start group and that for the control group. Each ellipse may be viewed as a scatterplot. Within each of these two groups, because the average pretest score and the average posttest score are equal, the center of each ellipse lies on the 45° line, which suggests no average gain for either group.

In contrast, the second statistician, upon close examination of the two ellipses, makes the observation that the control group scores higher than the Head Start group on average in the pretest and that the pretest–posttest association is essentially the same across the two groups. Hence, this statistician fits two parallel regression lines, one for each group, as shown in Figure 2.2. The difference in the intercepts of these two regression lines, which is the vertical distance between the two regression lines, represents the average difference in the posttest score between the Head Start children and the control children who scored the same on the pretest. After making this adjustment for the preexisting between-group difference in the pretest score, this second statistician finds that the control group displays a significantly higher posttest score than the Head Start group on average and therefore concludes that the control group has gained more than the Head Start group during the treatment year.

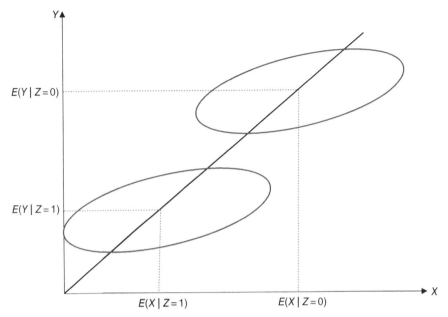

Figure 2.1   Lord's paradox: statistician 1's gain score analysis

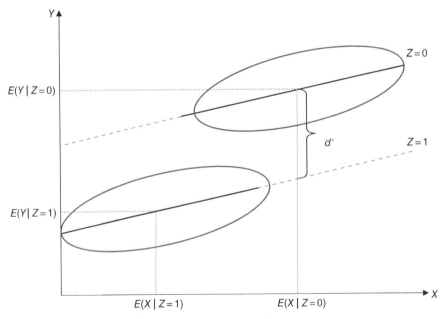

Figure 2.2   Lord's paradox: statistician 2's analysis of covariance

Neither of these two statisticians' conclusions would lend support to a legislation author-izing a continuation of federal funding for Head Start programs. Yet the conclusions drawn from analyzing the same data are vastly different. These seemingly contradictory results become less paradoxical when they are represented in terms of potential outcomes (Holland and Rubin, 1983; Weisberg, 1979). The causal question of interest is whether the federally funded Head Start programs targeted at children from low-income households pro-duce an intended positive effect on these children's posttest scores. The treatment effect on the treated (i.e., the effect of Head Start vs. no Head Start for the population of children who were actually enrolled in Head Start) is represented as

$$E[Y(1) - Y(0)|Z = 1]. \tag{2.7}$$

Statistician 1's conclusion is based on an analysis that compares the gain score computed as the pre-post difference in test score $Y - X$ between the Head Start group and the control group:

$$Y - X = a + dZ + e, \tag{2.8}$$

where a is the average gain score in the control group, d is the average between-group differ-ence in the gain score:

$$d = [E(Y|Z = 1) - E(X|Z = 1)] - [E(Y|Z = 0) - E(X|Z = 0)], \tag{2.9}$$

and e is the error term capturing the difference between an individual gain score and the group average gain score.

Statistician 2 conducts an analysis of covariance (ANCOVA) instead by modeling the covariance between pretest and posttest scores conditioning on group membership Z and letting the X–Y slope be the same between the two groups. One may represent the analysis in the form of a regression model:

$$Y = a' + bX + d'Z + e'. \tag{2.10}$$

In Figure 2.2, the relationship between X and Y is represented by the regression lines with slope b parallel between the Head Start group and the control group. Similar to Model (2.8), Model (2.10) makes a between-group comparison of the average posttest score with the between-group difference in the average pretest score removed:

$$Y - bX = a' + d'Z + e'.$$

The treatment effect is then identified as

$$d' = [E(Y|Z = 1) - E(Y|Z = 0)] - b[E(X|Z = 1) - E(X|Z = 0)]. \tag{2.11}$$

Clearly, the conclusions of the two statisticians would converge if b is 1.0.

However, neither statistician's result can be equated to the treatment effect on the treated defined in (2.7) without further assumptions about the Head Start children's average counterfac-tual outcome associated with the control condition $E(Y(0)|Z = 1)$. Statistician 1 has implicitly assumed that if the Head Start children had counterfactually not been enrolled in Head Start, they would have gained the same amount on average as would the control group children over the year. That is,

$$E(Y(0)|Z=1)-E(X|Z=1)=E(Y(0)|Z=0)-E(X|Z=0).$$

Statistician 2's conclusion is based on two assumptions. The first is that in the absence of Head Start, a subgroup of Head Start children and a subgroup of control children with the same pretest score would have performed the same on average on the posttest. The second assumption is that for the Head Start children, the linear association between pretest score X and the counterfactual posttest score in the absence of Head Start $Y(0)$ would be the same as that between pretest score X and the observed posttest score Y in the control group represented by the slope b. The assumptions can be summarized as follows:

$$E(Y(0)|Z=1, X=x)=E(Y(0)|Z=0, X=x)=a'+bx.$$

Under these two assumptions, the control group regression line extended to the left supplies information about the counterfactual outcome $Y(0)$ for the Head Start children.

Appendix 2.2 shows how statistician 1's assumption leads to (2.9) and how statistician 2's assumptions lead to (2.11) for identifying the average treatment effect on the treated. Once these assumptions required for identifying the causal effect are made transparent, each assumption can then be evaluated on scientific grounds. An empirical result is convincing only if its identification assumptions are highly plausible.

Given that the Head Start children are generally from a more disadvantaged background than the control children, Statistician 1's assumption that in the absence of Head Start, the Head Start children on average would nonetheless have gained the same amount as would the control group children over the year seems questionable. The assumption contradicts empirical knowledge accumulated in past research showing that lower-SES children learn at a lower rate than higher-SES children when the school is not in session. According to this reasoning, statistician 1 might have overestimated Head Start children's average counterfactual outcome and therefore underestimated the Head Start benefit.

Statistician 2's assumptions seem equally dubious. A subgroup of Head Start children and a subgroup of control children who score the same on the pretest are not necessarily "comparable" in every other aspect. One may argue that despite their disadvantaged family background, the subgroup of Head Start children who come on par with the control children in the pretest might have cognitive skills and personality traits that allow them to excel at a faster rate than their counterparts in the control group even in the absence of Head Start. Yet it also seems likely that during the treatment year, the same Head Start children might fail to overcome unpredictable new stressors in life that tend to plague low-income households and impede cognitive development in the absence of the Head Start intervention. Statistician 2 would have perhaps underestimated the potential negative impact of Head Start for this group of children under the former argument and would have perhaps overlooked the Head Start benefit if the latter argument were true.

Clearly, to estimate the average counterfactual outcome for Head Start children with least problematic assumptions, one would randomize children eligible for Head Start either to a Head Start program or to a control condition. The randomization could be conducted within subpopulations of children who score the same on the pretest not only for increasing the precision of treatment effect estimation but also for detecting possible heterogeneous treatment effect if there are theoretical reasons to hypothesize that disadvantaged children with higher pretest performance would benefit more (or less) from Head Start.

2.2.2 Simpson's paradox

Simpson's paradox, named after Edward H. Simpson, describes the paradoxical phenomenon that the relationship between a supposed cause and an outcome appearing in different sub-populations becomes nonexistent or the opposite when the subpopulations are combined. A classic example is the relationship between maternal smoking and infant mortality (Hernández-Díaz, Schisterman, and Hernán, 2006; Wilcox, 1993; Yerushalmy, 1971). It has been found that US infants born to mothers who smoked had higher risks of both low birth weight and infant mortality, suggesting a harmful effect of smoking on child health. However, among infants with low birth weight, mortality was lower for those born to smokers than non-smokers, which suggests, quite counterintuitively, a health benefit of maternal smoking for infants with low birth weight.

Some may view these paradoxical findings as an instance of moderation and conclude that the relationship between smoking and infant mortality depends on birth weight. Others may reason that low birth weight is at least partly a result of smoking and that birth weight can be regarded as a mediator on the causal pathway from smoking to mortality. According to this second point of view, although the total effect of smoking on infant mortality is detrimental, the direct effect of smoking on infant mortality not transmitted through birth weight could possibly be positive. Neither the reasoning of moderation nor that of mediation challenges the conclusion that maternal smoking appears to benefit infants with low birth weight.

The potential outcomes framework again sheds light on this paradoxical issue. The main controversy has to do with the causal effect of smoking versus nonsmoking on the mortality of low-birth-weight infants born to smokers. Here, maternal smoking is the treatment of interest and is denoted by $Z = 1$ if the mother smoked and 0 otherwise. The outcome Y takes value 1 if a child did not survive during the first year of life and 0 otherwise. We use M to denote a child's birth weight that takes value 1 if the birth weight was below 2000 g, a crossing point below which children born to smokers showed a lower mortality rate than those born to nonsmokers; $M = 0$ if the child's birth weight was at or above 2000 g (Hernández-Díaz, Schisterman, and Hernán, 2006). Because birth weight is generally regarded as an intermediate outcome of smoking, the observed birth weight of children born to smokers is equal to their potential inter-mediate outcome $M(1)$ associated with maternal smoking; in parallel, the observed birth weight of children born to nonsmokers is equal to their potential intermediate outcome $M(0)$ associated with nonsmoking. Child mortality can be written as a potential outcome of both maternal smoking and birth weight, denoted by $Y(z, M(z))$ for $z = 0, 1$. To be specific, the potential outcome is $Y(1, M(1) = 1)$ if a child was born to a smoker and recorded low birth weight; the potential outcome is $Y(0, M(0) = 1)$ if a child was born to a nonsmoker and recorded low birth weight; the potential outcome is $Y(1, M(1) = 0)$ if a child was born to a smoker and recorded normal birth weight; and the potential outcome is $Y(0, M(0) = 0)$ if a child was born to a nonsmoker and recorded normal birth weight.

For an individual child in the subpopulation of low-birth-weight children born to smokers, the causal effect of smoking versus nonsmoking is $Y(1, M(1)) - Y(0, M(0))$, which is equivalent to $Y(1) - Y(0)$ because the second argument $M(z)$ in this case is simply a function of the first argument z for $z = 0, 1$ and therefore provides no additional information in defining the potential outcome of maternal smoking and that of nonsmoking. Now, we can represent the subpopulation-specific average causal effect of smoking versus nonsmoking on infant mortality as the conditional expected difference between the two potential outcomes $Y(1)$ and $Y(0)$ for the subpopulation defined by the maternal smoking status $Z = 1$ and by the status of low birth weight $M(1) = 1$:

$$E[Y(1) - Y(0)|Z = 1,\ M(1) = 1].\tag{2.12}$$

Within this subpopulation, the mean difference between the two potential outcomes is always equal to the difference between the two means. Hence, the above is equal to

$$E[Y(1)|Z = 1,\ M(1) = 1] - E[Y(0)|Z = 1,\ M(1) = 1].$$

The first quantity $E[Y(1)|Z = 1,\ M(1) = 1]$ is equal to the observed mortality rate among low-birth-weight infants born to smokers; yet the second quantity $E[Y(0)|Z = 1,\ M(1) = 1]$ is counterfactual because it represents the mortality rate in the subpopulation of low-birth-weight infants born to smokers should these children have lived, counterfactually, in a fetal environment free of cigarette toxicities. It is also important to note that even though the birth weight of children born to smokers is low in this particular subpopulation, the same children's counterfactual birth weight $M(0)$ if their mothers had not smoked is not necessarily low. Rather, it is conceivable that without maternal smoking, a certain proportion of these children might have been born with normal weight. Hence, the second quantity is the average counterfactual mortality rate of children who would have been born with normal weight had their mothers not smoked and those who would still have been born with low weight even if their mothers had not smoked.

In past research that stratifies the population by birth weight (Wilcox, 1993, 2001; Yerushalmy, 1971), a comparison of the observed mortality rate was made between the aforementioned subpopulation of low-birth-weight children born to smokers and a different subpopulation of low-birth-weight children born to nonsmokers. This alternative subpopulation brought in for comparison is defined by their mothers' nonsmoking status $Z = 0$ and by the status of low birth weight $M(0) = 1$. The comparison can be represented in terms of potential outcomes as follows:

$$E[Y(1)|Z = 1,\ M(1) = 1] - E[Y(0)|Z = 0,\ M(0) = 1].\tag{2.13}$$

Simpson's paradox can now be unpacked by examining the identification assumption invoked when (2.13) is chosen for identifying the causal effect defined in (2.12). The identification assumption is simply

$$E[Y(0)|Z = 1,\ M(1) = 1] = E[Y(0)|Z = 0,\ M(0) = 1].\tag{2.14}$$

That is, the mortality rates of low-birth-weight children born to smokers and nonsmokers are expected to be the same should none of the mothers have smoked. This assumption is not highly plausible on the basis of the following reasoning.

Yerushalmy (1971) reported a higher rate of congenital defects among low-birth-weight infants born to nonsmokers than to smokers. It is well known that infants born with congenital defects have a significantly higher mortality rate than those born without such defects. Hence, the overrepresentation of infants with congenital defects in the subpopulation of low-birth-weight children born to nonsmokers would inevitably inflate the infant mortality rate in this sub-population. Moreover, as discussed earlier, one may expect that a certain proportion of the low-birth-weight children with smoking mothers would have been born with normal weight if their mothers had not smoked. Under this counterfactual condition, these children's mortality rate is expected to be lower than that of children actually born with low birth weight even though their mothers did not smoke. The theoretical reasoning suggests that should none of the mothers have smoked, the mortality rate of low-birth-weight children born to smokers would be lower than rather than equal to the mortality rate of low-birth-weight children born to nonsmokers. Assumption (2.14) leads to an overestimation of the counterfactual mortality rate of low-birth-weight children born to smokers should their mothers have not smoked, which is the second quantity in (2.12). Consequently, the average causal effect of smoking versus nonsmoking on the mortality rate of low-birth-weight children born to smokers is obtained with bias in favor of the smokers.

In conclusion, the comparison of mortality rate between low-birth-weight children born to smokers and nonsmokers does not address the causal question about the impact of smoking on infant mortality for the former subpopulation. In fact, when conditioning on low birth weight that could have at least partially been a result of smoking, the two subpopulations of children under comparison become less "comparable" given the higher rate of congenital defects in the latter subpopulation and given that a proportion of the former subpopulation would have been born with normal weight had their mothers not smoked. It is now well known in the causal inference literature that statistical adjustment for an intermediate outcome of the treatment (such as conditioning on birth weight that could have been affected by maternal smoking) has a tendency to introduce rather than reduce selection bias in estimating the treatment effect on a final outcome (Hernández-Díaz, Schisterman, and Hernán, 2006; Rosenbaum, 1984). Intermediate outcomes such as birth weight in this case are often called "posttreatment covariates." The posttreatment status indicates that these variables could have been affected by the treatment; they are associated with the outcome because they either are causes of the outcome or share other common causes with the outcome. In the previous example, congenital defects at birth are one of the common causes of low birth weight and infant mortality.

In another variant of Simpson's paradox, researchers may mistakenly stratify on a covariate that could have been affected by two unrelated causes. Such a covariate has been called "a collider" in causal graphical models (Elwert and Winship, 2014; Pearl, 2009). In Berkson's classic example (1946), two diseases that are initially unrelated in the general population become negatively associated in a hospitalized subpopulation. because patients become hospitalized perhaps due to one of the two diseases but not necessarily both. Suppose that patients with disease A in general have a higher mortality rate and are more likely to be hospitalized than patients with disease B. The treatment of interest is a radical procedure received by some patients with disease B that would in fact have no impact on mortality. An analyst who has no access to the information about each patient's disease type may decide to restrict his analysis to those who are hospitalized and may subsequently find that hospitalized patients not receiving the treatment have a higher mortality rate than those receiving the treatment and therefore wrongly conclude that the treatment is beneficial.

The previous examples of Lord's paradox and Simpson's paradox, along with numerous other similar examples, demonstrate the crucial importance of clarifying causal thinking by utilizing the potential outcomes framework. Unless a causal effect of interest is defined with no ambiguity and the identification assumptions are spelled out and closely scrutinized on scientific grounds, the public can easily be confused and misled by contradictory "empirical findings" that blur and disguise rather than disclose the truth.

2.3   Identification and estimation

Once a causal effect has been defined, this section makes a conceptual distinction between identification and estimation as two distinct inferential tasks. The coursework for students of statistics focuses primarily on estimation, that is, how to make inferences about population parameters on the basis of sample data. The importance of identification has received great emphasis only in the recent statistical literature on causal inference. Econometric training in its best form attends to identification along with estimation. Yet the econometric approach to identification traditionally does not state the assumptions in terms of potential outcomes but rather relies heavily on model-based assumptions—that is, assuming that the parametric models correctly represent the causal structure. In the following, we discuss a converging trend in these fields.

Heckman and Vytlacil (2007) delineated all three tasks arising in causal inference:

1. *Define* the potential outcomes and the causal effect of interest.

2. *Identify* the causal effect from observable population data.

3. *Estimate* the causal effect from observed sample data.

A causal effect may be defined for an individual, for a subpopulation of individuals, or for the entire population. It always involves some counterfactuals, that is, potential outcomes associated with treatment conditions different from what one was actually exposed to. In the example of the Head Start evaluation, the causal effect of interest is the population average effect of Head Start versus no Head Start for children eligible for and actually attending Head Start defined as $E[Y(1)-Y(0)|Z=1]$ in (2.7). Here, the potential outcome associated with not attending Head Start $Y(0)$ is counterfactual for Head Start children.

A causal effect defined for a given population is identifiable only if the counterfactual quantities can be equated with *observable* population data even if such data were not actually *observed* for all individuals in the population. To identify the average effect of Head Start for Head Start children, the average potential outcome of attending Head Start for these children can simply be related to the observable outcome of all those eligible and attending Head Start in the population, that is, $E[Y(1)|Z=1]=E[Y|Z=1]$. However, identification assumptions are required for relating the counterfactual quantity $E[Y(0)|Z=1]$ to observable population data. Identification assumptions are typically determined by research designs and analytic methods. In Lord's paradox, statistician 1 and statistician 2 related the aforementioned counterfactual quantity to different observable quantities by invoking two different sets of identification assumptions. Statistician 1 assumed that if the Head Start children had been assigned to the control condition instead, they would have displayed the same gain score on average as did those who were actually in the control group. Statistician 2 assumed that in the absence of Head Start, Head Start children and control children with the same pretest score would have the same potential posttest score and that the latter is the same linear function of the pretest score for the two groups. They therefore generated contradictory answers to the same causal question. We have explained that the identification assumption invoked by neither statistician is credible on scientific grounds. Simpson's paradox provides another illustration of how an implausible identification assumption could lead to a misleading "finding."

In the conceptual discussions of all the hypothetical examples up to this point, we have supposed that one has access to population data. Of course, this is generally not true. Rather, researchers typically collect information from individual units in a sample that has been drawn one way or another from a certain population. The observable population quantities are then estimated from the observed sample data. Conceivably, a random sample of children enrolled in Head Start programs can be drawn from the national Head Start population when a sampling frame is used of all children eligible for and attending Head Start. The random sampling procedure will ensure that the sample is representative of the US population of Head Start children, as was done in the recent Head Start Impact Study (U.S. Department of Health and Human Services, Administration for Children and Families, 2010). Therefore, the average observed outcome of the Head Start sample denoted by $\bar{Y}^{(E)}$, a sample statistic that is easy to obtain, estimates the average observable outcome of the Head Start population denoted by $E(Y|Z=1)$.

In summary, at the identification stage, the major challenge is to locate observable information that can be equated with the counterfactuals of interest without introducing selection bias. Selection bias often becomes inevitable when the identification assumptions—including parametric model-based assumptions—do not hold. At the estimation stage, the major concerns are sampling bias and estimation efficiency. Sampling bias arises when a sample is unrepresentative of the population or when the choice of a statistical procedure introduces bias. Estimation efficiency is of concern when a sample contains relatively limited information about the population or when a statistical procedure is chosen that does not make full use of the information in the sample. In the following, we discuss each of these three problems.

2.3.1 Selection bias

As before, we use δ to denote the average treatment effect for the Head Start population $E(Y(1)-Y(0)|Z=1)$. In most quasiexperimental evaluations of Head Start, the control group often consists of children either ineligible for Head Start or eligible yet having access to resources for alternative childcare. By invoking an identification assumption that equates the average counterfactual outcome associated with the control condition for the Head Start children $E[Y(0)|Z=1]$ with the average observable outcome of the control population $E(Y|Z=0)$, one attempts to identify the causal effect with $\delta_{PF}=E(Y|Z=1)-E(Y|Z=0)$ where $E(Y|Z=1)$ and $E(Y|Z=0)$ are both observable population quantities. As clarified by Holland (1986), the selection bias is $\delta_{PF}-\delta$ and can be attributed in the current example to the nonequivalence of the Head Start population and the control population. A treatment effect is identified without selection bias if $\delta_{PF}-\delta=0$. Section 2.1.3.1 discussed two potential sources of selection bias when the population average treatment effect rather than the average treatment effect on the treated is of interest: the first source of bias is due to the between-group difference in the average potential outcome in the absence of the treatment, and the other due to the between-group difference in the average gain from the treatment.

2.3.2 Sampling bias

Distinguished from selection bias, sampling bias often arises as a consequence of nonrandom sampling or of nonrandom sample attrition. For example, a convenience sample of Head Start children in an extremely poor urban neighborhood is clearly not representative of the national population of Head Start children and would generate a sample mean outcome $\bar{Y}^{(E)}$ that is likely a biased estimate of the average observable outcome of the Head Start population $E(Y|Z=1)$. To make a clear distinction between the selection bias and the sampling bias, we use $\widehat{\delta}_{PF} = \bar{Y}^{(E)} - \bar{Y}^{(C)}$ to denote a sample estimate of the observable population quantity δ_{PF}. Hence, the sampling bias in causal inference is defined as $E\left(\widehat{\delta}_{PF} - \delta_{PF}\right)$. In words, $\widehat{\delta}_{PF}$ is considered to be an unbiased estimate of δ_{PF} if over an infinite number of random samples of the same size drawn from the same population, the mean of $\widehat{\delta}_{PF}$ is equal to the population parameter δ_{PF}.

2.3.3   Estimation efficiency

In addition to unbiasedness, efficiency is another desired property of estimation because it enables researchers to assess the treatment effect with precision and to distinguish a nonzero effect from zero. Unless one analyzes population data, the sample mean difference in the observed outcome between the experimental group and the control group $\widehat{\delta}_{PF} = \bar{Y}^{(E)} - \bar{Y}^{(C)}$ always varies from sample to sample. Its sampling variability is quantified by $\mathrm{Var}\left(\widehat{\delta}_{PF}\right)$, the square root of which is usually reported as the standard error of the sample estimate in statistical software output. The smaller the sampling variability is, the smaller the standard error and the confidence interval and hence the higher the efficiency, in which case there is greater statistical power to detect a nonzero treatment effect. Efficiency can be improved through increasing the sample size, through equalizing the sample size of the experimental group and that of the control group, or through conditioning on important covariates. An unbiased sample estimate with an enormous amount of uncertainty is no more informative than a biased sample estimate with a modest amount of sampling error. Hence, when comparing the performance of different analytic methods for causal inference, bias reduction and efficiency both need to be taken into account.

Appendix 2.1: Potential bias in a prima facie effect

As defined in Section 2.1.3, model (2.2), the population average causal effect is $\delta = E(\Delta) = E[Y(1)-Y(0)]$, where $\Delta = Y(1)-Y(0)$ is the corresponding individual-specific causal effect, the value of which may vary from individual to individual. In contrast, the prima facie effect is $\delta_{PF} = E[Y|Z=1] - E[Y|Z=0]$, that is, the mean difference in the observed outcome between the experimental group and the control group. Here, we derive the two potential sources of bias when δ_{PF} is used to identify δ. We will show that $\delta_{PF} - \delta = Bias1 + Bias2$ where $Bias1 = E[Y(0)|Z=1] - E[Y(0)|Z=0]$ and $Bias2 = \{E[\Delta|Z=1] - E[\Delta|Z=0]\} \times \mathrm{pr}(Z=0)$.

First of all, the population average treatment effect δ can be written as the average of the treatment effect for the treated units $E[\Delta|Z=1]$ and the treatment effect for the control units $E[\Delta|Z=0]$ weighted by the proportion of the treated units $\mathrm{pr}(Z=1)$ and the proportion of the control units $\mathrm{pr}(Z=0)$, respectively, in the population. This is equal to the difference between the average treatment effect for the treated and $Bias2$:

$$\begin{aligned}
\delta &= E[\Delta|Z=1]\mathrm{pr}(Z=1) + E[\Delta|Z=0]\mathrm{pr}(Z=0) \\
&= E[\Delta|Z=1][1-\mathrm{pr}(Z=0)] + E[\Delta|Z=0]\mathrm{pr}(Z=0) \\
&= E[\Delta|Z=1] - \{E[\Delta|Z=1] - E[\Delta|Z=0]\}\mathrm{pr}(Z=0) \\
&= E[\Delta|Z=1] - Bias2.
\end{aligned}$$

Next, we can show that the average treatment effect for the treated is equal to the difference between the prima facie effect and *Bias*1:

$$E[\Delta|Z=1]=E[Y(1)|Z=1]-E[Y(0)|Z=1]$$
$$=\{E[Y(1)|Z=1]-E[Y(0)|Z=0]\}-\{E[Y(0)|Z=1]-E[Y(0)|Z=0]\}$$
$$=\delta_{PF}-Bias1.$$

Hence, $\delta=\delta_{PF}-Bias1-Bias2$.

In some studies, the researcher is interested in estimating the average treatment effect on the treated $E[\Delta|Z=1]$ rather than the average treatment effect for all units in the population δ. When this is the case, *Bias*1 is the only bias contained in δ_{PF} because $E[\Delta|Z=1]=\delta_{PF}-Bias1$.

Appendix 2.2: Application of the causal inference theory to Lord's paradox

The key distinction between the two statisticians is in their assumptions about what constitutes information useful for drawing causal inference about the Head Start children's counterfactual outcome had they not attended Head Start.

Statistician 1 has assumed that in the absence of Head Start, the average gain score of the Head Start children, represented in terms of potential outcomes as $E(Y(0)|Z=1)-E(X|Z=1)$, would be equal to the average gain score actually observed in the control group: $E(Y|Z=0)-E(X|Z=0)$. Hence, the counterfactual quantity $E(Y(0)|Z=1)$ is replaced by the observed quantity: $[E(Y|Z=0)-E(X|Z=0)]+E(X|Z=1)$. We then obtain the treatment effect on the treated as the average between-group difference in the gain score as shown in (2.9):

$$E(Y(1)|Z=1)-E(Y(0)|Z=1)=E(Y|Z=1)-\{[E(Y|Z=0)-E(X|Z=0)]+E(X|Z=1)\}.$$

Under statistician 2's assumptions, for Head Start children and control children with the same pretest score $X=x$, the average potential outcomes in the absence of Head Start are expected to be the same between the two groups, that is, $E(Y(0)|Z=1, X=x)=E(Y|Z=0, X=x)=a'+bx$. Taking the average of the counterfactual outcome for the Head Start children $Y(0)$ over their respective distribution of X, we obtain that

$$E(Y(0)|Z=1)=a'+bE(X|Z=1).$$

Similarly, the average observed outcome of the control children is $E(Y|Z=0)=a'+bE(X|Z=0)$. Because $E(Y|Z=0)$ and $E(X|Z=0)$ are both observed, we may compute the intercept: $a'=E(Y|Z=0)-bE(X|Z=0)$. One then derives

$$E(Y(0)|Z=1)=a'+bE(X|Z=1)=[E(Y|Z=0)-bE(X|Z=0)]+bE(X|Z=1).$$

Hence, the treatment effect on the treated

$$E(Y(1)|Z=1)-E(Y(0)|Z=1)=E(Y(1)|Z=1)-\{[E(Y|Z=0)-bE(X|Z=0)]+bE(X|Z=1)\}$$

is equal to d' as shown in (2.11), the coefficient for the treatment indicator in the ANCOVA model.

References

Berkson, J. (1946). Limitations of the application of fourfold table analysis to hospital data. *Biometrics Bulletin*, *2*, 47–53.

Bertrand, M. and Mullainathan, S. (2004). Are Emily and Greg more employable than Lakisha and Jamal? A field experiment on labor market discrimination. *American Economic Review*, *94*(4), 991–1013.

Cox, D.R. (1958). *Planning of Experiments*, Wiley, New York.

Elwert, F. & Winship, C. (2014). Endogenous selection bias: The problem of conditioning on a collider variable. *Annual Review of Sociology*, *40*, 31–53.

Fisher, R.A. (1935). *The Design of Experiments*, 9th edn, Macmillan Press, New York.

Haavelmo, T. (1943). The statistical implications of a system of simultaneous equations. *Econometrica*, *11*(1), 1–12.

Heckman, J.J. (1976). The common structure of statistical models of truncation, sample selection and limited dependent variables and a simple estimator for such models. *Annals of Economic and Social Measurement*, *5*(4), 475–492.

Heckman, J.J. (1979). Sample selection bias as a specification error. *Econometrica*, *47*(1), 153–162.

Heckman, J.J. and Vytlacil, E.J. (2007). Econometric evaluation of social programs, part I: causal models, structural models and econometric policy evaluation, in *Handbook of Econometrics*, vol. 6B (eds J.J. Heckman and E. Leamer), Elsevier, Amsterdam, pp. 4779–4874.

Hernández-Díaz, S., Schisterman, E.F. and Hernán, M.A. (2006). The birth weight "paradox" uncovered? *American Journal of Epidemiology*, *164*(11), 1115–1120.

Holland, P.W. (1986). Statistics and causal inference. *Journal of the American Statistical Association*, *81*(396), 945–960.

Holland, P. and Rubin, D. (1983). On lord's paradox, in *Principles of Modern Psychological Measurement: A Festschrift for Frederic M. Lord* (eds H. Wainer and S. Messick), Lawrence Erlbaum Associates, Hillsdale, NJ.

Lord, F.M. (1967). A paradox in the interpretation of group comparisons. *Psychological Bulletin*, *68*(5), 304–305.

Lord, F.M. (1975). Lord's paradox, in *Encyclopedia of Educational Evaluation* (eds S.B. Anderson, S. Ball and R.T. Murphy), Jossey-Bass, San Francisco, CA, pp. 232–236.

Neyman, J. (1923). *On the Application of Probability Theory to Agricultural Experiments. Essay on Principles*. Section 9. (Translated and edited by D. M. Dabrowska and T. P. Speed from the Polish original, which appeared in *Roczniki Nauk Rolniczych Tom X.*) *Annals of Agricultural Sciences*, 1–51. (The translation was published in 1990 in *Statistical Science*, *5*(4), 465–480.)

Neyman, J., with cooperation of K. Iwaskiewicz and St. Kolodziejczyk. (1935). Statistical problems in agricultural experimentation (with discussion). *Supplement to the Journal of the Royal Statistical Society, Series B*, *2*(2), 107–180.

Pearl, J. (2009). *Causality: Models, Reasoning, and Inference*, 2nd edn, Cambridge University Press, Cambridge, UK.

Rosenbaum, P.R. (1984). The consequences of adjustment for a concomitant variable that has been affected by the treatment. *Journal of the Royal Statistical Society, Series A (General)*, *147*(5), 656–666.

Roy, A. (1951). Some thoughts on the distribution of earnings. *Oxford Economic Papers*, *3*(2), 135–146.

Rubin, D.B. (1978). Bayesian inference for causal effects: the role of randomization. *Annals of Statistics*, *6*(1), 34–58.

Rubin, D.B. (1980). Discussion of "randomization analysis of experimental data: the Fisher randomization test," by D. Basu. *Journal of the American Statistical Association*, *75*, 591–593.

Rubin, D.B. (1986). Statistics and causal inference: comment: which ifs have causal answers. *Journal of the American Statistical Association*, *81*(396), 961–962.

US Department of Health and Human Services, Administration for Children and Families. (2010). *Head Start Impact Study*. Final Report. Washington, DC: U.S. Department of Health and Human Services, Administration for Children and Families.

VanderWeele, T.J. and Hernán, M.A. (2012). Results on differential and dependent measurement error of the exposure and the outcome using signed directed acyclic graphs. *American Journal of Epidemiology*, *175*(12), 1303–1310.

Weisberg, H.I. (1979). Statistical adjustments and uncontrolled studies. *Psychological Bulletin*, *86*(5), 1149–1164.

Westinghouse Learning Corporation and Ohio University (1969). *The Impact of Head Start Experience on Children's Cognitive and Affective Development*, Office of Economic Opportunity, Washington, DC.

Wilcox, A.J. (1993). Birth weight and perinatal mortality: the effect of maternal smoking. *American Journal of Epidemiology*, *137*(10), 1098–1104.

Wilcox, A.J. (2001). On the importance—and the unimportance—of birthweight. *International Journal of Epidemiology*, *30*, 1233–1241.

Willis, R.J. and Rosen, S. (1979). Education and self-selection. *Journal of Political Economy*, *87*(5), S7–S36.

Yerushalmy, J. (1971). The relationship of parents' cigarette smoking to outcome of pregnancy—implications as to the problem of inferring causation from observed associations. *American Journal of Sociology*, *93*(6), 443–456.

3

Review of causal inference designs and analytic methods

This chapter provides a brief review of research designs and statistical methods that researchers often employ for addressing causal questions, with a particular emphasis on revealing the identification assumptions in each case. A review of statistical adjustment through weighting will be provided in Chapter 4. For simplicity, this review is restricted to designs and methods for evaluating the causal effect of a binary treatment on a continuous outcome. Many of these designs and methods have limitations in evaluating multivalued treatments, a topic to be discussed further in Chapter 5. Readers who are familiar with these designs and methods may skip the remaining pages and move straight to the next chapter. Those who seek a more detailed introduction are referred to textbooks on experimental and quasiexperimental designs (e.g., Bloom, 2005; Murnane and Willett, 2011; Shadish, Cook, and Campbell, 2002).

3.1 Experimental designs

Experimental designs have been the "gold standard" for causal inference not just in natural science, engineering, agriculture, medicine, and psychological research but also increasingly in public policy and educational research. In social sciences, traditional laboratory-based small-sample experiments, often criticized for generating results lacking generalizability, are now supplemented with large-scale field experiments administered to random samples from well-defined populations. The trend promises to offer valid causal knowledge by eliminating both selection bias and sampling bias.

3.1.1 Completely randomized designs

Suppose that a random sample has been drawn from a well-defined population. In a completely randomized design, sampled individuals are assigned at random with the same probability to either an experimental condition or a control condition. The mean difference in the observed outcome between the experimental group and the control group $\widehat{\delta}_{PF} = \bar{Y}^{(E)} - \bar{Y}^{(C)}$

Causality in a Social World: Moderation, Meditation and Spill-over, First Edition. Guanglei Hong.
© 2015 John Wiley & Sons, Ltd. Published 2015 by John Wiley & Sons, Ltd.

estimates the population average treatment effect $\delta = E(Y(1) - Y(0))$ and the average treatment effect on the treated $\delta = E(Y(1) - Y(0)|Z=1)$.

As discussed in Section 2.1.3.2, a major strength of a completely randomized experiment is the prevention of selection bias by design. In analyses of experimental data, the ignorability assumption is the only identification assumption required. The randomization procedure guarantees that treatment assignment Z is independent of the potential outcomes $Y(1)$ and $Y(0)$. Hence, individuals in the population who may actually be assigned to the experimental condition are no different from those who may actually be assigned to the control condition on average in each of the two potential outcomes. This identification assumption is mathematically represented as

$$E[Y(z)|Z=1] = E[Y(z)|Z=0] = E[Y(z)], \text{ for } z = 0, 1.$$

Therefore, we have that

$$\begin{aligned}
\delta &= E[Y(1)] - E[Y(0)] \\
&= E[Y(1)|Z=1] - E[Y(0)|Z=0] \\
&= E[Y|Z=1] - E[Y|Z=0] \\
&= \delta_{\text{PF}}.
\end{aligned}$$

In the example of Head Start evaluation, the researcher may assign eligible children at random with a probability of 0.5 to Head Start programs. Note that the probability of treatment assignment does not have to be 0.5, though a balanced design with an equal number of sampled individuals in the experimental group and the control group maximizes efficiency in estimation. This is because when the sample size $N = n^{(E)} + n^{(C)}$ is given,

$$\text{Var}\left(\hat{\delta}_{\text{PF}}\right) = \text{Var}\left(\bar{Y}^{(E)} - \bar{Y}^{(C)}\right) = \sigma_{Y|Z}^2 \left(\frac{1}{n^{(E)}} + \frac{1}{n^{(C)}}\right) = \frac{\sigma_{Y|Z}^2}{NP(1-P)} \tag{3.1}$$

is minimized when $n^{(E)} = n^{(C)}$, that is, when $P = 0.5$. Here, $P = n^{(E)}/N$ is the probability of being assigned to Head Start; $\sigma_{Y|Z}^2$ denotes the within-group variance in the outcome that is assumed to be constant across the Head Start group and the control group.

3.1.2   Randomized block designs

In a randomized block design, randomization is conducted within blocks of relatively homogeneous individuals. Blocking serves two purposes: one is to reduce sampling variability within blocks and thereby improve efficiency in estimation; the other is to ease the investigation of heterogeneous treatment effects across blocks.

In the Head Start evaluation, the experimenter may use child race to create blocks and randomize within each racial group. Let $x = 0, 1, 2, 3, 4$ represent white, black, Hispanic, Asian, and Native Americans, respectively. Within-block randomization ensures that among children of the same racial background, treatment assignment Z is independent of the potential outcomes $Y(1)$ and $Y(0)$; and hence, the ignorability assumption is satisfied. This can be represented mathematically as follows:

$$E[Y(z)|Z=1,X=x] = E[Y(z)|Z=0,X=x] = E[Y(z)|X=x], \tag{3.2}$$

for $z=0$, 1 and for $x=0$, 1, 2, 3, 4. The aforementioned identification assumption relates the causal parameter of interest to the observable population data. In this case, the population average treatment effect $\delta = E[Y(1)-Y(0)]$ is a weighted average of block-specific treatment effects $E[Y(1)-Y(0)|X=x]$ where the weight is the proportion of the population in each block represented by $\mathrm{pr}(X=x)$:

$$\delta = E[Y(1)-Y(0)]$$
$$= \sum_{x=0}^{4}\{E[Y(1)-Y(0)|X=x]\mathrm{pr}(X=x)\}$$
$$= \sum_{x=0}^{4}\{[E(Y(1)|X=x)-E(Y(0)|X=x)]\mathrm{pr}(X=x)\}.$$

The ignorability assumption allows us to equate the mean potential outcomes in a racial group $E(Y(1)|X=x)$ and $E(Y(0)|X=x)$ with the respective mean observable outcomes in the same group $E(Y|Z=1,\ X=x)$ and $E(Y|Z=0,\ X=x)$, and hence, we have that

$$\delta = \sum_{x=0}^{4}\{[E(Y|Z=1,\ X=x)-E(Y|Z=0,\ X=x)]\mathrm{pr}(X=x)\} = \delta_{\mathrm{PF}}.$$

In a given sample, let the probability of treatment assignment be a constant P across all blocks. We use $\bar{Y}_x^{(E)}$ and $\bar{Y}_x^{(C)}$ to denote the sample mean observed outcome of the Head Start children and that of the control children, respectively, in racial group x. Analyzing the sample data within each block, one obtains an estimate of the treatment effect for each racial group: $\bar{Y}_x^{(E)} - \bar{Y}_x^{(C)}$ for $x=0$, 1, 2, 3, 4. The sample estimate of δ_{PF} is

$$\widehat{\delta}_{\mathrm{PF}} = \sum_{x=0}^{4}\left[\left(\bar{Y}_x^{(E)} - \bar{Y}_x^{(C)}\right)\mathrm{pr}(X=x)\right].$$

The sampling variability of the sample estimate is $\mathrm{Var}\left(\widehat{\delta}_{\mathrm{PF}}\right) = \left(\sigma_{Y|Z,X}^2\right)/(NP(1-P))$. Here, $\sigma_{Y|Z,X}^2$ is the constant variance in the outcome within a block and within each treatment. To the extent that race explains a portion of the outcome variance within each treatment, $\sigma_{Y|Z,X}^2$ is expected to be smaller than $\sigma_{Y|Z}^2$. For this reason, a sample estimate obtained from a randomized block design is expected to be more efficient than an estimate obtained from a completely randomized design.

To test whether the treatment effect depends on children's racial background, one may apply a two-way analysis of variance (ANOVA) to data from a randomized block design in which treatment group membership and block membership are the two explanatory factors. In the two-way ANOVA framework, the total variation in the observed outcome is partly due to the main effect of the treatment and the main effect of race and partly due to the treatment-by-race interaction effect. A significant F value for the interaction effect will indicate that the treatment effect varies by race.

3.1.3 Covariance adjustment for improving efficiency

In analyses of experimental data, the efficiency of estimation can often be improved through covariance adjustment if pretreatment covariates are available that have a strong linear association with the outcome. Let X be a pretreatment covariate such as pretest score measured on a continuous scale. An analysis of covariance (ANCOVA) as specified in the following allows one to remove within-treatment group variation in the outcome due to X while estimating the treatment effect:

$$Y = a' + bX + d'Z + e'. \tag{3.3}$$

Because in a randomized design the experimental group and the control group are not expected to have different distributions of X, we have that $\delta = \delta_{PF} = E[Y|Z=1] - E[Y|Z=0]$. As shown in Section 3.1.1, when no covariance adjustment is made, the variance of the treatment effect estimate is $\sigma^2_{Y|Z}/NP(1-P)$. The gain in precision from the covariance adjustment for X depends on the correlation between Y and X denoted by ρ. According to Cochran (1957), when the sample size approaches infinity, the adjustment reduces the variance of the treatment effect estimate by reducing $\sigma^2_{Y|Z}$ to $\sigma^2_{Y|Z}(1-\rho^2)$. In most education data, the correlation between the pretest and the posttest is often around 0.7. Hence, covariance adjustment for the pretest is expected to reduce the within-group variance of the outcome by half $(1 - 0.7^2 = 0.51)$, thereby doubling the precision of the treatment effect estimate.

3.1.4 Multilevel experimental designs

Because individuals are often nested in families, schools, firms, and neighborhoods and because an intervention may operate through group dynamics, these social clusters often constitute meaningful sampling units, units of treatment assignment, and units of analysis in a *cluster randomized design*, which allows the results to be generalized to a population of social clusters. Cluster randomized designs, however, tend to suffer from a lack of efficiency in estimation when the sample of clusters is limited in size (Bloom, Bos, and Lee, 1999; Moerbeek, van Breukelen, and Berger, 2000; Raudenbush, 1997; Schochet, 2008). In contrast, *multisite randomized trials* randomly assign individuals within geographical or organizational sites to individual-based interventions. By using the natural sites as blocks, multisite randomized trials remove noise due to between-site variability, thereby improving efficiency in estimation. Multisite randomized trials additionally enable researchers to investigate possible heterogeneity in treatment effect across sites (Fuller, Mattson, and Allen, 1994; Raudenbush and Liu, 2000). Yet within-site "contamination" or treatment "spill-over" from the experimental group to the control group is sometimes of concern as it compromises identification and often attenuates the treatment effect estimate. Viewing possible spill-overs of treatment effects as a social process of key interest, we have dedicated Chapters 14 and 15 to this particular issue.

Although a randomized experiment when combined with random sampling is expected to eliminate selection bias and sampling bias, nonrandom attrition from the experimental and control groups, often inevitable in studies that span a relatively long period of time, might introduce both sampling bias and selection bias. Nonresponse for various other reasons could occur at any time during a study. Multivariate imputation of missing data (Little and Rubin, 2002; Schaffer, 2010; Shin and Raudenbush, 2007, 2011; Su, Yajima, and Gelman, 2011), whenever feasible, may keep the initial sample intact though lose efficiency due to the additional uncertainty involved in the imputed data. The current imputation techniques typically require the assumption that the unobserved values are missing at random given all the observed information. Results tend to be robust even when this assumption fails if the observed information strongly predicts the missing information. However, such an assumption cannot be checked empirically. The missing at random assumption is not always warranted especially because covariates predictive of nonresponses are often unmeasured in experimental studies. Randomized experiments may also suffer from noncompliance when some of those assigned to the experimental condition manage to avoid exposure to the treatment, while some of those assigned to the control condition find access to the treatment, leaving the substantive meaning of the "treatment" dubious. In many of such cases, experimental studies with initial randomization often fall in the category of quasiexperimental studies.

3.2 Quasiexperimental designs

Quasiexperimental designs are aimed at addressing the same causal questions as experimental designs yet lacking the rigor of randomization. Researchers turn to quasiexperimental designs when randomization is impractical or unethical, when there are opportunities to take advantage of natural experiments, or when large-scale survey data collected from representative samples offer the advantage of generalizability. A wide variety of quasiexperimental designs have been used in evaluation research (Shadish, Cook, and Campbell, 2002). To be inclusive, the term "quasiexperimental designs" encompasses "nonexperimental studies" in psychology, "observational studies" in epidemiology, and "natural experiments" in economics and public policy research. In the following, we review commonly used quasiexperimental designs with a particular focus on nonequivalent comparison group designs.

3.2.1 Nonequivalent comparison group designs

In a nonequivalent comparison group design, researchers identify, often retrospectively, a group of individuals who have not been exposed to the treatment as a comparison group for the treated individuals, acknowledging that the two groups might differ in numerous ways in the absence of the treatment. Such preexisting differences between the treated group and the nonequivalent comparison group would likely confound the evaluation of the treatment effect and lead to selection bias. The previously mentioned Westinghouse evaluation of Head Start employed this type of design. As pointed out by Cochran (1965) and many others, setting up the comparisons is a major difficulty in planning quasiexperimental studies of human populations.

In the article "Choice as an alternative to control in observational studies," Rosenbaum (1999) gave examples of several well-planned nonequivalent comparison group designs. One of these examples is a clinical psychology study assessing the impact of bereavement on mental functioning over long periods of time. The researchers (Lehman, Wortman, and Williams, 1987) focused on individuals who experienced a sudden and unexpected loss of a spouse or a child in a motor vehicle crash in which the treated individuals were driving cars that were not at fault. They constructed a control group sampled from a pool of individuals seeking to renew their licenses by matching a control individual to a treated individual on the basis of gender, age, and pretreatment income. The treated and control groups were chosen such that the treatment—in this case a traumatic loss of a spouse or a child in a car crash for which one is not responsible—could as likely have happened to the controls as it did to the treated. The haphazardness of the treatment assignment greatly reduces even though not necessarily eliminating selection bias.

Most studies using the nonequivalent comparison group design have embedded in it a pre-post design; that is, measurements of individuals in the experimental group and the control group occur at the baseline as well as at the end of the treatment period. Researchers often hope that the baseline measurement, if comprehensive enough, would capture most if not all of the preexisting differences between the treated group and the control group. In this type of design, causal inference is crucially reliant on *the strong ignorability assumption*; that is, a subgroup of treated individuals and a subgroup of control individuals displaying the same baseline conditions are expected to have the same distribution of the potential outcome if they were all treated and to have the same distribution of the potential outcome if none were treated. This notion has been challenged by Bryk and Weisberg (1977) who revealed nonequivalent growth between treated individuals and control individuals equated on pretest information at the baseline, a problem to which the interrupted time-series design may offer a possible solution.

3.2.2   Other quasiexperimental designs

An interrupted time-series design tracks the treated individuals over time through repeated measurements of the outcome of interest both before and after an intervention takes place. In a traditional interrupted time-series design, every experimental unit serves as its own control as seen in single-subject studies. The temporal association between the introduction of an intervention and the change in outcome trend is taken as evidence for the intervention effect. With the availability of large-scale panel data, this design has been employed for evaluating the impact of a policy change (e.g., Deere, Murphy, and Welch, 1995). The traditional interrupted time-series design cannot rule out the confounding impact of other events concurrent to the intervention. This design is greatly strengthened when repeated measurements of individuals in a nonequivalent comparison group are collected over the same period of time (e.g., Webster, Vernick, and Hepburn, 2002). In a comparison of two time series of data between the treated group and the control group, an abrupt shift in an otherwise relatively smooth pattern of the time series for the treated group upon the introduction of the intervention is usually regarded as evidence for the intervention effect if a parallel change is not detected in the control group. However, because the treated group and the control group could be nonequivalent in possibly many other ways, one is not in a position of claiming that the design has eliminated selection bias.

Regression discontinuity designs take advantage of the fact that cutoffs have been used widely in determining individual eligibility for treatments. For example, some school districts require that students scoring below a certain cutoff on a test receive remedial education or be retained in grade (Jacob and Lefgren, 2004; Roderick and Nagaoka, 2005). In a sharp regression discontinuity design, participants are assigned to treatment or control based solely on whether a pretreatment measure falls above or below a cutoff point (Thistlewaite and Campbell, 1960). Unlike most other quasiexperimental designs, the selection process in a sharp regression discontinuity design is completely known to the researcher. To avoid unwarranted extrapolation in interpreting the estimated causal effect, Imbens and Lemieux (2008) argued that regression discontinuity designs, at best, provide estimates of the average treatment effect for a subpopulation of individuals whose pretreatment measure is at the cutoff score.

Challenges arise when treatment assignment is based only partly on the standing of an individual relative to the cutoff point. These are called fuzzy regression discontinuity designs because, even though the probability of receiving the treatment changes discontinuously at the cutoff, the selection process is no longer completely known to the researcher. One solution that has been proposed and implemented in the literature is to apply the instrumental variable (IV) method to the data within a certain bandwidth from the cutoff (Imbens and Lemieux, 2008; Jacob and Lefgren, 2004). The combination of a fuzzy regression discontinuity design with the IV method requires that data additionally satisfy all the identification assumptions needed in applications of the IV method. These assumptions will be reviewed in Section 3.3.3.

3.3   Statistical adjustment methods

Analyses of experimental data can benefit from statistical adjustment for strong predictors of an outcome because such adjustment reduces error variance and thereby improves the precision of treatment effect estimation. In analyses of quasiexperimental data, however, statistical adjustment becomes essential for reducing selection bias. This section briefly reviews the adjustment methods most often used in analyses of quasiexperimental data. Among them, ANCOVA and multiple regression rely on the strong ignorability assumption and additional model-based assumptions for identifying the treatment effect; matching and stratification similarly rely on the strong ignorability assumption yet are free of model-based assumptions; the IV method and difference-in-differences (DID) analysis each invokes a different set of identification assumptions that are often no weaker than the strong ignorability assumption.

3.3.1   ANCOVA and multiple regression

ANCOVA combines linear regression and ANOVA. This technique, frequently applied in nonequivalent control group designs, analyzes the covariance between a pretreatment covariate X and a continuous outcome Y within each treatment group. In Figure 2.1, an ANCOVA model $Y = a' + bX + d'Z + e'$ as specified in (3.3) has been fit to the hypothetical population data of preschool-aged children. The model assumes that the relationship between pretest X and posttest Y represented by a linear regression slope b is the same in the Head Start group and the control group. The intercept a' is the control group mean of the adjusted outcome when $X = 0$; the coefficient for the treatment indicator d' is the mean difference in the adjusted outcome between the Head Start group and the control group. Under the strong ignorability assumption and the assumption that the treatment effect does not depend on X, d' is equal to the population average treatment effect and the treatment effect on the treated.

3.3.1.1   ANCOVA for removing selection bias

In a quasiexperimental study, the experimental group and the control group may differ in the distribution of X. A pretreatment covariate X that predicts both treatment assignment Z and the outcome Y is called a confounder. This is because the mean difference in the observed

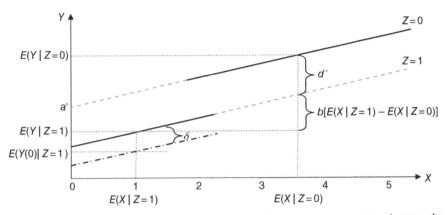

Figure 3.1   Treatment effect estimation through an analysis of covariance: $Y = a' + bX + d'Z + e'$

outcome between the experimental group and the control group may be at least partly due to the preexisting difference in X between the two groups. As illustrated in Figure 3.1 and consistent with Equation (2.11), the mean difference in the observed outcome between the Head Start group and the control group, $E(Y|Z=1) - E(Y|Z=0)$, is the sum of two quantities. The first is the bias term $b[E(X|Z=1) - E(X|Z=0)]$, a product of the within-group X–Y relationship and the between-group mean difference in X. The second is the adjusted between-group mean difference in the outcome d' with the bias associated with X removed through covariance adjustment. To the extent that the model-based assumptions are satisfied—that is, the within-group X–Y relationship is linear and is the same across the two treatment groups—ANCOVA removes all the confounding due to X.

3.3.1.2   Potential pitfalls of ANCOVA with a vast between-group difference

However, Cochran (1957) emphasized a major difficulty in covariance adjustment when the experimental group and the control group have little or no overlap in the distribution of X, in which case the ANCOVA model is based more or less on linear extrapolation. It is conceivable that should the Head Start children be assigned to the control condition instead, their counterfactual outcome in the absence of Head Start may be either equal to or less than their observed outcome. Figure 3.1 illustrates one possible scenario in which the dashed line on the bottom represents the relationship between Head Start children's pretest X and their counterfactual outcome under the control condition $Y(0)$. Head Start children's observed outcome corresponds to their potential outcome under Head Start $Y(1)$. The treatment effect on the treated defined as $\delta = E[Y(1) - Y(0)|Z=1]$ is positive in this scenario. Yet the ANCOVA model has assumed that the Head Start children, without actually attending Head Start programs, would have scored the same on the posttest as their control counterparts would when many of such "control counterparts" are nonexistent and are represented by the dotted line on the top purely on the basis of extrapolation. This scenario would arise because Head Start children and the control children would likely differ in many other measured and unmeasured pretreatment characteristics in addition to their difference in pretest scores. Relying on seemingly invalid assumptions about Head Start children's counterfactual outcome, an ANCOVA model would likely generate a biased result concluding that Head Start is detrimental to disadvantaged children, a result as misleading as that illustrated in Lord's paradox.

Cochran (1957) also pointed out a possible compromise in efficiency when the experimental group and the control group have little overlap in pretreatment covariate X. Intuitively, the larger the preexisting difference between the experimental group and the control group, the less information the data have to offer for estimating the treatment effect. Again, for simplicity, we assume equal sample size n in each treatment group. With covariance adjustment for X, Cochran has shown that the variance of the sample estimate of the treatment effect $\hat{d}'$ can be written as

$$\mathrm{Var}\left(\hat{d}'\right) = \sigma^2_{Y|Z,X}\left[\frac{2}{n} + \frac{[E(X|Z=1)-E(X|Z=0)]^2}{\sigma^2_{X|Z}}\right],$$

where $\sigma^2_{Y|Z,X}$ and $\sigma^2_{X|Z}$ are the within-group variance of Y conditioning on X and the within-group variance of X, respectively. Clearly, the standard error of the treatment effect estimate will increase as the between-group mean difference in X increases, which will lead to a loss of statistical power. As a result, one may fail to detect a nonzero treatment effect.

3.3.1.3 Bias due to model misspecification

As shown previously, a major constraint of ANCOVA is the assumption of linearity and additivity. That is, the relationship between pretreatment covariate X and outcome Y is assumed linear within each group; moreover, this linear relationship is assumed the same across treatment groups, and hence, the outcome model is additive. To relax this model-based assumption, Cochran (1965) suggested approximating the regression function of Y on X by a polynomial function that includes higher-order terms such as X^2 and X^3. If the X–Y relationship is nonlinear, omitting the nonlinear terms may lead to a biased estimation. Suppose that the true model is as follows:

$$Y = a + dZ + b_1 X + b_2 X^2 + e.$$

Hence, the treatment effect can be identified as

$$d = [E(Y|Z=1)-E(Y|Z=0)]-b_1[E(X|Z=1)-E(X|Z=0)]-b_2\left[E\left(X^2|Z=1\right)-E\left(X^2|Z=0\right)\right].$$

When the two treatment groups differ in the average X^2, omitting X^2 in the outcome model will result in the bias $b_2[E(X^2|Z=1)-E(X^2|Z=0)]$ in identifying the treatment effect. Even if the two treatment groups have the same mean of X, the bias remains to be $b_2[\mathrm{Var}(X|Z=1)-\mathrm{Var}(X|Z=0)]$. However, a nonlinear X–Y relationship sometimes cannot easily be approximated by a polynomial model. Correct model specification will depend on whether the analyst has the knowledge of various nonlinear functions and is skillful in analyzing nonparametric regression models.

Even when the X–Y relationship is linear in each treatment group, its slope in the experimental group may differ from that in the control group as illustrated in Figure 3.2. This is the case in which the treatment effect depends on one's pretreatment characteristic indicated by the value of X. Suppose that the correct outcome model is

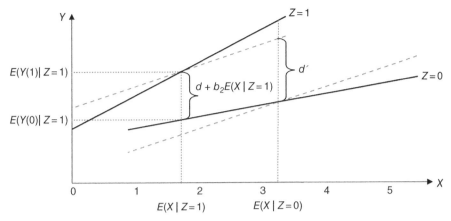

Figure 3.2   Biased estimation of the treatment effect due to the omission of a treatment-by-covariate interaction

$$Y = a + dZ + b_1 X + b_2 ZX + e. \tag{3.4}$$

For individuals with pretreatment covariate value $X = x$, the treatment effect is $d + b_2 x$. In a hypothetical evaluation of the Head Start program, suppose that d and b_2 are both positive. One may conclude that Head Start is more beneficial for eligible children with higher skills at the baseline. Averaging over the distribution of X in the Head Start population, the population average treatment effect on the treated is $d + b_2 E(X|Z = 1)$.

A commonly seen mistake in model specification is the omission of the interaction between the treatment indicator Z and a confounding covariate X. Analyzing the misspecified ANCOVA model (3.3) that omits ZX, one obtains the following coefficient for the treatment indicator:

$$d' = [E(Y|Z = 1) - E(Y|Z = 0)] - b[E(X|Z = 1) - E(X|Z = 0)].$$

Replacing Y with the linear nonadditive function of x as specified in the true model (3.4), we obtain that

$$E[Y|Z = 1] = E[a + d + b_1 X + b_2 X|Z = 1] = a + d + (b_1 + b_2)E[X|Z = 1]$$

and

$$E[Y|Z = 0] = E[a + b_1 X|Z = 0] = a + b_1 E[X|Z = 0].$$

Hence, as derived in Appendix 3.A, we have that

$$d' = d + b_2 E(X|Z = 1) + (b_1 - b)[E(X|Z = 1) - E(X|Z = 0)].$$

Clearly, $d' = d + b_2 E(X|Z = 1)$ only when $E(X|Z = 1) = E(X|Z = 0)$. In other words, the treatment effect on the treated is identified by the misspecified linear additive ANCOVA model only when the experimental group and the control group do not differ in the mean of X. Because this condition is always warranted in an experimental study, an omission of ZX interaction does not introduce bias in analyses of experimental data. However, when X is a confounding covariate in quasiexperimental data, the bias is $(b_1 - b)[E(X|Z=1) - E(X|Z=0)]$. The magnitude of the bias is proportional to the between-group mean difference in X (Figure 3.2).

In analyzing quasiexperimental data, researchers typically employ multiple regression to make covariance adjustment for one or more pretreatment covariates in estimating the treatment effect. The strengths and limitations of ANCOVA as described previously apply similarly to multiple regression. In particular, as the number of covariates increases in a regression model, the possibility of misspecifying the functional form of the model increases multiplicatively.

3.3.2 Matching and stratification

To completely avoid model-based assumptions including linear extrapolation in statistical adjustment for a pretreatment covariate X, one may choose to match the control units with the experimental units on the basis of X. Matching is recommendable especially when there is a large control reservoir and when the distributions of X are not vastly different between the two treatment groups. Because the difference in X between the experimental unit and the control unit in each matched pair is kept minimal, the experimental group and its matched control group are expected to show comparable distribution of X.

Alternatively, one may stratify the sample on the basis of X. For example, after ranking all the units in a sample by their observed values in X, one may subdivide the sample into S strata. Each stratum will contain some experimental units and some control units though the two groups are not necessarily of equal size. Notably, the between-group difference in X within a stratum will in general become smaller than that in the entire sample. Let $\bar{Y}_s^{(E)}$ and $\bar{Y}_s^{(C)}$ denote the mean outcome of the experimental group and that of the control group, respectively, in stratum s for $s = 1, \ldots, S$. The treatment effect estimate can be taken as the simple average of $\bar{Y}_s^{(E)} - \bar{Y}_s^{(C)}$ over all S strata that are of equal size. If the treatment effect on the treated is of interest, this is instead estimated as $\sum_{s=1}^{S} W_s \left(\bar{Y}_s^{(E)} - \bar{Y}_s^{(C)} \right)$, where W_s is the proportion of the treated in stratum s. Alternatively, one may subdivide the sample to have an equal number of treated units in each stratum. Under this stratification, a simple average of within stratum mean difference in the outcome between the experimental group and the control group over all the strata estimates the average treatment effect on the treated. In a sense, stratification can be viewed as a coarse way of matching due to the relatively small amount of remaining between-group difference in X within a stratum. As Cochran (1968) has shown, for monotonic relations between Y and X, creating five strata removes about 90% or more of the initial bias. The percentage of bias removal increases as the number of strata increases. Similar to matching, stratification improves the efficiency of the treatment effect estimation although not to the same degree as ANCOVA or multiple regression. For example, when X is normally distributed with a 0.7 correlation with Y, creating five strata reduces the variance of the treatment effect estimate by 44%. The degree of improvement in precision tends to increase as the number of strata increases.

Cochran (1965) concluded that "If the original x-distributions diverge widely, none of the methods can be trusted to remove all, or nearly all, the bias" (p. 247). Yet matching and stratification will enable the researcher to quickly detect this problem if some of the experimental units do not have matches in the control group or vice versa or if one or more strata contain only one of the treatment groups. Another important strength shared by matching and stratification is the opportunity that they provide for visually detecting heterogeneous treatment effects across matched pairs or across strata. Nonetheless, matching, stratification, and covariance adjustment are suitable only for removing bias associated with the measured pretreatment covariates. The results are unbiased only when the strong ignorability assumption holds, that is, when treatment assignment is independent of the potential outcomes within levels of the measured pretreatment covariates. Because this assumption is never warranted in a quasiexperimental study, many potential unmeasured confounders threaten to bias the treatment effect estimation. Additional difficulties arise when the pretreatment covariates are measured with error or even with bias, in which case none of the aforementioned methods can succeed in complete removal of bias.

3.3.3 Other statistical adjustment methods

The IV method and the DID method are popular especially among economists because they provide alternative strategies for identifying the treatment effect in the presence of unmeasured confounders. Yet each of these methods replaces the strong ignorability assumption with a different set of identification assumptions that are often hard to satisfy.

3.3.3.1 The IV method

Even in a randomized experiment, the treatment that one actually receives is not always in correspondence with the treatment that one has been assigned to due to noncompliance. A well-known example is the study of veteran status in the Vietnam era on civilian mortality in which an individual's draft status was determined by a randomly drawn lottery number related to one's date of birth. However, among those born in 1950 who were eligible to be drafted according to their lottery numbers, only 35% actually served in the army, while as many as 19% of those who would not be drafted actually served (Angrist, 1990; Hearst, Newman, and Hulley, 1986). Although the average effect of the treatment assigned, often called the "intent-to-treat" (ITT) effect, can be estimated without bias, the effect of the treatment actually received has more direct relevance to the causal theory. In this case, the theory suggests that wartime military service may elevate mortality even after one has been discharged from the army. Yet whether one actually served in the military despite one's draft status might be subjected to numerous selection factors, many of which have never been measured.

The IV method has often been employed for estimating the effect of the treatment received when the initial randomization of treatment assignment serves as the instrument variable. An IV is explicitly excluded from one regression equation but included in another regression equation. Let $Z = 1$ if an individual has been randomly assigned to be treated and 0 otherwise; let $D = 1$ if an individual has actually received the treatment and 0 otherwise. Of interest is the causal effect of D on Y defined as $E[Y(d=1) - Y(d=0)]$ irrespective of one's actual treatment assignment. Yet possible self-selection of the treatment received poses a major challenge to causal inference. The analysis, often conducted through either two-stage least squares or maximum likelihood estimation, involves two structural models as follows:

$$D = \alpha_0 + \alpha_1 Z + v,$$
$$Y = \beta_0 + \beta_1 D + \varepsilon. \tag{3.5}$$

The treatment received can be viewed as a potential outcome of the treatment assigned $D(z)$ for $z = 0$, 1. Because Z has been randomized, it is easy to obtain an unbiased estimate of α_1 representing the ITT effect of Z on D. However, because D was not randomized, an analysis of the second regression model alone does not generate an unbiased estimate of β_1. This second regression model is specified under a crucial assumption that Z can affect Y only through D, that is, being assigned to the treatment would have no impact on one's outcome unless one actually received the treatment. One may replace D in the outcome model with $\alpha_0 + \alpha_1 Z + v$ and have the reduced form of the outcome model as follows:

$$\begin{aligned} Y &= \beta_0 + \beta_1(\alpha_0 + \alpha_1 Z + v) + \varepsilon \\ &= (\beta_0 + \beta_1 \alpha_0) + \beta_1 \alpha_1 Z + (\beta_1 v + \varepsilon) \\ &= \gamma_0 + \gamma_1 Z + e, \end{aligned}$$

where $\gamma_1 = \beta_1 \alpha_1$ and can be estimated easily as the ITT effect of Z on Y, that is, the mean difference in the outcome between the experimental group and the control group. Therefore, the IV estimand for the causal effect of D on Y denoted by β_1^{IV} is simply

$$\beta_1^{IV} = \frac{\gamma_1}{\alpha_1} = \frac{\beta_1 \alpha_1}{\alpha_1}.$$

When D and Z are both binary, α_1 is the difference between the experimental group and the control group in the proportion of individuals who actually received the treatment. Angrist, Imbens, and Rubin (1996) viewed α_1 as the proportion of individuals in the population whose participation behavior could possibly be changed by the treatment assignment and labeled these individuals as "compliers." The compliers constitute a subpopulation of individuals for whom $D(1) = 1$ and $D(0) = 0$. In contrast, individuals who would always receive the treatment (i.e., $D(1) = D(0) = 1$) and those who would never receive the treatment (i.e., $D(1) = D(0) = 0$) regardless of their treatment assignment are labeled as "always-takers" and "never-takers," respectively.

Angrist, Imbens, and Rubin (1996) clarified the following identification assumptions under which the IV estimand has a causal interpretation:

1. SUTVA. When applied to the Vietnam veteran study, SUTVA implies that an individual's potential outcomes are not affected by other individuals' treatment assigned and treatment received.

2. Exogeneity of the treatment assignment. The random assignment of lottery numbers to dates of birth ensures that the treatment assignment is exogenous or, in other words, that Z is independent of all the potential outcomes, and hence, α_1 and γ_1 can be estimated without bias.

3. The exclusion restriction. Under this assumption, Z can affect Y only through changing one's participation behavior D. Hence, the effect of Z on Y is zero for "always-takers" and "never-takers" whose participation behavior cannot be changed by the treatment assignment.

4. A nonzero effect of Z on D. The effect of Z on D is nonzero if the treatment assignment changes at least one person's participation behavior such that $\alpha_1 \neq 0$.

5. Monotonicity. This assumption states that there is not a single "defier" in the population whose participation behavior would be opposite to the treatment assigned.

When D and Z are both binary and when all these five assumptions hold, Angrist, Imbens, and Rubin (1996) showed that the IV method estimates the "local average treatment effect" for the compliers defined as

$$E[Y(z=1, \ d=1) - Y(z=0, \ d=0)|D(1)=1, \ D(0)=0].$$

For the compliers, Z and D are always equivalent. Therefore, the causal effect of the treatment assignment is equivalent to that of the treatment received on the outcome Y in this subpopulation.

A major strength of the IV method is that it is not reliant on the measurement and adjustment of numerous pretreatment covariates. Hence, it is not vulnerable to potential bias associated with unobserved confounders or with measurement errors in the observed confounders. However, the IV method is vulnerable to possible violations of the aforementioned identification assumptions. Controversies often arise in particular when the values of the IV have not been randomized or when the exclusion restriction is questionable. For example, Altonji, Elder, and Taber (2005) analyzed Catholic school effects on educational attainment. One of their instrument variables was an individual's religious affiliation with the Catholic Church. Catholics and non-Catholics are different on average in many aspects including demographics and family background. These characteristics generally in favor of the Catholics are often found to be influential on educational attainment. Therefore, neither the exogeneity of the IV nor the exclusion restriction seems plausible in this case. The researchers attempted to remove potential selection bias by controlling for pretreatment covariates with the hope that the identification assumptions would become plausible within levels of these pretreatment characteristics. Yet further sensitivity analysis revealed a substantial amount of remaining bias and therefore ruled out Catholic religion as a qualified instrument.

The IV method has been utilized in many economics applications when D or Z is continuous, in which case the functional form of the Z–D relationship and the D–Y relationship needs to be correctly specified. Moreover, given that the treatment effects are unlikely to be constant in a heterogeneous population, when D and Z are not binary, the monotonicity assumption must be replaced with the assumption that the Z effect on D is independent of the D effect on Y (Reardon and Raudenbush, 2013). Specifically, let A_{1i}, B_{1i}, and G_{1i} denote, for individual i, the individual-specific causal effects of Z on D, of D on Y, and of Z on Y, respectively. Each of these individual-specific causal effects is a random variable that can take different values in the population. Analyzing the structural models specified in (3.5), one obtains the population average causal effects $\alpha_1 = E(A_1)$ and $\gamma_1 = E(G_1)$. Note that

$$\beta_1^{IV} = \frac{E(G_1)}{E(A_1)} = \frac{E(B_1 A_1)}{E(A_1)} = \frac{E(B_1)E(A_1) + \text{Cov}(B_1, A_1)}{E(A_1)}.$$

In general, the expected value of the product of two random variables such as $E(B_1 A_1)$ is the sum of two components. The first component is the product of the two expected values of these two respective random variables; and the second component is the covariance between these two random variables. According to the aforementioned derivation, the ratio of $E(G_1)$ to $E(A_1)$ is equal to $E(B_1)$, the causal effect of D on Y, only if $\text{Cov}(B_1, A_1) = 0$, that is, if the individual-specific causal effect of D on Y is independent of the individual-specific causal effect of Z on D. This assumption can easily be violated if individuals who are expected to benefit more from participation (i.e., those whose B_{1i} is relatively higher) make a rational decision to participate with a relatively higher rate when they are assigned to the treatment (i.e., those whose A_{1i} is also relatively higher). In general, the utility of the IV method is greatly constrained by some of the identification assumptions that may seem hard to satisfy.

3.3.3.2 DID analysis

To evaluate the causal effect of a policy that was introduced at a certain time point, researchers may be tempted to calculate the difference in the mean outcome between two cohorts, one cohort that never experienced the policy and the other that did. Yet because a policy of interest may be introduced in the midst of other contextual changes, the policy effect will likely be confounded by cohort difference and other concurrent events. The DID strategy resorts to a nonequivalent comparison group unaffected by the policy that is often located in the same jurisdiction or adjacent to the experimental group affected by the policy. When a longitudinal cohort in each group is followed before and after the onset of the new policy, one may argue that the average outcome change in the comparison group is solely attributable to the confounding effect of the concurrent events including maturation. Subtracting the average outcome change in the comparison group from that in the experimental group therefore produces an estimate of the policy effect of interest. For example, in a study by Card and Krueger (1993) assessing the causal effect of minimum wage increase on employment, the employment growth in New Jersey's fast-food industry before and after the statewide increase in minimum wage in 1992 was compared with the employment growth in fast-food restaurants in adjacent eastern Pennsylvania where the minimum wage was constant during the same time period.

A standard DID model invokes the strong assumption that the average confounding effect of concurrent events is the same for the comparison group and the experimental group and, in the case of an analysis involving prepolicy and postpolicy cohorts, that the two groups experience the same temporal changes in cohort composition. However, the confounding effect of concurrent events may vary by pretreatment characteristics that are distributed differently across the experimental group and the comparison group (Meyer, 1995) and across cohorts. To overcome this limitation, researchers have typically employed a DID model with linear covariance adjustment for a vector of observed pretreatment characteristics (e.g., Barnow, Cain, and Goldberger, 1980; Card and Kruger, 1993; Dynarski, 2003; Fitzpatrick, 2008). Similar to an ANCOVA model, linear covariance-adjusted DID additionally assumes that the covariate–outcome relationships are the same across the experimental group and the comparison group and are invariant over time. As the number of covariates increases and as nonlinearity arises, model misspecification becomes increasingly likely and is consequential for identification.

3.4   Propensity score

In analyses of quasiexperimental data that adjust for selection through covariance adjustment, matching, or stratification, the causal validity of the estimated treatment effects relies primarily on the plausibility of the strong ignorability assumption. This assumption states that the treatment assignment is independent of the potential outcomes conditioning on the observed pretreatment covariates. In other words, it is assumed that, within levels of these covariates, the assignment of individuals to different treatment conditions can be viewed "as if" random. For example, in the study assessing the long-term psychological impact of a sudden loss of a spouse or a child in a motor vehicle crash that one was not responsible for, individuals who experienced the loss were matched with control individuals on the basis of gender, age, and pretreatment income (Lehman, Wortman, and Williams, 1987). It was assumed that, within each matched pair, the traumatic loss could have happened with equal probability to either the treated or the control individual. In this natural experiment, the treated group and the control group may nonetheless differ systematically in some other aspects such as the physical or mental condition of the individual drivers. One may suspect that the drivers in the treated group were perhaps more likely than their counterparts in the control group to have a condition that lowered their alertness and swiftness in response to an unexpected traffic situation. Such between-group differences could possibly contribute to a higher probability of involvement in motor vehicle crash for the treated group than for the control group on average. Moreover, physical and mental conditions may likely predict an individual's long-term psychological well-being even in the absence of the traumatic event. Therefore, one may argue that an individual's physical and mental conditions are important potential confounders that the researchers ideally should have measured and used for forming matched pairs.

In most quasiexperimental studies, the number of pretreatment covariates that require statistical adjustment can be daunting. For example, in evaluating the effectiveness of retaining children in kindergarten rather than promoting them to the first grade, researchers identified more than 200 pretreatment covariates that could potentially confound the estimation of the retention effects on academic outcomes (Hong and Raudenbush, 2005). These include age, gender, race, preschool experience, health, disability, family structure, literacy resources at home, and maternal education, in addition to child cognitive and social development, instructional experience, and class and school environment during the first kindergarten year. Matching and stratification on more than a few variables are difficult to implement. Some analysts may opt for a "kitchen-sink" approach by entering all the measured covariates in a multiple regression model. Yet the more covariates that are included in regression, the higher the risk for misspecifying the functional relationships between the outcome and some of the covariates. It is known that one cannot succeed in removing all bias even if all the pretreatment covariates have been considered if the functional form of the outcome model is misspecified (Drake, 1993). An increasingly popular alternative, in evaluating a binary treatment, is to match or stratify cases on a single variable: the propensity score.

The propensity score is the conditional probability of assignment to a treatment given observed covariates. Rosenbaum and Rubin (1983a) were among the first to propose propensity score-based approaches to statistical adjustment when pretreatment covariates are high dimensional. For example, a group of kindergarten retainees and a group of promoted children with the same propensity score for retention are expected to have the same pretreatment composition in all the baseline characteristics described earlier. In the past several decades, applied researchers have increasingly employed propensity score-based matching, stratification, and covariance adjustment.

In the following, we discuss the balancing property of propensity scores and review the aforementioned statistical adjustment methods conditioning on propensity scores. These methods follow the rationale of estimating the treatment effect by comparing the mean outcome between the treated group and the control group for those who share the same or similar propensity score values and then pooling the treatment effect estimates over the entire distribution of the propensity score. These adjustment methods have the potential of removing the selection bias associated with the observed pretreatment covariates and arriving at a causal answer to the extent that the strong ignorability assumption holds.

3.4.1   What is a propensity score?

In a study contrasting an experimental condition with a control condition denoted by $z = 1$ and $z = 0$, respectively, the probability that individual i would be assigned to the experimental condition, denoted by $\theta_i = \mathrm{pr}(Z_i = 1)$, is the individual's propensity score. If the researcher has adopted a completely randomized design, the propensity score would be known and constant for all individuals. For example, suppose that in a sample of 100 individuals, 50 of them would be assigned at random to the experimental condition and 50 to the control condition. Then the propensity score is $\theta = 0.5$ for all the study participants. Alternatively, the researcher may conduct a randomized block design blocking on gender in light of past research suggesting a larger benefit of the treatment for boys than for girls. Suppose that the sample includes 100 boys and 100 girls. The researcher may choose to assign 70 boys and 30 girls at random to the experimental condition and the rest to the control condition. In this case, the probability of treatment assignment depends on gender because the conditional probability of being assigned to the experimental condition is 0.7 for boys and 0.3 for girls. Let $X_i = 1$ if individual i is a boy and 0 otherwise. The propensity score for individual i therefore is $\theta_i = \theta_i(x) = \mathrm{pr}(Z_i = 1|X_i = 1) = 0.7$ if the individual is a boy and is $\theta_i = \theta_i(x) = \mathrm{pr}(Z_i = 1|X_i = 0) = 0.3$ if the individual is a girl.

In a quasiexperimental study, an individual's probability of treatment assignment may depend on a wide range of pretreatment factors denoted by a vector $\mathbf{X}$. For example, in a study of the causal effect of regular exposure to secondhand smoke on individual health, whether a nonsmoking individual is regularly exposed to secondhand smoke is a function of, among other factors, the individual's age, gender, race, education, and poverty status. A vector of k measures of pretreatment characteristics, transposed from a column vector to a row vector with k elements, is written as $\mathbf{X}^{\mathrm{T}} = (X_1\ X_2\ \cdots\ X_k)$ and is read "X-transpose." Let $Z_i = 1$ if individual i is regularly exposed to secondhand smoke and 0 otherwise. The individual's propensity for exposure to secondhand smoke is then defined as the individual's conditional probability of exposure to secondhand smoke given all the pretreatment characteristics $\mathbf{X}_i$ that have taken particular values $\mathbf{x}$, that is, $\theta_i = \theta_i(\mathbf{x}) = \mathrm{pr}(Z_i = 1|\mathbf{X}_i = \mathbf{x})$. Here, the propensity score is a unidimensional index that summarizes all the pretreatment information indicating an individual's predisposition of being exposed to a given treatment (e.g., secondhand smoking). To simplify the notation, henceforth, we omit the subscript and use θ as a shorthand for $\theta(\mathbf{x})$.

In the earlier example, when the propensity score for an individual is unknown to the researcher, one may obtain a predicted propensity score according to the proportion of individuals with the same pretreatment characteristics $\mathbf{x}$ who are actually exposed to secondhand smoking. Let us suppose that among 40-year-old white women with high school education, 20% are exposed to secondhand smoking. The researcher may predict that the propensity score for exposure to secondhand smoke is 0.2 for a high school-educated 40-year-old white woman. The researcher would revise this individual's predicted propensity score if additional pretreatment information were to become available. Suppose that this particular individual lives below the poverty level; and suppose that 30% of the women with the same age, race, education, and poverty status have regular exposure to secondhand smoking. Then this individual's predicted propensity score might become 0.3. This example shows that, in quasiexperimental studies, an individual's predicted propensity score depends on the specific pretreatment information that the researcher has taken into consideration. The prediction is often based on analysis of a logit model. If we let X_1, X_2, X_3, X_4, and X_5 denote age, gender, race, education, and poverty status, respectively, then $\mathbf{X}^T = (1 \ \ X_1 \ \ X_2 \ \ X_3 \ \ X_4 \ \ X_5)$. A logit model can be specified as follows:

$$\ln\left(\frac{\theta}{1-\theta}\right) = \eta = \mathbf{X}^T\beta, \tag{3.6}$$

where $\beta = \begin{pmatrix} \beta_0 \\ \beta_1 \\ \beta_2 \\ \beta_3 \\ \beta_4 \\ \beta_5 \end{pmatrix}$ and hence $\mathbf{X}^T\beta = \beta_0 + \beta_1 X_1 + \beta_2 X_2 + \beta_3 X_3 + \beta_4 X_4 + \beta_5 X_5$. We use $\widehat{\beta}$ to

denote the sample estimates of β and obtain the estimated propensity score

$$\widehat{\theta} = \frac{e^{\widehat{\eta}}}{1+e^{\widehat{\eta}}} = \frac{e^{\mathbf{X}^T\widehat{\beta}}}{1+e^{\mathbf{X}^T\widehat{\beta}}}.$$

The logit of the estimated propensity score $\widehat{\eta} = \mathbf{X}^T\widehat{\beta}$ rather than the estimated propensity score itself is often used in further analysis.

3.4.2   Balancing property of the propensity score

In quasiexperimental studies, a major concern is selection bias in treatment effect identification due to preexisting differences between the treated group and the control group. Such between-group differences will lead to bias if they are also related to the potential outcome of interest. For example, if middle-aged white women with lower education and living in poverty tend to have more health problems than middle-aged minority women with higher education and living above the poverty level and if the former are more likely to be exposed to secondhand smoke than the latter, then the causal effect of secondhand smoke on the health of middle-aged women is confounded at least by race, education, and poverty status.

The propensity score is appealing primarily because of its property as a balancing score. According to Rosenbaum and Rubin (1983a), a balancing score θ is a function of the pretreatment covariates X such that for the treated units and the control units who share the same value of θ, the joint distribution of X is balanced conditional on θ. The propensity score estimated from a given sample, denoted by $\hat{\theta}$, has the same properties when applied to the sample data. In the previous example, the estimated propensity score is a function of age, gender, race, and education. A group with regular exposure to secondhand smoke and a group without such exposure are expected to have the same age distribution, gender distribution, racial distribution, and education distribution if the two groups have the same estimated propensity score for exposure to secondhand smoking. When the estimated propensity score is additionally a function of poverty status, the two treatment groups with the same estimated propensity score are expected to have the same distribution of poverty as well. Therefore, when the estimated propensity score is held constant, treatment assignment (secondhand smoking vs. no secondhand smoking in this case) becomes independent of all these pretreatment characteristics. Later, we use the true propensity score and the estimated propensity score interchangeably except when a distinction becomes necessary.

In an ideal world, the researcher would like to have collected all the pretreatment information X that are predictive of the potential outcomes $Y(1)$ and $Y(0)$ and are associated with treatment assignment Z such that, conditioning on X, treatment assignment becomes independent of the potential outcomes. Most importantly, once the pretreatment information has been summarized in a propensity score θ, conditioning on θ, treatment assignment is independent of the potential outcomes as well (Rosenbaum and Rubin, 1983a). This assumption is mathematically represented as follows:

$$E[Y(z)|Z=1, \theta] = E[Y(z)|Z=0, \theta] = E[Y(z)|\theta] \tag{3.7}$$

for $z = 0, 1$. Here, $E[Y(z)|Z=1, \theta]$ is the average potential outcome associated with treatment z for the treated units with propensity score value θ. When $z = 1$, $E[Y(1)|Z=1, \theta]$ is equal to the average observed outcome of the treated units with propensity score value θ written as $E[Y|Z=1, \theta]$. Similarly, $E[Y(z)|Z=0, \theta]$ is the average potential outcome associated with treatment z for the control units with propensity score value θ. When $z = 0$, this is equal to the average observed outcome of the control units with propensity score value θ written as $E[Y|Z=0, \theta]$. In quasiexperimental data, if treatment assignment is strongly ignorable given the observed pretreatment covariates X, then it is strongly ignorable given the propensity score θ. This assumption is relatively strong in the sense that it is not warranted by any particular research design. The strong ignorability assumption also implies that in a subpopulation characterized by a given propensity score value θ, the proportion of individual units being treated must be neither zero nor one, that is,

$$0 < pr(Z=1|\theta) < 1. \tag{3.8}$$

In this subpopulation, the strong ignorability assumption enables us to identify the average conditional treatment effect $E[Y(1)-Y(0)|\theta]$ by computing the mean difference in the observed outcome between the treated units and the control units with propensity score value θ:

$$E[Y(1)-Y(0)|\theta] = E[Y|Z=1, \theta] - E[Y|Z=0, \theta]. \tag{3.9}$$

(3.8) is a necessary condition because the conditional treatment effect cannot be identified if at a given propensity score value θ, either nobody is in the treated group or nobody is in the control group. The population average treatment effect is obtained by taking an integral over the distribution of θ:

$$\delta = \int_{\theta} \{E[Y|Z=1,\ \theta] - E[Y|Z=0,\ \theta]\} f(\theta) d\theta,$$

where $f(\theta)$ represents a probability density function of a continuous θ. Alternatively, suppose that all the pretreatment covariates are discrete and hence θ takes only discrete values. Then the integral can be replaced by the following summation:

$$\delta = \sum_{\theta} \{E[Y|Z=1,\ \theta] - E[Y|Z=0,\ \theta]\} \mathrm{pr}(\theta),$$

where $\mathrm{pr}(\theta)$ is the proportion of the population with propensity score value θ.

Table 3.1 illustrates the balancing property of the propensity score with a relatively simple evaluation of the causal effect of grade retention on academic achievement in a population of students at risk of repeating a grade. In this hypothetical example, suppose that whether a student repeats a grade is independent of the potential outcomes $Y(1)$ and $Y(0)$ when gender and SES are given. In this case, $Y(1)$ is a student's achievement score if retained, and $Y(0)$ is the same student's achievement score if promoted. Hence, a student's probability of being retained is a function of gender (X_1) and SES (X_2) represented by the propensity score $\theta = \mathrm{pr}(Z=1|X_1=x_1, X_2=x_2)$. To further simplify, we have dichotomized the SES measure into a "low-SES" category and a "high-SES" category. Suppose that in a population of 2000 students at risk of repeating a grade, low-SES boys, high-SES boys, low-SES girls, and high-SES girls each constitute a quarter of the population. Within each of these four subpopulations defined by gender and SES, the subpopulation average treatment effect can be identified as the mean difference in the observed outcome between the treated group and the control group with the same gender and SES:

Table 3.1 Propensity score and its balancing property

Gender (X_1)	Boys		Girls		
SES (X_2)	Low SES	High SES	Low SES	High SES	Total
Total n	500	500	500	500	$N = 2000$
n retained	400	250	250	100	$N_{Z=1} = 1000$
$E(Y|Z=1,X_1,X_2)$	10	45	35	70	$E(Y|Z=1) = 31$
n promoted	100	250	250	400	$N_{Z=0} = 1000$
$E(Y|Z=0,X_1,X_2)$	20	60	40	80	$E(Y|Z=0) = 59$
$\mathrm{pr}(X_1, X_2)$	0.25	0.25	0.25	0.25	
$E[Y(1)-Y(0)|X_1,X_2]$	-10	-15	-5	-10	
Propensity score (θ)	0.8	0.5	0.5	0.2	$\delta = -10$
$\mathrm{pr}(\theta)$	0.25		0.5	0.25	
$E[Y(1)-Y(0)|\theta]$	-10		-10	-10	

$$E(Y(1)-Y(0)|X_1,X_2) = E(Y|Z=1,X_1,X_2) - E(Y|Z=0,X_1,X_2).$$

The population average treatment effect is a weighted average of the subpopulation-specific treatment effect where the weight is the proportion of individuals in each subpopulation.

The propensity score for grade retention can be easily computed as the proportion of students retained in each subpopulation. For example, 400 out of the 500 low-SES boys are retained in grade. Hence, the propensity score for low-SES boys is $\theta = 400/500 = 0.8$. Similarly, one finds the propensity scores for high-SES boys, low-SES girls, and high-SES girls to be 0.5, 0.5, and 0.2, respectively. At each propensity score value, we can see that the retained group and the control group have the same gender distribution and SES distribution. Those at the propensity score value 0.8 are all low-SES boys. Hence, the retained group and the promoted group do not differ in gender and SES. Similarly, the two treatment groups are balanced at the propensity score value 0.2 as all those students are high-SES girls. It is most revealing when we compare the retained group and the promoted group at the propensity score value 0.5. Here, within each treatment group, 50% of the students are high-SES boys and 50% are low-SES girls. Therefore, the two treatment groups are again balanced in their gender composition and SES composition.

As shown in Table 3.1, the mean observed outcomes of the low-SES boys, high-SES boys, low-SES girls, and high-SES girls who are retained are 10, 45, 35, and 70, respectively; the mean observed outcomes of low-SES boys, high-SES boys, low-SES girls, and high-SES girls who are promoted are 20, 60, 40, and 80, respectively. A naïve analysis would compare the mean observed outcome of the retained students $E(Y|Z=1)$ and that of the promoted students $E(Y|Z=0)$ and generate a prima facie effect: 31–59 = –28. A negative selection bias seems likely because students with a relatively higher propensity of being retained tend to have relatively lower potential achievement scores under each treatment condition. In particular, among all four subpopulations, low-SES boys have the lowest potential achievement scores regardless of whether they are retained or promoted; and these students are most likely to be retained. In contrast, high-SES girls have the highest potential achievement scores and are least likely to be retained. The conditional average treatment effects are –10, –15, –5, and –10 for low-SES boys, high-SES boys, low-SES girls, and high-SES girls, respectively, all considerably smaller in magnitude than the prima facie effect. Averaging over the four subpopulations, we obtain the population average treatment effect $\delta = -10$. Instead of conditioning on gender and SES, we now condition on the propensity score instead. At each of the three distinct values of θ, that is, for $\theta = 0.2, 0.5$, and 0.8, the conditional treatment effect estimate is –10. The respective proportions of individuals with propensity score values 0.2, 0.5, and 0.8 are $pr(\theta=0.2)=0.25; pr(\theta=0.5)=0.5$; and $pr(\theta=0.8)=0.25$. The population average treatment effect over the distribution of the propensity score is a weighted average of the conditional treatment effect where the weight is the proportion of individuals taking each propensity score value: $\delta = (-10) \times 0.25 + (-10) \times 0.5 + (-10) \times 0.25 = -10$.

3.4.3   Pooling conditional treatment effect estimate: Matching, stratification, and covariance adjustment

The previous hypothetical example illustrates a basic rationale when propensity scores are used to adjust for selection bias, that is, to estimate an average of conditional treatment effect over the distribution of the propensity score. This is because under strong ignorability, treated and untreated units with the same propensity score value can act as controls for each other. Following this logic, propensity score matching, stratification, and covariance adjustment have been developed for assessing the effect of a binary treatment in quasiexperimental data. In each case, the propensity score is regarded as a single covariate, which greatly eases the implementation of these standard techniques for adjustment in quasiexperimental studies.

3.4.3.1 Propensity score matching

This method matches each treated unit with a control unit on the basis of the propensity score. For a treated unit with a given propensity score value θ, a control unit is randomly sampled from all the control units with the same propensity score value. The difference in the observed outcome between the treated unit and the matched control unit is, in expectation, equal to the average conditional treatment effect for the subpopulation at the propensity score value θ under the strong ignorability. If every treated unit has a match in the control group, the mean of the matched pair differences in the observed outcome is unbiased for the average treatment effect on the treated. This is because the distribution of pretreatment covariates is expected to be similar between the treated group and the matched control group.

Matching is particularly useful when there is a large reservoir of control units that supplies potential matches to the treated units, when the sample size of the treated group is relatively small, and when the number of confounding covariates is relatively large. An infinitely large control reservoir makes exact matching possible, in which case 100% of the initial bias is expected to be removed. Rubin's (1979) simulation study showed that as the ratio of the size of the control reservoir to the size of the treated group decreases, exact matching becomes increasingly difficult. The nearest available matching is imperfect and provides only a partial adjustment for the confounding factors. For example, when the control reservoir is four times the size of the treated group, matching removes at least 90% of the initial bias; when this ratio is reduced to two, matching may remove only 70% of the initial bias. Propensity score matching can be combined with covariance adjustment of matched pair differences to further reduce bias and improve precision. To date, researchers have written extensively on optimal matching methods (Hansen, 2004, 2006; Ho, Imai, King, and Stuart, 2007; Rosenbaum, 1988, 1989, 2002; Rosenbaum and Rubin, 1985a, 1985b; Rubin, 1973, 1976a, 1976b, 1979, 1980; Stuart, 2010; Stuart and Ialongo, 2010).

Rosenbaum (1986) illustrated propensity score matching with an application study evaluating the effect of dropping out of high school on cognitive achievement. In the High School and Beyond data, sophomores from more than 1000 high schools nationwide were observed in 1982 and then followed up in 1984. The number of the students who stayed in high school was about 10 times the number of the dropouts in the sample. A propensity score representing the conditional probability of dropping out was computed as a function of 32 sophomore-year predictors. Each dropout from a school was matched to a student who stayed in the same school on the estimated propensity score. Within-school matching effectively controlled for all observed and unobserved pretreatment covariates that were constant for all students within the same school. However, among those with relatively high propensity score values, the absolute number of students who stayed in high school was inadequate for matching with the dropouts. Even though the matching led to about 60–70% reduction in initial bias, a considerable amount of bias remained within matched pairs. The dropouts and those who stayed in high school showed large baseline differences in five tests as well as in seven other covariates. These baseline characteristics remained imbalanced between the two types of students in the matched pairs according to the t statistics. After additional adjustment for the within-pair differences through ANCOVA, the estimated effects of dropping out were negative in all five subjects with the largest decline in mathematics.

3.4.3.2   Propensity score stratification

Propensity score stratification divides the treated and control units into a number of strata denoted by $s = 1,…,S$ on the basis of the propensity score. When the propensity score takes discrete values, the treated and control units in the same stratum have the same propensity score value and therefore have the same distribution of pretreatment covariates. Under strong ignorability, the mean difference in the observed outcome between the treated units and the control units in stratum s is unbiased for the conditional treatment effect in stratum s. A weighted average of these conditional treatment effects with the weight equal to the proportion of the population in stratum s is unbiased for the population average treatment effect, that is,

$$\delta = \sum_{s=1}^{S} \{E[Y|Z=1,s] - E[Y|Z=0,s]\} \text{pr}(s). \tag{3.10}$$

In most cases, the estimated propensity score is on a continuum. The researcher sorts all the individuals in the analytic sample in an ascending order on the propensity score and then divides the sample into, say, five strata with equal proportions of individuals. Within each stratum, the treated and control units will have similar though not exactly the same estimated propensity score values. When the sample size is sufficiently large, the more strata created, the more homogeneous the treated and control units within a stratum will be in their propensity score values. Cochran's (1968) result for stratification on a single covariate is applicable in this case. This result indicates that with five strata of equal size, at least 90% of the initial bias is removed.

Rosenbaum and Rubin (1984) applied propensity score stratification to an evaluation of surgical treatment versus drug treatment for coronary artery disease. They reported that even though numerous pretreatment covariates displayed quite different distributions between the two treatment groups, within each of the five constructed propensity score strata, there was greater balance than would have been expected if treatments had been randomized. Rubin (1997) provided another example using propensity score stratification to compare the relative effectiveness of two different treatments for breast cancer and reported satisfactory between-treatment group balance in the distribution of all covariates within each of the five strata. In their evaluation of the effect of kindergarten retention versus promotion, Hong and Raudenbush (2005) constructed as many as 14 strata, which greatly reduced if not eliminated the preexisting differences in comparing the retained and promoted students within each stratum. These applications illustrated a nine-step analytic procedure for implementing propensity score stratification:

1. Collect a large set of pretreatment covariates. For example, the data analyzed by Rosenbaum and Rubin (1984) contained 74 covariates; Hong and Raudenbush (2005) considered as many as 207 pretreatment covariates including nearly all the factors that teachers and parents typically consider in making decisions with regard to grade retention.

2. Build a logistic regression model that uses the pretreatment covariates to predict treatment assignment. Save the estimated propensity score and convert it onto the logit scale. This is the balancing index denoted by η to be used in the rest of the analysis. When the number of covariates is large, researchers have used a stepwise procedure to remove covariates that have redundant information with those already in the model. Variable selection for the propensity score model is a separate topic that we leave to Appendix 3.B. The propensity score model may include the main effects of the selected covariates, nonlinear terms, and interactions. Rubin (1997) suggested that the propensity score model should include the linear and quadratic terms of a covariate if the treated group and the control group differ in the variance of this covariate; the model should include the product of two covariates if their correlation is different between the treated group and the control group.

3. Compare the distribution of the balancing index between the treated group and the control group and identify the range of propensity scores within which the two distributions overlap. This range constitutes the common support for causal inference. For example, in the kindergarten retention study, a large number of promoted students displayed propensity score values lower than that of all the retainees and therefore were excluded from the common support. In Figure 3.3, these students are concentrated in the left tail of the distribution in the upper panel. Clearly, for these promoted students who were essentially at no risk of being retained, the retention

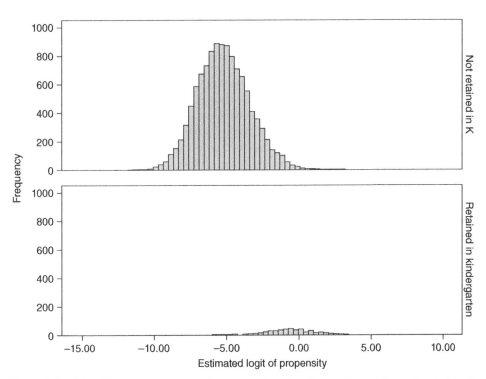

Figure 3.3   Identify common support by comparing the distribution of the estimated logit of propensity score between the treated group and the control group

effect is undefined and cannot be empirically identified. A small number of retainees were also excluded from the common support because their propensity score values were higher than those of all the promoted students. Step (3) empirically examines the second component of the strong ignorability assumption summarized in (3.8) and ensures that, within the common support, each treated unit and each control unit in the sample have their counterparts in the alternative treatment groups.

4. Stratify the sample on the balancing index η, typically starting with five or six strata of approximately equal size.

5. Check within-stratum balance between the treated group and the control group with respect to the balancing index η and the observed pretreatment covariates. Rosenbaum and Rubin (1984) subjected each covariate to a two-way ANOVA for examining the main effect of treatment assignment conditioning on stratum membership, the main effect of strata conditioning on treatment group membership, and the treatment-by-stratum interaction effect. Each F test for the treatment main effect and for the treatment-by-stratum interaction effect assesses the adequacy of the propensity score model and the adequacy of the current stratification. Ideally, the mean difference between the treated group and the control group is expected to disappear after stratification, not only in the balancing index but also in 95% of the pretreatment covariates. Some researchers (Stuart, 2007) have adopted an alternative criterion by computing the standardized difference (also called "standardized bias") in each covariate after stratification. This is the within-stratum mean difference between the treated group and the control group divided by the standard deviation of the entire control group and is expected to be less than 0.25 to prevent bias. Values greater than 0.5 are considered to be particularly problematic.

6. Restratify the sample or revise the propensity score model if necessary, and check balance again. For example, the researcher may divide the sample into a larger number of strata or subdivide some strata in which there is a lack of balance between the treated group and the control group. The propensity score model may be revised to include additional covariates remaining imbalanced or to incorporate nonlinear or interaction terms. In the study by Rosenbaum and Rubin (1984), the final propensity score model included 37 main effects, 7 interactions, and 1 quadratic term.

7. Examine the within-stratum mean difference in the outcome between the treated group and the control group. In this step, the researcher may estimate and tabulate stratum-specific treatment effects as that shown in Table 3.2. In the kindergarten retention study, the retained students would generally be lower achieving than the promoted students, as predicted by their pretreatment information, even if the former group were not retained. When the researcher simply compared the retainees with all the promoted students in the common support, this negative bias was contained in the naïve estimate of the retention effect on the reading outcome reported to be −18.51. When the retained students were instead compared with the promoted students with very similar propensity score values within each stratum, it was observed that, across the 14 propensity score strata, the retained students still consistently scored lower than their comparable promoted peers, suggesting a detrimental effect of retention across the board after

Table 3.2   Within-stratum mean difference in the reading outcome between retained and promoted students

Stratum	Retained			Promoted			Mean difference
	N	Mean	SD	N	Mean	SD	
1	9	47.41	11.73	3044	59.22	11.64	−11.80
2	12	47.25	15.50	1654	55.62	11.30	−8.37
3	14	42.66	15.92	977	52.49	11.95	−9.83
4	12	45.24	17.32	321	52.45	12.53	−7.21
5	24	44.07	14.43	440	50.05	12.15	−5.97
6	23	36.63	13.88	153	48.77	12.67	−12.14
7	47	37.41	10.92	211	45.89	12.40	−8.48
8	48	39.79	15.37	143	46.00	13.37	−6.21
9	45	37.78	12.27	85	43.28	11.52	−5.50
10	48	35.00	9.77	49	42.18	13.34	−7.18
11	47	34.30	9.05	35	40.45	13.08	−6.16
12	49	32.06	9.31	16	33.03	14.33	−0.96
13	45	31.09	11.96	8	35.27	8.25	−4.18
14	38	33.87	9.15	3	40.47	8.60	−6.60

Adapted with permission from Hong and Raudenbush (2005). © 2005 by Sage Publications.

removing the selection bias associated with the observed covariates. The magnitude of the negative effects, however, became considerably smaller than the naïve estimate, ranging from −12.14 to −0.96 among the 14 strata. This step additionally allows the researcher to detect possible systematic associations between stratum-specific treatment effects and the propensity score. For example, comparing the within-stratum survival rate between women who received breast conservation therapy and those who received mastectomy, Rubin (1997) observed similar performance of these two treatments in general. However, there was some indication that women and their physicians might be making optimal treatment choices: women who had a relatively higher propensity for receiving mastectomy rather than breast conservation therapy appeared to benefit more from mastectomy, while those who had a relatively lower propensity for receiving mastectomy apparently did not benefit from this particular treatment.

8. Pool the within-stratum treatment effect over all the strata. When all the strata are of approximately equal size, the population average treatment effect can be estimated as a simple average of the stratum-specific treatment effects. Otherwise, (3.10) can be applied to assign to each stratum-specific treatment effect a weight equal to the proportion of units in that stratum. A two-way ANOVA can again be employed for testing whether the sample estimate of the population average treatment effect is statistically significant. The remaining within-stratum difference between the treated group and the control group in the balancing index η can be further adjusted through covariance adjustment for the balancing index, in an example with five strata, through analyzing the following regression model:

$$Y = \beta_0 + \beta_1 Z + \sum_{s=2}^{5} \beta_s I(s) + \beta_6 \eta + e. \tag{3.11}$$

Here, Z is the treatment indicator, and hence, β_1 estimates the population average treatment effect; $I(s)$ is a dummy indicator for stratum s. To the extent that η explains a portion of the variance in the outcome conditioning on treatment group membership and stratum membership, the combination of propensity score stratification with covariance adjustment also improves precision in estimation.

9. Conduct sensitivity analysis. This important step will be discussed separately in Section 3.4.3.4.

3.4.3.3 Covariance adjustment for the propensity score

Rosenbaum and Rubin (1983a) proposed covariance adjustment for the propensity score as a third method. The researcher may simply fit an ANCOVA model that regresses the outcome on the treatment and the balancing index η, which is the logit of the estimated propensity score. This could be as effective as covariance adjustment for the observed pretreatment covariates. To avoid linear extrapolations and possible misspecifications of the functional relationship between the outcome and the balancing index η, Rosenbaum and Rubin (1983a) suggested inspecting residual plots for detecting nonlinear or nonparallel relationships. Yet as discussed earlier in this chapter, the linearity and additivity assumptions required by ANCOVA models are strong; and violations of these model-based assumptions have important consequences for treatment effect estimation. In contrast, propensity score matching and stratification are nonparametric in nature and hence are preferred to covariance adjustment for the propensity score. When propensity score matching and stratification are combined with covariance adjustment, however, model misspecifications become less of a concern because an ANCOVA model is fit within matched pairs or within strata in which the treated group and the control group have very similar propensity score values.

3.4.3.4 Sensitivity analysis

When applied to quasiexperimental data, propensity score-based adjustment methods generate unbiased estimates of average treatment effects only when the strong ignorability assumption holds. The results can be challenged on scientific grounds if important pretreatment covariates known to be potential confounders of the treatment effect are not observed and therefore possibly unadjusted. For example, in the kindergarten retention study, even though the data contain extensive information that educators typically rely on for making retention decisions, one cannot rule out the possibility that some parents might contribute to the decision-making on the basis of their observations of child behaviors that were perhaps unknown to the research analyst. The confounding effect of an unmeasured covariate is sometimes partially adjusted if the unmeasured covariate is correlated with a measured covariate and if the latter has been balanced between the treated group and the control group through statistical adjustment. Here, we consider the worst possible scenario in which the unmeasured covariate is unrelated to any of the measured covariates. A sensitivity analysis assesses the extent to which the estimated treatment effect would possibly be changed and the causal conclusion altered should additional adjustment be made of such an unmeasured covariate.

To reveal the basic logic, we consider the simplest case that involves only two potential confounders, one observed and one unobserved, that are independent of each other. Let X denote the observed confounder and U the unobserved one. The discussion can be generalized to the case in which U is a latent variable. As before, Y denotes a continuous outcome and Z a binary treatment that takes value 1 for the treated and 0 for the control units. To further simplify, we assume that X and U each have a linear association with Y that is constant across the treated group and the control group. The true model that generates the observed outcome is

$$Y = \alpha + \delta Z + \beta X + \pi U + e. \tag{3.12}$$

Here, the regression coefficients β and π represent the relationship between X and Y and that between U and Y, respectively; $\delta = E[Y(1) - Y(0)]$ is the average treatment effect of Z on Y. In other words, the adjustment for both X and U as shown in (3.12) is sufficient for removing all the selection bias, while adjustment for the observed covariate X only is not sufficient.

However, when U is unobserved, the researcher makes covariance adjustment for X only; and the model becomes

$$Y = \alpha^* + \delta_{PF} Z + \beta X + e^*. \tag{3.13}$$

This is a familiar ANCOVA model in which the prima facie effect is

$$\delta_{PF} = \sum_x [E(Y|Z=1, X=x) - E(Y|Z=0, X=x)] \times \mathrm{pr}(X=x).$$

At any given value of x, we have that

$$E(Y|Z=1, X=x) = E(\alpha + \delta + \beta x + \pi U + e|Z=1, X=x)$$
$$= \alpha + \delta + \beta x + \pi E(U|Z=1, X=x) + E(e|Z=1, X=x);$$
$$E(Y|Z=0, X=x) = E(\alpha + \beta x + \pi U + e|Z=0, X=x)$$
$$= \alpha + \beta x + \pi E(U|Z=0, X=x) + E(e|Z=0, X=x).$$

Here, e is a random error with mean 0 within each treatment group and is independent of X. In other words, there are no other unobserved covariates that are different in their means between the treated group and the control group and could possibly confound the treatment effect. Taking the difference between the earlier two equations, we obtain the prima facie effect as the sum of the average treatment effect and the bias:

$$\delta_{PF} = \sum_x [\delta + \pi E(U|Z=1, X=x) - \pi E(U|Z=0, X=x)] \times \mathrm{pr}(X=x)$$
$$= \delta + \pi [E(U|Z=1) - E(U|Z=0)].$$

Therefore, when U is omitted, the selection bias due to such an omission, denoted by $\text{BIAS}_{(U)}$, is

$$\text{BIAS}_{(U)} = \delta_{\text{PF}} - \delta = \pi[E(U|Z=1) - E(U|Z=0)]. \qquad (3.14)$$

The bias is a product of two quantities: the relationship between the outcome and the unobserved covariate denoted by π and the mean difference in the unobserved covariate between the treated group and the control group represented by $E(U|Z=1) - E(U|Z=0)$. Therefore, the impact of the omission of U is determined by its associations with the outcome and with the treatment assignment.

In a sensitivity analysis, the researcher speculates the possible existence of an unobserved covariate U and attempts to determine on the basis of prior scientific knowledge a plausible range of values that π and $E(U|Z=1) - E(U|Z=0)$ could each take. For example, in the kindergarten retention study (Hong and Raudenbush, 2005, 2006), the researchers adjusted for the observed pretreatment covariates including the pretest scores through propensity score stratification. The estimated causal effect of retention as opposed to promotion on a reading outcome was $\widehat{\delta}_{\text{PF}} = -8.86$ (standard error $= 1.38$). The researchers reasoned that, in studies of student learning, once the pretest scores have been controlled, additional covariates typically explain comparatively little of the variation in the outcome. For this reason, rather than speculating that an unobserved covariate U could rival the observed pretest scores, the researchers identified an observed covariate—teacher rating of a student's approach to learning—that demonstrated the second largest confounding effect. This covariate has the second largest absolute standardized regression coefficient in predicting the outcome within each treatment group and the second largest standardized difference between the treated group and the control group. It seemed relatively plausible that an unobserved covariate might have as strong associations with the outcome and with the treatment assignment as this referent observed covariate had when the pretest scores were controlled. The researchers incorporated its reference values for π and $E(U|Z=1) - E(U|Z=0)$ and, following (3.14), computed their product as the potential bias associated with the hypothetical unobserved covariate: $\text{BIAS}_{(U)} = -2.73$. Removing this amount of bias from the original estimate of the treatment effect, the researchers obtained a new estimate of the retention effect: $-8.86 - (-2.73) = -6.13$. Hence, with additional adjustment for U, the magnitude of the retention effect estimate became smaller. When the original standard error was applied, the 95% confidence interval for the new estimate of the retention effect was $(-6.95, -1.54)$ and did not contain zero or any positive values. The researchers concluded that if an unobserved covariate were to explain away the negative effect of kindergarten retention, that covariate would need to be a stronger confounder than the strongest observed confounder other than the pretest scores, which seemed unlikely given prior knowledge about predictors of student learning. Therefore, the empirical result with regard to the negative effect of kindergarten retention on the reading outcome did not seem particularly sensitive to potential omissions of unmeasured covariates.

The previously described procedure for sensitivity analysis is ANCOVA based and is relatively easy to implement by applied researchers. Rosenbaum (1986) applied a similar procedure to assess the sensitivity of the estimated effect of dropping out of high school initially obtained from propensity score matching combined with covariance adjustment. Others have developed procedures for assessing the sensitivity to an unobserved discrete covariate with a binary outcome (Rosenbaum, 2002; Rosenbaum and Rubin, 1983b) or with censored survival time data (Lin, Psaty, and Kronmal, 1998). Hosman, Hansen, and Holland (2010) further showed that additional adjustment for an omitted confounder may change not only the point estimate of the treatment effect but also the confidence interval. However, Rosenbaum (1986) cautioned that even if one concludes that the empirical results are insensitive to the impact of a hypothetical confounder assumed to be as influential as one of the most influential observed confounders, the sensitivity analysis generally does not rule out the possibility that unmeasured covariates with an even stronger joint confounding impact exist such that additional adjustment for all these unmeasured covariates might lead to qualitatively different causal conclusions.

In response to this concern, Frank and colleagues (Frank, 2000; Frank *et al.*, 2013) asked an alternative question: "How large must be the impact of a confounding variable to alter an inference?" (p. 176) Extending the earlier work by Mauro (1990) within the multiple linear regression framework, Frank (2000) developed a sensitivity index as the product of the correlation between the outcome and a hypothetical unobserved covariate and the correlation between the treatment assignment and the unobserved covariate. On the continuum of this index, he then identified the impact threshold or a "switching point" indicating the magnitude of a confounding impact necessary to change the result of hypothesis testing with regard to the treatment effect. The reference distribution of the sensitivity index then allows one to assess the probability of observing such an impact. An empirical result was considered to be insensitive to hypothetical unobserved confounding if the component correlations in the sensitivity index each have to be large by social science standards to alter the causal conclusion and if the probability is small that the impact of an unobserved covariate exceeds the threshold. In the kindergarten retention example, the result is considered to be insensitive because to change the conclusion with regard to the negative effect of retention, a hypothetical unobserved covariate must have a relatively strong correlation with the academic outcome and with the retention assignment. Yet it seems unlikely that such an unobserved covariate exists given all the observed covariates that the researchers have already made adjustment for.

In summary, this chapter has reviewed some of the most popular analytic methods each having its potential for generating valid causal evidence when the accompanying identification assumptions are valid. Comparing quasiexperimental results with causal estimates obtained from randomized experiments, some researchers have cautioned that many quasiexperimental evaluations do not replicate the experimental results (Agodini and Dynarski, 2004; Glazerman, Levy, and Myers, 2003; LaLonde, 1986), while others have found the results of some well-designed and carefully analyzed quasiexperimental studies to be promising (Cook, Shadish, and Wong, 2008; Dehejia and Wahba, 2002; Shadish, Clark, and Steiner, 2008). This book introduces innovative analytic methods in the subsequent chapters and compares them with existing methods, with a focus on examining identification assumptions as well as important properties including bias removal and efficiency. The reader is advised to always define the causal effect of interest, to opt for a research design that makes identification assumptions highly plausible, and to choose an analytic method that is efficient in estimation and requires as few additional model-based assumptions as possible.

Appendix 3.A: Potential bias due to the omission of treatment-by-covariate interaction

Many applied researchers are unclear about whether omitting a treatment-by-covariate interaction would bias the treatment effect estimate. As discussed in Section 3.3.1.3, if the experimental group and the control group differ in the mean of a pretreatment covariate X, omitting the ZX interaction would lead to a biased estimate of the average treatment effect on the treated. Suppose that the correct outcome model is

$$Y = a + dZ + b_1 X + b_2 ZX + e.$$

The population average treatment effect on the treated is $d + b_2 E(X|Z=1)$. Omitting the ZX interaction, the misspecified ANCOVA model is

$$Y = a' + d'Z + bX + e'.$$

When d' is interpreted as the average treatment effect on the treated, the bias is $(b_1 - b)[E(X|Z=1) - E(X|Z=0)]$. In the following, we derive this bias term.

From analyzing the ANCOVA model, one obtains the following estimate:

$$d' = \{E[Y|Z=1] - E[Y|Z=0]\} - b\{E[X|Z=1] - E[X|Z=0]\},$$

where $Y = a + d + b_1 X + b_2 X$ when $Z=1$ and where $Y = a + b_1 X$ when $Z=0$, and hence, $E[Y|Z=1] = a + d + (b_1 + b_2)E[X|Z=1]$ and $E[Y|Z=0] = a + b_1 E[X|Z=0]$. Substituting these terms in the expression for d', we have that

$$d' = d + b_2 E[X|Z=1] + (b_1 - b)\{E[X|Z=1] - E[X|Z=0]\}.$$

Hence, the bias term is $(b_1 - b)\{E[X|Z=1] - E[X|Z=0]\}$.

If the analyst is interested in estimating the population average treatment effect instead, from the true model, we can derive the population average treatment effect $d + b_2 E(X)$. The difference between d' and $d + b_2 E(X)$ is the bias:

$$
\begin{aligned}
& d' - [d + b_2 E(X)] \\
&= d + b_2 E[X|Z=1] + (b_1 - b)\{E[X|Z=1] - E[X|Z=0]\} - [d + b_2 E(X)] \\
&= (b_2 \mathrm{pr}(Z=0) + b_1 - b)\{E[X|Z=1] - E[X|Z=0]\}.
\end{aligned}
$$

The last equation holds because

$$
\begin{aligned}
& b_2 E[X|Z=1] - b_2 E(X) \\
&= b_2 \{E[X|Z=1] - E[X|Z=1]\mathrm{pr}(Z=1) - E[X|Z=0]\mathrm{pr}(Z=0)\} \\
&= b_2 \mathrm{pr}(Z=0)\{E[X|Z=1] - E[X|Z=0]\}.
\end{aligned}
$$

Appendix 3.B: Variable selection for the propensity score model

Researchers working with large-scale survey data or quasiexperimental data with multiple measurements are concerned with not only the hidden bias due to unobserved covariates but also the handling of an excessively large number of observed covariates. Even though propensity score models greatly ease statistical adjustment by summarizing the pretreatment information in multiple covariates into a unidimensional propensity score, a propensity score model cannot accommodate an extremely large number of covariates relative to the number of sampled units in each treatment group. In general, the smaller the sample size is, the smaller the number of covariates to be included in a propensity score model. Model overfitting is a primary concern here. It occurs when the parameter estimates in a propensity score model are driven by random noise in a given sample rather than reflecting the underlying selection mechanism in the population.

To avoid model overfitting, analysts must select pretreatment covariates for propensity score models first of all on scientific grounds. For example, following a checklist that educators use to screen students for grade retention, Hong and colleagues (Hong, 2004; Hong and Raudenbush, 2005, 2006; Hong and Yu, 2007, 2008) identified a relatively small number of variables that are essential for retention decision-making among more than 200 pretreatment covariates available in the data. The rest of the covariates were then subjected to a stepwise procedure in logistic regression as suggested by Rosenbaum and Rubin (1984). The final propensity score model included 39 covariates.

Should pretreatment covariates be selected into a propensity score model on the basis of their associations with the treatment, the outcome, or both? This has been a topic for discussion and debate in the causal inference literature. In theory, selection bias comes from pretreatment covariates that are associated with both the treatment and the outcome. The stronger these associations are, the larger the bias. Hence, such covariates are called "confounders." Variables that predict the treatment but not the outcome or predict the outcome but not the treatment do not contribute to bias. Including these two different types of variables in a propensity score model, though with no implication for bias removal, has different consequences for efficiency in estimation. Including nonconfounding treatment predictors will decrease the precision of the treatment effect estimate, while including nonconfounding outcome predictors will increase the precision. This is true regardless of whether the covariate–outcome relationship is linear or nonlinear (Brookhart et al., 2006). In light of these results, researchers would ignore pretreatment variables that are associated with the treatment only and would opt for a stepwise procedure that selects covariates according to the strength of their associations with the outcome within each treatment group.

Yet difficulties may arise in two scenarios. One is if it is particularly expensive to collect certain outcome information, propensity scores may be utilized at the design stage to match potential control units to the treated units, thereby reducing both selection bias and sampling error before the outcomes are measured. The propensity scores will be estimated as a function of the treatment predictors only (Rubin, 1997). Even when the outcome was already observed, Rubin (1997) was additionally concerned that covariates that are strongly related to the treatment and only weakly related to the outcome might contribute a considerable amount of bias that outweighs the gain in precision when such covariates are excluded. In the second scenario, the researcher is interested in evaluating the treatment effects on a wide range of outcomes. For example, Hong and Yu (2007, 2008) examined the effects of kindergarten retention on children's reading and math outcomes in first, third, and fifth grades; on their self-reported interest and competence in reading, math, and all school subjects; and on their interpersonal relationships, internalizing problem behaviors, and externalizing problem behaviors in all these posttreatment years. Specifying a single propensity score model on the basis of treatment predictors and applying it to these multiple outcomes seem more sensible than tailoring a separate propensity score model to each of the outcomes. When variables for the propensity score model are selected according to their associations with the treatment, additional covariance adjustment for strong predictors of the outcome can nonetheless be employed in the outcome model to improve efficiency (Rubin and Thomas, 2000).

References

Agodini, R. and Dynarski, M. (2004). Are experiments the only option? A look at dropout prevention programs. *The Review of Economics and Statistics*, *86*(1), 180–194.

Altonji, J.G., Elder, T.E. and Taber, C.R. (2005). An evaluation of instrumental variable strategies for estimating the effects of Catholic schooling. *Journal of Human Resources*, *40*(4), 791–821.

Angrist, J.D. (1990). Lifetime earnings and the Vietnam era draft lottery: evidence from social security administrative records. *American Economic Review*, *80*(3), 313–335.

Angrist, J.D., Imbens, G.W. and Rubin, D.B. (1996). Identification of causal effects using instrumental variables. *Journal of the American Statistical Association*, *91*(434), 444–472.

Barnow, B.S., Cain, G.G. and Goldberger, A.S. (1980). Issues in the analysis of selectivity bias. *Evaluation Studies*, *5*, 43–59.

Bloom, H.S. (2005). *Learning More from Social Experiments: Evolving Analytic Approaches*, Russell Sage Foundation, New York.

Bloom, H.S., Bos, J.M. and Lee, S.W. (1999). Using cluster random assignment to measure program impacts statistical implications for the evaluation of education programs. *Evaluation Review*, *23*(4), 445–469.

Brookhart, M.A., Schneeweiss, S., Rothman, K.J. *et al.* (2006). Variable selection for propensity score models. *American Journal of Epidemiology*, *163*(12), 1149–1156.

Bryk, A.S. and Weisberg, H.I. (1977). Use of the nonequivalent control group design when subjects are growing. *Psychological Bulletin*, *84*(5), 950–962.

Card, D. and Krueger, A. (1993). Minimum wages and employment: a case study of the fast-food industry in New Jersey and Pennsylvania. *American Economic Review*, *84*(4), 772–793.

Cochran, W.G. (1957). Analysis of covariance: its nature and uses. *Biometrics Special Issue on the Analysis of Covariance*, *13*(3), 261–281.

Cochran, W.G. (1965). The planning of observational studies of human populations (with discussion). *Journal of the Royal Statistical Society, Series A*, *128*, 234–266.

Cochran, W.G. (1968). The effectiveness of adjustment by subclassification in removing bias in observational studies. *Biometrics*, *24*(2), 295–313.

Cook, T.D., Shadish, W.R. and Wong, V.C. (2008). Three conditions under which experiments and observational studies produce comparable causal estimates: new findings from within-study comparisons. *Journal of Policy Analysis and Management*, *27*(4), 724–750.

Deere, D., Murphy, K. and Welch, F. (1995). Employment and the 1990–1991 minimum-wage hike. *American Economic Review*, *85*(2), 232–237.

Dehejia, R.H. and Wahba, S. (2002). Propensity score-matching methods for nonexperimental causal studies. *Review of Economics and Statistics, 84*(1), 151–161.

Drake, C. (1993). Effects of misspecification of the propensity score on estimators of treatments effects. *Biometrics, 49*, 1231–1236.

Dynarski, S. (2003). Does aid matter? Measuring the effect of student aid on college attendance and completion. *The American Economic Review, 93*(1), 279–288.

Fitzpatrick, M. D. (2008). Starting school at four: the effect of universal pre-kindergarten on children's academic achievement. *The B.E. Journal of Economic Analysis & Policy, 8*(1), 1–38 (Advances), Article 46.

Frank, K. (2000). Impact of a confounding variable on the inference of a regression coefficient. *Sociological Methods and Research, 29*(2), 147–194.

Frank, K.A., Maroulis, S., Duong, M. and Kelcey, B. (2013). What would it take to change an inference? Using Rubin's causal model to interpret the robustness of causal inferences. *Educational Evaluation and Policy Analysis, 35*(4), 437–460.

Fuller, R.K., Mattson, M.E., Allen, J.P. *et al.* (1994). Multisite clinical trials in alcoholism treatment research: organizational, methodological and management issues. *Journal of Studies in Alcohol and Drugs*, (Suppl. 12), 30–37.

Glazerman, S., Levy, D.M. and Myers, D. (2003). Nonexperimental versus experimental estimates of earnings impacts. *The Annals of the American Academy of Political and Social Science, 589*, 63–93.

Hansen, B.B. (2004). Full matching in an observational study of coaching for the SAT. *Journal of the American Statistical Association, 99*(467), 609–618.

Hansen, B.B. (2006). Optimal full matching and related designs via network flows. *Journal of Computational and Graphical Statistics, 15*(3), 609–627.

Hearst, N., Newman, T. and Hulley, S. (1986). Delayed effects of the military draft on mortality: a randomized natural experiment. *New England Journal of Medicine, 314*(10), 620–624.

Ho, D.E., Imai, K., King, G. and Stuart, E.A. (2007). Matching as nonparametric preprocessing for reducing model dependence in parametric causal inference. *Political Analysis, 15*, 199–236.

Hong, G. (2004). Causal inference for multi-level observational data with application to kindergarten retention, Unpublished doctoral dissertation. University of Michigan, School of Education.

Hong, G. and Raudenbush, S.W. (2005). Effects of kindergarten retention policy on children's cognitive growth in reading and mathematics. *Education Evaluation and Policy Analysis, 27*(3), 205–224.

Hong, G. and Raudenbush, S.W. (2006). Evaluating kindergarten retention policy: a case study of causal inference for multi-level observational data. *Journal of the American Statistical Association, 101* (475), 901–910.

Hong, G. and Yu, B. (2007). Early grade retention and children's reading and math learning in elementary years. *Educational Evaluation and Policy Analysis, 29*(4), 239–261.

Hong, G. and Yu, B. (2008). Effects of kindergarten retention on children's social-emotional development: an application of propensity score method to multivariate multi-level data. *Developmental Psychology, 44*(2), 407–421.

Hosman, C.A., Hansen, B.B. and Holland, P.W. (2010). The sensitivity of linear regression coefficients' confidence limits to the omission of a confounder. *Annals of Applied Statistics, 4*(2), 849–870.

Imbens, G.W. and Lemieux, T. (2008). Regression discontinuity designs: a guide to practice. *Journal of Econometrics, 142*(2), 615–635.

Jacob, B.A. and Lefgren, L. (2004). Remedial education and student achievement: a regression-discontinuity analysis. *Review of Economics and Statistics, 86*(1), 226–244.

Lalonde, R.J. (1986). Evaluating the econometric evaluations of training programs with experimental data. *The American Economic Review, 76*(4), 604–620.

Lehman, D., Wortman, C. and William, A. (1987). Long-term effects of losing a spouse or a child in a motor vehicle crash. *Journal of Personality and Social Psychology*, *52*, 218–231.

Lin, D.Y., Psaty, B.M. and Kronmal, R.A. (1998). Assessing the sensitivity of regression results to unmeasured confounders in observational studies. *Biometrics*, *54*, 948–963.

Little, R.J. and Rubin, D.B. (2002). *Statistical Analysis with Missing Data*, 2nd edn, Wiley, Hoboken, NJ.

Mauro, R. (1990). Understanding L.O.V.E. (left out variables error): a method of estimating the effects of omitted variables. *Psychological Bulletin*, *108*, 314–329.

Meyer, B. (1995). Natural and quasi-experiments in economics. *Journal of Business & Economic Statistics*, *13*, 151–161.

Moerbeek, M., van Breukelen, G.J. and Berger, M.P. (2000). Design issues for experiments in multilevel populations. *Journal of Educational and Behavioral Statistics*, *25*(3), 271–284.

Murnane, R.J. and Willett, J.B. (2011). *Methods Matter: Improving Causal Inference in Educational and Social Science Research*, Oxford University Press, New York.

Raudenbush, S.W. (1997). Statistical analysis and optimal design for cluster randomized trials. *Psychological Methods*, *2*(2), 173–185.

Raudenbush, S.W. and Liu, X. (2000). Statistical power and optimal design for multisite randomized trials. *Psychological Methods*, *5*(2), 199–213.

Reardon, S.F. and Raudenbush, S.W. (2013). Under what assumptions do site-by-treatment instruments identify average causal effects? *Sociological Methods & Research*, *42*(2), 143–163.

Roderick, M. and Nagaoka, J. (2005). Retention under Chicago's high-stakes testing program: helpful, harmful, or harmless? *Educational Evaluation and Policy Analysis*, *27*(4), 309–340.

Rosenbaum, P.R. (1986). Dropping out of high school in the United States: an observational study. *Journal of Educational Statistics*, *11*, 207–224.

Rosenbaum, P.R. (1988). Sensitivity analysis for matching with multiple controls. *Biometrika*, *75*, 577–581.

Rosenbaum, P.R. (1989). Optimal matching for observational studies. *Journal of the American Statistical Association*, *84*, 1024–1032.

Rosenbaum, P.R. (1999). Choice as an alternative to control in observational studies (with comments). *Statistical Science*, *14*(3), 259–278.

Rosenbaum, P.R. (2002). *Observational Studies*, 2nd edn, Springer, New York.

Rosenbaum, P.R. and Rubin, D.B. (1983a). The central role of the propensity score in observational studies for causal effects. *Biometrika*, *70*, 41–55.

Rosenbaum, P.R. and Rubin, D.B. (1983b). Assessing sensitivity to an unobserved binary covariate in an observational study with binary outcome. *Journal of the Royal Statistical Society, B*, *45*, 212–218.

Rosenbaum, P.R. and Rubin, D.B. (1984). Reducing bias in observational studies using subclassification on the propensity score. *Journal of the American Statistical Association*, *79*(387), 516–524.

Rosenbaum, P.R. and Rubin, D.B. (1985a). The bias due to incomplete matching. *Biometrics*, *41*, 106–116.

Rosenbaum, P.R. and Rubin, D.B. (1985b). Constructing a control group using multivariate matched sampling methods that incorporate the propensity score. *The American Statistician*, *39*, 33–38.

Rubin, D. B. (1973). Matching to remove bias in observational studies. *Biometrics*, *29*, 159–183. Correction (1974), 30, 728.

Rubin, D.B. (1976a). Matching methods that are equal percent bias reducing: some examples. *Biometrics*, *32*, 109–120.

Rubin, D.B. (1976b). Multivariate matching methods that are equal percent bias reducing: maximums on bias reduction for fixed sample sizes. *Biometrics*, *32*, 121–132.

Rubin, D.B. (1979). Using multivariate matched sampling and regression adjustment to control bias in observational studies. *Journal of the American Statistical Association*, *74*, 318–328.

Rubin, D.B. (1980). Discussion of "randomization analysis of experimental data: the Fisher randomization test," by D. Basu. *Journal of the American Statistical Association*, *75*, 591–593.

Rubin, D.B. (1997). Estimating causal effects from large data sets using propensity score. *Annuals of Internal Medicine*, *127*, 757–763.

Rubin, D.B. and Thomas, N. (2000). Combining propensity score matching with additional adjustments for prognostic covariates. *Journal of the American Statistical Association*, *95*, 573–585.

Schaffer, J.L. (2010). *Analysis of Incomplete Multivariate Data*, Chapman & Hall/CRC, Boca Raton, FL.

Schochet, P.Z. (2008). Statistical power for random assignment evaluations of education programs. *Journal of Educational and Behavioral Statistics*, *33*(1), 62–87.

Shadish, W.R., Cook, T.D. and Cambell, D.T. (2002). *Experimental and Quasi-Experimental Designs for Generalized Causal Inference*, Houghton Mifflin, Boston, MA.

Shadish, W.R., Clark, M.H. and Steiner, P.M. (2008). Can nonrandomized experiments yield accurate answers? A randomized experiment comparing random and nonrandom assignments. *Journal of the American Statistical Association*, *103*(484), 1334–1344.

Shin, Y. and Raudenbush, S.W. (2007). Just-identified versus overidentified two-level hierarchical linear models with missing data. *Biometrics*, *63*(4), 1262–1268.

Shin, Y. and Raudenbush, S.W. (2011). The causal effect of class size on academic achievement: multivariate instrumental variable estimators with data missing at random. *Journal of Educational and Behavioral Statistics*, *36*(2), 154–185.

Stuart, E.A. (2007). Estimating causal effects using school-level data sets. *Educational Researcher*, *36*(5), 187–198.

Stuart, E.A. (2010). Matching methods for causal inference: a review and a look forward. *Statistical Science*, *25*(1), 1–21.

Stuart, E.A. and Ialongo, N.S. (2010). Matching methods for selection of subjects for follow-up. *Multivariate Behavioral Research*, *45*(4), 746–765.

Su, Y.S., Yajima, M., Gelman, A.E. and Hill, J. (2011). Multiple imputation with diagnostics (mi) in R: opening windows into the black box. *Journal of Statistical Software*, *45*(2), 1–31.

Thistlewaite, D. and Campbell, D.T. (1960). Regression-discontinuity analysis: an alternative to the ex-post facto experiment. *Journal of Educational Psychology*, *51*, 309–317.

Webster, D.W., Vernick, J.S. and Hepburn, L.S. (2002). Effects of Maryland's law banning "Saturday night special" handguns on homicides. *American Journal of Epidemiology*, *155*(5), 406–412.

4

Adjustment for selection bias through weighting

This chapter introduces an alternative rationale for statistical adjustment that makes use of weighting. The idea originates from survey sampling in which the difference between sample composition and population composition is typically adjusted through weighting in estimating population parameters (Holt and Smith, 1979; Horvitz and Thompson, 1952; Kish, 1965; Little, 1982, 1986; Little and Vartivarian, 2003, 2005; Oh and Scheuren, 1983). This particular rationale, when applied to causal inference, focuses on obtaining the population average treatment effect as the difference between two population average potential outcomes, one associated with the experimental condition and the other with the control condition. In observational studies, the pretreatment composition of a certain treatment group is usually different from that of the population due to treatment selection. Propensity score-based weighting (Robins, 1986; Rosenbaum, 1987) estimates each population average potential outcome by transforming the pretreatment composition of the corresponding treatment group to resemble the population composition.

For example, to estimate the effect of grade retention for the population of retainees, one may use weighting to transform the composition of the group of promoted students such that the weighted promoted group will display the same pretreatment composition as the retained group. The weighted outcome of the promoted group will then estimate the population average potential outcome that one would obtain in a hypothetical world in which all the retainees were counterfactually promoted.

The chapter contrasts two propensity score-based weighting approaches, one parametric and the other nonparametric: the inverse-probability-of-treatment weighting (IPTW) method computes the weight as a direct function of the estimated propensity score (Robins, 1999), while marginal mean weighting through stratification (MMWS) computes the weight on the basis of propensity score stratification (Hong, 2010, 2012). The relative advantages of the nonparametric MMWS procedure are revealed through simulations. For simplicity, this chapter is restricted to a binary treatment as before in introducing the basic rationale of weighting. We will emphasize in later chapters the unique strengths of propensity score-based weighting for evaluating multivalued treatments, multiple concurrent treatments,

---

and time-varying treatments. In contrast, most other propensity score-based adjustment methods show limitations in these relatively complex treatment evaluations. Specifically, Chapter 5 will discuss evaluations of treatments with more than two categories; Chapter 7 will examine the joint effects of multiple concurrent treatments; and Chapter 8 will focus on the problem of evaluating time-varying treatments. In each of these cases, we will extend the nonparametric MMWS method to remove preexisting differences between treatment groups.

4.1 Weighted estimation of population parameters in survey sampling

The purpose of survey research is to gain knowledge of important population characteristics often captured by the population means of corresponding variables. Survey researchers employ probability sampling in which every individual in the population has a known probability of being selected into the sample. In the following, we discuss three types of probability sampling, the third of which generates data that require weighted analyses.

4.1.1 Simple random sample

The most basic form of survey sampling is simple random sampling that selects individuals at random independently with equal probability from the population. Even though the sample mean of a certain variable depends on which individuals happen to be selected into a given sample and hence may vary from sample to sample, in expectation, a sample obtained through the simple random sampling procedure is representative of the population. The mean of all the sample means, hypothetically obtained by replicating this sampling procedure an infinite number of times, is therefore equal to the population mean. Hence, the simple random sampling procedure ensures that a sample mean is an unbiased estimate of the population mean.

Language minority students (i.e., those whose primary home languages are not English) are a rapidly growing population in US schools. About 80% of these students are born in the United States; and the rest are born in other countries (Hernandez, Denton, and Macartney, 2007). Educators are interested in the distribution of English proficiency in this population, say, when the students enter middle school. In a simple random sample of size $n = 1000$, the sample mean outcome is simply

$$\bar{Y} = \frac{\sum_{i=1}^{1000} Y_i}{1000}. \tag{4.1}$$

It is essential to assess the amount of sampling error in using the above sample mean as an estimate of the population mean. Let S^2 denote the sample variance representing the degree of heterogeneity among students in their English proficiency. When the size of the population is very large, the variance of the sample mean is $\mathrm{Var}(\bar{Y}) = S^2/n$ representing the degree of uncertainty in estimation, that is, the extent to which the sample mean varies between different samples of 1000 students each. Its square root is reported as the standard error of the sample mean.

4.1.2   Proportionate sample

In general, foreign-born students tend to display a lower level of English proficiency on average than US-born students. With a particular interest in these two subpopulations, the researcher may stratify the population by native-born status and draw a simple random sample within each stratum with equal sampling probability across the strata. Let $X = 1$ if a student is foreign-born and $X = 0$ if US-born. In a proportionate sample as shown in the first column of Figure 4.1, the proportions of foreign-born students and US-born students are $pr(X = 1) = 0.2$ and $pr(X = 0) = 0.8$, equal to their respective proportions in the population. Therefore, a proportionate sample of 1000 students will include exactly 800 US-born students and 200 foreign-born students.

We may create two dummy indicators: $I_i(X = 1)$ is a dummy indicator that takes value 1 if student i is foreign-born and 0 otherwise; $I_i(X = 0)$ takes value 1 if student i is US-born and 0 otherwise. The subsample mean for the foreign-born students is

$$\bar{Y}_{X=1} = \frac{\sum_{i=1}^{1000} Y_i I_i(X = 1)}{\sum_{i=1}^{1000} I_i(X = 1)},$$

while that for the US-born students is

$$\bar{Y}_{X=0} = \frac{\sum_{i=1}^{1000} Y_i I_i(X = 0)}{\sum_{i=1}^{1000} I_i(X = 0)}.$$

The mean for the entire sample is

$$\bar{Y} = \frac{\sum_{i=1}^{1000} Y_i I_i(X = 1) + \sum_{i=1}^{1000} Y_i I_i(X = 0)}{\sum_{i=1}^{1000} I_i(X = 1) + \sum_{i=1}^{1000} I_i(X = 0)}. \tag{4.2}$$

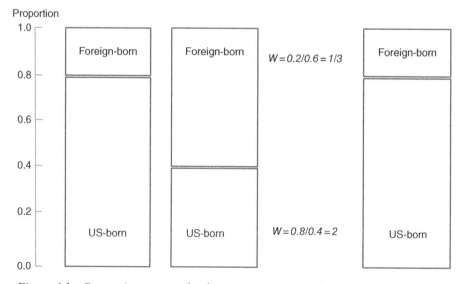

Figure 4.1   Proportionate sample, disproportionate sample, and weighted sample

(4.2) and (4.1) are equal in expectation. The variance of the sample mean (i.e., the standard error squared) obtained from the proportionate sample tends to be smaller than the variance of the sample mean obtained from a simple random sample. This is because when the foreign-born students and the US-born students have unequal means in the population, the variance within each subsample tends to be smaller than the variance in the entire sample.

4.1.3 Disproportionate sample

Because the foreign-born students differ in their national origins and their ages of arrival in the United States and because these two factors are strongly associated with a student's exposure to English, a much greater variation in English proficiency may exist among foreign-born students than among US-born students. A simple random sample or a proportionate sample might fail to include, by chance, foreign-born students from certain countries who arrived at certain ages. To allow for meaningful statistical inference with regard to the subpopulation of foreign-born students, a survey researcher would not only stratify the population but would also differentiate the sampling probability by increasing the sampling fraction of foreign-born students and accordingly decreasing the sampling fraction of US-born students. For example, the researcher may decide to include 600 foreign-born students and 400 US-born students in the sample as shown in the second column of Figure 4.1. The sampling probability for each foreign-born student becomes considerably higher than that for each US-born student though these probabilities are still known to the researcher. Because the foreign-born students and the US-born students are now disproportionately represented in the sample, if we simply apply (4.1) or (4.2) in estimating the population mean, the sample mean obtained from the dispro-portionate sample will be biased. Given that foreign-born students tend to have relatively lower English proficiency than US-born students, the higher representation of the former group in the sample than in the population will lead to an underestimation of the population mean.

Nonetheless, an unbiased estimate of the population mean can be obtained from such a sample through appropriate weighting to correct for the overrepresentation of the foreign-born students and the underrepresentation of the US-born students. To be specific, the weight for each foreign-born student is $0.2/0.6 = 1/3$; and the weight for each US-born student is $0.8/0.4 = 2$. In each case, the numerator is the proportion of such students in the population, and the denominator is the proportion of such students in the current sample. Intuitively speaking, we let every foreign-born student in the current sample represent a third of such a student and let every US-born student in the current sample double his or her representation. Note that the sample before weighting has $\sum_{i=1}^{1000} I_i(X=1) = 600$ foreign-born students and $\sum_{i=1}^{1000} I_i(X=0) = 400$ US-born students; the weighted sample now consists of $\sum_{i=1}^{1000} W_i I_i(X=1) = 200$ foreign-born students and $\sum_{i=1}^{1000} W_i I_i(X=0) = 800$ US-born students, the same as what one would expect in a proportionate sample. The transformation of the sample composition through weighting is shown in the third column of Figure 4.1. The sum of the weight over all the individuals in a sample is equal to the sample size; or in other words, the mean of the weight is equal to 1.0.

The weighted sample mean is

$$\bar{Y}_W = \frac{\sum_{i=1}^{1,000} W_i Y_i I_i(X=1) + \sum_{i=1}^{1,000} W_i Y_i I_i(X=0)}{\sum_{i=1}^{1,000} W_i I_i(X=1) + \sum_{i=1}^{1,000} W_i I_i(X=0)}. \tag{4.3}$$

Here, the denominator is $\sum_{i=1}^{1000} W_i I_i(X=1) + \sum_{i=1}^{1000} W_i I_i(X=0) = 200 + 800 = 1000$; the numerator is the sum of the observed outcome of the foreign-born students and the US-born students in the weighted sample. The weighted sample mean is an unbiased estimate of the population mean. Yet the unbiasedness is acquired through weighting often at the expense of reduced efficiency in estimation.

4.2 Weighting adjustment for selection bias in causal inference

The weighting strategy used to transform sample composition in survey sampling and the weighting adjustment for selection bias in causal inference share the same logic. In quasiexperimental studies, even though the sample might be representative of a given population, due to treatment selection, typically, the treated group and the control group differ in pretreatment composition and therefore also differ in the distribution of each potential outcome. The two treatment groups can be viewed as two different disproportionate samples of the same population. Hence, the mean outcome of neither treatment group is an unbiased estimate of the population average potential outcome under the corresponding treatment condition. This section explains how propensity score-based weighting transforms the pretreatment composition of each treatment group to resemble that of the entire sample representative of the population.

Hong (2012) described a hypothetical example that reveals the rationale for removing selection bias through weighting. The treatment of interest is English language learning (ELL) services for language minority students, denoted by $Z=1$ if a child received ELL services and $Z=0$ if a child did not receive such services. The outcome is a child's English vocabulary at age 6. We define the population average treatment effect on child vocabulary as $\delta = E[Y(1)] - E[Y(0)]$. Here, $E[Y(1)]$ is the population average potential outcome if, in a hypothetical world, the entire population of language minority students were to receive ELL services; and $E[Y(0)]$ is the population average potential outcome if ELL services were not available to any student in the population. Understandably, foreign-born students tend to acquire a lower English vocabulary in comparison with US-born students at age 6 with or without ELL services. As before, let $X=1$ if a student is foreign-born and $X=0$ if US-born. Suppose that if all the foreign-born students in the population would receive ELL services, their average potential outcome would be 5000 English words at age 6, that is, $E[Y(1)|X=1] = 5000$; in the absence of ELL services, their average potential outcome would still be 5000 words, that is, $E[Y(0)|X=1] = 5000$. Suppose also that the average potential outcome of US-born students would be 10 000 words at age 6 regardless of whether they would receive ELL services, that is, $E[Y(1)|X=0] = E[Y(0)|X=0] = 10\,000$. As mentioned in the previous section, 20% of the language minority student population are foreign-born and 80% are US-born. Hence, if all students in the language minority population would receive ELL services, the population average potential outcome averaging over the two subpopulations would be

$$E[Y(1)] = E[Y(1)|X = 1] \times \text{pr}(X = 1) + E[Y(1)|X = 0] \times \text{pr}(X = 0)$$
$$= 5000 \times 0.2 + 10\,000 \times 0.8$$
$$= 9000.$$

Similarly, if ELL services were not available to anyone in the population, the population average potential outcome would be $E[Y(0)] = 9000$. Hence, in this hypothetical example, the population average causal effect of ELL services on child vocabulary at 6 years of age is null:

$$\delta = E[Y(1)] - E[Y(0)] = 9000 - 9000 = 0.$$

4.2.1 Experimental result

A proportionate sample from this population would consist of 200 foreign-born students and 800 US-born students. If the students were sampled at kindergarten entry, the researcher might conduct a randomized block experiment with the probability of treatment assignment $\text{pr}(Z = 1) = 0.5$ within the block of foreign-born students and also within the block of US-born students. The treated group and the control group would each consist of a random sample of 100 foreign-born students and 400 US-born students, representing the composition of the sample as well as the population. The sample mean outcome of each treatment group would be an unbiased estimate of the population average potential outcome associated with the corresponding treatment. Hence, the sample mean difference in the observed outcome between the treated group and the control group would be an unbiased estimate of the population average treatment effect.

4.2.2 Quasiexperimental result

Alternatively, the researcher might choose not to intervene but let the schools assign the treatment as they normally do. Suppose that ELL services can accommodate 50% of the language minority population. Understandably, schools are much more likely to assign foreign-born students than US-born students to ELL services and may let the probability of receiving ELL services (i.e., the propensity score) be 0.9 for a foreign-born student and 0.4 for a US-born student in the population as well as in the current sample, that is,

$$\text{pr}(Z = 1|X = 1) = 0.9;$$
$$\text{pr}(Z = 1|X = 0) = 0.4.$$

As a result, the sample would have 180 foreign-born students and 320 US-born students in the treated group, while in the control group, there would be only 20 foreign-born students and as many as 480 US-born students. Given that in the population as well as in the entire sample, 20% of the students are foreign-born and 80% are US-born, foreign-born students are clearly overrepresented in the treated group and underrepresented in the control group.

In fact, because foreign-born students are overrepresented and constitute 36% rather than 20% of the treated group, while the US-born students are underrepresented and constitute 64% rather than 80% of the treated group, the outcome of the treated group is expected to be 8200 rather than 9000 English words:

$$E(Y|Z=1) = E(Y|X=1,Z=1) \times \mathrm{pr}(X=1|Z=1) + E(Y|X=0,Z=1) \times \mathrm{pr}(X=0|Z=1)$$
$$= 5000 \times 0.36 + 10\,000 \times 0.64$$
$$= 8200.$$

In the above computation, $E(Y|X=1, Z=1)$ is the expected outcome of the foreign-born students in the treated group, and $E(Y|X=0, Z=1)$ is the expected outcome of the US-born students in the treated group; $\mathrm{pr}(X=1|Z=1)$ is the proportion of foreign-born students in the treated group, and $\mathrm{pr}(X=0|Z=1)$ is the proportion of US-born students in the treated group.

In contrast, the outcome of the control group is expected to be 9800 English words. This is because foreign-born students are underrepresented and constitute only 4% rather than 20% of the control group, while the US-born students are overrepresented and constitute as many as 96% rather than 80% of the control group:

$$E(Y|Z=0) = E(Y|X=1,Z=0) \times \mathrm{pr}(X=1|Z=0) + E(Y|X=0,Z=0) \times \mathrm{pr}(X=0|Z=0)$$
$$= 5000 \times 0.04 + 10000 \times 0.96$$
$$= 9800.$$

Taking the expected difference between the treated group mean and the control group mean, we obtain a prima facie effect:

$$\delta_{PF} = E(Y|Z=1) - E(Y|Z=0) = 8200 - 9800 = -1600.$$

The selection bias in this case is as large as $\delta_{PF} - \delta = -1600$, which will lead to a misleading conclusion that ELL services are detrimental to language minority students' English vocabulary learning.

4.2.3   Sample weight for bias removal

From the perspective of survey sampling, it seems straightforward that one may remove the selection bias by transforming the composition of each treatment group. As shown in the first row of Figure 4.2, should the entire population be treated in a hypothetical world, 20% of this hypothetical treated population would be foreign-born and 80% US-born. Yet the current treated group in which 36% are foreign-born and 64% US-born is a disproportionate sample of the hypothetical treated population. We may transform the composition of this disproportionate sample by assigning the weight $0.2/0.36 = 5/9$ to each of the 180 treated foreign-born students and the weight $0.8/0.64 = 1\frac{1}{4}$ to each of the 320 treated US-born students. As before, the weight is a ratio of the population fraction to the sample fraction for each subgroup of students. The treated group after weighting would then have the same composition as the hypothetical treated population. The weighted mean outcome of the treated group is now an unbiased estimate of the population average potential outcome should the entire population be treated and is expected to be equal to 9000 English words.

Similarly, viewing the current control group in which 4% are foreign-born and 96% US-born as a disproportionate sample of the hypothetical control population, we

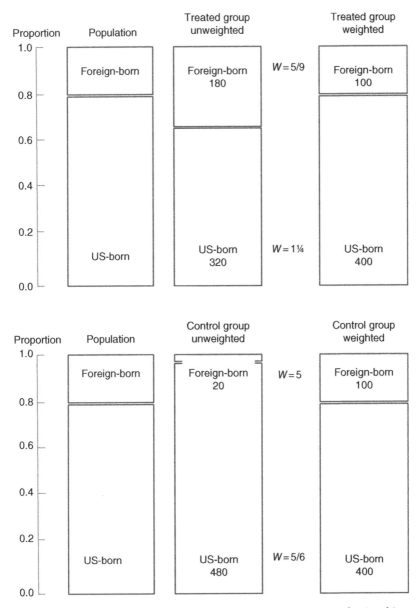

Figure 4.2   Propensity score-based weighting for removing selection bias

assign the weight $0.2/0.04 = 5$ to each of the 20 foreign-born students and the weight $0.8/0.96 = 5/6$ to each of the 480 US-born students in the control group. The transformation of the control group composition is shown in the second row in Figure 4.2. The weighted mean outcome of the control group is expected to be 9000 English words, unbiased for the population average potential outcome should the entire population be assigned to the control condition.

4.2.4   IPTW for bias removal

Alternatively, one may use weighting to transform the experimental group composition and the control group composition such that the probability of treatment assignment in the weighted sample would resemble that in a hypothetical complete randomized design or a randomized block design with equal probability of treatment assignment for all individuals. The probability of treatment assignment in the hypothetical experimental design reflects the resource constraint in the natural world. Suppose that ELL services can accommodate only 50% of the language minority students in the population and hence approximately 50% of the students in a simple random sample are treated, the target probability of receiving ELL services is set to be 0.5 for all students in a weighted sample as well. Yet in the unweighted sample, the actual probabilities of receiving ELL services are 0.9 for the foreign-born students and 0.4 for the US-born students. Comparing the treated group and the control group in the middle column of Figure 4.2, we can see that these two groups do not have "comparable" pretreatment composition due to the unequal probability of treatment assignment for foreign-born and US-born students.

For every treated student, the weight can be computed as a ratio of the target probability to the actual probability of receiving ELL services, that is, $0.5/0.9 = 5/9$ for a treated foreign-born student and $0.5/0.4 = 1\frac{1}{4}$ for a treated US-born student. The target probability of being assigned to the control condition is $1 - 0.5 = 0.5$. Yet the actual probabilities of being assigned to the control condition are 0.1 for the foreign-born students and 0.6 for the US-born students. Hence, for every control student, the weight is a ratio of the target probability to the actual probability of being assigned to the control condition, that is, $0.5/0.1 = 5$ for a foreign-born student and $0.5/0.6 = 5/6$ for a US-born student in the control group.

In this weighting scheme, the denominator of the weight is always an individual's probability of receiving the treatment that the individual actually received, that is, the probability of being treated for a treated unit and the probability of being untreated for a control unit. Hence, this strategy has been called "inverse-probability-of-treatment weighting" (IPTW) in epidemiology (Hernan, Brumbeck, and Robins, 2000; Robins, 1987, 1999; Robins, Brumbeck, and Hernan, 2000). The numerator is the target probability for the corresponding treatment expected in an experimental design. Once all individuals in the weighted sample display the same probability of treatment assignment regardless of their different pretreatment background, as shown in the last column of Figure 4.2, the treated group and the control group will become comparable in pretreatment composition, the same as one would expect to see in a randomized experiment.

The computed values of the sample weight for the treated foreign-born, the treated US-born, the untreated foreign-born, and the untreated US-born are 5/9, 1¼, 5, and 5/6, respectively, the same as their IPTW values. However, the computation of the sample weight becomes extremely challenging if the sample composition deviates from the population composition in multiple dimensions. For example, in assigning language minority students to ELL services, schools may decide to give priority to not only foreign-born students but also students who are male; who have had no preschool education; who display limited literacy, math, and general knowledge; who are reluctant to participate in class; and whose parents have relatively low socioeconomic status. It may become impractical to stratify a sample on these multiple dimensions all at once. A ready solution is to estimate a student's propensity score of receiving (or not receiving) ELL services as a function of these multiple covariates and use the propensity score in the denominator for computing IPTW.

Let $\theta = \theta(x) = \text{pr}(Z = 1 | X = x)$ denote a student's propensity of receiving ELL services, which is a function of the student's native-born status X. As in the previous chapter, we use θ as a shorthand for $\theta(x)$ henceforth. The same student's propensity of being assigned to the control condition is simply $1 - \theta = \text{pr}(Z = 0 | X = x)$. Let $\text{pr}(Z = 1)$ be the target probability of receiving ELL services and $\text{pr}(Z = 0)$ the target probability of being assigned to the control condition. These target probabilities are usually determined by the respective fractions of the sample treated and untreated. If a student is in the treated group, we have

$$\text{IPTW} = \frac{\text{pr}(Z = 1)}{\theta}; \tag{4.4}$$

if, instead, the student is in the control group, then

$$\text{IPTW} = \frac{\text{pr}(Z = 0)}{1 - \theta}. \tag{4.5}$$

Robins (1999) proved that when the treatment assignment is strongly ignorable given the observed pretreatment covariates and is thus strongly ignorable given the propensity score as defined in (3.7) and (3.8) in the previous chapter, the IPTW-adjusted sample mean outcome of each treatment group consistently estimates the corresponding population average potential outcome. Therefore, the IPTW-adjusted sample mean difference between the treated group and the control group consistently estimates the population average treatment effect. In statistics, a sample estimate of the treatment effect is consistent if it converges in probability to the population average treatment effect as the same size grows to infinity.

Rosenbaum (1987) called the propensity score-based weighting method "model-based direct adjustment" because the weight is completely determined by the fitted propensity score model and is estimated parametrically. This was contrasted with the direct adjustment by sample weight computed on the basis of sample proportions. In particular, if a population has been stratified into many subclasses, due to sampling error, individuals from some relatively small subclasses may be assigned large sample weights often in an erratic manner. In an extreme case, such small subclasses may not even be represented in a particular sample; their sample weights computed as a ratio of the population proportion to the sample proportion will become infinite in theory. In contrast, as long as the propensity score model is correctly specified, the estimated propensity scores for individuals from such small subclasses tend to be relatively stable. The weight is simply zero for those not represented in a given sample.

However, the recent literature has sometimes reported unsatisfactory performance of the IPTW adjustment in causal inference (Imbens, 2000; Kang and Schafer, 2007; Schafer and Kang, 2008; Waernbaum, 2012). Hong (2010) pointed out that IPTW estimation of treatment effects can potentially be distorted when units that have no counterfactual information under an alternative treatment condition receive a nonzero weight. In addition, when the functional form of the propensity model is misspecified, IPTW adjustment typically leads to a bias proportional to the amount of confounding associated with the nonlinearity.

4.3   MMWS

MMWS is an alternative weighting method that makes use of propensity scores. This method has recently been illustrated in epidemiological (Huang *et al.*, 2005) and educational research (Hong, 2010, 2012; Hong and Hong, 2009; Hong *et al.*, 2012). The idea is related to poststratification adjustment (Horvitz and Thompson, 1952), direct standardization (Little, 1982; Little and Pullum, 1979), and adjustment for nonresponse through weighting (Little, 1986; Little and Vartivarian, 2003, 2005; Oh and Scheuren, 1983) in survey methodology. For example, to adjust for nonresponse in surveys, Little and Vartivarian (2003, 2005) suggested computing a propensity score for response and creating adjustment cells based on this score. A similar strategy can be applied to adjust for nonrandom selection in treatment assignment. The major difference between MMWS and IPTW is that, instead of computing a weight as a parametric function of the propensity score, MMWS *stratifies* a sample on the basis of the propensity score for each treatment and then computes the weight according to the proportion of the individual units in a given stratum assigned to the corresponding treatment. As a result, each weighted treatment group resembles the composition of the target population, thereby enabling pairwise comparisons between the treatment groups.

When applied to the hypothetical example of ELL evaluation, the logic of MMWS is as follows: we have noted that the target probability of receiving ELL services is 0.5 in the language minority population. Among the 200 foreign-born students in the sample, we expect $200 \times 0.5 = 100$ of them to be assigned at random to receive ELL services in an experiment. Yet as many as 180 foreign-born students actually received ELL services. Hence, the weight for each of these 180 foreign-born students in the treated group is $(200 \times 0.5)/180 = 5/9$. After weighting, these 180 foreign-born students would represent 100 such children. Here, the numerator of the weight is equal to the expected number of foreign-born students in the treated group in a randomized trial; the denominator is the actual number of such children in the treated group. The 20 foreign-born students in the control group would each have a weight equal to $(200 \times 0.5)/20 = 5$ and would represent 100 such children after weighting. The numerator of this weight is equal to the expected number of foreign-born students in the control group in a randomized trial; the denominator is the actual number of such children in the control group. Similarly, the 320 US-born students in the treated group would each have a weight equal to $(800 \times 0.5)/320 = 1\frac{1}{4}$, whereas the 480 US-born students in the control group would each have a weight equal to $(800 \times 0.5)/480 = 5/6$. The weighted sample would eventually have 100 foreign-born children and 400 US-born children in each treatment group, therefore resembling data from a randomized trial.

MMWS and IPTW are identical only when the pretreatment covariates are discrete, in which case the propensity scores are discrete as well. Most applications involve continuous pretreatment covariates and therefore require specification assumptions with regard to the functional form of the propensity score model. MMWS and IPTW tend to diverge especially when the functional form of a propensity score model is misspecified.

4.3.1   Theoretical rationale

Hong (2010) proved that, in population data, the MMWS-adjusted mean observed outcome of a treatment group is unbiased for the population average potential outcome (i.e., the marginal mean outcome) associated with that treatment under the strong ignorability assumption.

The proof is formally presented in Appendix 4.A. The following is an intuitive explanation of the rationale.

4.3.1.1 MMWS for a discrete propensity score

In an evaluation of a binary treatment such as ELL services versus the control condition, we have used θ to denote a student's propensity of receiving ELL services. When θ takes discrete values such as $\theta = 0.9$ for the foreign-born students and $\theta = 0.4$ for the US-born students, $\mathrm{pr}(\theta)$ simply represents the proportion of individuals in the population sharing the particular propensity score value θ. In the current example, we have that $\mathrm{pr}(\theta = 0.9) = 0.2$ and $\mathrm{pr}(\theta = 0.4) = 0.8$ given that 20% of the language minority population are foreign-born and 80% are US-born. In a randomized experiment, the distribution of θ in each treatment group is expected to be the same as that in the entire population. Yet in a quasiexperimental study, those with a higher propensity score for receiving ELL services are more likely to be assigned to the treated group. Hence, the distribution of θ in each treatment group becomes different from its distribution in the population. We have seen that the proportions of foreign-born students and US-born students in the treated group are $\mathrm{pr}(\theta = 0.9|Z = 1) = 0.36$ and $\mathrm{pr}(\theta = 0.4|Z = 1) = 0.64$, respectively; and their proportions in the control group are $\mathrm{pr}(\theta = 0.9|Z = 0) = 0.04$ and $\mathrm{pr}(\theta = 0.4|Z = 0) = 0.96$, respectively. To adjust for selection bias, we have used the sample weight now represented in a general form as

$$\frac{\mathrm{pr}(\theta)}{\mathrm{pr}(\theta|Z=z)}, \text{ for } z = 0,1.$$

The treated group and the control group with the same propensity score are expected to have the same joint distribution of all the observed pretreatment covariates, as proved by Rosenbaum and Rubin (1983). When there are no unmeasured confounders, the treated group and the control group with the same propensity score will have the same average potential outcomes $Y(1)$ and $Y(0)$. Under this assumption, the population average potential outcome associated with the treated condition $E[Y(1)]$ is equal to the weighted mean outcome of the treated group $E(Y_W|Z = 1)$ where the weight is $\mathrm{pr}(\theta)/\mathrm{pr}(\theta|Z = 1)$. The weight can be derived as follows:

$$E[Y(1)] = \sum_{\theta}[E(Y(1)|\theta)\mathrm{pr}(\theta)]$$
$$= \sum_{\theta}[E(Y||Z = 1,\theta)\mathrm{pr}(\theta)]$$
$$= \sum_{\theta}\left[E\left(\frac{\mathrm{pr}(\theta)}{\mathrm{pr}(\theta|Z=1)}Y|Z = 1,\theta\right)\mathrm{pr}(\theta|Z = 1)\right]$$
$$= \sum_{\theta}[E(Y_W|Z = 1,\theta)\mathrm{pr}(\theta|Z = 1)]$$
$$= E(Y_W|Z = 1).$$

The first equation represents the population average potential outcome as the sum of the average potential outcome of the foreign-born students $E(Y(1)|\theta = 0.9)$ and that of the

US-born students $E(Y(1)|\theta=0.4)$ each multiplied by the respective proportion of such students in the population—that is, $\mathrm{pr}(\theta=0.9)=0.2$ for the foreign-born and $\mathrm{pr}(\theta=0.4)=0.8$ for the US-born. The second equation holds because $E(Y(1)|\theta)=E(Y|Z=1,\theta)$ under the strong ignorability. In the third equation, we simply multiply and divide the initial quantity by $\mathrm{pr}(\theta|Z=1)$. Averaging over the distribution of θ, this is then equal to the weighted mean outcome of the treated group. The population average potential outcome associated with the control condition can be estimated similarly by the weighted mean outcome of the control group:

$$E[Y(0)]=E(Y_W|Z=0),$$

where the weight is $\mathrm{pr}(\theta)/\mathrm{pr}(\theta|Z=0)$ for every control unit.

4.3.1.2   MMWS for a continuous propensity score

When the pretreatment covariates are continuous and multidimensional, however, the propensity score θ is typically continuous, in which case $\mathrm{pr}(\theta)$, typically written in its density form, represents the relative likelihood that θ takes on a given value. For each set of θ values close together on the continuum, $\mathrm{pr}(\theta)$ can be empirically approximated through stratification. To proceed, one may divide a simple random sample or a proportionate sample of N units from the population into H strata such that the treated group and the control group within each stratum will have the same or very similar propensity score values. When the strong ignorability holds, we expect that the average observed outcome of the treated units within each stratum be approximately equal to the average potential outcome associated with the treated condition for all the units in that stratum regardless of their actual treatment assignment, that is, $E(Y|Z=1,\ S=s)=E[Y(1)|S=s]$, for all H strata denoted by $s=1,\dots,H$. Hence, to estimate the population average potential outcome $E[Y(1)]$, we multiply the sample estimate of the within-stratum average observed outcome of the treated units $E(Y|Z=1,\ S=s)$ by the sample proportion estimating the population proportion in that stratum $\mathrm{pr}(S=s)$. We then sum over all the H strata. This is equal to the weighted mean outcome of the treated group:

$$E[Y(1)]\approx\sum_{s=1}^{H}[E(Y|Z=1,S=s)\mathrm{pr}(S=s)]$$

$$=\sum_{s=1}^{H}[E(Y_W|Z=1,S=s)\mathrm{pr}(S=s|Z=1)],$$

where the weight is $\mathrm{pr}(S=s)/\mathrm{pr}(S=s|Z=1)$. Here, $\mathrm{pr}(S=s)$ is an approximation of $\mathrm{pr}(\theta)$, representing the population proportion of individual units in stratum s; $\mathrm{pr}(S=s|Z=1)$ is an approximation of $\mathrm{pr}(\theta|Z=1)$, representing the proportion of treated individuals in stratum s. In a given sample with size N, we may compute $\mathrm{pr}(S=s)=n_s/N$, where n_s denotes the number of units in stratum s; we may compute $\mathrm{pr}(S=s|Z=1)=n_{z=1,s}/n_{z=1}$, where $n_{z=1,s}$ denotes the number of treated individuals in stratum s and where $n_{z=1}$ denotes the total number of individuals in the treated group. Hence, the aforementioned weight can be computed as:

$$\frac{n_s}{N}\times\frac{n_{z=1}}{n_{z=1,s}}.$$

Similarly, the weighted mean outcome of the control group estimates the population average potential outcome associated with the control condition:

$$E[Y(0)] \approx \sum_{s=1}^{H} [E(Y|Z=0, \ S=s)\text{pr}(S=s)]$$

$$= \sum_{s=1}^{H} [E(Y_W|Z=0, \ S=s)\text{pr}(S=s|Z=0)],$$

where the weight is $\text{Pr}(S=s)/\text{Pr}(S=s|Z=0) = (n_s/N) \times (n_{z=0}/n_{z=0,s})$. Here, $n_{z=0}$ denotes the total number of control individuals; and $n_{z=0,s}$ denotes the number of control individuals in stratum s.

In general, for an individual unit in propensity score stratum s who has been assigned to treatment group z, the weight is

$$\text{MMWS} = \frac{n_s}{N} \times \frac{n_z}{n_{z,s}} \tag{4.6}$$

for $z=0, 1$ when the treatment is binary. We will show in the next chapter that the same weight applies to evaluations of multivalued treatments.

Hong (2010) has shown that MMWS adjustment is equivalent to propensity score stratification in terms of bias removal. For example, when a sample is divided into five strata, MMWS adjustment is expected to remove at least 90% of the initial bias. Increasing the number of strata generally increases the percentage of bias reduction in large samples. However, when the sample size is given, increasing the number of strata leads to a reduction of sample size within a stratum. With a relatively small number of units per stratum, the proportion of treated units in the stratum may become less stable, and hence, the marginal mean weight computed in the next step may become more volatile. This may compromise the precision of the treatment effect estimate. In practice, a rule of thumb is to choose a minimal number of strata such that the experimental group and the control group in each stratum display the same distribution of the logit propensity score. After starting with five or six strata, one may increase the number of strata by restratifying the entire analytic sample or subdividing certain strata when necessary. Weighting on the basis of propensity score stratification can be easily generalized to applications with discrete propensity scores because when a propensity score takes a relatively small number of discrete values, the sample may simply be stratified by the propensity score values, in which case 100% of the bias is expected to be removed under the strong ignorability. In general, the MMWS estimator of the treatment effect is consistent because, as the sample size increases, the number of strata can increase as well, and the sample estimate will converge to the population parameter value.

4.3.1.3   MMWS for estimating the treatment effect on the treated

For the population of the treated individuals indicated by $Z=1$, the treatment effect on the treated is defined as

$$\delta = E[Y(1)-Y(0)|Z=1] = E[Y(1)|Z=1] - E[Y(0)|Z=1]. \tag{4.7}$$

For example, in evaluating the effect of kindergarten retention, the population of students retained in kindergarten is represented by the sampled students who were retained; the population of those promoted to first grade is represented by the sampled students who were promoted.

Figure 3.3 clearly shows that these two populations are different in pretreatment composition due to the concentration of students with a relatively high propensity of retention among the retained. Moreover, neither the population of retainees nor the population of promoted students has the same pretreatment composition as does the population at risk of being retained. This last population consists of students who had a nonzero probability of being retained, among them some were actually retained and some were promoted. Because the retention effect is unlikely to be constant for all students, the average treatment effects for these three populations are expected to be different. Given that many schools and school districts have been debating a social promotion policy promoting all kindergartners to first grade, such a policy would arguably influence the population of retainees more than it would with the other two populations.

When evaluating the treatment effect on the treated, we take an average of each potential outcome over the propensity score distribution of the treated population. For those in the treated population, the observed outcome Y is equal to the potential outcome of being treated $Y(1)$. Hence, we have that $E(Y|Z=1)=E[Y(1)|Z=1]$. The mean observed outcome of the treated group therefore is unbiased for the population average potential outcome of being treated for the treated population. For this reason, we simply set the weight to be 1.0 for every treated individual. The population average potential outcome associated with the control condition for the treated population $E[Y(0)|Z=1]$ is to be estimated from the observed outcome of those in the control group after we use weighting to transform the control group composition to resemble that of the treated population.

Because the first element of the causal effect $E[Y(1)|Z=1]$ can be estimated without bias, we are concerned only with the potential bias in estimating the second element $E[Y(0)|Z=1]$. As shown in Appendix 4.B, the bias would come from the mean difference in $Y(0)$ between the treated group and the control group: $E[Y(0)|Z=1]-E[Y(0)|Z=0]$. We therefore summarize pretreatment information predictive of $Y(0)$ in a propensity score θ. Under the assumption that, within subpopulations with the same propensity score θ, the mean of $Y(0)$ is the same between the treated units and the control units, that is,

$$E[Y(0)|Z=1,\theta]=E[Y(0)|Z=0,\theta],$$
$$0<\text{pr}(Z=1|\theta)<1 \tag{4.8}$$

Appendix 4.B derives the following weight for the control units, which enables the researcher to estimate $E[Y(0)|Z=1]$ without bias:

$$\text{MMWS}=\frac{\text{pr}(\theta|Z=1)}{\text{pr}(\theta|Z=0)}.$$

For individuals with propensity score θ, the numerator of the weight is the proportion of the treated group displaying this propensity score, while the denominator is the proportion of the control group displaying the same propensity score.

When θ is continuous, its distribution can be approximated by the proportion of individuals in each stratum who were actually treated. We compute the weight after dividing the sample into $s=1,\ldots,H$ strata on θ. An application of Bayes theorem generates the following result:

$$\text{MMWS}=\frac{\text{pr}(\theta|Z=1)}{\text{pr}(\theta|Z=0)}=\frac{\text{pr}(\theta,\ Z=1)}{\text{pr}(\theta,\ Z=0)}\times\frac{\text{pr}(Z=0)}{\text{pr}(Z=1)},$$

which can be obtained nonparametrically by using the sample counts for every individual in the control group in stratum s:

$$\text{MMWS} = \frac{n_{z=1,s}}{n_{z=0,s}} \times \frac{n_{z=0}}{n_{z=1}}. \tag{4.9}$$

Here, $n_{z=1,s}$ is the number of treated individuals in stratum s; $n_{z=0,s}$ is the number of control units in stratum s; $n_{z=0}$ is the total number of control individuals; and $n_{z=1}$ is the total number of treated individuals in the sample. When the strong ignorability assumption holds, the mean observed outcome of the weighted control group is approximately equal to the average potential outcome associated with the control condition for the treated population. One thereby obtains an estimate of the treatment effect on the treated.

4.3.2 MMWS analytic procedure

When MMWS is applied to an evaluation of a binary treatment, the analysis involves eight steps listed below. The first four steps and the last step of the analytic procedure are identical to those in propensity score stratification:

1. Collect a large set of pretreatment covariates.
2. Build a logistic regression model for estimating the propensity score.
3. Identify the common support for causal inference.
4. Stratify the sample on the propensity score.

An MMWS analysis then proceeds with the following four steps:

5. Compute MMWS for each individual on the basis of stratum membership and treatment group membership by applying (4.6). Individuals in the same stratum and the same treatment group receive the same weight.
6. Check balance in the observed pretreatment composition between the treated group and the control group in the weighted sample; if necessary, restratify the sample or revise the propensity score model, recompute the weight, and check balance again.
7. Analyze a weighted outcome model to estimate the treatment effect. After applying the marginal mean weight to the data in the same way as with a sample weight, one can analyze the data within the ANOVA framework. When the treatment is binary, the analysis simply involves computing the weighted mean difference between the treated group and the control group and estimating its standard error through weighted least squares. To improve precision, one may include strong predictors of the outcome in the weighted analysis of the treatment effect, a strategy analogous to ANCOVA.
8. Conduct sensitivity analysis.

These steps can be carried out in any standard statistical software. However, a stand-alone MMWS program greatly eases the computation especially when the number of pretreatment covariates becomes daunting. The MMWS program and the SPSS, SAS, Stata, and R syntax files along with data example are all freely available online at the publisher's website: http://www.wiley.com/go/social_world.

Hong and colleagues (Hong, 2010; Hong and Hong, 2009) illustrated the procedure with an evaluation of the effect of within-class homogeneous grouping according to ability versus no

Table 4.1 Propensity score stratification and MMWS computation

Stratum	Size	Unweighted sample						MMWS		Weighted sample	
		Grouping ($Z = 1$)			No grouping ($Z = 0$)						
		N	M	SD	N	M	SD	$Z = 1$	$Z = 0$	$Z = 1$	$Z = 0$
$s = 1$	311	38	−1.96	0.32	273	−1.97	0.33	3.62	0.64	138	173
$s = 2$	652	158	−1.14	0.21	494	−1.17	0.21	1.82	0.74	288	364
$s = 3$	652	241	−0.51	0.16	411	−0.53	0.16	1.20	0.89	288	364
$s = 4$	587	281	−0.03	0.12	306	−0.04	0.12	0.92	1.07	260	327
$s = 5$	391	241	0.35	0.10	150	0.35	0.10	0.72	1.45	173	218
$s = 6$	587	414	0.95	0.31	173	0.89	0.30	0.63	1.89	259	328
$s = 7$	65	61	2.09	0.30	4	2.08	0.39	0.47	9.07	29	36
Total	3245	1434	0.15	0.88	1811	−0.64	0.91	—	—	1434	1817

Adapted with permission from Hong and Hong (2009). © 2009 by Sage Publications.

grouping in kindergarten literacy instruction on students' literacy achievement at the end of the year. In the Early Childhood Longitudinal Study–Kindergarten (ECLS-K) cohort data, a sample of 3260 kindergarten classes in 1044 schools had approximately 44% of the classes using within-class grouping. The researchers identified 146 pretreatment measures of class composition of student characteristics, teacher background, school type, and school climate. With 15 classes displaying the lowest propensity score values excluded from the common support due to a lack of counterparts in the treated group, the sample was divided into seven strata on the basis of the estimated propensity score. Less than 5% of the 146 observed covariates showed within-stratum differences between the treated group and the control group at the significance level of 0.05.

The sample sizes under the stratification, as listed in Table 4.1, provide sufficient information for computing the marginal mean weight for each kindergarten class. As shown in the last two columns of this table, the composition of the treated group and that of the control group after weighting both resemble the composition of the entire sample in terms of the distribution of the propensity score. For example, the proportion of the sample in the first propensity score stratum is $311/3245 = 0.096$. Before weighting, the proportion of the treated group in this stratum is $38/1434 = 0.026$, while the proportion of the control group in this stratum is $273/1811 = 0.151$. After weighting, these proportions become $137/1434 = 0.096$ and $174/1811 = 0.096$ for the treated and the control, respectively. Similar transformations as a result of weighting can be seen in each of the other six strata.

If the outcome had been measured at the class level, one could have simply computed the weighted mean difference in the outcome between the treated classes and the control classes as an estimate of the population average treatment effect. However, ECLS-K employed multistage sampling with about 5–10 students sampled from each kindergarten class. A class-level analysis of aggregated child outcomes would be less efficient than a multilevel analysis that takes into account the precision of class mean outcome. Using HLM 6.4 for the weighted multilevel analysis, the researchers reported an estimated effect of grouping versus no grouping, $\widehat{\delta} = 0.57$, SE $= 0.27$, $t = 2.13$, $p < 0.05$, indicating a statistically significant small positive effect of grouping students by ability for kindergarten literacy instruction.

4.3.3   Inherent connection and major distinctions between MMWS and IPTW

MMWS and IPTW are theoretically equivalent. Hong (2010) revealed their inherent connection by resorting to Bayes theorem. The derivation is shown in Appendix 4.C. However, IPTW is directly a function of the estimated propensity score, while MMWS computes the weight nonparametrically on the basis of sample proportions under propensity score stratification. To make this clear, we may rewrite (4.6) as follows:

$$\text{MMWS} = \frac{n_s}{n_{z,s}} \times \frac{n_z}{N}$$
$$= \frac{n_s}{n_{z,s}} \times \text{pr}(Z = z) \tag{4.10}$$

for $z = 0, 1$. Hence, MMWS is proportional to a ratio of the number of units in stratum s to the number of units in stratum s who were actually assigned to treatment z: $n_s/n_{z,s}$. For the treated units in stratum s, the inverse of this ratio is a nonparametric estimate of their propensity score for being treated, $\widehat{\theta} = n_{z=1,s}/n_s$, while for the control units in stratum s, the inverse of this ratio estimates nonparametrically their propensity score for being assigned to the control group, $1 - \widehat{\theta} = n_{z=0,s}/n_s$. Therefore, MMWS can be viewed as a nonparametric version of IPTW.

In general, the numerator $n_s \times \text{pr}(Z = z)$ estimates the expected number of individuals assigned to treatment z in stratum s in a randomized experiment. In such an experiment with equal probability of treatment assignment, the expected value of MMWS would be 1.0 for all units. In a quasiexperimental study, MMWS would be greater than 1.0 for the individual units underrepresented and would be smaller than 1.0 for those overrepresented in each treatment group. As illustrated in the previous section, with 44% of the ECLS-K sample of kindergarten classes in the treated group, the goal of MMWS adjustment is to approximate a randomized trial with equal probability of treatment assignment: $\text{pr}(Z = 1) = 0.44$. However, in the first stratum, with 38 classes in the treated group and as many as 273 classes in the control group, the average probability of being treated is only $38/311 = 0.12$, while the average probability of being untreated is $273/311 = 0.88$. The expected number of classes to be treated in this stratum is $311 \times 0.44 = 137$; and the expected number of classes to be untreated is $311 \times 0.56 = 174$. Once we have assigned the weight 3.62 to each of the 38 treated classes and the weight 0.64 to each of the 273 control classes, the probability of being treated becomes $138/311 = 0.44$ and the probability of being untreated becomes $173/311 = 0.56$ in this first stratum. The same probability of treatment assignment has been obtained in all the six other strata in the weighted sample. Hence, the classes can be viewed as if randomized to be treated in the weighted sample under strong ignorability.

Hong (2010) has shown that the IPTW method is expected to slightly outperform the MMWS method in bias reduction when the propensity score model is correctly specified or when the omitted covariates are linear predictors of the logit of propensity and of the outcome. However, IPTW adjustment tends to introduce additional bias when only a portion of the population provides support for causal inference. This problem can be effectively avoided in MMWS adjustment due to its built-in procedure in Step (3) that identifies the common support and excludes the units without counterfactual information in the observed data. More importantly, Hong (2010) provided simulation results showing that, in typical applications in which a nonlinear or nonadditive propensity score model is misspecified as a linear and additive one, MMWS estimates of treatment effects display a much higher level of robustness when compared with IPTW estimates. This is because, with its nonparametric approach on the basis of stratification, the MMWS method usually provides a better approximation of nonlinear or nonadditive relationships between treatment assignment and pretreatment covariates. In contrast, because IPTW is computed as a direct function of the estimated propensity score, the estimated IPTW tends to deviate systematically from the true weight when the functional form of the propensity score model is misspecified, leading to a bias in treatment effect estimation. The same pattern has been observed across a range of outcome distributions. The simulation design and the results are presented in Appendix 4.D. In practice, it is advisable to compare IPTW with MMWS estimation results, as a major discrepancy between the two may implicate a possible misspecification of the functional form of the propensity score model. The nonparametric MMWS strategy demonstrates an additional gain in efficiency when compared with the IPTW strategy. This is consistent with the results in survey research methodology showing that constructing weights on the basis of a nonparametric spline function increases the efficiency in estimation by reducing the variation in weight values (Elliott and Little, 2000).

Appendix 4.A: Proof of MMWS-adjusted mean observed outcome being unbiased for the population average potential outcome

Suppose that individual units in a population can possibly be assigned to either treatment z or alternative treatment z'. We prove that the MMWS-adjusted mean observed outcome of those assigned to treatment z is unbiased for the population average potential outcome associated with treatment z. This population average potential outcome can be represented as

$$E[Y(z)] \equiv E\{E[Y(z)|\theta_z]\} = E[E(Y|Z=z,\ \theta_z)] = \iint yf(y|Z=z,\ \theta_z)\mathrm{pr}(\theta_z)dyd\theta_z.$$

The second equality is warranted in a randomized experiment and holds in a nonexperimental study under the strong ignorability assumption. Here, $\mathrm{pr}(\theta_z)$ is a density or probability mass function for individuals in the population whose pretreatment characteristics are summarized in θ_z. In a randomized experiment, the mean observed outcome of each treatment group is unbiased for the corresponding population average potential outcome simply because $\mathrm{pr}(\theta_z|Z=z) = \mathrm{pr}(\theta_z)$, that is, the distribution of the propensity score in treatment group z is the same as that in the entire population. Hence, for treatment z, we have that

$$E[Y(z)] = E[E(Y|Z=z,\theta_z)]$$
$$= \iint yf(y|Z=z,\theta_z)\mathrm{pr}(\theta_z)dyd\theta_z$$
$$= \iint yf(y|Z=z,\theta_z)\mathrm{pr}(\theta_z|Z=z)dyd\theta_z$$
$$= E(Y|Z=z).$$

However, when individuals are selected into treatment groups on the basis of their pretreatment characteristics θ_z, the distribution of θ_z in treatment group z will become different from its distribution in the population. To adjust for selection, we apply the marginal mean weight $\mathrm{MMWS} = \mathrm{pr}(\theta_z)/\mathrm{pr}(\theta_z|Z=z)$ to the individuals in treatment group z. The weight can be derived as follows:

$$E[Y(z)] = E[E(Y|Z=z,\theta_z)]$$
$$= \iint yf(y|Z=z,\theta_z)\mathrm{pr}(\theta_z)dyd\theta_z$$
$$= \iint \frac{\mathrm{pr}(\theta_z)}{\mathrm{pr}(\theta_z|Z=z)} yf(y|Z=z,\theta_z)\mathrm{pr}(\theta_z|Z=z)dyd\theta_z$$
$$= E\left\{ E\left[\frac{\mathrm{pr}(\theta_z)}{\mathrm{pr}(\theta_z|Z=z)} Y|Z=z,\theta_z \right] |Z=z \right\}$$
$$= E\{Y_W|Z=z\}.$$

The expectations are taken first over the distribution of the propensity score and then over the distribution of the weighted outcome. This concludes the proof.

Appendix 4.B: Derivation of MMWS for estimating the treatment effect on the treated

Here, we evaluate the treatment effect on the treated defined in (4.7) as

$$\delta = E[Y(1) - Y(0)|Z = 1] = E[Y(1)|Z = 1] - E[Y(0)|Z = 1].$$

The mean difference in the observed outcome between the treated and the control, which we have called the prima facie effect δ_{PF} in Chapter 2, now contains only the first source of bias $E[Y(0)|Z = 1] - E[Y(0)|Z = 0]$. This is because

$$
\begin{aligned}
\delta_{PF} &= E(Y|Z = 1) - E(Y|Z = 0) \\
&= E[Y(1)|Z = 1] - E[Y(0)|Z = 0] \\
&= \{E[Y(1)|Z = 1] - E[Y(0)|Z = 1]\} - \{E[Y(0)|Z = 1] - E[Y(0)|Z = 0]\} \\
&= \delta + Bias1.
\end{aligned}
$$

This bias is the mean difference in $Y(0)$ between the treated group and the control group.

To adjust for selection bias in estimating $E[Y(0)|Z = 1]$, we apply the marginal mean weight to the individuals in the control group. Below, we derive the weight under the identification assumption stated in (4.8):

$$
\begin{aligned}
&E[Y(0)|Z = 1] \\
&= E\{E[Y(0)|Z = 1, \theta]\} \\
&= \int_{\theta} E[Y(0)|Z = 1, \theta] \mathrm{pr}(\theta|Z = 1) d\theta.
\end{aligned}
$$

When the identification assumption holds, the above is equal to

$$\int_{\theta} E[Y(0)|Z = 0, \theta] \mathrm{pr}(\theta|Z = 1) d\theta$$

$$= \int_{\theta} E(Y|Z = 0, \theta) \mathrm{pr}(\theta|Z = 1) d\theta$$

$$= \int_{\theta} \left[\int_{y} yf(y|Z = 0, \theta) dy \right] \mathrm{pr}(\theta|Z = 1) d\theta$$

$$= \int_{\theta} \int_{y} \frac{\mathrm{pr}(\theta|Z = 1)}{\mathrm{pr}(\theta|Z = 0)} yf(y|Z = 0, \theta) \mathrm{pr}(\theta|Z = 0) dy d\theta$$

$$= E\left\{ E\left[\frac{\mathrm{pr}(\theta|Z = 1)}{\mathrm{pr}(\theta|Z = 0)} Y|Z = 0, \theta \right] |Z = 0 \right\}$$

$$= E(Y_W|Z = 0).$$

Appendix 4.C: Theoretical equivalence of MMWS and IPTW

We can show that, for the treated units,

$$\text{MMWS} = \frac{\text{pr}(\theta)}{\text{pr}(\theta|Z=1)}$$

$$= \frac{\text{pr}(\theta)}{\text{pr}(Z=1|\theta)\text{pr}(\theta)} \times \text{pr}(Z=1)$$

$$= \frac{\text{pr}(Z=1)}{\text{pr}(Z=1|\theta)}$$

$$= \frac{\text{pr}(Z=1)}{\theta} = \text{IPTW}.$$

Similarly, for the control units,

$$\text{MMWS} = \frac{\text{pr}(\theta)}{\text{pr}(\theta|Z=0)}$$

$$= \frac{\text{pr}(\theta)}{\text{pr}(Z=0|\theta)\text{pr}(\theta)} \times \text{pr}(Z=0)$$

$$= \frac{\text{pr}(Z=0)}{\text{pr}(Z=0|\theta)}$$

$$= \frac{\text{pr}(Z=0)}{1-\theta} = \text{IPTW}.$$

Appendix 4.D: Simulations comparing MMWS and IPTW under misspecifications of the functional form of a propensity score model

When the estimated propensity score model fails to represent nonlinear relationships between pretreatment covariates and treatment assignment, treatment effect estimation on the basis of IPTW adjustment tends to be less robust than that based on MMWS adjustment. Hong (2010) provided some examples as follows.

Suppose that a true propensity score model is $\eta = \beta_0 + \beta_1 X_1 + \beta_2 X_2 + \beta_3 X_1 X_2$, where X_1 and X_2 are both binary. In practice, researchers often fail to include interaction terms such as $X_1 X_2$ when specifying a propensity score model. As a result, the estimated logit of θ will systematically deviate from its true score by the amount $\beta_3 X_1 X_2$. The estimated IPTW will consequently deviate from the true weight except for units whose observed values of either X_1 or X_2 are zero, leading to a bias in the IPTW estimate of treatment effect. Such an omission in the propensity score model, however, generally does not change *propensity stratum membership* for units in either treatment group. Because MMWS is estimated as a ratio of the sample sizes within each stratum, the MMWS estimate of the treatment effect will remain robust.

A similar pattern can be observed for continuous covariates that are nonlinearly associated with the logit of θ. Suppose that the true propensity score model is polynomial $\eta = \beta_0 + \beta_1 X + \beta_2 X^2$. A misspecified propensity score model omits the quadratic term, as is often the case in practice. IPTW will be wrongly estimated except for those whose observed values of X are zero. The IPTW estimate of treatment effect will be biased as a result. MMWS adjustment is much less susceptible to such distortion. This is because stratification on the basis of the estimated linear function of X will divide the sample into segments along the dimension of X. In large samples, as the number of strata increases, MMWS will likely converge to the true weight despite the misspecification of the propensity score model.

In general, when a nonlinear or nonadditive propensity score model $\eta = f(\mathbf{X})$ is misspecified as a linear and additive one $\eta' = f'(\mathbf{X})$, as long as the linear relationship is nonzero, stratification on the basis of η' will nonetheless enable us to approximate the probability of treatment assignment for any given values of $\mathbf{X}$ in a piecemeal manner. This nonparametric approach increases the relative robustness of MMWS estimation of treatment effect when compared with IPTW adjustment.

Hong (2010) presented three series of simulations in support of this general conclusion. In each case, the true propensity score model was nonlinear, while the misspecified propensity score model was linear. In the first series of simulations, treatment assignment was based on a dichotomized version of a continuous covariate; in the second series, treatment assignment was a quadratic function of a continuous covariate; and in the third series, treatment assignment was a logarithmic function of a continuous covariate. In each simulated random sample, units with no counterfactual information were identified and were assigned a zero weight in both IPTW adjustment and MMWS adjustment. The MMWS adjustment was then based on six strata with equal proportions. The simulation results consistently show that as the amount of nonlinearity increases, the amount of bias and the mean square error increase in the IPTW estimate of treatment. In comparison, the MMWS adjustment based on six strata generates robust estimates of the treatment effect with a relatively small amount of bias. The mean square error of MMWS estimate never exceeds that of IPTW estimate. This pattern holds across a variety of commonly seen outcome distributions including a normal outcome as a linear function of the covariate, a normal outcome as a polynomial function of the covariate, a Poisson outcome, and a Bernoulli outcome. These results generally favor the MMWS method when the functional form of a propensity score model is unknown, as is usually the case in practice.

References

Elliott, M.R. and Little, R.J.A. (2000). Model-based alternatives to trimming survey weights. *Journal of Official Statistics*, *16*(3), 191–209.

Hernán, M.A., Brumback, B. and Robins, J. (2000). Marginal structural models and causal inference in epidemiology. *Epidemiology*, *11*(5), 550–560.

Hernandez, D.J., Denton, N.A. and Macartney, S.E. (2007). *Children in Immigrant Families, the U.S. and 50 States: National Origins, Language, and Early Education*, Research Brief Series, Publication No. 2007–11, Child Trends Data Bank, Washington, DC.

Holt, D. and Smith, T.M.F. (1979). Post stratification. *Journal of the Royal Statistical Society, Series A (General)*, *142*(1), 33–46.

Hong, G. (2010). Marginal mean weighting through stratification: adjustment for selection bias in multilevel data. *Journal of Educational and Behavioral Statistics*, *35*(5), 499–531.

Hong, G. (2012). Marginal mean weighting through stratification: a generalized method for evaluating multi-valued and multiple treatments with non-experimental data. *Psychological Methods*, *17*(1), 44–60.

Hong, G. and Hong, Y. (2009). Reading instruction time and homogeneous grouping in kindergarten: an application of marginal mean weighting through stratification. *Educational Evaluation and Policy Analysis*, *31*(1), 54–81.

Hong, G., Corter, C., Hong, Y. and Pelletier, J. (2012). Differential effects of literacy instruction time and homogeneous grouping in kindergarten: who will benefit? Who will suffer? *Educational Evaluation and Policy Analysis.*, *34*(1), 69–88.

Horvitz, D.G. and Thompson, D.J. (1952). A generalization of sampling without replacement from a finite universe. *Journal of the American Statistical Association*, *47*(260), 663–685.

Huang, I.C., Diette, G.B., Dominici, F. *et al.* (2005). Variations of physician group profiling indicators for asthma care. *American Journal of Managed Care*, *11*(1), 38–44.

Imbens, G.W. (2000). The role of the propensity score in estimating dose-response functions. *Biometrika*, *87*(3), 706–710.

Kang, J.D. and Schafer, J.L. (2007). Demystifying double robustness: a comparison of alternative strategies for estimating a population mean from incomplete data. *Statistical Science*, *22*(4), 523–539.

Kish, L. (1965). *Survey Sampling*, John Wiley & Sons, Inc., New York.

Little, R. (1982). Models for nonresponse in sample surveys. *Journal of the American Statistical Association*, *77*(378), 237–250.

Little, R. (1986). Survey nonresponse adjustments. *International Statistical Review*, *54*(1), 1–10.

Little, R. and Pullum, T. (1979). *The General Linear Model and Direct Standardization: A Comparison.* International Statistical Institute, Voorburg. Occasional Papers No. 20.

Little, R.J. and Vartivarian, S. (2003). On weighting the rates in non-response weights. *Statistics in Medicine*, *22*(9), 1589–1599.

Little, R.J. and Vartivarian, S. (2005). Does weighting for nonresponse increase the variance of survey means? *Survey Methodology*, *31*(2), 161–168.

Oh, H.L. and Scheuren, F.S. (1983). Weighting adjustments for unit nonresponse, in *Incomplete Data in Sample Surveys*, vol. 2 (eds W.G. Madow, I. Olkin and D.B. Rubin), Academic Press, New York.

Robins, J.M. (1986). Sparse data and large strata limiting models. *Biometrics*, *42*(2), 311–323.

Robins, J.M. (1987). Addendum to "A new approach to causal inference in mortality studies with sustained exposure periods—application to control of the healthy worker survivor effect." *Computers and Mathematics with Applications*, *14*, 923–945.

Robins, J. (1999). Marginal structural models versus structural nested models as tools for causal inference, in *Statistical Models in Epidemiology, the Environment, and Clinical Trials* (eds M.E. Halloran and D. Berry), Springer, New York.

Robins, J.M., Hernan, M.A. and Brumback, B. (2000). Marginal structural models and causal inference in epidemiology. *Epidemiology*, *11*(5), 550–560.

Rosenbaum, P.R. (1987). Model based direct adjustment. *Journal of the American Statistical Association*, *82*(398), 387–394.

Rosenbaum, P.R. and Rubin (1983). The central role of the propensity score in observational studies for causal effects. *Biometrika*, *70*, 41–55.

Schafer, J.L. and Kang, J. (2008). Average causal effects from nonrandomized studies: a practical guide and simulated example. *Psychological Methods*, *13*(4), 279–313.

Waernbaum, I. (2012). Model misspecification and robustness in causal inference: comparing matching with doubly robust estimation. *Statistics in Medicine*, *31*, 1572–1581.

5

Evaluations of multivalued treatments

Chapters 2 and 3 have reviewed causal inference concepts and methods in the simple case of a binary treatment. The causal effect simply contrasts the potential outcome under the experimental condition with that under the control condition. The current chapter extends the previous discussion of a binary treatment to a more general case that compares three or more treatment conditions. In addition to making multiple pairwise comparisons, researchers will need to address the global question of whether *any* of the treatments differ in their impacts. The methods discussed in the previous chapters for analyzing quasiexperimental data, while of some use, fall short. In particular, propensity score matching and stratification are limited in their applicability when there are more than two treatment conditions.

This chapter will start with defining the causal effects for multivalued treatments in a one-way classification (as opposed to two-way or multiway classifications of multiple treatments to be discussed in section "Moderation"). It will then briefly review the existing research designs and analytic methods for identifying and estimating the causal effects. Analysis of such data in a randomized study is simply via the one-way ANOVA. Yet challenges arise in analyses of quasiexperimental data. These challenges can be addressed by the propensity score-based weighting strategies including inverse-probability-of-treatment weighting (IPTW) and marginal mean weighting through stratification (MMWS). IPTW, while dramatically increasing generality, can be made more robust by means of MMWS. The chapter will focus on explaining the rationale and procedure of MMWS applications.

5.1 Defining the causal effects of multivalued treatments

A binary treatment indicator is sometimes an oversimplification when, in fact, there are multiple versions of the experimental condition representing distinct intervention theories. The control condition, often defined as "the status quo" or "business as usual," is not necessarily a uniform condition either. In general, distinguishing between these different versions can greatly sharpen a theoretical argument or enhance a policy position.

For example, in evaluating the impact of providing English language learning (ELL) services to language minority students, one may contrast the potential outcome of receiving ELL services with that of not receiving ELL services. Yet in reality, ELL services vary in format and intensity. Some schools *pull out* language minority students from their regular classes for intensive English-as-a-second-language (ESL) instruction; some leave language minority students in the regular classes and offer *in-class assistance* when a subject is taught in English to the entire class; other schools provide a *bilingual education* program, teaching academic subjects in both English and the students' home language. The relative effectiveness of these different ELL services has been highly debated for years. There is also a question about the optimal intensity of ELL services, that is, the amount of time per day or per week that should be spent on ESL or bilingual education, in which case further distinctions would need to be made between different dosage levels. Identifying the optimal format and intensity of ELL services has direct implications for ELL policy and practice.

Evaluations of multiple treatment conditions typically involve a designated reference condition. For example, to determine the relative effectiveness of various formats of ELL services, the researcher may contrast each type of ELL services with the control condition in which none of the ELL services is offered. This does not preclude contrasts between different types of ELL services such as a contrast between pull-out ESL instruction and bilingual instruction. In general, we may use a categorical variable Z with multiple values $z = 1,\ldots,K$ to denote K treatment conditions. For example, let $z = 1$ if a language minority student is assigned to pull-out ESL instruction; let $z = 2$ if the student is assigned to in-class ESL assistance; let $z = 3$ if the student is assigned to bilingual education; and let $z = 4$ if the student receives no ELL services. A causal effect contrasts the potential outcome of a given treatment condition z with the potential outcome of an alternative treatment condition z'. The population average treatment effect can be written as

$$\delta_{z-z'} = E[Y(z) - Y(z')]. \tag{5.1}$$

Here, $Y(z)$ is an individual's potential outcome if assigned to treatment condition z; and $Y(z')$ is the individual's potential outcome if assigned to treatment condition z' instead.

With four possible treatment conditions, we may have six pairwise comparisons:

1. The causal effect of pull-out ESL versus no ELL services, $\delta_{1-4} = E[Y(1) - Y(4)]$

2. The causal effect of in-class ESL versus no ELL services, $\delta_{2-4} = E[Y(2) - Y(4)]$

3. The causal effect of bilingual education versus no ELL services, $\delta_{3-4} = E[Y(3) - Y(4)]$

4. The causal effect of pull-out ESL versus bilingual education, $\delta_{1-3} = E[Y(1) - Y(3)]$

5. The causal effect of in-class ESL versus bilingual education, $\delta_{2-3} = E[Y(2) - Y(3)]$

6. The causal effect of pull-out ESL versus in-class ESL, $\delta_{1-2} = E[Y(1) - Y(2)]$

5.2   Existing designs and analytic methods for evaluating multivalued treatments

5.2.1   Experimental designs and analysis

The experimental designs for evaluating multivalued treatments are straightforward. In general, participants are assigned at random to as many treatment arms as the number of treatment conditions under study. The analysis of variance (ANOVA) is the standard method for analyzing experimental data that involve two or more treatment conditions. Under randomization, the observed mean outcome of each treatment group provides an unbiased estimate of the population mean outcome associated with that treatment. A comparison of the mean observed outcome between any two treatment groups is an unbiased estimate of the corresponding population average causal effect. ANOVA provides an omnibus test of the null hypothesis that, in comparing between multiple treatment conditions, at least one of the causal effects is nonzero in the population.

5.2.1.1   Randomized experiments with multiple treatment arms

Formal evaluations of multivalued treatments in social sciences have been carried out in either laboratory-based experiments or field experiments. For example, in the midst of the policy debate with regard to promoting economic independence among welfare recipients in the United States, one of the central questions was: What strategy works best? Is it investing in education programs for building basic skills that might enable participants to secure stable jobs in the long run? Or is it providing job search assistance that might bring immediate employment opportunities albeit low paying or temporary? To address this question, the National Evaluation of Welfare-to-Work Strategies (NEWWS) included a series of randomized experiments around the country in the mid-1990s. At some of the experimental sites, welfare applicants were assigned at random to one of three treatment conditions: a control condition that continued to provide cash assistance to those eligible for welfare without a work mandate, a human capital development (HCD) program that emphasized basic education and long-term skill building before entering the labor market with the goal of reducing future welfare dependence, and a labor force attachment (LFA) program that offered short-term job search assistance and emphasized quick entry into the labor market with the goal of reducing reliance on welfare as soon as possible. At the end of the fifth year after randomization, about three-quarters of the control group members found jobs and more than half left the welfare rolls. Consistent with the intervention theories, it was found that HCD programs and LFA programs both increased the participants' 5-year earnings by a higher amount and reduced welfare payments by a greater amount when compared with the control condition. In contrary to the anticipation of some advocates, however, HCD programs did not produce additional long-term economic benefits when compared with LFA programs (Hamilton *et al.*, 2001).

5.2.1.2   Identification under ignorability

In NEWWS, each participant has three potential outcomes corresponding to the three treatment conditions. The randomization procedure assured that which treatment condition one was assigned to was independent of these potential outcomes. In the causal inference literature, the treatment assignment is said to be *ignorable* if it is independent of all the potential outcomes. Let $z = 1$ if a welfare applicant was assigned to an LFA program; let $z = 2$ if the applicant was assigned to an HCD program; and let $z = 3$ if the applicant was assigned to the control condition. The ignorability assumption can be formally represented as follows:

$$E[Y(z)] = E(Y|Z=z),\tag{5.2}$$

for $z = 1, 2, 3$. Suppose that the entire population is included in a randomized study. The left-hand side of the equation $E[Y(z)]$ represents the population average potential outcome should the entire population have been assigned to treatment condition z; the right-hand side of the equation $E(Y|Z=z)$ represents the average observed outcome of the individuals actually assigned to treatment condition z. In the current example, when $z = 1$, $E[Y(1)]$ is the population average potential earnings if all the welfare applicants were assigned to LFA programs; $E(Y|Z=1)$ represents the average observed earnings of the individuals actually assigned to LFA. Similarly, the population average potential earnings associated with the HCD programs $E[Y(2)]$ and that associated with the control condition $E[Y(3)]$ are equal to the average observed earnings of the HCD group members and that of the control group members denoted by $E(Y|Z=2)$ and $E(Y|Z=3)$, respectively.

Because the ignorability assumption is guaranteed by the random treatment assignment, one can easily identify each of the causal effects through pairwise comparisons of the average observed outcomes between the treatment groups. To be specific, the causal effect of the LFA programs versus the control condition is identified by

$$E[Y(1) - Y(3)] = E(Y|Z=1) - E(Y|Z=3);$$

the causal effect of the HCD programs versus the control condition is identified by

$$E[Y(2) - Y(3)] = E(Y|Z=2) - E(Y|Z=3);$$

and finally, the causal effect of the LFA programs versus the HCD programs is identified by

$$E[Y(1) - Y(2)] = E(Y|Z=1) - E(Y|Z=2).$$

5.2.1.3   ANOVA

A randomized experiment typically involves a sample rather than the entire population. When a sample is representative of the population, the sample mean difference in the observed outcome between two treatment groups is an unbiased estimate of the corresponding mean difference in the population, which in turn is unbiased for the population average causal effect that contrasts the two treatment conditions under the ignorability assumption. However, due to sampling variability, the sample mean difference is probably nonzero even though the population mean difference is zero. The standard statistical procedure for testing mean differences between multiple groups is ANOVA familiar to most social scientists. An ANOVA model for the population can be written as follows:

$$E(Y|Z=z) = \mu + d_z,\tag{5.3}$$

for all possible values of z. Here, $E(Y|Z=z)$ is the mean observed outcome of all individuals in the population who have been assigned to treatment z; μ is the mean observed outcome of all individuals in the population regardless of their actual treatment assignment; and d_z is the discrepancy between the mean observed outcome in treatment group z and the population mean observed outcome μ.

In essence, ANOVA compares the between-group variability to the within-group variability in the observed outcome in a given sample. The ratio of the between-group variability to the within-group variability computed as an F value is expected to be 1.0 if all the groups have equal means in the population, that is, if $d_1 = d_2 = d_3 = 0$ in the example involving three treatment conditions. If the between-group variability is considerably greater than the within-group variability in a sample to the extent that it becomes highly improbable (as indicated by a p value < 0.05) to obtain such an F value under the null hypothesis that the treatment group means are all equal in the population, one would then reject the null hypothesis and conclude that the mean difference in the observed outcome between at least one pair of the treatment groups is significantly different from zero. In multigroup comparisons, a global F test is preferred over multiple t tests because the former provides a safeguard against inflated type I errors—that is, rejecting a null hypothesis when it is true—when multiple tests are conducted with data from the same sample.

In the analysis of NEWWS data collected from Riverside, California, for example, the researchers reported that the LFA group members earned $2459 more on average than the control group members over 5 years, that the HCD group members earned $2017 more on average than the control group members, and that the LFA group members earned $442 more on average than the HCD group members. The results from an F test indicated significant between-group mean differences in the population. Post hoc comparisons further indicated that both the LFA program and the HCD program increased 5-year earnings relative to the control condition. However, even though the LFA participants appeared to earn more than the HCD participants on average, this mean difference of $442 in earnings was not statistically significant. Therefore, one cannot conclude that LFA was more effective than HCD in raising the earnings of welfare recipients (Hamilton *et al.*, 2001).

If we use dummy indicators to represent $K-1$ treatment groups (i.e., all the treatment groups except for one) in an evaluation of K treatment conditions, Equation 5.3 is equivalent to the following multiple regression model:

$$Y = a + \sum_{z=1}^{K-1} \gamma_z D_z + e. \tag{5.4}$$

Here, D_z takes value 1 if an individual was actually assigned to treatment condition z and 0 otherwise. In the above example, the model would contain two dummy indicators as follows:

$$Y = a + \gamma_1 D_1 + \gamma_2 D_2 + e,$$

where D_1 and D_2 are dummy indicators for being assigned to HCD and LFA, respectively. Hence, γ_1 and γ_2 represent the respective causal effects of HCD and LFA in contrast with the control condition. This multiple regression model is equivalent to the ANOVA model in (5.3) because $\gamma_1 = d_1 - d_3$ and $\gamma_2 = d_2 - d_3$.

5.2.2   Quasiexperimental designs and analysis

Evaluations of multivalued treatments are often conducted with quasiexperimental data including large-scale survey data. The causal validity of the results is called into question again when special care has not been taken to remove selection bias. In the following, we illustrate with an application study evaluating multiple formats of ELL services. In the Early Childhood Longitudinal Study–Kindergarten (ECLS-K) cohort, language minority kindergartners received pull-out ESL instruction, in-class ESL instruction, bilingual education, or none of these services. We use teacher rating of child approaches to learning at the end of the kindergarten year as the outcome assessing a child's engagement and concentration as perceived by the teacher. Earlier research has shown that a child's approaches to learning constitute an important domain of school readiness development and is an important prerequisite for achievement during kindergarten (Heaviside and Farris, 1993; Ladd, Birch, and Buhs, 1999). Language minority students experiencing difficulties in comprehending others and in communicating their own ideas may receive a relatively low teacher rating unless appropriate language support is provided to facilitate their participation in classroom instruction.

At the end of the year, children who did not have any ELL services received the highest rating in approaches to learning, followed by those with in-class ESL instruction and those with pull-out ESL instruction. Children in bilingual education programs had the lowest rating, almost a quarter of a standard deviation lower than the average rating received by those with no ELL services. These between-group mean differences in the observed outcome largely reflect the preexisting differences between these four groups at the beginning of the kindergarten year. The outcome is associated with at least 54 pretreatment measures covering important domains including child oral language proficiency in English and Spanish (among Spanish speakers); child demographic characteristics; disability status; both parent and teacher ratings of child social–emotional development; child literacy, math, and general knowledge; home literacy; family structure; family socioeconomic status; and school composition of minority and free-lunch students. These pretreatment measures are mostly predictive of a relatively higher teacher rating for the language minority students receiving no ELL services; students who were assigned to bilingual education generally displayed pretreatment characteristics predictive of a relatively lower teacher rating.

We now review the existing analytic approaches to evaluating multivalued treatments that have been applied to quasiexperimental data and discuss their limitations in the context of the current example. The existing methods include analysis of covariance (ANCOVA) and multiple regression; propensity score-based matching, stratification, and IPTW; and the instrumental variable (IV) method.

5.2.2.1   ANCOVA and multiple regression

Chapter 3 has reviewed ANCOVA for evaluating binary treatments. We now extend the ANCOVA model to multivalued treatments with covariance adjustment for a single pretreatment covariate X. For example, in evaluating the different formats of ELL services, let X be a student's oral English proficiency at kindergarten entry. For simplicity, suppose that one analyzes population data rather than sample data. The population model can be written as follows:

$$E(Y|Z=z,\ X=x)=\mu+b(x-E(X))+d_z. \qquad (5.5)$$

Here, $E(Y|Z=z, X=x)$ is the mean observed outcome of all individuals in the subpopulation with oral English proficiency value x who have been assigned to treatment z. $E(X)$ is the population mean of X regardless of actual treatment assignment; and hence, $x-E(X)$ is the individual observed value x centered at the population mean. To be specific, $x-E(X)$ is negative, zero, or positive if a student's oral English proficiency at the baseline was below, at, or above the population mean. Due to the centering, μ is equal to $E(Y)$, the population mean of Y regardless of actual treatment assignment. For students in treatment group z whose oral English proficiency at the baseline x was equal to the population mean, that is, $x-E(X)=0$, d_z is the discrepancy between their average observed outcome and the population mean outcome. The model has assumed that d_z does not depend on the value of x. To be concrete, this assumption states that, for students with low English proficiency at the baseline, the mean difference in the observed outcome between two treatment groups is the same as that for children with high English proficiency at the baseline. This is an alternative way of stating the standard ANCOVA assumption that the X–Y relationship is the same across all the treatment groups as shown in Figure 5.1. In this hypothetical data example, d_1 and d_4 are positive, while d_2 and d_3 are negative.

The ANCOVA model allows one to compare the mean observed outcomes of students from different treatment groups who had the same baseline oral English proficiency and thereby removes the preexisting differences in average oral English proficiency between the treatment groups. We can see this by moving $b(x-E(X))$ from the right-hand side to the left-hand side of the equation. The aforementioned model then becomes equivalent to an ANOVA model applied to a subpopulation of individuals with the same baseline oral English proficiency x. Analyzing the adjusted outcome $Y-b[x-E(X)]$, a global F test for the main effect of treatment tests $d_1=d_2=d_3=d_4=0$. As is the case in ANOVA, the F value

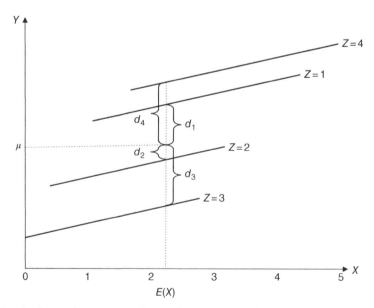

Figure 5.1   Analysis of covariance for evaluating multivalued treatments: $E(Y|Z=z,X=x)=\mu+b(x-E(X))+d_z$

is expected to be 1.0 if the mean observed outcomes are the same across the four treatment groups after adjusting for the pre-existing differences in baseline oral English proficiency.

Yet to draw causal conclusions with regard to the relative effectiveness of different types of ELL services, one needs to remove preexisting between-group differences not only in child oral English proficiency but also in all the other baseline characteristics that are predictive of the outcome, such that the strong ignorability assumption becomes plausible. The strong ignorability assumption, as explained in the previous chapters, states that the assignment to a given treatment condition z is independent of the corresponding potential outcome $Y(z)$ given the observed pretreatment covariates denoted by a vector $\mathbf{X} = \mathbf{x}$. In other words, within each subpopulation defined by the baseline characteristics $\mathbf{x}$, the assignment of individuals to different treatment conditions can be viewed "as if random." The strong ignorability assumption can be formally represented as follows:

$$E[Y(z)|\mathbf{X}=\mathbf{x}] = E(Y|Z=z, \ \mathbf{X}=\mathbf{x}),$$
$$0 < \mathrm{pr}(Z=z|\mathbf{X}=\mathbf{x}) < 1 \tag{5.6}$$

for all possible values of z.

To satisfy the strong ignorability assumption, the data would need to contain all the pretreatment information that characterizes preexisting between-group differences and is predictive of the outcome such that no unmeasured covariates would confound the treatment effects given the observed covariates. Moreover, all the observed covariates, should they be independent of one another, would need to be entered into the ANCOVA model and satisfy the model-based assumption that the relationship between each covariate and the outcome is linear and that the linear relationship is the same across all the treatment groups. Chapter 3 has discussed potential pitfalls that could lead to bias when an ANCOVA model is misspecified due to omissions of nonlinear or interaction terms. The same concerns apply when ANCOVA is employed for evaluating multivalued treatments.

If we use dummy indicators to represent $K-1$ treatment groups in an evaluation of K treatment conditions, Equation 5.5 is equivalent to the following multiple regression model:

$$Y = a + bX + \sum_{z=1}^{K-1} \gamma_z D_z + e. \tag{5.7}$$

As before, D_z takes value 1 if an individual was actually assigned to treatment condition z and 0 otherwise. In the above example, the model would contain three dummy indicators as follows:

$$Y = a + bX + \gamma_1 D_1 + \gamma_2 D_2 + \gamma_3 D_3 + e,$$

where D_1, D_2, and D_3 are dummy indicators for being assigned to pull-out ESL, in-class ESL, and bilingual education, respectively. Hence, γ_1, γ_2, and γ_3 represent the respective effectiveness of pull-out ESL, in-class ESL, and bilingual education as opposed to no ELL services. This multiple regression model is equivalent to the ANCOVA model in (5.5) because $\gamma_1 = d_1 - d_4$, $\gamma_2 = d_2 - d_4$, and $\gamma_3 = d_3 - d_4$. The researcher may expand the number of covariates in the multiple regression model the same way as that in the ANCOVA model. The two models invoke exactly the same model-based assumptions. When all these assumptions hold,

γ_1 identifies the causal effect of pull-out ESL versus no ELL services denoted by δ_{1-4} at the beginning of this chapter; γ_2 and γ_3 identify δ_{2-4} and δ_{3-4}, respectively.

5.2.2.2 Propensity score-based adjustment

When the number of observed pretreatment covariates that require statistical adjustment is large, as is often the case in quasiexperimental studies, the risk of misspecifying covariate–outcome relationships in an ANCOVA model or a multiple regression model increases. Propensity score-based methods enable researchers to make statistical adjustment for a large number of covariates without having to specify the functional form of each covariate–outcome relationship. The key is to summarize multidimensional pretreatment information associated with the selection into an experimental rather than a control condition in a unidimensional index called the propensity score. Yet their applications to evaluations of multivalued treatments are limited. According to Hong's (2012) report, a survey of the applications of propensity score-based techniques by searching PsycINFO located 202 studies published before May 2010. Among them, only four studies used propensity scores to adjust for selection bias in comparing three or four treatment groups. Two of these studies used propensity score matching (Citrome *et al.*, 2009; Hill *et al.*, 2005), one used propensity score stratification (Tiihonen *et al.*, 2006), and one used IPTW adjustment (Zhu *et al.*, 2007). In the following, we assess the suitability of each of these three propensity score-based methods for evaluating multivalued treatments.

Propensity score matching

Propensity score matching was designed to evaluate the effect of a binary treatment by matching the untreated units to the treated units on a single propensity score. The propensity score for receiving treatment z as a function of pretreatment covariates $\mathbf{x}$ is $\theta_z(\mathbf{x}) = \mathrm{pr}(Z = z | \mathbf{X} = \mathbf{x})$. To simplify the notation, henceforth, we use θ_z as a shorthand for $\theta_z(\mathbf{x})$. The propensity of being untreated denoted by θ_0 and the propensity of being treated denoted by θ_1 are linearly dependent as the two propensity scores add up to 1.0 for each individual. Matching on θ_0 therefore is equivalent to matching on θ_1. In studies of multivalued treatments, however, treatment assignment becomes multidimensional, as do the propensity scores. For instance, the covariates that predict whether a student would be assigned to pull-out ESL rather than the control condition do not necessarily predict whether the same student would be assigned to in-class ESL rather than the control condition.

Applications of propensity score matching to multivalued treatments is challenging. For example, in a study investigating the effect of maternal employment on child development, Hill and colleagues (Hill *et al.*, 2005) compared three employment groups with a control group of mothers who did not work in the first 3 years after child birth denoted by $z = 4$. The three employment groups, denoted here by $z = 3$, 2, and 1, consisted of mothers who worked only after the first year, who worked part time during the first year, and who worked full time during the first year, respectively. These treatment groups were conceptualized as lying on a continuum from the least to the most amount of work time. The researchers then made pairwise comparisons between selected treatment conditions in separate analyses. These included a comparison between full-time work and part-time work in the first year ($z = 1$ vs. $z = 2$), a comparison between full-time work in the first year and not working until after the first year ($z = 1$ vs. $z = 3$), a comparison between part-time work in the first year and not working until after the first year ($z = 2$ vs. $z = 3$), and a comparison between working after first year and never working ($z = 3$ vs. $z = 4$). The researchers argued that these comparisons correspond most closely to employment options that mothers might consider in the absence of a strong intervention.

In essence, these researchers transformed a multiple comparison problem into a sequence of binary comparisons. For example, in comparing full-time work with part-time work in the first year, a logistic regression estimated the propensity score for full-time work. Children whose mothers worked full time in the first year were then matched with those whose mothers worked only part time on the basis of the estimated propensity score. This strategy seems reasonable if a particular comparison is of central interest. The results are especially meaningful if the population of mothers who would choose between full-time work and part-time work in the first year is quite different from the population of mothers who could choose between working after the first year and not working at all.

However, this strategy made it impossible to conduct a global test for the mean differences between the four treatment groups in the outcome. This is because different pairwise comparisons were based on different matched samples for these separate analyses and were to be generalized to possibly different populations. If there is a common population of interest for the entire set of inferences, that is, if a mother who never worked in the 3 years after child birth would conceivably work part time or full time in the first year perhaps due to a policy intervention, then without a global test, the inflated type I error would become a major concern.

Propensity score stratification

Imbens (2000) generalized Rosenbaum and Rubin's (1983) results to multivalued treatments and defined the generalized propensity score as the conditional probability of being assigned to a particular treatment condition z given a set of pretreatment covariates $\mathbf{X}$:

$$\theta_z = \mathrm{pr}(Z = z | \mathbf{X} = \mathbf{x}) \tag{5.8}$$

for all possible values of z. As before, let D_z be a dummy indicator that takes value 1 if an individual was assigned to treatment condition z and 0 otherwise. The propensity score defined in (5.8) is equal to $E[D_z | \mathbf{X} = \mathbf{x}]$ representing, among individuals with pretreatment characteristics $\mathbf{x}$ in the population, the proportion that is expected to be assigned to treatment condition z. With four possible treatment conditions, each individual has four corresponding propensity scores. In the example of ELL services evaluation, θ_1, θ_2, θ_3, and θ_4 are the propensity scores for being assigned to pull-out ESL, in-class ESL, bilingual education, and no ELL services, respectively. Three of these four propensity scores are linearly independent, while the fourth can be represented as a linear combination of the other three given that $\theta_1 + \theta_2 + \theta_3 + \theta_4 = 1$. In an evaluation of four treatment conditions, a global test is possible if statistical adjustment is made for three of the four propensity scores.

When the treatment conditions are qualitatively distinct and without a logical ordering, one may estimate the propensity scores through analyzing a multinomial logistic regression model. Unlike the binary logistic regression employed by Hill and colleagues (2005) that analyzes data from a pair of treatment groups each time, a multinomial logistic regression analyzes all the treatment groups at once. In an evaluation of four treatment conditions with the last treatment condition being the reference group, a multinomial logistic regression model analyzes the following three equations simultaneously:

$$\ln\left(\frac{\theta_1}{\theta_4}\right) = \eta_1 = \mathbf{X}^T \boldsymbol{\beta}_{(1)};$$

$$\ln\left(\frac{\theta_2}{\theta_4}\right) = \eta_2 = \mathbf{X}^T \boldsymbol{\beta}_{(2)}; \qquad (5.9)$$

$$\ln\left(\frac{\theta_3}{\theta_4}\right) = \eta_3 = \mathbf{X}^T \boldsymbol{\beta}_{(3)}.$$

Here, $\boldsymbol{\beta}_{(z)}$ for $z = 1, 2, 3$ is a vector of coefficients corresponding to the elements in vector $\mathbf{X}$ predicting how probable an individual would be assigned to the treatment condition $Z = z$ rather than to the control condition $Z = 4$. Hence, each vector of coefficients indicates how the pretreatment covariates distinguish treatment group z from the reference group. Once obtaining the coefficient estimates, the researcher may then estimate all four propensity scores for each individual as follows:

$$\widehat{\theta}_1 = \frac{e^{\widehat{\eta}_1}}{1 + e^{\widehat{\eta}_1} + e^{\widehat{\eta}_2} + e^{\widehat{\eta}_3}};$$

$$\widehat{\theta}_2 = \frac{e^{\widehat{\eta}_2}}{1 + e^{\widehat{\eta}_1} + e^{\widehat{\eta}_2} + e^{\widehat{\eta}_3}};$$

$$\widehat{\theta}_3 = \frac{e^{\widehat{\eta}_3}}{1 + e^{\widehat{\eta}_1} + e^{\widehat{\eta}_2} + e^{\widehat{\eta}_3}};$$

$$\widehat{\theta}_4 = \frac{1}{1 + e^{\widehat{\eta}_1} + e^{\widehat{\eta}_2} + e^{\widehat{\eta}_3}}.$$

For example, a child might have a 0.13 probability of receiving pull-out ESL, a 0.32 probability of receiving in-class ESL, a 0.44 probability of receiving bilingual education, and a 0.11 probability of not receiving any ELL services.

This procedure summarizes in $K - 1$ propensity scores the pretreatment information associated with individual selection into K different treatment conditions. Given the balancing property of the propensity scores as discussed in Chapter 3, propensity score stratification enables the researcher to identify the causal effect of treatment condition z versus an alternative treatment condition z' under the strong ignorability assumption taking the following form:

$$E[Y(z)|\theta_1, \ldots, \theta_{K-1}] = E(Y|Z = z, \theta_1, \ldots, \theta_{K-1}),$$

$$0 < \mathrm{pr}(Z = z|\theta_1, \ldots, \theta_{K-1}) < 1 \qquad (5.10)$$

for all K possible values of z. This assumption states that, within a subpopulation defined by the $K - 1$ propensity scores, the assignment of individuals to each of the K different treatment conditions is independent of the corresponding potential outcome.

Matching on multiple propensity scores becomes extremely challenging. If employing the propensity score stratification method instead, after dividing the sample into five strata along the dimension of each of the three propensity scores in an evaluation of three different treatment conditions relative to a control condition, one would need to identify individuals that are homogeneous in all three propensity scores by generating at least $5 \times 5 \times 5 = 125$ cells. Within each cell, one could estimate cell-specific treatment effects through pairwise comparisons among the four treatment categories and then pool the results over all the cells to estimate the average treatment effects. This procedure would be not only cumbersome but also complicated by the sparseness of data in many cells that might have only two or one treatment categories represented. The number of cells would easily grow out of hand as the number of treatment conditions increases.

IPTW

IPTW is particularly flexible for evaluating multivalued treatments. As discussed in great detail in Chapter 4, the rationale for IPTW is not to estimate the conditional treatment effect for those with the same propensity score value and then pool the conditional treatment effects over the distribution of the propensity score. Rather, IPTW focuses on estimating the population average potential outcome associated with each treatment condition. Weighting is used to transform the observed pretreatment composition of each treatment group such that it resembles the pretreatment composition of the entire sample representative of the target population.

To apply IPTW to the example of ELL services evaluation, one starts with estimating the four propensity scores for each individual through analyzing a multinomial logistic regression model specified in (5.9). For individuals assigned to treatment z, the weight is

$$\text{IPTW} = \frac{\text{pr}(Z=z)}{\hat{\theta}_z}; \qquad (5.11)$$

for $z = 1, 2, 3, 4$. Here, $\text{pr}(Z=z)$ is the proportion of individuals in the sample who were actually assigned to treatment z and represents the target probability of random assignment to treatment z in a hypothetical experiment. For example, if the estimated propensity score for receiving pull-out ESL services is 0.8 for a student actually in the pull-out group and if 20% of the language minority students received pull-out services, the weight for this student is $0.2/0.8 = 0.25$. One then analyzes a weighted outcome model comparing the weighted mean observed outcome between the treatment groups. The model may take the form of a one-way ANOVA as in (5.3) or its equivalent form in multiple regression as in (5.4).

The identification requires an ignorability assumption that is considerably weaker than that stated in (5.10) such that Imbens (2000) called it the weak ignorability assumption:

$$E[Y(z)|\theta_z] = E(Y|Z=z,\theta_z),$$
$$0 < \text{pr}(Z=z|\theta_z) < 1 \qquad (5.12)$$

for all possible values of z. The assumption in (5.12) is relatively weaker in the sense that the adjustment for selection into treatment group z involves only one propensity score θ_z and does not involve the propensity scores for other treatment conditions. In addition, the assignment to treatment condition z needs to be independent of only the corresponding potential outcome $Y(z)$. For example, the assignment to pull-out services is not required to be independent of one's potential outcome of receiving in-class services or that of receiving bilingual education. When the weak ignorability assumption holds, the weighted data approximate data from an experiment that assigns individuals at random to each of the multiple treatment conditions. Hence, the weighted mean observed outcome of each treatment group consistently estimates the population average potential outcome associated with the corresponding treatment condition (Imbens, 2000; Robins, 1999).

IPTW shows clear advantages over matching and stratification. Unlike propensity score stratification, IPTW does not require simultaneous conditioning on multiple propensity scores. And unlike the pairwise matching method, IPTW allows for a global test and for tests of a priori or post hoc contrasts among the treatment groups. Such tests use standard methods for controlling type I errors in ANOVA. However, as discussed in Chapter 4, because IPTW is computed as a direct function of the estimated propensity score, this method is often sensitive to misspecifications of the functional form of the propensity score model. Recent research has shown that in typical applications in which a nonlinear or nonadditive propensity score model is misspecified as a linear additive one, IPTW adjustment leads to a bias and a rapid rise in standard error proportional to the amount of confounding associated with nonlinearity (Hong, 2010). Moreover, earlier research has shown that a lack of overlap in the covariate distributions between treatment groups may not only reduce precision (Cochran, 1957; Imbens, 2004; Rubin, 1997) but also lead to bias in treatment effect estimation when the treatment effect is not constant (Hong, 2010). IPTW adjustment is particularly susceptible to this problem. It tends to introduce additional bias when only a portion of the population provides empirical support for causal inference, that is, when individuals whose true propensity score for being assigned to a certain treatment condition is essentially zero receive a nonzero weight.

5.2.2.3   Other adjustment methods

Methods that are employed relatively less often in evaluations of multivalued treatments include the use of IV. Chapter 3 reviewed the IV method in the case of a binary treatment—veteran status from the Vietnam era. This method identifies the treatment effect on the outcome by utilizing an IV for the treatment. The IV must have an unconfounded effect on the treatment; its effect on the outcome must also be unconfounded and must be entirely channeled through the treatment. Additional assumptions include that the effect of the IV on the treatment must be independent of the treatment effect on the outcome. These assumptions are generally hard to meet.

To evaluate three different formats of ELL services in contrast with the control condition, the researcher would need to use at least three IV. Supposing that all the identification assumptions are met, Appendix 5.A shows how the IV method identifies each of the three treatment effects. Clearly, the major difficulty is in obtaining multiple IVs that satisfy all the identification assumptions. The IVs used in many past applications have been dubious either because their effects on the treatments and the outcomes are likely confounded or because they could have affected the outcome arguably through pathways other than the treatments of interest.

5.3   MMWS for evaluating multivalued treatments

Chapter 4 has introduced the MMWS method. This is a newly emerging nonparametric strategy of statistical adjustment that combines key elements of propensity score stratification and IPTW (Hong, 2010; Hong and Hong, 2009; Huang *et al.*, 2005; Zanutto, Lu, and Hornik,

2005). MMWS overcomes important limitations of the existing methods when applied to multivalued treatments. Unlike the propensity score stratification method, MMWS does not require simultaneous conditioning on all the propensity scores. And unlike the pairwise matching method, MMWS involves identifying a common analytic sample for a global test in ANOVA and for multiple comparisons among treatment groups while controlling type I errors. While IPTW shares these same flexibilities, MMWS brings unique strengths by having a built-in procedure of excluding from the analysis individuals that do not have counterfactual information in the observed data. As shown in the kindergarten retention study, the MMWS procedure identified and removed from the subsequent analysis students who were not at risk of repeating a grade. Moreover, because MMWS has adopted a nonparametric procedure by computing the weight on the basis of propensity score stratification, its estimation of treatment effects remains robust across a range of outcome distributions and tends to be more efficient than IPTW estimation (Hong, 2010). Later, we clarify the rationale and delineate the analytic procedure in the context of the ELL services evaluation.

5.3.1 Basic rationale

In quasiexperimental data, the observed mean outcome of each treatment group is often a biased estimate of the population average potential outcome associated with the corresponding treatment. For example, the mean observed outcome of students receiving bilingual education is likely an underestimate of the population average potential outcome if the entire population of language minority students had hypothetically received bilingual education. This is due to the concentration of students in bilingual education from low socioeconomic backgrounds that are predictive of low teacher ratings of student approaches to learning. Earlier, we used $z = 3$ to denote the assignment to bilingual education. The selection problem at hand is $E(Y|Z = 3) \neq E[Y(3)]$. The basic rationale of MMWS, as explained in Chapter 4, is to directly estimate the population average potential outcome (i.e., the marginal mean outcome) of each treatment condition by transforming the pretreatment composition of the corresponding treatment group to resemble that of the entire sample representative of the target population.

In an MMWS application, one uses the estimated propensity score for a given treatment to stratify the sample. To estimate the population average potential outcome associated with bilingual education $E[Y(3)]$, for example, the entire sample would be stratified on the estimated propensity score for receiving bilingual education denoted by θ_3. Suppose that six strata denoted by $S_3 = 1, \ldots, 6$ are created for this purpose. Rather than comparing all four treatment groups within strata, we focus in this case on examining (i) the proportion of students receiving bilingual education in a given stratum s represented by $\mathrm{pr}(Z = 3|S_3 = s)$ relative to (ii) the proportion of students receiving bilingual education in the entire sample represented by $\mathrm{pr}(Z = 3)$. The nonparametric weight for each bilingual education student is a ratio of (ii) to (i). Bilingual education students underrepresented in a certain stratum will receive a relatively high weight to increase their representation, while those overrepresented in another stratum will receive a relatively low weight to reduce their representation. After weighting, the proportion of students receiving bilingual education in each stratum will be equal to the proportion of students receiving bilingual education in the entire sample.

This procedure is applied to one treatment group at a time. In general, in order to estimate the population average potential outcome $E[Y(z)]$ for $z = 1, 2, 3, 4$ in the evaluation of ELL services, the sample is stratified on the basis of the estimated propensity score $\widehat{\theta}_z$, which creates H_z strata denoted by $S_z = 1,\ldots,H_z$. Under this stratification, the nonparametric weight for students in stratum s who were assigned to treatment z is

$$
\begin{aligned}
\text{MMWS} &= \frac{\text{pr}(Z=z)}{\text{pr}(Z=z|S_z=s)} \\
&= \frac{n_{S_z=s}}{n_{z,S_z=s}} \times \text{pr}(Z=z).
\end{aligned}
\tag{5.13}
$$

Here, the numerator $\text{pr}(Z=z)$ is the proportion of students in the entire analytic sample assigned to treatment z; the denominator $\text{pr}(Z=z|S_z=s)$ is the proportion of students in stratum s who were assigned to treatment z. The latter is estimated as a ratio of two sample counts. The first is the number of students in stratum s represented by $n_{S_z=s}$; the second is the number of students in stratum s who were assigned to treatment z represented by $n_{z,S_z=s}$. This ratio could be viewed as a nonparametric representation of the average propensity of being assigned to treatment z in stratum s. When one focuses on the assignment to treatment z only, (5.13) is equivalent to (4.6) applied to a binary treatment in Chapter 4. After completing this procedure with each of the four treatment groups, the MMWS method effectively equates the observed pretreatment composition across all four treatment groups.

5.3.2   Analytic procedure

As discussed in Chapter 4, the general analytic procedure of using MMWS adjustment involves eight major steps which similarly apply to evaluations of multivalued treatments:

1. Collect a large set of pretreatment covariates.

2. Estimate the propensity scores.

3. Identify the common support.

4. Stratify the analytic sample.

5. Compute the nonparametric weight.

6. Check balance in pretreatment composition across the treatment groups in the weighted sample.

7. Analyze a weighted outcome model to estimate the treatment effects.

8. Conduct sensitivity analysis.

The procedures differ, however, between multinomial treatment measures and ordinal treatment measures. In the following, we first highlight the analytic procedure for multinomial treatment measures that is clearly different from that for binary treatment measures. We then discuss the procedure for ordinal treatment measures that could be relatively simpler under certain conditions.

We omit the discussions of Steps 1 and 8 as they are similar to the corresponding steps for binary treatment measures.

5.3.2.1 MMWS for a multinomial treatment measure

Step 2. Estimate the propensity scores. A multinomial treatment measure indicates different treatment categories on a multivalued nominal scale. In other words, indicators such as $z = 1$, $z = 2$, $z = 3$, and $z = 4$ are arbitrarily chosen to represent four treatment conditions without a particular order. In evaluating different formats of ELL services, after collecting pretreatment measures, one analyzes a multinomial logistic regression model as shown in (5.9) and estimates four propensity score θ_1, θ_2, θ_3, and θ_4 for each child corresponding to the four possible treatment conditions.

Step 3. Identify the common support. To identify the common support that contains counterfactual information needed for causal inference, one may use the minimum of the maximum values of a logit propensity score among all treatment groups as the upper bound and may use the maximum of the minimum values of the same logit propensity score as the lower bound. A predefined propensity score radius (or caliper) of width no more than 0.2 standard deviations of the logit propensity score is generally acceptable (Austin, 2011). Figure 5.2 compares the distributions of the four logit propensity scores between the four treatment groups. The dashed lines mark the lower and upper bounds within which there is common support across all four treatment groups. Students whose four logit propensity scores all fall between the corresponding lower and upper bounds constitute the analytic sample that provides an empirical basis for the causal inference. In other words, a student who does not have counterparts in an alternative treatment group is excluded from the analytic sample because such a student has inadequate counterfactual information in the observed data. Here, the analytic sample includes 2819 kindergartners, roughly 94% of the original sample. Among them, 1167 children had no ELL services, 421 had pull-out ESL instruction, 406 had in-class ESL instruction, and 825 had bilingual education. A comparison between the analytic sample and the original full sample may reveal whether the two samples represent somewhat different populations to which the causal inference results can be generalized.

Step 4. Stratify the analytic sample. With four possible treatment conditions, one needs to stratify the analytic sample on each of the four estimated logit propensity scores one at a time. Specifically, after sorting and stratifying the sample on the logit of θ_1, let $S_1 = 1, \ldots, H_1$ denote stratum membership under this first stratification. The goal of this stratification is to balance the within-stratum distribution of the logit of θ_1 between those assigned to pull-out ESL and those in the rest of the sample. Subsequently, one sorts and stratifies the analytic sample on the logit of θ_2. Repeat this procedure until the sample has been stratified on the last logit propensity score. Every individual will have four indicators for stratum membership corresponding to the four sets of stratification. The number of strata does not have to be the same across the four sets of stratification. As shown in Table 5.1, seven strata are created for balancing the logit propensity score for pull-out ESL between the pull-out ESL group and the rest of the sample; seven strata for balancing the logit propensity score for in-class ESL between the in-class ESL group and the rest of the sample; and eight strata for balancing between the bilingual education group and the rest of the sample. As many as 10 strata are needed for balancing between the group receiving no ELL services and the rest of the sample.

Step 5. Compute the nonparametric weight. Under the first set of stratification shown in Table 5.1, we can compute the nonparametric weight defined in (5.13) for every student receiving pull-out ESL; under the second set of stratification, the nonparametric weight for every student receiving in-class ESL is computed; and so on and so forth. The table displays, under each set of stratification created on the basis of propensity score $\widehat{\theta}_z$ for treatment z, the

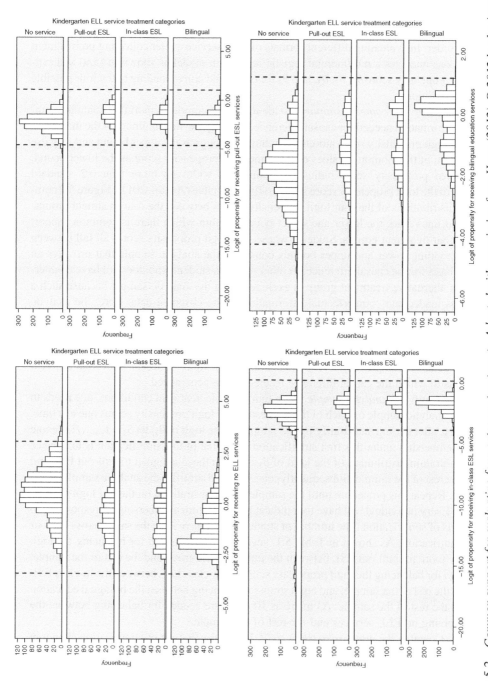

Figure 5.2   Common support for evaluating four-category treatments. Adapted with permission from Hong (2012). © 2012 by the American Psychological Association. The use of this information does not imply endorsement by the publisher.

Table 5.1 Computation of marginal mean weight through stratification for four treatment groups

Pull-out ESL instruction				In-class ESL instruction				Bilingual education				No ELL services			
Stratum	n_{s_1}	n_{1,s_1}	MMWS	Stratum	n_{s_2}	n_{2,s_2}	MMWS	Stratum	n_{s_1}	n_{3,s_3}	MMWS	Stratum	n_{s_4}	n_{4,s_4}	MMWS
1	563	25	3.36	1	352	11	4.61	1	563	37	4.45	1	352	33	4.42
2	564	51	1.65	2	352	23	2.20	2	564	75	2.20	2	352	46	3.17
3	564	69	1.22	3	705	85	1.19	3	564	131	1.26	3	353	85	1.72
4	282	35	1.20	4	705	114	0.89	4	376	137	0.80	4	352	97	1.50
5	282	60	0.70	5	352	71	0.71	5	376	202	0.54	5	353	168	0.87
6	282	69	0.61	6	177	47	0.54	6	125	74	0.49	6	352	206	0.71
7	282	112	0.38	7	176	55	0.46	7	188	120	0.46	7	235	154	0.63
								8	63	49	0.38	8	235	177	0.55
												9	117	94	0.52
												10	118	107	0.46
Total	2819	421		Total	2819	406		Total	2819	825		Total	2819	1167	

Adapted with permission from Hong (2012). © 2012 by the American Psychological Association. The use of this information does not imply endorsement by the publisher.

frequency distribution of stratum membership $n_{S_z = s}$, the frequency distribution of students in the focal treatment group by strata $n_{z, S_z = s}$, and the nonparametric weight for the students in the focal treatment group in each stratum. For example, under the stratification for pull-out ESL, the weight for the 25 children in stratum 1 who received pull-out ESL is computed as $(563/25) \times (421/2819) = 3.36$. Here, $421/2819 = 0.15$ is the proportion of students in the entire sample that received pull-out ESL; 563 is the total number of students in stratum 1; and 25 is the number of students in stratum 1 who received pull-out ESL.

Step 6. Check balance in the weighted sample. The attempt to remove selection bias through weighting might fail if important pretreatment covariates are omitted from the multinomial logistic regression model or if the sample is stratified inappropriately. Hence, an important step is to check the balance in pretreatment composition between the weighted treatment groups through a weighted global test of between-group mean difference in each logit propensity score. Also, compare between treatment groups the distributions of all the observed pretreatment covariates in the weighted sample. In theory, 5% of the pretreatment covariates could show statistically significant differences between the treatment groups at the significance level of 0.05 even in a completely randomized experiment. If the proportion of covariates remaining significantly different between the treatment groups exceeds 5%, researchers may need to modify the propensity score model or restratify the sample. Table 5.2 compares the between treatment group differences in the distribution of each of the four logit propensity scores before and after weighting. Results from weighed one-way ANOVA show that, in each case, the pretreatment differences between the four treatment groups are mostly eliminated after weighting. Hence, the weighting strategy has effectively balanced across all four treatment groups the pretreatment composition indexed by the four logit propensity scores. Further analysis shows that the same results hold for approximately 95% of the observed pretreatment covariates.

Step 7. Estimate the treatment effects. If the weighted sample successfully approximates a randomized experiment, the weighted mean differences between treatment groups will provide consistent estimates of the treatment effects. After applying the marginal mean weight to the data in the same way as with a sample weight, one can analyze the data within the ANOVA framework. Rubin and Thomas (2000) suggested combining propensity score matching with covariance adjustment for a set of strong predictors of the outcome to further reduce bias and improve precision in estimation. Applying the same logic here, one may include strong predictors of the outcome in the weighted analysis of treatment effects, a strategy analogous to ANCOVA. As discussed in Chapter 3, covariance adjustment has many potential pitfalls when the treatment groups differ in the distributions of pretreatment covariates. Because the weighted treatment groups display very similar pretreatment composition, the potential bias due to possible misspecifications of the covariate–outcome relationships becomes minimal.

In the weighted ECLS-K data, children receiving pull-out ESL appear to have the highest teacher rating of approaches to learning, while those receiving bilingual education show the lowest rating. The weighted mean difference in the outcome between these two treatment groups is 0.10, $SE = 0.04$, $t = 2.48$, $p < 0.050$. However, the result of a global F test, $F(3, 2,816) = 2.15$, $p = 0.09$, indicates that the earlier mean difference could have occurred by chance under the null hypothesis at the significance level of 0.5. Subsequent inclusion of the baseline measure of child approaches to learning as a covariate in the same model improves the precision of the estimation and alters the result of the global test, $F(3, 2816) = 2.69$, $p < 0.05$. The positive impact of pull-out ESL instruction over bilingual education

Table 5.2 Between treatment group differences in logit propensity scores before and after weighting

(a) Logit propensity score for pull-out ESL instruction

Treatment groups	Before weighting			After weighting		
	N	Mean	SD	N	Mean	SD
Pull-out ESL	431	−1.42	0.84	421	−1.93	0.78
In-class ESL	410	−1.86	1.16	406	−1.89	0.75
Bilingual education	858	−2.13	1.36	825	−1.94	0.80
No ELL services	1320	−2.43	1.52	1167	−1.95	0.83
Total	3019	−2.12	1.36	2819	−1.94	0.80
F test of mean differences		70.65***			0.47	

(b) Logit propensity score for in-class ESL instruction

Treatment groups	Before weighting			After weighting		
	N	Mean	SD	N	Mean	SD
Pull-out ESL	431	−1.94	1.61	421	−1.92	0.61
In-class ESL	410	−1.61	0.57	406	−1.93	0.65
Bilingual education	858	−1.85	1.12	825	−1.92	0.67
No ELL services	1320	−2.94	2.83	1167	−1.93	0.68
Total	3019	−2.31	2.14	2819	−1.92	0.66
F test of mean differences		75.63***			0.05	

(c) Logit propensity score for bilingual education

Treatment groups	Before weighting			After weighting		
	N	Mean	SD	N	Mean	SD
Pull-out ESL	431	−1.26	1.05	421	−1.15	1.14
In-class ESL	410	−0.91	1.07	406	−1.12	1.09
Bilingual education	858	−0.36	1.00	825	−1.09	1.12
No ELL services	1320	−1.88	1.13	1167	−1.16	1.14
Total	3019	−1.23	1.25	2819	−1.13	1.13
F test of mean differences		360.32***			0.72	

(d) Logit propensity score for no ELL services

Treatment groups	Before weighting			After weighting		
	N	Mean	SD	N	Mean	SD
Pull-out ESL	431	−0.73	1.11	421	−0.45	1.22
In-class ESL	410	−0.91	1.12	406	−0.48	1.22
Bilingual education	858	−1.19	1.18	825	−0.51	1.27
No ELL services	1320	0.53	1.25	1167	−0.46	1.27
Total	3019	−0.33	1.42	2819	−0.47	1.25
F test of mean differences		70.65***			0.47	

Adapted with permission from Hong (2012). © 2012 by the American Psychological Association. The use of this information does not imply endorsement by the publisher. ***$p < 0.001$.

(coefficient $= 0.09$, SE $= 0.03$, $t = 2.83$, $p < 0.01$) shows an effect size equivalent to about 13% of a standard deviation of the outcome. A further weighted analysis contrasting pull-out ESL instruction with a combination of the other three treatment categories again indicates a positive effect of pull-out ESL instruction. This result holds with or without covariance adjustment for child approaches to learning at the baseline even though the estimated effect with covariance adjustment is again relatively more precise.

5.3.2.2   MMWS for an ordinal treatment measure

In large-scale surveys, a treatment with potentially continuous values is often measured on an ordinal scale. The measurement of ESL instruction time provides such an example. ECLS-K researchers asked every kindergarten teacher how many times a week and how much time a day students worked on ESL lessons as a whole class, in small groups, or in individualized arrangements. Combining teacher responses to these two survey items along with information about whether an individual student received ELL services, one may create six levels of time exposure to ESL instruction defined as follows:

$z = 0$: Never

$z = 1$: No more than 1–2 times a week and no more than 30 minutes each time

$z = 2$: 3–4 times a week with no more than 30 minutes each time or 1–2 times a week with 30–60 minutes each time

$z = 3$: Daily with no more than 30 minutes each time or 3–4 times a week with 30–60 minutes each time or 1–2 times a week with 60–90 minutes each time

$z = 4$: Daily with 31–60 minutes each time or 3–4 times a week with 61–90 minutes each time

$z = 5$: Daily with at least 61–90 minutes each time or 3–4 times a week with at least 90 minutes each time

When treatments are measured on an ordinal scale, researchers may take advantage of the systematic relationships between pretreatment covariates and treatment dosage (Joffe and Rosenbaum, 1999; Imai and van Dyke, 2004). A conventional strategy is to model an ordinal treatment measure Z as a function of pretreatment covariates $\mathbf{X}$ through an ordinal logistic regression specified as follows:

$$\ln\left(\frac{\text{pr}(Z \le z|\mathbf{X} = \mathbf{x})}{\text{pr}(Z > z|\mathbf{X} = \mathbf{x})}\right) = \eta_z = \alpha_z + \mathbf{X}^{\mathsf{T}}\boldsymbol{\beta} \tag{5.14}$$

for $z = 0$, 1, 2, 3, 4. An ordinal logistic regression model with the six-level measure of ESL instruction time as the outcome assumes that, for any given language minority student in the example, the log odds of ESL instruction time exposure differ between any two dosage levels z and z' only in the intercepts α_z and $\alpha_{z'}$. In other words, it assumes that the partial association between each pretreatment covariate and the odds of being assigned to a certain dosage or lower is the same across all dosage levels. If this assumption does not hold in the given data, a multinomial logistic regression will be more suitable for estimating the propensity scores for different dosages. Lu and colleagues (Lu *et al.*, 2001; Zanutto, Lu, and Hornik, 2005) suggested that when the aforementioned assumption holds, there is a single balancing score $\mathbf{X}^T\boldsymbol{\beta}$ for all dosage levels. The logit scores estimated from an ordinal logistic regression model are monotonic across the dosage levels. In other words, across the range of $\mathbf{X}$, η_z is either always greater than or always smaller than $\eta_{z'}$. For this reason, in applications of propensity score matching, stratification, or MMWS, one may simply use the estimated logit propensity of being assigned to the first dosage level $\eta_0 = \alpha_0 + \mathbf{X}^T\boldsymbol{\beta}$ as the balancing score.

To identify the common support for causal inference, one then compares the distribution of the balancing score across the six dosage groups. In the ECLS-K data, 81 students who displayed relatively extreme values in the balancing score apparently had no counterparts at an alternative dosage level and are removed from the analytic sample. Table 5.3 shows the result of dividing the analytic sample into eight strata on the balancing score. The computation of MMWS is similar to that for a binary or multinomial treatment. For students in stratum s who were assigned to dosage level z, the weight is

$$\text{MMWS} = \frac{\text{pr}(Z=z)}{\text{pr}(Z=z|S=s)}$$

$$= \frac{n_s}{n_{z,s}} \times \text{pr}(Z=z).$$

Here, the numerator $\text{pr}(Z=z)$ is the proportion of students in the entire analytic sample assigned to dosage level z; the denominator $\text{pr}(Z=z|S=s)$ is the proportion of students in stratum s who were assigned to dosage level z. The inverse of the latter is estimated as a ratio of two sample counts. The first is the number of students in stratum s represented by n_s; the second is the number of students in stratum s who were assigned to dosage level z represented by $n_{z,s}$. This ratio could be viewed as a nonparametric representation of the average propensity of being assigned to dosage level z in stratum s. For example, in stratum 1, the average propensity of being assigned to the lowest dosage level is $84/573 = 0.15$, while the marginal probability of being assigned to the lowest dosage level is $1395/2865 = 0.49$ in the entire analytic sample. Hence, for the 84 students in stratum 1 who were actually assigned to the lowest dosage level, the weight is $(573/84) \times (1395/2865) = 3.32$.

After weighting, the mean logit propensity score becomes equal across the six dosage groups. The same result holds for approximately 95% of the observed pretreatment covariates. The next step is to estimate the weighted mean outcome of each dosage group and compare across the dosage levels. The six dosage groups show significant differences in the outcome before weighting, $F(5, 2859) = 4.25$, $p = 0.001$. Without adjustment for pretreatment differences, language minority children who never received ESL instruction displayed the best approaches to learning at the end of the kindergarten year. The between-group differences disappeared after weighting, $F(5, 2859) = 0.51$, $p = 0.77$. The same result holds even after controlling for child approaches to learning at the baseline, $F(5, 2859) = 0.40$, $p = 0.85$. If the ANOVA result would instead suggest significant between-group mean differences, one would then assess the expected incremental change in the outcome when the treatment dosage increases at each level.

5.3.3   Identification assumptions

MMWS invokes the same weak ignorability assumption stated in (5.12) as does IPTW. Conditioning on a propensity score such as that for bilingual education θ_3, the assignment

Table 5.3 MMWS for a multidosage treatment

Stratum	0		1		2		3		4		5		Total
	$n_{z=0,s}$	MMWS	$n_{z=1,s}$	MMWS	$n_{z=2,s}$	MMWS	$n_{z=3,s}$	MMWS	$n_{z=4,s}$	MMWS	$n_{z=5,s}$	MMWS	n_s
$s=1$	84	3.32	22	1.43	23	0.97	143	0.62	143	0.50	158	0.50	573
$s=2$	45	2.75	17	0.82	15	0.66	64	0.62	56	0.57	57	0.62	254
$s=3$	81	1.53	16	0.87	17	0.58	48	0.83	46	0.69	47	0.75	255
$s=4$	108	1.15	17	0.82	21	0.47	56	0.71	25	1.27	28	1.27	255
$s=5$	201	0.93	26	0.81	12	1.23	56	1.06	42	1.14	45	1.18	382
$s=6$	247	0.75	23	0.91	9	1.64	45	1.32	23	2.08	35	1.52	382
$s=7$	286	0.65	20	1.05	10	1.48	26	2.29	17	2.81	23	2.31	382
$s=8$	343	0.54	16	1.31	4	3.70	8	7.43	6	7.96	5	10.61	382
Total	1395		157		111		446		358		398		2865

Dosage of ESL instruction time

to bilingual education can be viewed as if random if this assignment is independent of the potential outcome associated with bilingual education. The weighted data therefore approximates an experiment that assigns students at random to bilingual education. By the same token, the distribution of the propensity score θ_3 in the weighted bilingual education group will approximate the distribution of this propensity score in the entire sample. In a sufficiently large sample, the distribution of θ_3 in the entire sample approximates its population distribution. The weighted mean observed outcome of the bilingual education group therefore is averaged over the marginal distribution of θ_3 and consistently estimates the population average potential outcome of bilingual education. If a series of four weak ignorability assumptions holds for the assignment to each of the four treatment conditions, the weighted data will approximate an experiment that assigns language minority students at random to the four treatment conditions. The weighted mean difference in the outcome between any two treatment groups therefore estimates the causal effect of one treatment versus the other.

5.4   Summary

This chapter has shown applications of the MMWS method to multivalued treatments measured on an ordinal or nominal scale with the intention of approximating a completely randomized experiment. In general, researchers need to estimate for each unit as many propensity scores as the number of treatment categories and construct a separate set of strata on the basis of each estimated propensity score. One exception is when the treatment categories represent different dosage levels on an underlying continuum and when the association between treatment dosage and each pretreatment covariate is additive. In this special case, stratification on a single balancing score obtained from an ordinal logistic regression becomes sufficient despite the increase of the number of dosage levels.

Because the MMWS method adjusts the pretreatment composition of each treatment group to approximate a randomized experiment under the weak ignorability assumption (Imbens, 2000), a weighted outcome model directly estimates the treatment effects of interest, which allows researchers to apply familiar analytic tools within the ANOVA framework. The analysis typically involves a global test followed by post hoc pairwise comparisons between treatment groups. The precision of the treatment effects estimated from a weighted sample can often be improved by employing ANCOVA-like strategies. In a series of applications evaluating ELL services as described in this chapter, child approaches to learning at kindergarten entry is arguably the strongest predictor of child approaches to learning at the end of kindergarten. Covariance adjustment for the former typically increases the statistical power for detecting treatment effects on the latter. Given that the MMWS adjustment has successfully removed most of the preexisting differences between the treatment groups, possible misspecifications of the functional relationship between the outcome and the covariates in the outcome model will become much less consequential than is the case in ANCOVA and multiple regression.

Similar to IPTW, the MMWS method invokes fewer and weaker identification assumptions than those required by ANCOVA and multiple regression. Both weighting methods are more suitable than propensity score matching and stratification for evaluating multivalued treatments. Moreover, MMWS results are much more robust than IPTW results when the propensity score models are misspecified in functional forms. Nonetheless, causal inferences on the basis of all these methods require that, given the observed covariates, there is no additional confounding of treatment–outcome relationships. The causal validity of the analytic results requires close scrutiny in light of possible violations of the aforementioned assumption. In general, the success of statistical adjustment for the observed covariates depends heavily on the quality of data. The MMWS method is no exception in that regard.

Appendix 5.A: Multiple IV for evaluating multivalued treatments

To evaluate the effects of three different formats of ELL services in contrast with the control condition, we use dummies D_1, D_2, and D_3 to indicate whether a student was assigned to pull-out ESL, in-class ESL, or bilingual education. Suppose that the researcher has obtained three IVs denoted by Z_1, Z_2, and Z_3. Each treatment indicator is to be regressed on all three IVs, while the outcome is to be regressed on the predicted values of the three treatment indicators:

$$D_1 = \alpha_0 + \alpha_1 Z_1 + \alpha_2 Z_2 + \alpha_3 Z_3 + v_1;$$

$$D_2 = \beta_0 + \beta_1 Z_1 + \beta_2 Z_2 + \beta_3 Z_3 + v_2;$$

$$D_3 = \gamma_0 + \gamma_1 Z_1 + \gamma_2 Z_2 + \gamma_3 Z_3 + v_3;$$

$$Y = \delta_0 + \delta_1 \hat{D}_1 + \delta_2 \hat{D}_2 + \delta_3 \hat{D}_3 + e.$$

The reduced form of the outcome model is

$$Y = \delta_0 + \delta_1 \hat{D}_1 + \delta_2 \hat{D}_2 + \delta_3 \hat{D}_3 + e$$
$$= \delta_0 + \delta_1 (\alpha_0 + \alpha_1 Z_1 + \alpha_2 Z_2 + \alpha_3 Z_3) +$$
$$\delta_2 (\beta_0 + \beta_1 Z_1 + \beta_2 Z_2 + \beta_3 Z_3) +$$
$$\delta_3 (\gamma_0 + \gamma_1 Z_1 + \gamma_2 Z_2 + \gamma_3 Z_3) + e$$
$$= \lambda_0 + \lambda_1 Z_1 + \lambda_2 Z_2 + \lambda_3 Z_3 + e.$$

The causal effects of interest are identified by δ_1, δ_2, and δ_3 that can each be derived as a function of the coefficients estimated from the three treatment models and those from the reduced form outcome model. The causal effects are just identified because we can solve the following three simultaneous equations for δ_1, δ_2, and δ_3:

$$\begin{cases} \delta_1 \alpha_1 + \delta_2 \beta_1 + \delta_3 \gamma_1 = \lambda_1 \\ \delta_1 \alpha_2 + \delta_2 \beta_2 + \delta_3 \gamma_2 = \lambda_2 . \\ \delta_1 \alpha_3 + \delta_2 \beta_3 + \delta_3 \gamma_3 = \lambda_3 \end{cases}$$

References

Austin, P.C. (2011). Optimal caliper widths for propensity-score matching when estimating differences in means and differences in proportions in observational studies. *Pharmaceutical Statistics, 10*(2), 150–161.

Citrome, L., Reist, C., Palmer, L. *et al.* (2009). Impact of real-world ziprasidone dosing on treatment discontinuation rates in patients with schizophrenia or bipolar disorder. *Schizophrenia Research, 115*(2–3), 115–120.

Cochran, W.G. (1957). Analysis of covariance: its nature and uses. *Biometrics, 13*(3), 261–281.

Hamilton, G., United States, Administration for Children and Families *et al.* (2001). *How Effective Are Different Welfare-to-Work Approaches: Five-Year Adult and Child Impacts for Eleven Programs*, U.S. Department of Health and Human Services, and U.S. Department of Education, Washington, DC.

Heaviside, S. and Farris, E. (1993). *Public School Kindergarten Teachers' Views on Children's Readiness for School*. U.S. Department of Education, Office of Educational Research and Improvement, Washington, DC. National Center for Education Statistics No. 93-410.

Hill, J.L., Waldfogel, J., Brooks-Gunn, J. and Han, W.J. (2005). Maternal employment and child development: a fresh look using newer methods. *Developmental Psychology, 41*(6), 833–850.

Hong, G. (2010). Marginal mean weighting through stratification: adjustment for selection bias in multilevel data. *Journal of Educational and Behavioral Statistics, 35*(5), 499–531.

Hong, G. (2012). Marginal mean weighting through stratification: a generalized method for evaluating multi-valued and multiple treatments with non-experimental data. *Psychological Methods, 17*(1), 44–60.

Hong, G. and Hong, Y. (2009). Reading instruction time and homogeneous grouping in kindergarten: an application of marginal mean weighting through stratification. *Educational Evaluation and Policy Analysis, 31*(1), 54–81.

Huang, I.-C., Frangakis, C., Dominici, F. *et al.* (2005). Approach for risk adjustment in profiling multiple physician groups on asthma care. *Health Services Research, 40*, 253–278.

Imai, K. and van Dyke, D.A. (2004). Causal inference with general treatment regimes: generalizing the propensity score. *Journal of the American Statistical Association, 99*(467), 854–866.

Imbens, G.W. (2000). The role of the propensity score in estimating dose–response functions. *Biometrika, 87*, 706–710.

Imbens, G.W. (2004). Nonparametric estimation of average treatment effects under exogeneity: a review. *Review of Economics and Statistics, 86*(1), 4–29.

Joffe, M.M. and Rosenbaum, P.R. (1999). Invited commentary: propensity scores. *American Journal of Epidemiology, 150*(4), 327–333.

Ladd, G.W., Birch, S.H. and Buhs, E.S. (1999). Children's social and scholastic lives in kindergarten: related spheres of influence? *Child Development, 70*(6), 1373–1400.

Lu, B., Zanutto, E., Hornik, R. and Rosenbaum, P.R. (2001). Matching with doses in an observational study of a media campaign against drug abuse. *Journal of the American Statistical Association, 96*(456), 1245–1253.

Robins, J.M. (1999). Marginal structural models versus structural nested models as tools for causal inference. In M.E. Halloran and D. Berry (Eds.), *Statistical Models in Epidemiology, the Environment, and Clinical Trials* (pp. 95–134). New York: Springer.

Rosenbaum, P.R. and Rubin, D.B. (1983). The central role of the propensity score in observational studies for causal effects. *Biometrika, 70*, 41–55.

Rubin, D.B. (1997). "Estimating causal effects from large data sets using propensity score," *Annals of Internal Medicine, 127*, 757–763.

Rubin, D.B. and Thomas, N. (2000). Combining propensity score matching with additional adjustments for prognostic covariates. *Journal of the American Statistical Association*, 95(450), 573–585.

Tiihonen, J., Lönnqvist, J., Wahlbeck, K. *et al.* (2006). Antidepressants and the risk of suicide, attempted suicide, and overall mortality in a nationwide cohort. *Archives of General Psychiatry*, 63, 1358–1367.

Zanutto, E., Lu, B. and Hornik, R. (2005). Using propensity score subclassification for multiple treatment doses to evaluate a national antidrug media campaign. *Journal of Educational and Behavioral Statistics*, 30(1), 59–73.

Zhu, B., Kulkarni, P.M., Stensland, M.D. and Ascher-Svanum, H. (2007). Medication patterns and costs associated with olanzapine and other atypical antipsychotics in the treatment of bipolar disorder. *Current Medical Research and Opinion*, 23(11), 2805–2814.

Part II

MODERATION

6

Moderated treatment effects: concepts and existing analytic methods

This section of the book includes three chapters devoted to the discussion of "moderated treatment effects." This chapter reveals the conceptual confusion in the past literature and aims to clarify the definitions of "moderated treatment effects." It then reviews the existing research designs and analytic methods for investigating explicit and implicit moderators. Chapters 7 and 8 extend the marginal mean weighting through stratification (MMWS) method to moderation studies with quasiexperimental data. Chapter 7 focuses on MMWS estimation of subpopulation-specific treatment effects in which identifiers for subpopulation membership serve as moderators. The chapter also examines cases in which the effect of one treatment is moderated by a second concurrent treatment. In addition, the joint effects of two concurrent treatments may vary across subpopulations or across organizational contexts. Chapter 8 focuses on treatments delivered in a temporal sequence. When the effect of an earlier treatment depends on a later treatment, each treatment moderates the effect of the other. The search for an optimal treatment sequence becomes particularly relevant. The chapter explains challenges that arise in evaluating the cumulative effects of time-varying treatments and shows how to use MMWS to reduce selection bias associated with time-varying covariates. Such covariates are intermediate outcomes of the earlier treatments and may confound the relationship between the later treatments and the outcome. Moreover, what is optimal for one subpopulation may not be optimal for another. The chapter discusses strategies for identifying subpopulation-specific optimal treatment sequences.

6.1 What is moderation?

Moderation becomes a focus in social sciences when the researcher is interested not only in average treatment effects but also in the possible heterogeneity of treatment effects. Such heterogeneity typically has important implications for theoretical generalizability. For example, it has been found across a wide range of populations that uncontrollable negative life events

Causality in a Social World: Moderation, Meditation and Spill-over, First Edition. Guanglei Hong.
© 2015 John Wiley & Sons, Ltd. Published 2015 by John Wiley & Sons, Ltd.

such as illness, injuries, or parent unemployment contribute to adolescents' psychological distress; yet some adolescents appear to be more vulnerable than others to life stress (Ge, Conger, and Elder, 2001; Ge et al., 1994). This has led to a search for protective factors that may buffer the detrimental impact of negative life events for adolescents (Cohen and Wills, 1985). It has been found, for example, that girls are more vulnerable than boys to the impacts of negative life events; yet support from parents, peers, and school can mitigate the impact of such events. In addition, positive life events have been found to alleviate the impact of negative life events (Cohen and Hoberman, 1983; DuBois et al., 1992; Ystgaard, Tambs, and Dalgard, 1999). Interestingly, some studies have suggested that the buffering role of family environment appeared to be more pronounced for adolescents with physical disability than for physically normal adolescents. The researchers reasoned that families remaining cohesive when confronted with child disability were better equipped to deal with future uncontrollable negative life events than were families with able-bodied adolescents (Burt, Cohen, and Bjorck, 1988; Murch and Cohen, 1989).

In the above example, an uncontrollable negative life event is the treatment of interest, producing an average effect on the outcome—psychological distress. Among a large number of moderators examined in the past research, one type of moderator defines *subpopulations* of individuals (e.g., boys vs. girls) who may respond differently to the treatment. The second type of moderator characterizes organizational or social–cultural *contexts* (e.g., family environment) that may shape individual responses to the treatment. The third type of moderator includes *concurrent or consecutive treatments* (e.g., positive life events) that may change individual responses to the focal treatment. The importance of a contextual moderator (e.g., family cohesion) may differ across subpopulations defined by another moderator (e.g., normal vs. disabled adolescents).

Chapter 2 has clarified that defining, identifying, and estimating a causal effect are three distinct tasks in causal inference. In this chapter, after revealing past controversies in moderation research, Section 6.1.2 defines moderated treatment effects in terms of potential outcomes for various types of moderators. A causal effect is identifiable only if the counterfactual quantities can be equated with observable data under certain identifying assumptions. Section 6.2 reviews a number of experimental designs including randomized block designs, factorial designs, and multisite randomized trials that render identification assumptions plausible. The same section discusses the corresponding analytic models for experimental data.

6.1.1   Past discussions of moderation

6.1.1.1   Purpose of moderation research

Baron and Kenny (1986) generally defined moderation as existing when a moderator, either qualitative or quantitative, "affects the direction and/or strength of the relation between an independent or predictor variable and a dependent or criterion variable" (p. 1174). This definition was generated within a correlational analysis framework in which a moderator affects the zero-order correlation between two other variables. The predictor is not necessarily a treatment; and the relation between the predictor and the outcome is not necessarily causal in nature. In fact, some of the earliest moderation studies were primarily concerned with the utility of psychometric measurement such as using aptitude to predict job performance. The researchers intended to discriminate the more predictable individuals from the less predictable ones (Ghiselli, 1956). This tradition has continued in psychometrics with an emphasis on differentiating measurement models for different subpopulations of individuals.

Others have argued that even though the moderator is not necessarily a cause of the predictor or of the outcome, a causal relationship may be moderated (Rose *et al.*, 2004; Stolzenberg, 1980). Wu and Zumbo (2008) made the strong claim that the term "moderation effect" should be reserved for models in which the relationship between the predictor and the outcome is causal in nature.

In risk research focusing on identifying risk factors for the purpose of prevention, a moderator modifies the relationship between a risk factor and the outcome. Studies examining the impacts of negative life events on psychological distress are examples from this field. Questions about moderation are also prevalent in evaluation research. In addition to knowledge about whether a particular intervention has an effect on average, knowledge about for whom the intervention works and under what conditions it works can be particularly valuable. Judd, Kenny, and McClelland (2001) gave an example that students who are taught with a new curriculum might perform higher on a subsequent test than students taught with an old curriculum. The performance difference, however, "might be larger or smaller for different types of students or in different types of classrooms or when taught by different kinds of teachers. All of these then are potential moderators of the treatment effect" (p. 115). Kraemer and colleagues (2002) further elaborated the rationale for moderation studies in the field of psychiatric research. Treatment moderators suggest to clinicians which patients might be most responsive to a particular treatment and under what circumstances the treatment might have the most clinically significant effects even if the total effect of the treatment on the outcome is zero. Moreover, in randomized clinical trials (RCTs), understanding treatment moderators may help investigators clarify how to classify patients, which patients to include, and how to improve the circumstances for the treatment in subsequent studies.

Consistent with Baron and Kenny's approach, many researchers (Abrahams and Alf, 1972; Fairchild and MacKinnon, 2009; Holmbek, 1997; James and Brett, 1984) have formulated moderation typically in an analysis of variance (ANOVA) or linear regression framework. When the theoretical relationship between a predictor and an outcome is hypothesized to vary according to the level or value of the moderator, a moderator model specifies this relationship as a linear function of a third variable. For example, let δ_1 denote the relationship between negative life events and psychological distress for girls and let δ_0 denote the relationship for boys. If negative life events result in greater distress for girls than for boys, that is, if $\delta_1 > \delta_0$, then gender is a moderator and $\delta_{1-0} = \delta_1 - \delta_0$ is its moderating effect. The moderated relationship in this theoretical model can be represented in a linear form:

$$\delta_0 + \delta_{1-0} \times \text{Gender}$$

where Gender is an indicator taking value 1 for girls and 0 for boys. In this simple case, a theoretical model would represent the outcome as a linear function of the treatment, gender, and the cross-product of treatment and gender; hence, δ_{1-0} is also called the treatment-by-gender interaction effect. However, the moderated relationship does not have to be linear. Baron and Kenny (1986) presented a number of cases in which the moderator is continuous and the relationship between the predictor and the outcome is a nonlinear function of the moderator.

A major campaign searching for moderated treatment effects was launched in the field of educational psychology about half a century ago. In his 1957 presidential address at the American Psychological Association annual meeting, Lee Cronbach called for investigations of aptitude–treatment interactions (ATIs), that is, how treatment effects differ by individual

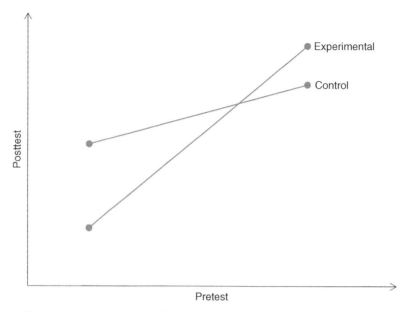

Figure 6.1   A hypothetical example of aptitude–treatment interaction

characteristics. A typical ATI study includes a comparison of two or more alternative treatments and one or more psychological variables. A comparison between alternative treatments is made for individuals at different levels of a psychological variable such as aptitude (Bracht, 1970). Since then, many psychologists have looked for evidence for ATIs that can be graphically represented as shown in Figure 6.1. Here, the hypothetical aptitude–outcome relationship represented by a regression line under one treatment condition crosses the relationship in the alternative treatment condition. The vertical distance between the two regression lines at a given level of aptitude is viewed as the treatment effect for individuals with that particular level of aptitude. In this hypothetical example, the treatment effect is positive for students with a high aptitude and is negative for those with a low aptitude. A case is then made for developing alternative instructional programs for students with different aptitudes so as to optimize educational payoff.

6.1.1.2   What qualifies a variable as a moderator?

In practice, there are numerous examples in which the term "moderation" has been misused and often confused with the term "mediation" (Holmbeck, 1997). In order to draw a clear line of demarcation between "moderators" and "mediators," some researchers have placed additional constraints on whether a variable can be viewed as a moderator. Following Abrahams and Alf (1972), James and Brett (1984) stated the desirability of having "minimal covariation between the moderator and both the independent and dependent variables" and contrasted this with the desirability of having "high degrees of covariation between the mediator and both the antecedent(s) and consequence(s)" (p. 310). More recently, Kraemer and colleagues (2008, 2001) proposed a set of eligibility criteria for determining whether a variable could be a potential moderator. This so-called MacArthur approach stipulates that, in risk research,

(i) a moderator must precede the risk factor and that (ii) these two variables must be independent. These researchers argued that the imposition of temporal precedence is necessary for determining whether variable A moderates the effect of variable B on outcome Y or whether, on the contrary, B moderates the effect of A on Y. To identify moderators of treatment effects on outcomes in RCTs, Kraemer and colleagues (2002) again emphasized that the moderator must be a baseline characteristic and hence by definition is uncorrelated with the random treatment assignment yet must have an interactive effect with the treatment on the outcome.

The aforementioned two sets of eligibility criteria are inconsistent in that James and Brett (1984) did not require temporal precedence of the moderator, while the "MacArthur approach" said nothing about a minimal association between the moderator and the outcome. Kraemer and her colleagues (2002) acknowledged that there is no consensus among methodologists with regard to these eligibility criteria. When any one or all three of these criteria are applied, however, many of the theoretically meaningful questions about moderation would become illegitimate. For example, gender might not qualify as a moderator of the effect of negative life events on adolescent psychological distress because its covariation with the outcome is more than "minimal." In fact, it is known that girls display a higher level of depression than boys on average (Ge, Conger, and Elder, 2001). Although scientific questions often arise regarding whether the effect of one treatment depends on the co-occurrence of another treatment, positive life events would not qualify as a moderator if independently happening at about the same time as the negative life events due to the violation of temporal precedence. In medical treatments of chronic disorder, the effect of initial treatment decisions often depends on subsequent decisions at later stages (Lavori and Dawson, 2008). Such moderation questions have also been ruled out by the aforementioned criteria. Finally, in many clinical trials, the investigator often deliberately sets the probability of treatment assignment to be a function of a potential moderator such as disease severity in order to maximize compliance. Even though the moderator and the treatment are clearly associated by design in these experiments, the association does not nullify the causal question about the moderated treatment effect. Hence, I conclude that these eligibility criteria are arbitrary and could potentially impede the development of scientific understanding.

6.1.2   Definition of moderated treatment effects

The previous chapters have defined the causal effect of a treatment on an outcome in terms of the potential outcomes associated with alternative treatment conditions. A moderated treatment effect is to be defined according to the nature of the moderator. We have discussed examples in which the moderator is a subpopulation identifier, a contextual characteristic, a concurrent treatment, or a preceding or succeeding treatment. In the following, we define the moderated treatment effect in each case except for a preceding or succeeding treatment, which will be defined and discussed at great length in Chapter 8.

6.1.2.1   Treatment effects moderated by individual or contextual characteristics

As before, we use Z to denote the treatment indicator that takes value 1 if an individual is assigned to the experimental condition and 0 if the individual is assigned to the control condition. The potential outcome that the individual would display under the experimental condition is denoted by $Y(1)$; the same individual, if assigned to the control condition, would display the potential outcome $Y(0)$. Hence, the individual-specific treatment effect can be defined as the difference between the two potential outcomes $Y(1)-Y(0)$.

In asking whether the treatment effect is moderated by a certain individual characteristic such as gender, the researcher is primarily interested in the treatment effects for different subpopulations defined by gender. Moderation occurs when, for example, the treatment effect for the subpopulation of girls is different from that for the subpopulation of boys. In the following, we use R to denote a moderator. In the current example, $R=1$ indicates a girl, and $R=0$ a boy.

The population average treatment effect has been represented as the expected value of the individual-specific treatment effect $E[Y(1)-Y(0)]$ where $E[\cdot]$ denotes the marginal mean of a random variable for the entire population. The average treatment effect for the subpopulation of girls is represented as the conditional expected value of the individual-specific treatment effect, that is, the mean treatment effect conditioning on gender being girl:

$$E[Y(1)-Y(0)|R=1].$$

Here, $E[\cdot|R=1]$ denotes the conditional mean of a random variable; the condition statement to the right of "|" defines the subpopulation of interest. Similarly, the average treatment effect for the subpopulation of boys is represented as

$$E[Y(1)-Y(0)|R=0].$$

The moderated treatment effect in this example is the between-gender difference in the average treatment effect:

$$E[Y(1)-Y(0)|R=1]-E[Y(1)-Y(0)|R=0]. \tag{6.1}$$

The treatment is said to have differential effects for boys and girls if (6.1) is nonzero.

The effect of a treatment may depend on the context in which the treatment is received. A contextual moderator may characterize distinct cultures, institutional history, organizational structure or climate, demographic composition of a community, or the density of a social network. A contextual moderator may also be defined by characteristics of the agents who deliver the treatment. For example, in comparing between an innovative curriculum denoted by $Z=1$ and a standard curriculum denoted by $Z=0$, one may hypothesize that the treatment effect depends on the quality of the teacher who delivers the curriculum. Let $R=1$ if the teacher has high quality and 0 otherwise. The treatment effect may be maximized if the students are taught by a high-quality teacher rather than by a low-quality teacher, that is,

$$E[Y(1)-Y(0)|R=1]-E[Y(1)-Y(0)|R=0]>0.$$

Even though these definitions have been given for the simple case of a binary treatment and a binary moderator, they can be extended easily to multivalued treatments and multivalued moderators.

6.1.2.2 Joint effects of concurrent treatments

Meta-analyses have often reported mixed findings with regard to the causal impact of a given treatment. This tends to be true even when the analysts attempt to summarize the results from a collection of studies of similar subpopulations. An important question to ask is whether the treatment was administered jointly with other concurrent treatments and whether the effect of the focal treatment depends on what other treatments are in place at the same time.

Grouping students by ability within an elementary class for differentiated literacy instruction is a practice that has been examined, challenged, debated, and reexamined for decades. Extensive reviews and meta-analyses have generated results ranging from negative to positive estimates of the overall effects of within-class homogeneous grouping versus no such grouping (Kulik and Kulik, 1987; Lou, Abrami, and Spence, 2000; Lou et al., 1996; Slavin, 1987). Most of the past researchers, however, have singled out ability grouping as an isolated practice without taking into account its mutual contingency with other concurrent instructional decisions. Hong and Hong (2009) argued that instruction must be viewed as a multifaceted intervention program because every single element needs to operate in concert with other parts of the program to produce a joint impact. The effects of various elements therefore cannot be assumed to be additive. They reasoned that the impact of grouping might depend on literacy instruction time which they considered to be a concurrent treatment for kindergartners. This was because vast differences in literacy instruction time exist between kindergarten classes. According to Hong and Hong (2009), in classes that have allocated only a limited amount of time to literacy instruction and intensively use homogeneous grouping, a relatively large proportion of the instructional time is perhaps spent on transitioning between different instructional activities and on managing student behavior, with less time left for every student to have direct instructional interactions with the teacher. Hence, the time constraint will likely offset the potential benefit of homogeneous grouping. Student learning will likely be optimized if the student receives a substantial amount of literacy instruction time and if homogeneous grouping sustains the student's engagement in meaningful learning tasks matched to his or her current ability.

In Hong and Hong's (2009) study, kindergarten classes that devoted more than 1 hour/day to literacy instruction were contrasted with those devoting at most 1 hour/day. High-intensity grouping was defined as spending nearly half of the literacy instruction time in homogeneous groups, which was contrasted with low-intensity grouping and with no grouping. To examine literacy instruction time as a potential moderator of the grouping impact, the researchers investigated (i) the effects of high-intensity grouping in comparison with low-intensity grouping or no grouping on kindergartners' literacy growth when a relatively high amount of time was allocated to literacy instruction, (ii) the effects of grouping under a relatively low amount of literacy instruction time, and (iii) whether the grouping effects changed as literacy instruction time increased.

In this study, a student's potential outcomes are each a function of two concurrent treatments—time and grouping. We use random variable Z_1 to denote the first treatment and let $z_1 = 1$ if a student received a high amount of literacy instruction time and $z_1 = 0$ otherwise; we use Z_2 to denote the second treatment and let $z_2 = 0, 1, 2$ represent no grouping, low-intensity grouping, and high-intensity grouping, respectively. With six possible combinations of time and grouping, every kindergartner may have six potential outcomes. As shown in Table 6.1, the potential outcomes are denoted by $Y(z_1, z_2)$. For example, $Y(0, 0)$ is a student's potential outcome if the kindergarten class that the student was attending allocated a relatively low amount of time to literacy instruction and never used grouping. The last column in Table 6.1 lists the causal effects of low-intensity grouping and high-intensity grouping as opposed to no grouping under each level of literacy instruction time. Specifically, $Y(0,1) - Y(0,0)$ defines the causal effect of low-intensity grouping versus no grouping, while

Table 6.1   Potential outcomes and joint effects of two concurrent treatments

Time, Z_1	Grouping, Z_2	Potential outcome, $Y(z_1, z_2)$	Causal effect
Low amount, $z_1 = 0$	No, $z_2 = 0$	$Y(0, 0)$	$Y(0,1) - Y(0,0)$
	Low intensity, $z_2 = 1$	$Y(0, 1)$	$Y(0,2) - Y(0,0)$
	High intensity, $z_2 = 2$	$Y(0, 2)$	
High amount, $z_1 = 1$	No, $z_2 = 0$	$Y(1, 0)$	$Y(1,1) - Y(1,0)$
	Low intensity, $z_2 = 1$	$Y(1, 1)$	$Y(1,2) - Y(1,0)$
	High intensity, $z_2 = 2$	$Y(1, 2)$	

$Y(0,2) - Y(0,0)$ defines the causal effect of high-intensity grouping versus no grouping under a low amount of time. In parallel to these, $Y(1,1) - Y(1,0)$ and $Y(1,2) - Y(1,0)$ define the two grouping effects under a high amount of time. The moderation exists if the effect of low-intensity grouping or high-intensity grouping versus no grouping depends on time. It was hypothesized that the potential benefit of grouping would become manifest only under a high amount of time and that high-intensity grouping would be especially detrimental under a low amount of time, that is,

$$E[Y(1,1) - Y(1,0)] - E[Y(0,1) - Y(0,0)] > 0;$$
$$E[Y(1,2) - Y(1,0)] - E[Y(0,2) - Y(0,0)] > 0. \tag{6.2}$$

The causal effect of grouping and that of time are said to be additive if each of the two grouping effects is a constant regardless of the time level.

This section has distinguished two types of moderation in causal inference: treatment effects moderated by individual or contextual characteristics and treatment effects moderated by a concurrent treatment. We can see a key difference between these two types of moderated treatment effects. When the effect of a single treatment indicator Z is moderated by a pretreatment measure R, the potential outcomes are functions of the values of Z only, while R defines the subpopulations across which the treatment effects are to be compared. In contrast, when a concurrent treatment Z_1 moderates the effect of a focal treatment Z_2, the potential outcomes are functions of the values of both Z_1 and Z_2. For a given individual, the pretreatment characteristic $R = r$ is fixed, while the concurrent treatment Z_1 may possibly take different values. For example, the researcher cannot easily change someone's gender but can conceivably manipulate the time allocation to literacy instruction.

6.2   Experimental designs and analytic methods for investigating explicit moderators

When the moderators have been explicitly hypothesized, corresponding to the two types of moderated treatment effects defined in the previous section, different experimental designs have been developed to identify the causal effects under assumptions that are often warranted by the designs. Section 6.2.1 discusses randomized block designs suitable for investigating the moderation of treatment effects by individual or contextual pretreatment characteristics.

Factorial designs are prevalent in psychological research for examining the joint effects of two or more concurrent treatments, which will be discussed in Section 6.2.2. We will clarify the identification assumptions and explain the analytic methods applicable to each of these designs.

6.2.1 Randomized block designs

Randomized block designs were initially developed to reduce chance error in evaluating treatment effects. Each block is composed of individual units as homogeneous as possible in their response to each treatment condition such that the variation in each potential outcome is accounted for as much as possible by between-block differences. For example, a sample of students may be divided into blocks according to their pretest scores. The treatment is then assigned at random to individual units within each block. If the pretest strongly predicts the potential outcomes, such blocking will increase the statistical precision of the experiment. A matched-pair design may be viewed as a special case of a randomized block design in which each matched pair is a block.

Generalized randomized block designs are employed not only for reducing error but, more importantly, for testing the generalizability of the inference with regard to the treatment effects (Hinkelmann and Kempthorne, 1994). When the experimenter suspects that the treatment may have differential effects for different subpopulations, these subpopulations may be deliberately introduced as blocks. The treatment is then randomized independently within each subpopulation. The between-subpopulation variation in treatment effect can be separated from the random error variation in a generalized randomized block design but cannot in a matched-pair design.

A study by Marshall (1969), representative of many other studies under the ATI framework, examined the educational aspect of a child's social environment, such as parent education, as a potential moderator of the relationship between mode of reinforcement and kindergartners' task performance. The modes of reinforcement to be contrasted included immediate verbal feedback and delayed verbal feedback. The outcome was the number of trials a child took before reaching the criterion of a fixed number of consecutive correct responses. The researcher characterized every child's educational environment (EE) to be at either a high level or a low level. Within each of these two blocks, children were randomly assigned to reinforcement conditions. The researcher reported differential effects of reinforcement for high EE children and low EE children. There was no significant difference in the performance of high EE children between the delayed verbal and the immediate verbal conditions. However, low EE children performed considerably better when the adult immediately told them whether they were correct than when the information was given after a delay. In the following, we discuss the assumptions required for identifying the moderated treatment effects and evaluate the plausibility of the assumptions under the randomized block design.

6.2.1.1 Identification assumptions

As before, we use Z to denote the binary treatment and use R to denote the moderator. Let $z = 1$ if a child received immediate verbal feedback and $z = 0$ if the child received delayed verbal feedback. Hence, $Y(1)$ and $Y(0)$ denote a kindergartners' task performance under immediate verbal feedback and delayed verbal feedback, respectively. We use $R = 1$ to indicate high EE children and $R = 0$ for low EE children. The treatment effect moderated by kindergartners' EE background takes a similar form as that defined in (6.1) and can be identified if the treatment effect within each EE subpopulation is identified. The identification assumption states that, within each subpopulation, the treatment assignment is independent of the potential outcomes $Y(1)$ and $Y(0)$. We can represent this assumption in terms of mean independence as follows for the high EE children:

$$
\begin{aligned}
E[Y(1)|R=1] &= E[Y(1)|Z=1,\ R=1] = E[Y|Z=1,\ R=1]; \\
E[Y(0)|R=1] &= E[Y(0)|Z=0,\ R=1] = E[Y|Z=0,\ R=1].
\end{aligned}
\tag{6.3}
$$

Under this assumption, the mean observed outcome of high EE children assigned to immediate verbal feedback $E[Y|Z=1,\ R=1]$ is unbiased for the subpopulation average potential outcome associated with immediate verbal feedback $E[Y(1)|R=1]$ should the entire subpopulation of high EE children have been assigned to receive immediate verbal feedback. Similarly, the mean observed outcome of high EE children assigned to delayed verbal feedback $E[Y|Z=0,\ R=1]$ is unbiased for the subpopulation average potential outcome associated with delayed verbal feedback $E[Y(0)|R=1]$. By the same token, for low EE children, the observed mean outcome of those assigned to immediate verbal feedback $E[Y|Z=1,\ R=0]$ and the observed mean outcome of those assigned to delayed verbal feedback $E[Y|Z=0,\ R=0]$ are unbiased for the corresponding subpopulation average potential outcomes $E[Y(1)|R=0]$ and $E[Y(0)|R=0]$.

We therefore derive the following identification result for the moderated treatment effect:

$$
\begin{aligned}
&E[Y(1)-Y(0)|R=1] - E[Y(1)-Y(0)|R=0] \\
&= \{E[Y(1)|R=1]-E[Y(0)|R=1]\} - \{E[Y(1)|R=0]-E[Y(0)|R=0]\} \\
&= \{E[Y|Z=1,\ R=1]-E[Y|Z=0,\ R=1]\} - \{E[Y|Z=1,\ R=0]-E[Y|Z=0,\ R=0]\}.
\end{aligned}
\tag{6.4}
$$

In the randomized block design conducted by Marshall (1969), the randomization of treatment within each block guaranteed that the treatment assignment was independent of the potential outcomes for children in a given block. The estimated treatment effects within each block (i.e., the sample estimates of $E[Y|Z=1,\ R=1]-E[Y|Z=0,\ R=1]$ and $E[Y|Z=1,\ R=0]-E[Y|Z=0,\ R=0]$) were unbiased. Hence, the estimation of the moderated treatment effect (i.e., the estimated difference in the treatment effect between the two blocks) was also unbiased.

6.2.1.2   Two-way ANOVA

Chapter 5 has introduced one-way ANOVA for analyzing data obtained from a completely randomized design. One-way ANOVA assesses the variation in the mean observed outcome between the treatment groups relative to the within-group variation in the outcome and thereby determines the probability of obtaining the sample results under the null hypothesis that the population average observed outcomes are equal across different treatment conditions. Under the general framework of ANOVA, two-way ANOVA has been typically employed for analyzing data from randomized block designs investigating whether the treatment effect is moderated by the blocks.

When applied to data from a randomized block design, a two-way ANOVA computes (i) the between-treatment variation conditioning on block membership, (ii) the between-block variation conditioning on treatment group membership, (iii) the remaining variation between treatment-by-block cells, and (iv) the variation within treatment-by-block cells due to the random sampling error. In the study by Marshall (1969), the between-treatment variation in the outcome was considerably greater than the variation of random sampling error according to an F value representing the ratio of (i) to (iv), indicating a statistically significant "main effect" (i.e., average effect) of the treatment favoring immediate verbal feedback over delayed verbal feedback. The main effect of block membership in this study was also statistically significant according to an F value representing the ratio of (ii) to (iv), indicating a clear advantage of the high EE children over the low EE children. After the main effect of the treatment and the main effect of block membership have been accounted for, the remaining variation within treatment-by-block cells reveals the extent to which the treatment effect is moderated by block membership. In this example, the F value representing the ratio of (iii) to (iv) is sufficiently large to indicate a statistically significant treatment-by-block interaction effect. Note that when the identification assumption (6.3) is met, only the main effect of the treatment and the treatment-by-block interaction effect are causal in nature; the main effect of block membership is purely descriptive indicating the mean difference in the observed outcome between subpopulations conditioning on treatment assignment.

In the two-way ANOVA framework, block membership does not have to be a fixed factor. If the experimenter has created more than a few blocks, modeling the random effects rather than the fixed effects of these blocks is sometimes preferred especially when the sample size within a block becomes relatively small and constrains the precision of the block-specific treatment effect estimate. Multisite randomized trials are a special type of randomized block design in which the sites serve as blocks. Data from multisite randomized trials are typically analyzed via mixed-effects models involving fixed effects of treatment and random effects of site membership. We will discuss such models in Section 6.3.1.

6.2.1.3   Multiple regression

Multiple regression provides an alternative way of analyzing experimental data from a randomized block design. When the treatment indicator Z and the block indicator R are both binary, one may simply regress the observed outcome Y on Z, R, and their cross-product ZR:

$$Y = \beta_0 + \beta_1 Z + \beta_2 R + \beta_3 ZR + \varepsilon.$$

In the current example, $\beta_0 = E(Y|Z=0, R=0)$ is the mean outcome of the low EE children receiving delayed verbal feedback; $\beta_1 = E(Y|Z=1, R=0) - E(Y|Z=0, R=0)$ is the mean difference in the outcome between low EE children receiving immediate verbal feedback and low EE children receiving delayed feedback; $\beta_2 = E(Y|Z=0, R=1) - E(Y|Z=0, R=0)$ is the mean difference in the outcome between high EE children and low EE children receiving delayed verbal feedback; $\beta_1 + \beta_3 = E(Y|Z=1, R=1) - E(Y|Z=0, R=1)$ is the mean difference in the outcome between high EE children receiving immediate verbal feedback and high EE children receiving delayed feedback. Because the treatment was randomized within each block, assumption (6.3) is met. Hence, β_1 is unbiased for the treatment effect for low EE children, while $\beta_1 + \beta_3$ is unbiased for the treatment effect for high EE children. Applying the identification result in (6.4), we can interpret β_3 as the moderating impact of children's EE background on the treatment effect.

Some researchers (Aiken and West, 1991; Kraemer *et al.*, 2008) have proposed that to reduce multicollinearity between each of the predictors Z and R and the product ZR in the multiple regression model, Z and R should be centered at their respective means before they are entered into the regression model. Let $\bar{Z}$ denote the sample mean of Z and let $\bar{R}$ denote the sample mean of R. The regression model then becomes

$$Y = \alpha_0 + \alpha_1 (Z - \bar{Z}) + \alpha_2 (R - \bar{R}) + \alpha_3 (Z - \bar{Z})(R - \bar{R}) + \varepsilon.$$

Here, α_0 is the population average outcome; α_1 is the average effect of treatment assignment conditioning on block membership; and α_2 is the average effect of block membership conditioning on treatment. Therefore, α_0, α_1, and α_2 are generally different from β_0, β_1, and β_2. Nonetheless, α_3 is the treatment effect moderated by block membership and remains equivalent to β_3.

6.2.2   Factorial designs

We made a case earlier for examining the joint effects of two concurrent treatments. The example involved a focal treatment (within-class homogeneous grouping denoted by Z_2) and a concurrent treatment (literacy instruction time denoted by Z_1). To examine whether time moderates the effect of grouping on students' literacy growth as hypothesized in (6.2), in an ideal world, one would design a factorial randomized experiment with two factors. As shown in Table 6.1, factor Z_1 has two levels, a high amount of time ($z_1 = 1$) and a low amount of time ($z_1 = 0$); factor Z_2 contains three levels, no grouping ($z_2 = 0$), low-intensity grouping ($z_2 = 1$), and high-intensity grouping ($z_2 = 2$). The two-by-three combinations define six distinct treatment conditions. Correspondingly, each individual student has six potential outcomes listed in the last column of Table 6.1. In a factorial design, the experimenter would assign kindergarten classes at random to one of these six treatment conditions.

6.2.2.1   Identification assumptions

The assumptions required for identifying the joint effects of two concurrent treatments in a factorial design may seem to be different from those required for identifying block-specific treatment effects in a randomized block design. This is because one's block membership is typically viewed as given, while one's assignment to a concurrent treatment has causal implications. In essence, identification requires that the assignment to each treatment condition must be independent of the corresponding potential outcome such that the average observed outcome of those who are actually assigned to the treatment condition is unbiased for the population average potential outcome associated with that treatment condition.

Let $D_{z_1 z_2}$ be a dummy indicator that takes value 1 if the class that a kindergartner attended was assigned to the treatment condition defined by (z_1, z_2); $D_{z_1 z_2}$ takes value 0 otherwise. For example, if the class was assigned to a high amount of time with high-intensity grouping, that is, $(z_1, z_2) = (1, 2)$, then we have that $D_{12} = 1$ and that $D_{00} = D_{01} = D_{02} = D_{10} = D_{11} = 0$. Due to the fundamental problem of causal inference (Holland, 1986) discussed in Chapter 2, the potential outcome $Y(1, 2)$ can be observed only for students whose classes were actually assigned to a high amount of time with high-intensity grouping and cannot be observed for students whose classes were assigned to any of the other five treatment conditions. Hence, the population average potential outcome associated with a high amount of time with high-intensity grouping $E[Y(1,2)]$ cannot be computed from the observable data.

Yet the factorial design generates data that may allow the researcher to make valid inferences about the population average potential outcomes. In the randomized factorial design described earlier, the mean observed outcome of students whose classes were actually assigned to a high amount of time with high-intensity grouping, denoted by $E[Y|D_{12} = 1]$, is unbiased for the population average potential outcome $E[Y(1,2)]$. This is because the data from such a randomized factorial design satisfies the identification assumption that the assignment to a high amount of time with high-intensity grouping is independent of the potential outcome $Y(1,2)$. Intuitively speaking, under the randomized factorial design, those who were assigned to a given treatment condition and those who were not assigned to this particular treatment condition are "comparable" in how they would have responded if they had all been assigned to this same treatment condition. In general, the identification assumption can be represented as

$$E[Y(z_1,z_2)] = E[Y(z_1,z_2)|D_{z_1 z_2} = 1] = E[Y|D_{z_1 z_2} = 1] \qquad (6.4)$$

for all possible values of z_1 and z_2. Because this assumption is guaranteed by randomization, the moderated treatment effects defined in (6.2) can each be identified. To determine whether the causal effect of low-intensity grouping versus no grouping depends on literacy instruction time, we compute the mean difference in the observed outcome between students assigned to low time with low-intensity grouping $E[Y|D_{01} = 1]$ and those assigned to low time with no grouping $E[Y|D_{00} = 1]$ and subtract this quantity from the mean difference in the observed outcome between students assigned to high time with low-intensity grouping $E[Y|D_{11} = 1]$ and those assigned to high time with no grouping $E[Y|D_{10} = 1]$:

$$
\begin{aligned}
&E[Y(1,1)-Y(1,0)]-E[Y(0,1)-Y(0,0)] \\
&= \{E[Y(1,1)]-E[Y(1,0)]\}-\{E[Y(0,1)]-E[Y(0,0)]\} \\
&= \{E[Y|D_{11} = 1]-E[Y|D_{10} = 1]\}-\{E[Y|D_{01} = 1]-E[Y|D_{00} = 1]\}.
\end{aligned}
$$

Similarly, we can generate empirical evidence with regard to whether the causal effect of high-intensity group versus no grouping differs between a high amount of literacy instruction time and a low amount of time.

Hinkelmann and Kempthorne (1994) discussed a potential limitation of factorial designs: as the number of factors and the number of levels in each factor increase, the number of treatment conditions increases rapidly. For example, to detect a nonlinear trend in the causal effect of literacy instruction time on literacy learning, one might wish to divide the continuous measure of time into three or more levels, which would then be crossed by the three levels of grouping and lead to at least nine treatment conditions. One might also wish to examine the joint effects of more than two treatments. Even though the causal effects are always identifiable in theory under a randomized factorial design, the estimation would become imprecise due to the limit of sample size. Yet the researcher may focus on a subset of all possible treatment conditions on the basis of theory or in a systematic manner as suggested by Hinkelmann and Kempthorne (1994).

The moderation of the causal effect of a focal treatment by a concurrent treatment may differ across subpopulations as well. For example, in evaluating the grouping effects on literacy growth, one may additionally investigate whether the moderating role of literacy instruction time depends on a student's prior ability level. Hong and colleagues (2012) challenged the belief that homogeneous grouping benefits high-ability students at the expense of their low-ability peers. They hypothesized that homogeneous grouping would benefit kindergartners at all ability levels under adequate instructional time and that low-ability children would suffer most from high-intensity grouping when instructional time is limited, suggesting a greater degree of moderation of the grouping effects by time for low-ability students than for medium- and high-ability students. The researchers reasoned that high-intensity grouping under time constraint could make it difficult for the teacher to have extensive interactions with and continually monitor the students concentrated in low-ability groups who are most in need of frequent assistance. The researchers empirically examined the joint effects of time and grouping within each subpopulation defined by students' prior ability and compared the time-by-grouping effects across the subpopulations. Even though the data were quasiexperimental in nature, the researchers made an attempt to approximate a two-by-three factorial design for the time and grouping factors within each of three blocks defined by student prior ability. We will review their adjustment strategy in the next chapter.

6.2.2.2 Analytic strategies

For data from randomized factorial designs evaluating the joint effects of two concurrent treatments, two-way ANOVA can be used to test whether the main effect of each treatment is nonzero and whether the effect of one treatment is contingent upon the other. A general linear model further quantifies each causal effect and its estimation error. The result can be reproduced by a multiple regression analysis when the treatment indicators are appropriately centered. These models take the same form as those discussed in Section 6.2.1.2.

6.3 Existing research designs and analytic methods for investigating implicit moderators

Heterogeneous treatment effects are important to detect in experimental data even if specific moderators of treatment effects are yet to be hypothesized and measured. Exploratory strategies for detecting possible moderations include comparing the outcome variances between the treated group and the control group (Bryk and Raudenbush, 1988; Kim and Seltzer, 2011), assessing between-site variation in the treatment effect in a multisite randomized design (Weiss, Bloom, and Brock, 2013), and evaluating the treatment effect for latent subpopulations of individuals characterized by posttreatment covariates through principal stratification analysis (Frangakis and Rubin, 2002).

In analyzing data from a completely randomized experiment, some researchers have emphasized the importance of comparing the entire distribution of the outcome between the experimental group and the comparison group (Bryk and Raudenbush, 1988; Kim and Seltzer, 2011). In other words, the focus is on not just the mean comparison but also the variance comparison. For example, when comparing the achievement of students receiving ability-grouped instruction with that of students receiving whole-class instruction, if one finds that the average achievement is the same across the two treatment conditions yet the variance of achievement for students receiving ability grouping is larger than the variance for students receiving whole-class instruction, the evidence might suggest that ability grouping increases the achievement gap between higher- and lower-ability students. One might further hypothesize that grouping might have been beneficial to the higher-ability students and detrimental to the lower-ability students. Hence, this type of investigation can be useful at the preliminary stage of a moderation analysis. However, finding equal variance of the outcome across the treatment groups does not rule out the possibility of moderation. This is because a treatment may change the rank order of students; yet the upward mobility for some and the downward mobility for others may lead to a zero sum at the end, leaving the variance of the outcome possibly unchanged.

Although estimating the entire distribution of the individual-specific treatment effect is extremely difficult (Heckman, Urzua, and Vytlacil, 2006), estimating the distribution of the site-specific treatment effect is possible in a multisite randomized trial. Such data may provide particularly useful information as researchers explore the generalizability of causal inference results across a population of sites and search for moderators that may explain between-site heterogeneity in the treatment effect. Section 6.3.1 defines the parameters characterizing the population distribution of site-specific treatment effects. It then clarifies the identification assumptions and compares various strategies for analyzing multisite experimental data.

Section 6.3.2 briefly reviews principal stratification, a distinct framework for defining, identifying, and estimating treatment effects for latent subpopulations of individuals characterized by their intermediate responses to the initial treatment assignment. Because there has been a considerable amount of confusion with regard to whether principal stratification analyses are moderation or mediation analyses, we will revisit this topic again in the part on mediation in Chapter 10.

6.3.1   Multisite randomized trials

Social scientists study individuals who reside in families, neighborhoods, towns, and cities. Many of them are additionally clustered in schools, churches, hospitals, or business firms. These social environments often influence how a treatment is implemented and shape individual responses to the treatment. Even though a treatment is found effective in one location, the result does not necessarily enable the researcher to predict its effectiveness in a different location.

This was the case in the National Evaluation of Welfare-to-Work Strategies (NEWWS) in the United States before these strategies were legislated in the late 1990s. The evaluation involved randomized experiments conducted in seven cities across the country comparing two different welfare programs. The old program entitled welfare recipients to public assistance without working; the new program made active participation in the job market a requirement and provided services and incentives along the way. Within each of the seven experimental sites, thousands of eligible welfare applicants were assigned at random either to an experimental group subjected to the work mandate or to a control group receiving welfare under the old program. Participants in the experimental group generally reported higher earnings than their control group counterparts on average over the 5 years after randomization. Yet there was considerable variation across the seven sites. For example, the program's impact on earnings was as large as $5000 in Portland, Oregon, in contrast with an impact of $1500 in Grand Rapids, Michigan (Hamilton, 2002). Additionally, the new program was found to increase depressive symptoms among participants with young children in Grand Rapids; however, the program's impact on depression was not uniformly positive or negative at the other sites (McGroder *et al.*, 2000). The between-site differences in the treatment impacts, possibly attributable to between-site differences in the welfare population, the local labor market, and the implementation of programs by local welfare offices, were informative to policy-making as the new program started to replace the old program nationwide (Morris, 2008).

A multisite randomized trial, as described earlier, provides suggestive evidence with regard to the generalizability of the treatment effect across the different sites. These sites were deliberately selected to represent a range of settings in which the policy change was to take place. One may view each site as a case study. Even though the total number of sites was limited, the relatively large sample size at each site enabled researchers to compare the treatment effect between two or more sites with distinct features and to explore the causal mechanism within each site.

In some other multisite designs, a relatively large sample of sites is drawn at random from a well-defined population of sites so that the estimated distribution of the treatment effect can be generalized to the entire population. For example, the goal of the Head Start Impact Study (HSIS) commissioned by the US Congress was to determine if access to federally funded Head Start programs caused better developmental and parenting outcomes for participating children and families living in poverty. The study was conducted in 2002 with a nationally representative sample of 84 federal grantee/delegate agencies, 383 randomly selected Head Start centers, and a total of 4667 newly entering 3-year-old and 4-year-old children. At each Head Start center, applicants were assigned at random either to a Head Start group or to a control group that could receive any non-Head Start services chosen by the parents. Even though the site-specific treatment effect may not be estimated with high precision in this case due to the relatively small number of individuals sampled at each site, the estimated distribution of the treatment effect over all the sites may provide valuable information with regard to how effective (or ineffective) a Head Start program could be. Moreover, the data may have adequate power for detecting whether the site-specific treatment effect is associated with certain site characteristics such as demographic composition, program implementation, or other childcare resources available to poor children.

6.3.1.1   Causal parameters and identification assumptions

A major strength of a multisite randomized trial is that the treatment effect can be estimated without bias at each site regardless of whether the number of participants at a site is only a handful as in HSIS or in the thousands as in NEWWS. HSIS involved $j = 1,\dots,$ 383 sites. Let $z = 1$ if an applicant at a given site was assigned to Head Start and 0 otherwise. The site-specific treatment effect for the subpopulation of children applying for Head Start at site j is defined as

$$\delta_j = E[Y(1) - Y(0)|j] = E[Y(1)|j] - E[Y(0)|j].$$

Here, $E[Y(1)|j]$ is the subpopulation average potential outcome if all the applicants at site j would be enrolled in Head Start; $E[Y(0)|j]$ is the subpopulation average potential outcome if all the applicants at site j would be denied access to Head Start. Due to the randomization at site j, the average observed outcome of Head Start children at site j, denoted by $E[Y|Z = 1, j]$, is unbiased for $E[Y(1)|j]$; similarly, the average observed outcome of control children at site j, denoted by $E[Y|Z=0, j]$, is unbiased for $E[Y(0)|j]$. Hence, the random treatment assignment at site j satisfies the identification assumption:

$$E[Y(1)|j] = E[Y(1)|Z=1,j] = E[Y|Z=1,j];$$
$$E[Y(0)|j] = E[Y(0)|Z=0,j] = E[Y|Z=0,j]. \tag{6.5}$$

If the site-specific treatment effect is obtained for every site in the population, one would then have information about the distribution of the treatment effect across all the sites and would be able to determine, in addition to the population average treatment effect, the amount of between-site heterogeneity in the treatment effect. One may further develop hypotheses regarding important site features that may explain why the treatment is more effective at some sites than at others. The population average treatment effect δ is simply the mean of the site-specific treatment effect, denoted by δ_j at site j, over all the sites in the population,

$$\delta \equiv E(\delta_j) = E\{E[Y(1) - Y(0)|j]\}, \tag{6.6}$$

and can be identified, in a multisite randomized trial, by the observable data

$$\delta = E\{E[Y|Z=1,j] - E[Y|Z=0,j]\}.$$

The right hand side of the equation is the average of the site-specific treatment effect identified as the difference between the site mean observable outcome of Head Start children and the site mean observable outcome of control children. The amount of variation in the site-specific treatment effect can be represented as the average squared difference between the treatment effect at site j and the population average treatment effect:

$$\mathrm{Var}(\delta_j) = E\left[(\delta_j - \delta)^2\right], \tag{6.7}$$

the squared root of which is the standard deviation of the population distribution of the site-specific treatment effect.

6.3.1.2   Analytic strategies

The NEWWS experiment and HSIS are two distinct representations of multisite randomized trials. The former features a relatively small number of sites and a large sample at each site; in contrast, the latter has a relatively large number of sites yet a relatively small sample at each site. Researchers typically employ different strategies for analyzing these two types of data. When the focus is on studying between-site heterogeneity in the treatment effect, two-way ANOVA/multiple regression is a suitable option for analyzing NEWWS data, while mixed-effects models rather than fixed-effects models, when used with great caution, can be applied to the HSIS data for estimating $\mathrm{Var}(\delta_j)$. In the following, we discuss each of these options.

Two-way ANOVA and multiple regression
In a multisite randomized trial, the experimental sites can be viewed as blocks within which the treatment is randomized. Accordingly, the data can be analyzed via two-way ANOVA or multiple regression as discussed in Section 6.2.1. In addition to estimating the main effect of

the treatment and the main effect of site membership, the researcher gains important information by examining the treatment-by-site interaction effect. In the NEWWS data, with seven experimental sites and two treatment groups per site generating 14 cells in total, the interaction effect is the difference between a cell mean outcome and the overall mean outcome unexplained by treatment group membership and site membership. If the magnitude of this difference is considerably greater than the amount attributable to sampling variation, the result of an F test will indicate that the treatment effect varies across the sites.

Equivalent to this, one may analyze a multiple regression model after creating dummy indicators D_j for sites $j = 2, \ldots, J$. The regression model represents an individual's observed outcome as a function of treatment assignment, site membership, and treatment-by-site interaction as follows for individual i in site j:

$$Y_{ij} = \pi_1 + \delta_1 Z_{ij} + \sum_{j=2}^{J} \pi_j D_j + \sum_{j=2}^{J} \gamma_j D_j Z_{ij} + \varepsilon_{ij}. \tag{6.8}$$

Here, π_1 is the control group mean and δ_1 the treatment effect at site 1, the reference site in the current setup; π_j for $j = 2, \ldots, J$ is the difference in the control group mean outcome between each of the other $J-1$ sites and the reference site; and γ_j for $j = 2, \ldots, J$ is the difference in the treatment effect between each of the other $J-1$ sites and the reference site. Therefore, $\delta_j = \delta_1 + \gamma_j$ represents the treatment effect at site j. To conduct a global F test for the $J-1$ interaction effects, we compare the complete model specified in (6.8) with an incomplete model that assumes a constant treatment effect across all the sites by omitting the interaction terms:

$$Y_{ij} = \pi_1 + \beta Z_{ij} + \sum_{j=2}^{J} \pi_j D_j + e_{ij}. \tag{6.9}$$

Note that model (6.9) is nested in model (6.8). An F statistic represents the proportion of error variation reduced by including the interaction terms, that is,

$$F = \frac{\left(\text{SSE}_{\text{incomp}} - \text{SSE}_{\text{comp}}\right)/df_1}{\text{SSE}_{\text{comp}}/df_2},$$

where $df_1 = J-1$ is the number of interaction terms that are entered in the complete model (6.8) but not in the incomplete model (6.9), $df_2 = N-2J$ is the degree of freedom in the complete model, and $\text{SSE}_{\text{incomp}}$ and SSE_{comp} are the sum of squares of error in the incomplete model and that in the complete model, respectively.

As we discussed in Section 6.2.1, the researcher may modify model (6.8) by centering the treatment indicator and the site indicators at their respective means. After centering, the coefficient for Z_{ij} is unbiased for the population average treatment effect δ defined in (6.6). Centering the predictors in the incomplete model (6.9), however, does not render β unbiased for δ. Model (6.9) is called a fixed-effects model in the econometrics literature (Angrist and Pischke, 2008; Greene, 2003; Woodridge, 2002, 2006). Appendix 6.A derives the bias in the fixed-effects model parameter when the treatment effect is heterogeneous across the sites.

Mixed-effects models

When the number of sites (or blocks) is relatively large while the sample size per site (or per block) is relatively small, researchers have often employed mixed-effects models—also called random-effects models, multilevel models, or hierarchical linear models. The models may be written separately for individuals at level 1 and sites at level 2. The level 1 model is a within-site model that includes a site-specific random intercept A_j representing the average outcome of the control group in site j and a site-specific random treatment effect B_j. The within-site random error r_{ij} is typically assumed to have mean zero and variance σ^2:

$$Y_{ij} = A_j + B_j Z_{ij} + r_{ij}, \quad r_{ij} \sim N(0, \sigma^2).$$

The level 2 model in this case has two simultaneous equations, one for each level 1 coefficient:

$$A_j = \alpha + a_j;$$
$$B_j = \beta + b_j;$$
$$\begin{pmatrix} a_j \\ b_j \end{pmatrix} \sim N\left(\begin{pmatrix} 0 \\ 0 \end{pmatrix}, \begin{pmatrix} \tau_a^2 & \\ \tau_{ab} & \tau_b^2 \end{pmatrix} \right).$$

When the model is correctly specified, α is the population average outcome of the control group over all the sites; and β is the population average treatment effect. The discrepancy between the control group mean at site j and the population mean of the control group over all the sites, denoted by a_j, is typically specified to have a normal distribution with mean zero and variance τ_a^2; similarly, the discrepancy between the site-specific treatment effect and the population average treatment effect, denoted by b_j, is specified to have a normal distribution with mean zero and variance τ_b^2. The model additionally generates the covariance between a_j and b_j denoted by τ_{ab}. The covariance indicates whether the treatment is more beneficial in sites where the average outcome tends to be lower (or higher) in the absence of the treatment. The level 1 model and the level 2 model can be combined into a mixed-effects model as follows:

$$Y_{ij} = \alpha + a_j + \beta Z_{ij} + b_j Z_{ij} + r_{ij}. \tag{6.10}$$

A major challenge, as pointed out by Raudenbush (2014), is that an estimator of β will be biased for the population average treatment effect δ if the random intercept a_j and the random slope b_j are not independent of the treatment indicator Z_{ij}. The bias is

$$\left[E\left(a_j | Z_{ij} = 0\right) - E\left(a_j | Z_{ij} = 1\right) \right] - E\left(b_j | Z_{ij} = 1\right). \tag{6.11}$$

Appendix 6.B derives the bias, which can arise when the probability of treatment assignment varies from site to site. We can show that a_j and b_j are likely not independent of Z_{ij} and therefore the potential bias is likely nonzero because

$$\text{Cov}\left(a_j, Z_{ij}\right) = \text{Cov}\left(a_j, Z_{ij} - P_j + P_j\right) = \text{Cov}\left(a_j, Z_{ij} - P_j\right) + \text{Cov}\left(a_j, P_j\right) = \text{Cov}\left(a_j, P_j\right);$$
$$\text{Cov}\left(b_j, Z_{ij}\right) = \text{Cov}\left(b_j, Z_{ij} - P_j + P_j\right) = \text{Cov}\left(b_j, Z_{ij} - P_j\right) + \text{Cov}\left(b_j, P\right) = \text{Cov}\left(b_j, P_j\right).$$

As defined earlier, P_j is the probability of treatment assignment at site j. Even though the treatment assignment within a site $Z_{ij}-P_j$ is independent of the site-specific increment to the control group mean a_j and of the site-specific increment to the treatment effect b_j due to the within-site randomization, the probability of treatment assignment at a given site P_j is not necessarily independent of the control group mean at the site; nor is it necessarily independent of the site-specific treatment effect. It is possible that in the sites where the average outcome would be relatively lower in the absence of the treatment in comparison with other sites, the demand for the treatment would be higher. In response to the political pressure in these sites, a relatively large proportion of participants might be assigned to the experimental group, which would lead to a positive covariance between a_j and P_j. It is also possible that at the sites where the treatment was known to be relatively more effective than those at other sites, the demand for the treatment would become higher accordingly, leading to a positive covariance between b_j and P_j. Nonetheless, researchers (Raudenbush, 2014; Woodridge, 2006) have shown that in comparison with the potential bias in the fixed-effects estimator, the bias in the random-effects estimator tends to be attenuated when the between-site variation is relatively large.

A key question is how to adjust for possible between-site variation in the probability of treatment assignment P_j. One may choose to center the observed outcome and the treatment indicator at their respective site means and thereby change model (6.10) to the following:

$$Y_{ij}-\mu_j = \beta(Z_{ij}-P_j) + b_j(Z_{ij}-P_j) + r_{ij}. \tag{6.12}$$

Here, μ_j is the mean of the observed outcome at site j and is equal to $\alpha+a_j+(\beta+b_j)P_j$. This strategy, however, does not eliminate the bias in (6.11).

Another solution that has been proposed in the literature is to include P_j as a covariate in the model (Hill, 2013):

$$Y_{ij} = \alpha+a_j+\beta Z_{ij}+b_j Z_{ij}+\gamma P_j+r_{ij}. \tag{6.13}$$

When P_j is held constant, treatment assignment Z_{ij} becomes independent of a_j and b_j. The rationale is to estimate the average treatment effect conditioning on P_j and then pool the treatment effects over the distribution of P_j. It can be shown that β is again biased for the population average treatment effect δ with the bias equal to $-E\left[E(b_j|P_j)\right]$ unless b_j is independent of P_j. Another issue is that conditioning on P_j changes the definitions of the random intercept and the random slope and their joint distribution. In model (6.13), a_j is now the discrepancy between the control group mean at site j and the average control group mean in the subpopulation of sites sharing the same probability of treatment assignment P_j. Similarly, b_j is now the discrepancy between the treatment effect at site j and the average treatment effect in the subpopulation of sites sharing the same probability of treatment assignment. The conditional variance of a_j and that of b_j are expected to be smaller than their respective marginal variances. The conditional covariance between a_j and b_j will likely be different from the marginal covariance as well.

A third strategy is to transform the relative proportions of the treated and the untreated within each site through weighting, such that in the weighted data, the probability of treatment assignment becomes constant across all the sites (Raudenbush, 2014). Let P be the average probability of treatment assignment in the population of sites. In site j, in which the probability of treatment assignment P_j could possibly be different from P, the weight for the treated units is P/P_j and that for the untreated units is $1-P/1-P_j$. In general, one may represent the weight for individual i at site j as

$$W_{ij} = Z_{ij} \times \frac{P}{P_j} + \left(1 - Z_{ij}\right) \times \frac{1-P}{1-P_j}. \tag{6.14}$$

Appendix 6.C shows the derivation and proves the identification result that when model (6.10) is analyzed under the aforementioned weighting, the coefficient for the treatment assignment β is unbiased for the population average treatment effect δ.

To investigate the heterogeneity of treatment effect, the variance of the site-specific treatment effect across all the sites in the population as defined in (6.7) is of key interest.

The estimation of this variance is often a preliminary step that researchers take before examining the moderating effects of specific individual or site characteristics. One may employ weighted maximum likelihood or generalized least squares for estimation (Hong and Raudenbush, 2008; Pfefferman et al., 1998; Raudenbush, 2014). These strategies involve partitioning the marginal weight specified in (6.14) into a level 1 weight and a level 2 weight in a multilevel modeling framework. Analyzing a weighted mixed-effects model as specified in (6.10) generates not only an estimate of the population average treatment effect but also estimates of the variance of the site-specific treatment effect, of the variance of the site-specific control group mean, and of their covariance.

In summary, multisite randomized trials offer important opportunities for investigating heterogeneous treatment effects and for assessing the generalizability of causal inference results over different sites in a population. Yet, as shown previously, when the treatment effect at a given site is associated with the probability of treatment assignment at the site, conventional use of fixed-effects models and mixed-effects models often generates estimates of population average treatment effects or between-site variance in treatment effects that are prone to bias. Several adjustment strategies have been proposed in the literature to correct for bias in analyses of mixed-effects models. Neither site mean centering nor covariance adjustment for the site-specific probability of treatment assignment removes bias. In contrast, adjustment through weighting by the inverse probability of treatment assignment at each site appears to be a viable solution. Analyzing a weighted mixed-effects model, one obtains an estimated distribution of the treatment effect over all the sites and its joint distribution with the site-specific control group mean.

6.3.2 Principal stratification

The principal stratification framework, formalized by Frangakis and Rubin (2002) and reviewed by Joffe, Small, and Hsu (2007), has been exploited by some researchers for investigating the heterogeneity in treatment effects across subpopulations (Barnard et al., 2003; Gallop et al., 2009; Jin and Rubin, 2008, 2009; Jo, 2008; Page, 2012). Principal strata are defined according to a set of potential intermediate outcome values under alternative treatment conditions. Individual units within a principal stratum are homogeneous in how they would respond to each of the treatment conditions at the intermediate stage and hence constitute a meaningful subpopulation. Principal stratification therefore provides a unique tool for investigating heterogeneity across subpopulations.

Barnard and colleagues' (2003) study is illustrative in this regard. In this example, public school students who applied for the New York City School Choice Scholarship Program were randomized to receive a voucher. Most students who were offered vouchers ($z = 1$) switched to private schools, while most of those who were not ($z = 0$) remained in public schools. The potential intermediate outcomes indicate whether a student would switch to a private school ($M(z) = 1$) or stay in a public school ($M(z) = 0$) as a result of the initial treatment assignment z for $z = 0$, 1. Three principal strata are plausible representing three subpopulations of students, each expected to display a unique set of potential compliance behavior:

1. *Always-takers* would always attend private schools regardless of the voucher offer ($M(1) = M(0) = 1$).

2. *Compliers* would attend private schools if offered vouchers and would attend public schools if not offered vouchers ($M(1) = 1$; $M(0) = 0$).

3. *Never-takers* would always attend public schools regardless of the voucher offer ($M(1) = M(0) = 0$).

In this randomized experiment, because $M(1)$ and $M(0)$ are potential intermediate outcomes that are independent of the treatment assignment, a student's principal stratum membership defined jointly by $M(1)$ and $M(0)$ is pretreatment in nature. The goal of the analysis is to estimate the intent-to-treat (ITT) effect of treatment assignment within each principal stratum. The ITT effects on the academic outcome of always-takers, compliers, and never-takers are defined as $E[Y(1) - Y(0)|M(1) = M(0) = 1]$, $E[Y(1) - Y(0)|M(1) = 1, M(0) = 0]$, and $E[Y(1) - Y(0)|M(1) = M(0) = 0]$, respectively.

In addition to the ignorability assumption assured by the random treatment assignment, another key identification assumption, named "monotonicity," states that being offered voucher does not reduce one's probability of switching to a private school. This assumption is seemingly plausible in the current study and rules out a principal stratum of individuals who would stay in public school if offered a voucher and would switch to private school if not offered a voucher.

Although students who attended private schools without voucher offers can be identified as always-takers while those who stayed in public schools despite being offered vouchers can be identified as never-takers, the principal stratum memberships of all other students remain unknown. Nonetheless, by utilizing pretreatment covariate information **X**, researchers can specify parametric models for predicting a student's joint distribution of $M(1)$ and $M(0)$ as a function of **x** as well as models for predicting the distributions of a student's potential outcomes $Y(1)$ and $Y(0)$ each as a function of principal stratum membership and **x**. A Bayesian inference framework is then employed to estimate the model parameters and generate the posterior distribution of the causal effect within each principal stratum. In the current application, the ITT effects for the compliers were found to be generally greater than the ITT effects for the entire population, suggesting that the treatment effect was perhaps greater for the compliers than for the always-takers and never-takers as one would anticipate.

Appendix 6.A: Derivation of bias in the fixed-effects estimator when the treatment effect is heterogeneous in multisite randomized trials

In analyses of data from multisite randomized trials, a fixed-effects model as specified in (6.9) can be recast as follows for individual i in site j:

$$Y_{ij} = \alpha_j + \beta Z_{ij} + e_{ij}. \tag{A6.1}$$

Here, Z_{ij} is the treatment assignment for individual i in site j; β is the fixed-effects parameter for the population average treatment effect δ as defined in (6.6); α_j is the fixed effect of site j on the outcome, that is, the control group mean at the site, previously represented in model (6.9) by π_1 for site 1 and by $\pi_1 + \pi_j$ for sites $j = 2, \ldots, J$; and e_{ij} is the within-site random error assumed to have a zero mean. One may center Y_{ij} at its site mean denoted by μ_j and center the treatment assignment Z_{ij} at its site mean denoted by P_j, the latter representing the proportion of participants at site j assigned to the experimental condition. The centered model eliminates α_j:

$$Y_{ij} - \mu_j = \beta (Z_{ij} - P_j) + e_{ij}. \tag{A6.2}$$

One then computes the covariance of $Z_{ij} - P_j$ with the elements on each side of the equation in (A6.2):

$$\mathrm{Cov}(Z_{ij} - P_j, Y_{ij} - \mu_j) = \beta \mathrm{Var}(Z_{ij} - P_j) + \mathrm{Cov}(Z_{ij} - P_j, e_{ij}). \tag{A6.3}$$

Under the assumption that e_{ij} is a random error and therefore has a zero covariance with $Z_{ij} - P_j$, the coefficient representing the average treatment effect in the aforementioned fixed-effects model is

$$\beta = \frac{\mathrm{Cov}(Z_{ij} - P_j, Y_{ij} - \mu_j)}{\mathrm{Var}(Z_{ij} - P_j)}.$$

This approach, however, invokes the seemingly implausible assumption that the treatment effect is constant across all the sites. When this assumption does not hold, as pointed out by Raudenbush (2014), β is possibly biased for the population average treatment effect δ. We derive the bias in the following.

When the treatment effect at site j takes a unique value δ_j, model (A6.2) becomes

$$Y_{ij} - \mu_j = \delta_j (Z_{ij} - P_j) + \varepsilon_{ij},$$

which is equivalent to the following:

$$Y_{ij} - \mu_j = \delta (Z_{ij} - P_j) + (\delta_j - \delta)(Z_{ij} - P_j) + \varepsilon_{ij}. \tag{A6.4}$$

The covariance of $Z_{ij} - P_j$ with the elements on each side of the equation in (A6.4) is

$$\mathrm{Cov}(Z_{ij} - P_j, Y_{ij} - \mu_j) = \delta \mathrm{Var}(Z_{ij} - P_j) + \mathrm{Cov}(Z_{ij} - P_j, (\delta_j - \delta)(Z_{ij} - P_j) + \varepsilon_{ij}), \tag{A6.5}$$

where ε_{ij} is a pure random error. Ignoring the fact that the treatment effect is heterogeneous across the sites, one would mistakenly analyze model (A6.2) rather than (A6.4). We can see that the error term e_{ij} in (A6.2) is equivalent to $(\delta_j - \delta)(Z_{ij} - P_j) + \varepsilon_{ij}$. Clearly, $Z_{ij} - P_j$ and e_{ij} are not independent unless there is a constant treatment effect $\delta_j = \delta$ over all the sites. Dividing each side of the equation (A6.5) by $\mathrm{Var}(Z_{ij} - P_j)$, we obtain the following:

$$\delta = \frac{\mathrm{Cov}(Z_{ij} - P_j, Y_{ij} - \mu_j)}{\mathrm{Var}(Z_{ij} - P_j)} - \frac{\mathrm{Cov}(Z_{ij} - P_j, (\delta_j - \delta)(Z_{ij} - P_j) + \varepsilon_{ij})}{\mathrm{Var}(Z_{ij} - P_j)}.$$

The difference between β and δ is the bias contained in β and is equal to

$$\frac{\mathrm{Cov}(P_j(1 - P_j), \delta_j)}{\mathrm{Var}(Z_{ij} - P_j)}.$$

The bias can be derived as follows:

$$\beta - \delta = \frac{\mathrm{Cov}(Z_{ij} - P_j, (\delta_j - \delta)(Z_{ij} - P_j) + \varepsilon_{ij})}{\mathrm{Var}(Z_{ij} - P_j)}$$

$$= \frac{E\left[(Z_{ij} - P_j)^2 (\delta_j - \delta)\right] - E\{(Z_{ij} - P_j)E\left[(\delta_j - \delta)(Z_{ij} - P_j)\right]\}}{\mathrm{Var}(Z_{ij} - P_j)}$$

$$= \frac{E\{(\delta_j - \delta)E\left[(Z_{ij} - P_j)^2 | j\right]\}}{\mathrm{Var}(Z_{ij} - P)}$$

$$= \frac{E\left[P_j(1 - P_j)(\delta_j - \delta)\right]}{\mathrm{Var}(Z_{ij} - P_j)}$$

$$= \frac{\mathrm{Cov}(P_j(1 - P_j), \delta_j)}{\mathrm{Var}(Z_{ij} - P_j)}.$$

The third equation holds because

$$E\{(Z_{ij} - P_j)E\left[(\delta_j - \delta)(Z_{ij} - P_j)\right]\} = E\left[(\delta_j - \delta)(Z_{ij} - P_j)\right]E(Z_{ij} - P_j | j) = 0.$$

The bias is positive, for example, if the site-specific treatment effect increases when the probability of treatment assignment at a given site approaches 0.5. This is not unlikely in the following scenarios. Suppose that the treatment of interest is an intervention program aimed at reducing antisocial behaviors among adolescents. When too few individuals are treated at a site, the treated individuals will be surrounded overwhelmingly by the untreated peers and will likely succumb to the antisocial norm, whereas when too many individuals are treated at a site with resource scarcity, the intervention will likely be spread too thin, which will then compromise the potential benefit of the program. The treatment benefit is expected to be maximized when the proportion of individuals at the site receiving the treatment is neither too small nor too large.

In conclusion, the fixed-effects models do not support the investigation of heterogeneous treatment effects across sites in a multisite randomized trial. Even the estimator of the population average treatment effect can be biased when the homogeneous treatment effect assumption does not hold.

Appendix 6.B: Derivation of bias in the mixed-effects estimator when the probability of treatment assignment varies across sites

Raudenbush (2014) revealed that when the probability of treatment assignment varies across sites in a multisite trial, the mixed-effects model estimator of the population average treatment effect can be biased. Here, we derive the bias. Taking the site-specific mean on each side of (6.10) for each treatment group, we have that

$$
\begin{aligned}
E\left(Y_{ij}|Z_{ij}=1, j\right) &= \alpha + \beta + E\left(a_j + b_j|Z_{ij}=1, j\right) + E\left(r_{ij}|Z_{ij}=1, j\right); \\
E\left(Y_{ij}|Z_{ij}=0, j\right) &= \alpha + E\left(a_j|Z_{ij}=0, j\right) + E\left(r_{ij}|Z_{ij}=0, j\right).
\end{aligned}
\tag{A6.6}
$$

Due to the randomization of treatment assignment within each site, $E\left(r_{ij}|Z_{ij}=1, j\right)$ and $E\left(r_{ij}|Z_{ij}=0, j\right)$ are both zero. Subtracting the site-specific control group mean from the experimental group mean in (A6.6) and then averaging over all the sites, we have that

$$
E\left[E(Y_{ij}|Z_{ij}=1,j)-E\left(Y_{ij}|Z_{ij}=0, j\right)\right] = \beta + \left[E\left(a_j|Z_{ij}=1\right)-E\left(a_j|Z_{ij}=0\right)\right] + E\left(b_j|Z_{ij}=1\right).
$$

Due to the randomization within each site, the left hand side of the equation is unbiased for the population average treatment effect δ, and therefore,

$$
\delta = \beta + \left[E\left(a_j|Z_{ij}=1\right)-E\left(a_j|Z_{ij}=0\right)\right] + E\left(b_j|Z_{ij}=1\right).
\tag{A6.7}
$$

We can see that β and δ are equal only if $E\left(a_j|Z_{ij}=1\right)=E\left(a_j|Z_{ij}=0\right)$ and if $E\left(b_j|Z_{ij}=1\right)=0$.

Hence, the potential bias in β is $\left[E\left(a_j|Z_{ij}=0\right)-E\left(a_j|Z_{ij}=1\right)\right]-E\left(b_j|Z_{ij}=1\right)$. As shown in Section 6.3.1.2.2, the within-site treatment assignment Z_{ij} is associated with the site-specific random effects a_j and b_j only through the site-specific probability of treatment assignment P_j. In other words, the bias will disappear if P_j is a constant across all the sites.

Appendix 6.C: Derivation and proof of the population weight applied to mixed-effects models for eliminating bias in multisite randomized trials

As shown in Appendix 6.B, an analysis of the mixed-effects model specified in (6.10)

$$
Y_{ij} = \alpha + a_j + \beta Z_{ij} + b_j Z_{ij} + r_{ij}
$$

will generate a biased estimate of the population average treatment effect if the probability of treatment assignment P_j varies across the sites. Here, we show that the bias can be eliminated through inverse-probability-of-treatment weighting.

Due to the randomization of treatment assignment within each site, the mean difference between the treated and the untreated in each site, averaged over all the sites, is unbiased for the population average treatment effect, that is, $E\big[E(Y_{ij}|Z_{ij}=1,j)-E\big(Y_{ij}|Z_{ij}=0,j\big)\big]=\delta$. Under model (6.10), the former is equal to

$$E\big[E\big(\alpha+a_j+\beta Z_{ij}+b_j Z_{ij}+r_{ij}|Z_{ij}=1,j\big)-E\big(\alpha+a_j+\beta Z_{ij}+b_j Z_{ij}+r_{ij}|Z_{ij}=0,j\big)\big]$$
$$=\alpha+\beta+E\big(a_j|Z_{ij}=1\big)+E\big(b_j|Z_{ij}=1\big)-\alpha-E\big(a_j|Z_{ij}=0\big)$$
$$=\beta+\int a_j f\big(a_j|Z_{ij}=1\big)da_j+\int b_j f\big(b_j|Z_{ij}=1\big)db_j-\int a_j f\big(a_j|Z_{ij}=0\big)da_j.$$

Note that

$$f\big(a_j|Z_{ij}=1\big)=\frac{\mathrm{pr}\big(Z_{ij}=1|a_j\big)f\big(a_j\big)}{\mathrm{pr}\big(Z_{ij}=1\big)}=\frac{P_j}{P}f\big(a_j\big)$$

and, similarly, we have that

$$f\big(a_j|Z_{ij}=0\big)=\frac{\mathrm{pr}\big(Z_{ij}=0|a_j\big)f\big(a_j\big)}{\mathrm{pr}\big(Z_{ij}=0\big)}=\frac{1-P_j}{1-P}f\big(a_j\big)$$

and that

$$f\big(b_j|Z_{ij}=1\big)=\frac{\mathrm{pr}\big(Z_{ij}=1|b_j\big)f\big(b_j\big)}{\mathrm{pr}\big(Z_{ij}=1\big)}=\frac{P_j}{P}f\big(b_j\big)$$

.

Hence, $E\big[E(Y_{ij}|Z_{ij}=1,j)-E\big(Y_{ij}|Z_{ij}=0,j\big)\big]$ is equal to the following:

$$\beta+\int a_j\times\frac{P_j}{P}f\big(a_j\big)da_j+\int b_j\times\frac{P_j}{P}f\big(b_j\big)db_j-\int a_j\times\frac{1-P_j}{1-P}f\big(a_j\big)da_j.$$

Once applying the population weight P/P_j to the treated units and applying $(1-P)/(1-P_j)$ to the untreated units, we can show that the weighted mean difference in the outcome between the treated and the untreated is unbiased for the population average treatment effect δ:

$$E\left[E\left(\frac{P}{P_j}\times Y_{ij}|Z_{ij}=1,j\right)-E\left(\frac{1-P}{1-P_j}\times Y_{ij}|Z_{ij}=0,j\right)\right]$$
$$=E\left[\frac{P}{P_j}E\big(Y_{ij}|Z_{ij}=1,j\big)|Z_{ij}=1\right]-E\left[\frac{1-P}{1-P_j}E\big(Y_{ij}|Z_{ij}=0,j\big)|Z_{ij}=0\right]$$
$$=\int\frac{P}{P_j}f\big(P_j|Z_{ij}=1\big)\left[\int yf\big(y|Z_{ij}=1,j\big)dy dP_j\right]-\int\frac{1-P}{1-P_j}f\big(P_j|Z_{ij}=0\big)\left[\int yf\big(y|Z_{ij}=0,j\big)dy dP_j\right]$$
$$=\int\frac{P}{P_j}\times\frac{\mathrm{pr}\big(Z_{ij}=1|P\big)}{\mathrm{pr}\big(Z_{ij}=1\big)}f\big(P_j\big)\left[\int yf\big(y|Z_{ij}=1,j\big)dy dP_j\right]-\int\frac{1-P}{1-P_j}\times\frac{\mathrm{pr}\big(Z_{ij}=0|P_j\big)}{\mathrm{pr}\big(Z_{ij}=0\big)}f\big(P_j\big)\left[\int yf\big(y|Z_{ij}=0,j\big)dy dP_j\right]$$
$$=\int\int yf\big(y|Z_{ij}=1,j\big)dyf\big(P_j\big)dP_j-\int\int yf\big(y|Z_{ij}=0,j\big)dyf\big(P_j\big)dP_j$$
$$=E\big[E\big(Y_{ij}(1)|P_j\big)\big]-E\big[E\big(Y_{ij}(0)|P_j\big)\big]$$
$$=\delta.$$

Analyzing model (6.10) under the same weighting scheme, we will find that β becomes unbiased for δ. This is because

$$E\left[E\left(\frac{P}{P_j}\times Y_{ij}|Z_{ij}=1,j\right)-E\left(\frac{1-P}{1-P_j}\times Y_{ij}|Z_{ij}=0,j\right)\right]$$

$$=E\left\{E\left[\frac{P}{P_j}\times\left(\alpha+a_j+\beta Z_{ij}+b_jZ_{ij}+r_{ij}\right)|Z_{ij}=1,j\right]-E\left[\frac{1-P}{1-P_j}\times\left(\alpha+a_j+\beta Z_{ij}+b_jZ_{ij}+r_{ij}\right)|Z_{ij}=0,j\right]\right\}.$$

Note that

$$E\left[E\left(\frac{P}{P_j}\times\alpha|Z_{ij}=1,j\right)\right]=\alpha\int\frac{P}{P_j}f\left(P_j|Z_{ij}=1\right)dP_j=\alpha\int\frac{P}{P_j}\times\frac{\text{pr}\left(Z_{ij}=1|P_j\right)}{\text{pr}\left(Z_{ij}=1\right)}f\left(P_j\right)dP_j=\alpha\int f\left(P_j\right)dP_j=\alpha;$$

$$E\left[E\left(\frac{P}{P_j}\times\beta|Z_{ij}=1,j\right)\right]=\beta\int\frac{P}{P_j}f\left(P_j|Z_{ij}=1\right)dP_j=\beta\int\frac{P}{P_j}\times\frac{\text{pr}\left(Z_{ij}=1|P_j\right)}{\text{pr}\left(Z_{ij}=1\right)}f\left(P_j\right)dP_j=\beta\int f\left(P_j\right)dP_j=\beta;$$

$$E\left[E\left(\frac{1-P}{1-P_j}\times\alpha|Z_{ij}=0,j\right)\right]=\alpha\int\frac{1-P}{1-P_j}f\left(P_j|Z_{ij}=0\right)dP_j=\alpha\int\frac{1-P}{1-P_j}\times\frac{\text{pr}\left(Z_{ij}=0|P_j\right)}{\text{pr}\left(Z_{ij}=0\right)}f\left(P_j\right)dP_j=\alpha\int f\left(P_j\right)dP_j=\alpha.$$

Also note that

$$E\left[\frac{P}{P_j}E\left(r_{ij}|Z_{ij}=1,j\right)\right]=0;$$

$$E\left[\frac{1-P}{1-P_j}E\left(r_{ij}|Z_{ij}=0,j\right)\right]=0.$$

This is because, by definition, $E\left(r_{ij}|Z_{ij}=1,j\right)=E\left(r_{ij}|Z_{ij}=0,j\right)=0$. Hence, the weighted parameter is

$$\alpha+\beta+E\left(\frac{P}{P_j}\times a_j|Z_{ij}=1\right)+E\left(\frac{P}{P_j}\times b_j|Z_{ij}=1\right)-\alpha-E\left(\frac{1-P}{1-P_j}\times a_j|Z_{ij}=0\right)$$

$$=\beta+\int\frac{P}{P_j}\times a_j\times\frac{P_j}{P}f\left(a_j\right)da_j+\int\frac{P}{P_j}\times b_j\times\frac{P_j}{P}f\left(b_j\right)db_j-\int\frac{1-P}{1-P_j}\times a_j\times\frac{1-P_j}{1-P}f\left(a_j\right)da_j$$

$$=\beta+\int a_jf\left(a_j\right)da_j+\int b_jf\left(b_j\right)db_j-\int a_jf\left(a_j\right)da_j$$

$$=\beta.$$

The last equation holds because $\int b_jf\left(b_j\right)db_j=E\left(b_j\right)=0$ by definition. This completes the proof.

References

Abrahams, N.M. and Alf, E., Jr (1972). Pratfalls in moderator research. *Journal of Applied Psychology*, *56*, 245–251.

Aiken, L.S. and West, S.G. (1991). *Multiple Regression: Testing and Interpreting Interactions*, Sage Publications, Newbury Park, CA.

Angrist, J.D. and Pischke, J. (2008). *Mostly Harmless Econometrics: An Empiricist's Companion*, Princeton University Press, Princeton, NJ.

Barnard, J., Frangakis, C., Hill, J. and Rubin, D.B. (2003). Principal stratification approach to broken randomized experiments: a case study of school choice vouchers in New York City (with discussion). *Journal of the American Statistical Association*, *98*, 299–323.

Baron, R.M. and Kenny, D.A. (1986). The moderator-mediator variable distinction in social psychological research: conceptual, strategic, and statistical considerations. *Journal of Personality and Social Psychology*, *51*(6), 1173–1182.

Bracht, G.H. (1970). Experimental factors related to aptitude-treatment interactions. *Review of Educational Research*, *40*(5), 627–645.

Bryk, A.S. and Raudenbush, S.W. (1988). Heterogeneity of variance in experimental studies: a challenge to conventional interpretations. *Psychological Bulletin*, *104*(3), 396–404.

Burt, C., Cohen, L. and Bjorck, J. (1988). Perceived family environment as a moderator of young adolescents' life stress adjustment. *American Journal of Community Psychology*, *16*(1), 101–122.

Cohen, S. and Hoberman, H.M. (1983). Positive events and social supports as buffers of life change stress. *Journal of Applied Social Psychology*, *13*(2), 99–125.

Cohen, S. and Wills, T.A. (1985). Stress, social support, and the buffering hypothesis. *Psychological Bulletin*, *98*, 310–357.

DuBois, D.L., Felner, R.D., Brand, S. *et al.* (1992). A prospective study of life stress, social support, and adaptation in early adolescence. *Child Development*, *63*(3), 542–557.

Fairchild, A.J. and MacKinnon, D.P. (2009). A general model for testing mediation and moderation effects. *Prevention Science*, *10*(2), 87–99.

Frangakis, C.E. and Rubin, D.B. (2002). Principal stratification in causal inference. *Biometrics*, *58*, 21–29.

Gallop, R., Small, D.S., Lin, J.Y., Elliott, M.R., Joffe, M.T. and Have, T.R. (2009). Mediation analysis with principal stratification. *Statistics in Medicine*, *28*, 1108–1130.

Ge, X., Lorenz, F.O., Conger, R.D. *et al.* (1994). Trajectories of stressful life events and depressive symptoms during adolescence. *Developmental Psychology*, *30*(4), 467.

Ge, X., Conger, R.D. and Elder, G.H., Jr (2001). Pubertal transition, stressful life events, and the emergence of gender differences in adolescent depressive symptoms. *Developmental Psychology*, *37*(3), 404.

Ghiselli, E.E. (1956). Differentiation of individuals in terms of their predictability. *Journal of Applied Psychology*, *40*(6), 374–377.

Greene, W.H. (2003). *Econometric Analysis*, Pearson Education Inc, Upper Saddle River, NJ.

Hamilton, G. (2002). *Moving People from Welfare to Work: Lessons from the National Evaluation of Welfare-to-Work Strategies*, U.S. Department of Health and Human Services, and U.S. Department of Education, Washington, DC.

Heckman, J.J., Urzua, S. and Vytlacil, E. (2006). Understanding instrumental variables in models with essential heterogeneity. *The Review of Economics and Statistics*, *88*(3), 389–432.

Hill, J. (2013). Multilevel models and causal inference, in M. Scott, J. Simonoff, and B. Marx (eds), *The SAGE Handbook of Multilevel Modeling* (pp. 201–221). London: Sage Publications, Ltd.

Hinkelmann, K. and Kempthorne, O. (1994). *Design and Analysis of Experiments, Volume I: Introduction to Experimental Design*, Wiley, New York.

Holland, P.W. (1986). Statistics and causal inference. *Journal of the American Statistical Association*, *81*(396), 945–960.

Holmbeck, G.N. (1997). Toward terminological, conceptual, and statistical clarity in the study of mediators and moderators: examples from the child-clinical and pediatric psychology literatures. *Journal of Consulting and Clinical Psychology*, *65*(4), 599–610.

Hong, G. and Hong, Y. (2009). Reading instruction time and homogeneous grouping in kindergarten: an application of marginal mean weighting through stratification. *Educational Evaluation and Policy Analysis*, *31*(1), 54–81.

Hong, G. and Raudenbush, S.W. (2008). Causal inference for time-varying instructional treatments. *Journal of Educational and Behavioral Statistics*, *33*(3), 333–362.

Hong, G., Corter, C., Hong, Y. and Pelletier, J. (2012). Differential effects of literacy instruction time and homogeneous grouping in kindergarten: who will benefit? Who will suffer? *Educational Evaluation and Policy Analysis*, *34*(1), 69–88.

James, L.R. and Brett, J.M. (1984). Mediators, moderators, and test for mediation. *Journal of Applied Psychology*, *69*(2), 307–321.

Jin, H. and Rubin, D.B. (2008). Principal stratification for causal inference with extended partial compliance. *Journal of the American Statistical Association*, *103*, 101–111.

Jin, H. and Rubin, D.B. (2009). Public schools versus private schools: causal inference with partial compliance. *Journal of Educational and Behavioral Statistics*, *34*, 24–45.

Jo, B. (2008). Causal inference in randomized experiments with mediational processes. *Psychological Methods*, *13*(4), 314–336.

Joffe, M.M., Small, D. and Hsu, C.-H. (2007). Defining and estimating intervention effects for groups that will develop an auxiliary outcome. *Statistical Science*, *22*(1), 74–97.

Judd, C.M., Kenny, D.A. and McClelland, G.H. (2001). Estimating and testing mediation and moderation in within-subject designs. *Psychological Methods*, *6*(2), 115–134.

Kim, J. and Seltzer, M. (2011). Examining heterogeneity in residual variance to detect differential response to treatments. *Psychological Methods*, *16*(2), 192–208.

Kraemer, H.C., Stice, E., Kazdin, A. *et al.* (2001). How do risk factors work together? Mediators, moderators, and independent, overlapping, and proxy risk factors. *American Journal of Psychiatry*, *158*, 848–856.

Kraemer, H.C., Wilson, G.T., Fairburn, C.G. and Agras, W.S. (2002). Mediators and moderators of treatment effects in randomized clinical trials. *Archives of General Psychiatry*, *59*, 877–883.

Kraemer, H., Kiernan, M., Essex, M. and Kupfer, D. (2008). How and why criteria defining moderators and mediators differ between the Baron & Kenny and MacArthur approaches. *Health Psychology*, *27*, 1–14.

Kulik, J.A. and Kulik, C.L. (1987). Effects of ability grouping on student achievement. *Equity and Excellence*, *23*, 22–30.

Lavori, P.W. and Dawson, R. (2008). Adaptive treatment strategies in chronic disease. *Annual Review of Medicine*, *59*, 443–453.

Lou, Y., Abrami, P.C., Spence, J.C. *et al.* (1996). Within-class grouping: a meta-analysis. *Review of Educational Research*, *66*(4), 423–458.

Lou, Y., Abrami, P.C. and Spence, J. (2000). Effects of within-class grouping on student achievement: an exploratory model. *Journal of Educational Research*, *94*(2), 101–112.

Marshall, H.H. (1969). Learning as a function of task interest, reinforcement, and social class variables. *Journal of Educational Psychology*, *60*(2), 133–137.

McGroder, S.M., Zaslow, M.J., Moore, K.A., and LeMenestrel, S. (2000). Impacts on Young Children and Their Families Two Years after Enrollment: Findings from the Child Outcomes Study. National Evaluation of Welfare-to-Work Strategies Report. U.S. Department of Health and Human Services, Office of the Assistant Secretary for Planning and Evaluation, Administration for Children and Families, Washington, DC.

Morris, P.A. (2008). Welfare program implementation and parents' depression. *Social Service Review*, 82(4), 579–614.

Murch, R.L. and Cohen, L.H. (1989). Relationships among life stress, perceived family environment, and the psychological distress of spina bifida adolescents. *Journal of Pediatric Psychology*, 14(2), 193–214.

Page, L.C. (2012). Principal stratification as a framework for investigating mediational processes in experimental settings. *Journal of Research on Educational Effectiveness*, special issue on the statistical approaches to studying mediator effects in education research, 5(3), 215–244.

Pfefferman, C.J., Skinner, D.J., Holmes, H. *et al.* (1998). Weighting for unequal selection probabilities in multilevel analysis. *Journal of the Royal Statistical Society, Series B*, 60(1), 23–40.

Raudenbush, S.W. (2014). *Random coefficient models for multi-site randomized trials with inverse probability of treatment weighting*. University of Chicago working paper, Chicago, IL.

Rose, B.M., Holmbeck, G.N., Coakley, R.M. and Franks, E.A. (2004). Mediator and moderator effects in developmental and behavioral pediatric research. *Developmental and Behavioral Pediatrics*, 25, 58–67.

Slavin, R.E. (1987). Ability grouping and student achievement in elementary schools: a best-evidence synthesis. *Review of Educational Research*, 57(3), 293–336.

Stolzenberg, R. (1980). The measurement and decomposition of causal effects in nonlinear and nonadditive models, in *Sociological Methodology* (ed K. Schuessler), Jossey-Bass for the American Sociological Association, San Francisco, CA.

Weiss, M.J., Bloom, H.S. and Brock, T. (2014). A conceptual framework for studying sources of variation in program effects. *Journal of Policy Analysis and Management*, 33(3), 778–808.

Woodridge, J.M. (2002). *Econometric Analysis of Cross Section and Panel Data*, MIT Press, Cambridge, MA.

Woodridge, J.M. (2006). *Introductory Econometrics: A Modern Approach*, Thomson, Mason, OH.

Wu, A.D. and Zumbo, B.D. (2008). Understanding and using mediators and moderators. *Social Indicator Research*, 87, 367–392.

Ystgaard, M., Tambs, K. and Dalgard, O.S. (1999). Life stress, social support and psychological distress in late adolescence: a longitudinal study. *Social Psychiatry and Psychiatric Epidemiology*, 34(1), 12–19.

7

Marginal mean weighting through stratification for investigating moderated treatment effects

In Chapter 5, we defined moderated treatment effects and discussed various experimental designs for identifying moderated treatment effects. The statistical methods reviewed in Chapter 5 were developed primarily for analyzing experimental data. This chapter extends the discussions to analyses of quasiexperimental data and focuses on analytic strategies for removing selection bias in such data. We will review the existing methods of statistical adjustment including ANCOVA, regression-based methods, and propensity score-based methods. The chapter then explains the rationale and the procedure for using marginal mean weighting through stratification (MMWS) to remove selection bias in moderation analyses and provides application examples. Specifically, to examine whether the treatment effect is moderated by individual or contextual characteristics, MMWS transforms the data to approximate a randomized block design. To investigate whether the effect of one treatment is moderated by a concurrent treatment, the goal of MMWS adjustment is to approximate data from a factorial randomized design. The weighted data can be conveniently analyzed within the two-way ANOVA framework.

7.1 Existing methods for moderation analyses with quasiexperimental data

In quasiexperimental research, individuals select themselves or are selected into certain treatments often on the basis of a prior belief that the treatments were designed for and would benefit those with certain pretreatment characteristics. Lower-achieving students, for example, are more likely to be retained in grade. Families with a stronger desire for upward mobility are more likely to take advantage of the opportunity to move to a low-poverty neighborhood. Without adjusting for selection bias, analysts of quasiexperimental data would most likely overestimate the harm (or underestimate the benefit) of grade retention and would most likely

Causality in a Social World: Moderation, Mediation and Spill-over, First Edition. Guanglei Hong.
© 2015 John Wiley & Sons, Ltd. Published 2015 by John Wiley & Sons, Ltd.

overestimate the benefit (or underestimate the harm) of moving to a low-poverty neighborhood.

One of the most influential findings in epidemiological research, obtained from the Nurses' Health Study that surveyed 122 000 nurses every other year in the 1980s, was that women taking estrogen as a treatment for menopausal symptoms had only a third as many heart attacks as women who had never taken the drug (Grodstein *et al.*, 1996, 2000). By the mid-1990s, the American Heart Association, the American College of Physicians, and the American College of Obstetricians and Gynecologists had all concluded that estrogen therapy could be recommended to older women as a means of preventing heart disease (Taubes, 2007). Yet in a double-blind randomized control trial a few years later, the Women's Health Initiative (WHI) tested estrogen therapy against a placebo in 16 500 healthy women and concluded that estrogen therapy would potentially increase the risk of heart attacks and stroke for all postmenopausal women (Writing Group for the Women's Health Initiative Investigators, 2002).

To interpret these contradictory findings, some researchers developed a hypothesis of moderation: the impact of estrogen therapy on cardiovascular health is perhaps moderated by the age of initiation. Estrogen therapy may protect younger women against heart disease if they start taking it during menopause yet increase the risk for heart disease for women who begin estrogen therapy after menopause (Manson *et al.*, 2007; Rossouw *et al.*, 2007). These researchers noted that the WHI experimental results were based on a sample of women whose average age was about 63 years at the baseline, whereas the average age of menopause in the United States is 51 years. In contrast, most estrogen therapy users in the Nurses' Health Study initiated use within a 2- to 3-year period after the start of menopause (Manson and Bassuk, 2007).

However, there are numerous concerns with the internal validity of the results obtained from observational studies. In the Nurses' Health Study, it was found that, in comparison with women who did not take estrogen therapy, those who used the therapy on average were less obese, more educated, wealthier, and more health conscious; exercised more; and displayed fewer risk factors for heart disease. All these differences suggest a possible bias due to the overrepresentation of healthy users in the treated group that would likely lead to an overestimation of the potential benefit or an underestimation of the potential harm of estrogen therapy. The initial researchers argued that they adjusted for the healthy user bias and still saw a substantial benefit of estrogen therapy. Nonetheless, the adjustment might be incomplete due to unobserved confounders and measurement error in the observed confounders.

Even in randomized trials, the experimenter often has limited control over the extent to which patients adhere to the treatment assigned. The low adherence rate in the WHI study raised concerns with possible misclassification of treatment users especially because some of those who were assigned at random to estrogen therapy might stop its use at any time before the end of the study. An intent-to-treat (ITT) analysis that compares the mean observed outcome between the treated group and the control group may not be highly informative because the experimenter's initial interest was in the average effect when everyone would adhere to the assigned treatment throughout the duration of the study (Hernán *et al.*, 2008). The lower the adherence rate, the larger the possible discrepancy between the ITT effect and the effect of actual treatment use.

There have been recent attempts at reanalyzing the Nurses' Health Study sample and the WHI sample by the age of initiation. Within each sample, it appears that evidence for the potential benefit of estrogen therapy is weaker, while evidence for its potential harm is greater for cardiovascular health among women who initiated use later rather than earlier after the onset of menopause (Grodstein, Manson, and Stampfer, 2006; Rossouw *et al.*, 2007). Selection bias can nonetheless be prevalent in these moderation analyses. In general, people who comply with their doctors' prescription tend to be different in the absence of the treatment, often healthier (or unhealthier and therefore more desperate) than those who do not. Those who comply also tend to respond to the treatment differently than those who do not should all individuals be assigned to the same treatment. For example, the adherence rate would likely be higher among earlier users than among later users due to the relief that estrogen therapy provides to menopausal symptoms experienced by earlier users; the adherence rate might be particularly low among later users who experienced side effects and would have suffered more health consequences had they instead adhered to the assigned treatment. When the selection mechanism is different between the early users and the late users, measured and unmeasured bias is not completely canceled out when the treatment effect is compared between these subpopulations. This could be true even in the experimental WHI study: the degree to which the ITT effect is moderated by age of initiation may not correctly indicate the degree to which the effect of the actual use of treatment is moderated. Because controversies over selection bias and moderation can become intertwined, methods are needed to clarify the assumptions under which findings of moderation are free of selection bias.

This section reviews the existing methods for removing selection bias in moderation analyses with quasiexperimental data. As in Chapter 6, we focus on two types of moderation questions: (i) treatment effects moderated by individual or contextual characteristics and (ii) treatment effects moderated by a concurrent treatment. The issue of moderation by consecutive treatments is left to Chapter 8. Let us first consider whether the causal effect of a treatment varies across subpopulations defined by theoretically important individual or contextual characteristics.

7.1.1   Analysis of covariance and regression-based adjustment

When a pretreatment covariate is suspected of confounding treatment effect estimation, the conventional strategy is to make covariance adjustment for the covariate. This is done through analysis of covariance (ANCOVA) or multiple regression on the basis of the same rationale: the treatment effect is estimated within each level of the covariate and is then pooled over the entire distribution of the covariate under the model-based assumption with regard to the covariate–outcome relationship. The same strategy has been extended to moderation analyses. In the following, we consider its applications to the two types of moderation questions.

7.1.1.1   Treatment effects moderated by subpopulation membership

In the hypothesis that estrogen therapy affects cardiovascular health differently for older and younger women, the subpopulations of interest have been defined by the age of initiation of estrogen therapy. To test the hypothesis with quasiexperimental data, the researcher may specify a two-way ANOVA model with treatment and age group as two fixed factors and make covariance adjustment for pretreatment covariates (e.g., SES) that may confound the treatment–outcome relationship within each age group. Under the assumption that the observed pretreatment covariates are sufficient for removing confounding of the treatment effect on cardiovascular health within each age group, the results would provide a test of the main effect of the treatment on health conditioning on age of initiation, a test of the main effect of age group on health conditioning on treatment, and a treatment-by-age interaction effect indicating whether the treatment effect on health differs between older and younger women.

This is equivalent to a multiple regression model with treatment and age group centered at their respective sample means. Let a random variable Z take value 1 if a woman used estrogen therapy and 0 otherwise; and let $\bar{Z}$ be the mean of this dummy indicator representing the sample proportion of estrogen therapy users. Let a random variable R take value 1 for an older woman above age 60 and 0 for a younger one in her 50s. Thus, $\bar{R}$ is the mean representing the sample proportion of older women. For simplicity, we use X to denote a single covariate SES typically measured on a continuous scale. The multiple regression model may be written as

$$Y = \beta_0 + \beta_1(Z-\bar{Z}) + \beta_2(R-\bar{R}) + \beta_3(Z-\bar{Z})(R-\bar{R}) + \beta_4 X + \varepsilon. \tag{7.1}$$

With linear covariance adjustment for SES, the above model defines the main effect of the treatment denoted by β_1, the main effect of age group membership by β_2, and the average difference in the treatment effect on health between older and younger women by β_3. Similar to ANCOVA, model (7.1) invokes a key assumption that the X–Y relationship—that is, the association between SES and the health outcome in the current example—is linear and constant across all four treatment-by-age combinations. The discussion of potential pitfalls of ANCOVA in Chapter 3 applies here. The estimation of the treatment effects will be biased if the covariate–outcome relationship differs between the treated group and the control group—for example, if health disparity between high-SES and low-SES older women is more severe among the treated than among the untreated. The adjustment would also fail if a nonlinear covariate–outcome relationship has been misspecified to be linear.

Some researchers may alternatively conduct subpopulation analysis that models the outcome as a function of the treatment and the covariates within each age subpopulation and then compares the treatment effect between the age subpopulations. Structural equation modeling (SEM) software can be used for such multigroup comparisons. Unlike model (7.1) that assumes a constant covariate–outcome relationship across all the age-by-treatment combinations, multigroup comparisons allow the covariate–outcome relationship to be different across different age groups, thereby relaxing the additivity assumption to some extent. Specifically, one analyzes two parallel regression models, one for the older women and the other for the younger women:

$$Y_{R=1} = \beta_{0(R=1)} + \beta_{1(R=1)}Z + \beta_{2(R=1)}X + \varepsilon_{(R=1)};$$

$$Y_{R=0} = \beta_{0(R=0)} + \beta_{1(R=0)}Z + \beta_{2(R=0)}X + \varepsilon_{(R=0)}.$$

One then tests the null hypothesis: $\beta_{1(R=1)} = \beta_{1(R=0)}$. A rejection of this null hypothesis will indicate that the treatment effect is moderated by age when the confounding effect of X has been adjusted for within each age group. The amount of moderation is estimated by $\widehat{\beta}_{1(R=1)} - \widehat{\beta}_{1(R=0)}$. Yet analytic decisions that remain crucial for causal inferences include which covariates require statistical adjustment and how to model the functional relationship between each covariate and the outcome within each age subpopulation.

In general, moderation analyses with quasiexperimental data can successfully remove selection bias only if, at each level of the moderator $R=r$, treatment assignment Z of individuals with the same observed pretreatment characteristics $\mathbf{X}$ is independent of the potential outcomes $Y(z)$ for all possible values of z. Here, $\mathbf{X}$ denotes a vector of covariates. In the aforementioned example, this assumption states that, within each age group, the treated and the control individuals with the same pretreatment characteristics $\mathbf{X}$ would have the same average health outcome in a hypothetical world in which they all used estrogen therapy; they would also have the same average health outcome if no one used estrogen therapy. The identification assumption can be formally represented as follows for a binary treatment:

$$E[Y(1)|R=r,\mathbf{X}] = E[Y(1)|Z=1,R=r,\mathbf{X}] = E[Y|Z=1,R=r,\mathbf{X}];$$
$$E[Y(0)R=r,\mathbf{X}] = E[Y(0)|Z=0,R=r,\mathbf{X}] = E[Y|Z=0,R=r,\mathbf{X}].$$

$$(7.2)$$

Under the above assumption, among those with the same pretreatment characteristics, the average observed health outcome of the older women who took estrogen would be unbiased for the average potential outcome of taking estrogen for the entire subpopulation of older women; and the average observed health outcome of the older women who did not take estrogen would be unbiased for the average potential outcome of not taking estrogen for the entire subpopulation of older women. The same logic would apply to the subpopulation of younger women. Clearly, assumption (7.2) would be made more plausible if $\mathbf{X}$ includes the entire set of pretreatment covariates that, according to theory or past empirical evidence, are related to cardiovascular health and are predictors of whether one would use estrogen therapy during a given age range. However, the risk of misspecifying the functional form of the covariate–outcome relationship in an ANCOVA or regression model increases as the number of covariates increases.

In moderation analyses of a binary treatment with quasiexperimental data, the identification assumption (7.2) additionally implies that, within each subpopulation defined by a moderator value and at each level of the pretreatment covariates, each individual must have a nonzero probability of adopting the alternative treatment. That is,

$$0 < \mathrm{pr}(Z=z|R=r,\mathbf{X}) < 1 \qquad (7.3)$$

for $z=0, 1$ and $r=0, 1$ in the current example. Suppose that among older women, none of those with a history of heart attack were given estrogen. Then the causal effect of estrogen on cardiovascular health cannot be defined and identified for such women. This consideration has important implications not only for the generalizability (i.e., external validity) of the analytic results but also for the internal validity of the results obtained from ANCOVA- and regression-based analyses. This is because older women with a history of heart attack are expected to have more cardiovascular problems than other older women on average; the former group may have an undue impact on the estimation of the slope relating heart attack history to cardiovascular health. As discussed in Chapter 3 and illustrated in Figure 3.1, when the experimental group and the control group have little or no overlap in the distribution of a covariate X, covariance adjustment for X relies heavily on linear extrapolation and may lead to a considerable amount of bias in the estimation of treatment effects. Applications of ANCOVA- and regression-based methods generally do not involve an empirical examination of the covariate space across different treatment groups. Later in this chapter, we will discuss the possibility of overcoming this limitation when researchers opt for propensity score-based adjustment methods.

7.1.1.2   Treatment effects moderated by a concurrent treatment

A second type of moderation questions involves the effect of one treatment on the outcome moderated by a concurrent treatment. Chapter 6 gave the example that, in theory, the potential benefit of grouping students in a kindergarten class by ability for literacy instruction is contingent upon whether the teacher has allocated adequate time for extensive instructional interactions with students in each group. Researchers have reasoned that ability grouping in a kindergarten class involves a considerable amount of teacher transitioning between groups and behavioral management. Hence, when the instructional time is constrained, a large share of the time will be spent on management, leaving little time for teaching and learning. Ability grouping was therefore hypothesized to be beneficial for literacy learning under a relatively high amount of time and detrimental under a low amount of time (Hong and Hong, 2009; Hong et al., 2012). The same researchers found that kindergarten classes with more literacy instruction time or more intensive grouping had, on average, a higher concentration of disadvantaged students, were more likely to be in schools with no selective criteria for kindergarten enrollment, and had limited educational resources. Yet, at the same time, these classes were under the greatest pressure to raise student achievement.

The study therefore involved two concurrent treatments: literacy instruction time denoted by Z_1 and within-class homogeneous grouping denoted by Z_2. The first treatment indicator Z_1 takes value 1 for a high amount of time and 0 for a low amount of time; the second treatment indicator Z_2 has three levels: $Z_2 = 0$ for no grouping, $Z_2 = 1$ for low-intensity grouping, and $Z_2 = 2$ for high-intensity grouping. To identify the causal effect of ability grouping at each level of instructional time in quasiexperimental data, the researchers invoked the assumption that, among kindergartners whose classes, teachers, and schools share the same observed pretreatment characteristics $\mathbf{X}$, the level of instructional time and intensity of grouping are independent of one's potential outcomes. With $2 \times 3 = 6$ treatment conditions, a kindergartner could have six potential outcomes, one associated with each of the six treatment conditions. For example, $Y(0, 0)$ is the potential outcome if the kindergarten class attended by a child adopted low instructional time with no grouping, $Y(1, 0)$ is the potential outcome if the same class adopted high instructional time with no grouping, and so on. The identification assumption can be formally represented as

$$E[Y(z_1, z_2)|\mathbf{X}] = E[Y(z_1, z_2)|Z_1 = z_1, Z_2 = z_2, \mathbf{X}] = E[Y|Z_1 = z_1, Z_2 = z_2, \mathbf{X}] \qquad (7.4)$$

for $z_1 = 0, 1$ and $z_2 = 0, 1, 2$.

A second part of the identification assumption states that, among kindergartners sharing the same class, teacher, and school characteristics $\mathbf{X}$, the probability is nonzero for every student that his or her class would adopt any one of the six treatment conditions:

$$0 < \mathrm{pr}(Z_1 = z_1, Z_2 = z_2|\mathbf{X}) < 1 \qquad (7.5)$$

for $z_1 = 0, 1$ and $z_2 = 0, 1, 2$.

When the above identification assumption holds, the main effect of each treatment and their interaction effect can be estimated and tested through a two-way ANOVA with covariance adjustment for the pretreatment covariates including student composition in class and in school, teacher characteristics, and school characteristics. Similar evidence can be generated through analyzing a multiple regression model. The two levels of time ($z_1 = 0, 1$) and the three levels of grouping ($z_2 = 0, 1, 2$) define six distinct treatment conditions. We may let D_1 be a dummy indicator for low-intensity grouping and D_2 a dummy indicator for high-intensity grouping each having an interaction with Z_1, the dummy indicator for high time as opposed to low time. With low instruction time and no grouping being the reference treatment condition, the following regression model directly tests the time-by-grouping interaction effects:

$$Y = \beta_0 + \beta_1 D_1 + \beta_2 D_2 + \beta_3 Z_1 + \beta_4 Z_1 D_1 + \beta_5 Z_1 D_2 + \gamma^T X + \varepsilon. \tag{7.6}$$

Here, γ^T is the transpose of a vector of regression coefficients. In a relatively simple case in which there are only three pretreatment covariates X_1, X_2, and X_3, $\gamma^T X$ is a shorthand for

$$(\gamma_1 \, \gamma_2 \, \gamma_3) \begin{pmatrix} X_1 \\ X_2 \\ X_3 \end{pmatrix} = \gamma_1 X_1 + \gamma_2 X_2 + \gamma_3 X_3.$$ The shorthand is particularly useful when the number

of covariates is relatively large. The regression coefficients β_1 and β_2 represent the effects of low-intensity grouping and high-intensity grouping, respectively, relative to no grouping under low time. β_3 represents the effect of high time versus low time when no grouping is employed in instruction. The grouping effects are said to depend on instructional time if $\beta_4 \neq 0$ and/or $\beta_5 \neq 0$.

Even though covariance adjustment for the potential confounders through ANCOVA provides a familiar approach to most researchers for investigating whether the effect of one treatment is moderated by a concurrent treatment, the potential pitfalls of this approach remain. As explained in Chapter 3, major risks include model misspecification and linear extrapolation. Model (7.6) assumes a constant relationship between the outcome and each covariate across all six treatment conditions. The results will be biased if the covariate–outcome relationship varies across different treatment conditions. Covariance adjustment becomes especially cumbersome, and the risk of model misspecification increases, when the selection bias is associated with a large number of pretreatment covariates, which is often the case in quasiexperimental research. An alternative analysis of two parallel regression models, one for each level of instructional time, however, would not allow for appropriate covariance adjustment for X that may confound the time effect on the outcome and therefore is inapplicable. Moreover, according to identification assumption (7.5), the causal inference applies to a population of individual units each having a nonzero probability of being assigned to each of the six treatment conditions. Yet the analytic procedure described here does not involve detection of a possible lack of counterfactual information for some individuals who do not have counterparts under some alternative treatment conditions. When the lack of counterfactual information is a problem for a certain portion of the population, the generalizability of the results becomes dubious, and bias often sneaks into the estimation.

7.1.2   Propensity score-based adjustment

In the recent years, applied researchers have started to use propensity scores to summarize pretreatment information in analyses of quasiexperimental data. As explained in Chapter 3, a propensity score represents an individual's conditional probability of being assigned to a certain treatment given the individual's pretreatment characteristics. This can be estimated by determining the proportion of individuals with the same or similar pretreatment characteristics who have actually been assigned to the given treatment. In practice, the estimated propensity score is a function of a relatively large number of observed pretreatment covariates $\mathbf{X}$. Rosenbaum and Rubin (1983) showed that the joint distribution of $\mathbf{X}$ is balanced between the treated units and the control units who share the same estimated propensity score value. Therefore, if the treatment assignment is "strongly ignorable"—that is, if the treatment assignment is independent of the potential outcomes given the observed pretreatment covariates $\mathbf{X}$—it is also "strongly ignorable" given the propensity score. Hence, the treated units and the control units with the same propensity score value can act as controls for each other.

7.1.2.1   Propensity score matching and stratification

Matching or stratifying the treated and control units on the estimated propensity scores provides important alternatives to covariance adjustment. As discussed in Chapter 3, in evaluating a binary treatment, one may estimate each individual's propensity score through analyzing a logistic regression model. For each treated unit, if the researcher is able to find a control unit who can be matched on the basis of the estimated propensity score, the difference in the observed outcome between the treated unit and the control unit, averaged over all the matched pairs, estimates the average treatment effect on the treated. Propensity score stratification makes use of nearly all the observations except for individuals who have extremely high or extremely low propensity scores who do not have counterparts in the alternative treatment group. The sample is sorted on the basis of the estimated propensity score and is then divided into a number of strata with the goal of minimizing the difference in the distribution of pretreatment covariates between the treated group and the control group within each stratum. The within-stratum mean difference in the observed outcome between the treated group and the control group, averaged over all the strata, estimates the population average treatment effect.

Propensity score matching and stratification are nonparametric in nature because they avoid model-based assumptions with regard to the functional relationship between each covariate and the outcome. Researchers often combine propensity score matching and stratification with covariance adjustment for the estimated propensity score or covariance adjustment for other strong predictors of the outcome. Their aim is to further reduce bias and increase precision. In this case, the risk of model misspecification is greatly reduced within matched pairs or within strata. Matching or stratification on the estimated propensity score also enables the researcher to easily detect and discard individual observations that do not have counterparts under an alternative treatment condition.

Matching and stratification may also provide useful information with regard to possible moderation by pretreatment characteristics. One can assess whether the within-matched pair or within-stratum treatment effect changes with a clear pattern across the range of the estimated propensity score. When the hypothesized moderator is an observed pretreatment covariate R that defines subpopulations and when the interest is in estimating the causal effect of a binary treatment Z moderated by subpopulation membership, propensity score matching or stratification can be conducted within each subpopulation to estimate the subpopulation-specific treatment effect. Specifically, in a subpopulation defined by $R = r$, the propensity of treatment assignment $\theta_{1|r}(\mathbf{x}) = \mathrm{pr}(Z = 1 | R = r, \mathbf{X} = \mathbf{x})$ is estimated for each individual as a function of pretreatment characteristics $\mathbf{x}$. To simplify the notation, we use $\theta_{1|r}$ in the following as a shorthand for $\theta_{1|r}(\mathbf{x})$. The treated individuals and the control individuals in the same subpopulation r are then matched or stratified on the propensity score $\theta_{1|r}$. The mean differences in the observed outcome between the treated individuals and the control individuals within matched pairs or strata are then pooled over all the matched pairs or all the strata, generating an estimate of the subpopulation-specific treatment effect. The estimated difference in the treatment effect between different subpopulations, if statistically significant, is taken as an indication for moderation. The analysis invokes the identification assumptions represented by (7.2) and (7.3).

However, as explained in Chapter 5, propensity score matching and stratification are limited in evaluating multivalued treatments and multiple concurrent treatments. This is because an evaluation of K treatment conditions when $K > 2$ requires simultaneous adjustment for $K-1$ propensity scores. One of the application examples in Chapter 5 was an evaluation of three distinct types of educational services for English language learners (ELLs) in comparison with no ELL services. In that case, simultaneous adjustment for three propensity scores became necessary. In another study investigating whether the effect of within-class grouping depends on instructional time, adjustment is required for removing pretreatment differences between six treatment conditions. Statistical adjustment is required for as many as five propensity scores. Matching and stratification on multiple propensity scores are generally difficult to implement. If the researcher conducts pairwise comparisons instead, by matching or stratification with a pair of treatment groups each time, a global test cannot be applied to avoid an inflated type I error in multiple pairwise comparisons.

7.1.2.2 Inverse-probability-of-treatment weighting

Inverse-probability-of-treatment weighting (IPTW) is relatively flexible for handling multivalued and multiple treatments. This strategy focuses on estimating the population average potential outcome associated with each treatment condition and compares the estimated population average potential outcomes across the treatment conditions. In addressing the first type of moderation question, the weighting adjustment is conducted within each subpopulation. For individuals in subpopulation r assigned to treatment z, the weight is inverse to each individual's propensity score of being assigned to that treatment:

$$\text{IPTW} = \frac{\text{pr}(Z=z|R=r)}{\text{pr}(Z=z|R=r,\mathbf{X}=\mathbf{x})} = \frac{\text{pr}(Z=z|R=r)}{\theta_{z|r}}.$$

When the identification assumptions in (7.2) and (7.3) hold, the weighted data approximate a randomized block design. The weighted outcome model takes a similar form as that in (7.1) except that it becomes unnecessary to make covariance adjustment for any pretreatment covariate.

The second type of moderation questions may involve two concurrent treatments denoted by Z_1 and Z_2. For individuals assigned to a treatment condition jointly defined by z_1 and z_2, the weight is inverse to one's propensity score of being assigned to that joint treatment condition denoted by $\theta_{z_1,z_2}(\mathbf{x}) = \text{pr}(Z_1=z_1,Z_2=z_2|\mathbf{X}=\mathbf{x})$. In the following, we use θ_{z_1,z_2} as a shorthand for $\theta_{z_1,z_2}(\mathbf{x})$:

$$\text{IPTW} = \frac{\text{pr}(Z_1=z_1,Z_2=z_2)}{\text{pr}(Z_1=z_1,Z_2=z_2|\mathbf{X}=\mathbf{x})} = \frac{\text{pr}(Z_1=z_1,Z_2=z_2)}{\theta_{z_1,z_2}}.$$

Under assumptions (7.4) and (7.5), the weighted data approximate a factorial randomized design. The weighted outcome model can be similar to (7.6) yet again with no need to make covariance adjustment for pretreatment covariates.

Reanalyzing data from the Nurses' Health Study, Hernán and colleagues (2008) made an attempt to mimic the WHI sample by restricting the study population to healthy postmenopausal women who were first-time users or reinitiators of estrogen therapy in each 2-year period. Employing IPTW, they adjusted for baseline covariates that predicted whether one would initiate or reinitiate estrogen therapy. Because some women discontinued their baseline treatment before experiencing a health event such as a heart attack or mortality, additional adjustment was made for noncompliance. The researchers reported a higher hazard ratio in women whose initiation or reinitiation occurred more than 10 years after menopause than in women within 10 years after menopause (p value for interaction = 0.01). These were provided as evidence suggesting that the effect of estrogen therapy on heart disease depends on the lapse of time since menopause.

However, Chapter 4 has discussed some major limitations of IPTW including its lack of robustness and efficiency (Hong, 2010). In particular, IPTW results are sensitive to misspecifications of the functional form of the propensity score models; it tends to introduce additional bias when only a portion of the population or subpopulation provides support for causal inference; and in comparison with other adjustment methods, IPTW results are generally less precise especially when the estimated weights become volatile.

Chapter 4 has introduced MMWS, an alternative weighting method that makes use of propensity scores. Later, we review its rationale and then extend the strategy to moderation analyses. For each type of moderation questions, we describe a case study that illustrates the application of MMWS.

7.2   MMWS estimation of treatment effects moderated by individual or contextual characteristics

MMWS builds on the rationale and techniques similar to poststratification adjustment (Horvitz and Thompson, 1952), direct standardization (Little, 1982; Little and Pullum, 1979), and adjustment for nonresponse through weighting (Little, 1986; Little and Vartivarian, 2003, 2005; Oh and Scheuren, 1983) in survey sampling. Rather than applying sample weight to adjust the sample composition such that it becomes representative of the population, MMWS computes and applies a weight to individuals in each treatment group in the sample such that the pretreatment composition of each group becomes similar to that of the entire sample. The goal is to transform the sample data to approximate experimental data. In that sense, MMWS and IPTW share the same logic: when individuals with certain pretreatment characteristics are overrepresented in one treatment group and underrepresented in another treatment group, their unequal representations make the treatment groups "incomparable" and will likely lead to selection bias in estimating the treatment effect. Such individuals need to be down-weighted in the first group and up-weighted in the second group. As a result of weighting, individuals with the same pretreatment characteristics will have equal representations in different treatment groups. Therefore, it becomes possible to attribute the between-group difference in the average outcome to the treatment rather than to the pretreatment characteristics that initially differed between the treatment groups.

Yet unlike IPTW, MMWS is not computed as a direct function of the estimated propensity score. Rather, MMWS uses the estimated propensity score to stratify a sample. In the case of a binary treatment, individuals in the same stratum are relatively homogeneous in their propensity to be treated regardless of their actual treatment status. The sample proportion of individuals in a given stratum that were actually treated is to be compared with the overall proportion of individuals in the entire sample that were treated. The inverse of the ratio of these two proportions is the marginal mean weight for the treated individuals in the given stratum. For example, suppose that 40% of the individuals in the entire sample are treated. Yet in the first stratum, only 20% of the individuals are treated, while 80% of the individuals are untreated. One may compute the weight $0.4/0.2 = 2$ for each treated individual and $(1-0.4)/(1-0.2) = 0.75$ for each untreated individual in the first stratum. A similar strategy is applied to each of the other strata. In the weighted data, 40% of the individuals will be treated and 60% will be untreated in every stratum.

In essence, MMWS uses the observed sample proportion of individuals treated in each stratum to replace the estimated propensity score. The sample proportion within a stratum is computed without resorting to any parametric models; hence, the MMWS procedure is nonparametric in nature. For this reason, MMWS reduces the potential pitfalls associated with misspecifications of the functional form of the propensity score model and generates results that tend to be more robust than the IPTW results. In addition, MMWS has a built-in procedure that identifies and excludes individuals who do not have counterparts in an alternative treatment group. In contrast, IPTW would assign a nonzero weight to such individuals as they often display an extremely high propensity of receiving the treatment that they actually received. The IPTW estimate of the treatment effect would become biased as a result (Hong, 2010).

Similar to IPTW, MMWS is particularly suitable for evaluating multivalued treatments (Hong, 2010; Hong and Hong, 2009; Huang et al., 2005; Zanutto, Lu, and Hornik, 2005). Chapter 5 provided an application example in which language minority students were assigned to one of three types of ELL services—pull-out ESL, in-class ESL, and bilingual education—or received no ELL services at all (Hong, 2012). Every language minority child therefore had four propensity scores corresponding to these four treatment conditions. Each propensity score is a summary of the observed pretreatment information associated with the assignment to the corresponding treatment condition. The identification assumption states that holding constant the propensity score associated with a given treatment condition z, the potential outcome associated with that treatment condition denoted by $Y(z)$ is no different on average between the individuals actually assigned to treatment condition z and the individuals who had not been assigned to this particular treatment condition. Under this assumption, once the pretreatment composition of treatment group z has been transformed through weighting to resemble that of the entire sample representative of a target population, the weighted mean outcome of the individuals in treatment group z consistently estimates the population average potential outcome if the entire population had hypothetically been assigned to treatment condition z. The estimate is consistent because as the sample size increases, the number of strata can grow as well, and the sample estimate will converge to the population parameter. The weighted data can be analyzed conveniently within an ANOVA framework.

The MMWS method can be readily extended to investigations of treatment effects moderated by individual or contextual characteristics in quasiexperimental data. Instead of attempting to approximate a completely randomized experiment, the goal is now to approximate a randomized block design in which the blocks are defined by values of a hypothesized moderator. Extending the previous example, Hong (2012) analyzed whether the impact of ELL services differs between Hispanic and non-Hispanic language minority students.

7.2.1   Application example

Nearly 80% of language minority students in the United States are Spanish speaking. The rest are from other non-English language backgrounds (Kindler, 2002). A great majority of Spanish-speaking students are of Hispanic origin. When compared with language minority students from most other racial/ethnic backgrounds, Hispanic language minority students tend to be more disadvantaged socioeconomically and display a lower oral English proficiency and a lower level of school readiness behaviors at kindergarten entry. Some researchers have considered Hispanic children as a subpopulation of unique educational needs (Reardon and Galindo, 2009; Robinson, 2008). To determine whether Hispanic and non-Hispanic language minority children may respond differently to ELL instruction, Hong (2012) compared the relative effectiveness of pull-out ESL instruction, in-class ESL assistance, bilingual education, and no provision of ELL services in fostering desirable learning approaches between Hispanic and non-Hispanic language minority kindergartners. The outcome is the teacher rating of child approaches to learning at the end of the kindergarten year. Effective ELL support is expected to remove important barriers to class participation and thereby improve a language minority student's engagement in learning activities.

As shown in the nationally representative Early Childhood Longitudinal Study–Kindergarten (ECLS-K) cohort data, Hispanic language minority kindergartners had a significantly higher likelihood of receiving bilingual education, while non-Hispanic language minority children were more likely to be pulled out for ESL instruction or receive no ELL services at all. The uneven allocation of the number of students to the treatment groups within each language subpopulation does not interfere with causal inference. Challenges arise only because the four treatment groups are "incomparable" in their pretreatment composition within each subpopulation. For example, in the ECLS-K data, non-Hispanic children from relatively higher-SES families were more likely to receive pull-out or in-class ESL instruction or have no ELL services rather than receiving bilingual education. In contrast, among Hispanic children, SES did not predict whether one was to receive pull-out or in-class ESL instruction or bilingual education. A higher SES was only associated with a higher likelihood of not receiving any ELL services.

In a hypothetical randomized block design ideal for evaluating the treatment effects moderated by Hispanic background, Hispanic language minority students and non-Hispanic language minority students would constitute two blocks. Within each block, students would be assigned at random to pull-out ESL, in-class ESL, bilingual education, and no ELL services. The probabilities of treatment assignment within each subpopulation would be consistent with the observed proportions of students in different treatment groups in the observed data, providing a realistic estimation of the sampling variation.

7.2.2   Analytic procedure

To approximate such a randomized block design, Hong (2012) recommended the following analytic procedure. The first five analytic steps need to be conducted within each of the two subpopulations. Step 6 applies to the combination of data from the two subpopulations.

Step 1. *Propensity score estimation.* Because the selection mechanisms are apparently different between Hispanic and non-Hispanic language minority students when they are assigned to various ELL services, a separate multinomial logistic regression model is specified for each subpopulation that estimates the propensity scores for receiving pull-out ESL ($z = 1$), in-class

ESL ($z=2$), bilingual education ($z=3$), or no ELL services ($z=0$). Let $r=1$ denote Hispanic and $r=0$ for non-Hispanic. Let $\theta_{z|r}$ denote the propensity for receiving treatment z for a student in subpopulation r. Every student therefore has four propensity scores that add up to 1.0. Chapter 5 explained how to specify a multinomial logistic model and obtain the propensity scores. Each propensity score is converted to a logit score $\eta_{z|r} = \theta_{z|r}/(1-\theta_{z|r})$ for subsequent analyses.

Step 2. *Common support identification*. Within each subpopulation, we compare the distribution of each logit score across the four treatment groups and identify the range of the logit score distribution within which the four treatment groups overlap. This range defines the common support for causal comparisons across the four treatment groups. Students whose four logit scores all fall within their corresponding common support constitute the analytic sample. Chapter 5 explained how to identify common support across multiple treatment groups.

Step 3. *Propensity score stratification*. Within each subpopulation, we then stratify the sample on the basis of each of the four logit scores one at a time. Tables 7.1 and 7.2 display the stratification results for Hispanic students and non-Hispanic students, respectively. In these two tables, $s_{z|r}$ denote the stratum in which a child in subpopulation r is located when the sample is stratified on $\eta_{z|r}$ for $z=0, 1, 2, 3$ and $r=0, 1$; $n_{s_{z|r}}$ is the number of students in stratum $s_{z|r}$; and $n_{z,s_{z|r}}$ is the number of students in stratum $s_{z|r}$ who actually received treatment z.

Step 4. *MMWS computation*. For example, after stratifying the sample of Hispanic language minority students on the logit propensity score for pull-out ESL services $\eta_{1|1}$, we compute the proportion of students in stratum $s_{1|1} = 1$ who actually received pull-out ESL services:

$$\text{pr}(Z=1|R=1, s_{1|1}=1) = \frac{n_{s_{1|1}=1}}{n_{z=1, s_{1|1}=1}} = \frac{4}{92}.$$

This is to be contrasted with the total proportion of Hispanic students who received pull-out ESL services:

$$\text{pr}(Z=1|R=1) = \frac{185}{1626}.$$

The marginal mean weight for the four Hispanic students in this stratum who actually received pull-out ESL services is the ratio of these two proportions:

$$\text{MMWS} = \frac{\text{pr}(Z=1|R=1)}{\text{pr}(Z=1|R=1, s_{1|1}=1)} = \frac{185}{1626} \times \frac{92}{4} = 2.62.$$

In general, when the sample of Hispanic students are stratified on the logit score for treatment z, if the Hispanic students receiving treatment z are underrepresented in their stratum relative to the total proportion of Hispanic students who received treatment z, as is the case in the previous example, these students would receive a weight above 1.0. If the students receiving treatment z are overrepresented in their stratum, they would receive a weight below 1.0. After weighting, all four treatment groups are expected to have the same pretreatment composition for Hispanic children. The same logic and procedure apply to the non-Hispanic students under a separate series of stratification as shown in Table 7.2.

Table 7.1 Marginal mean weight through stratification for four treatment groups: Hispanic

	No ELL services				Pull-out ESL instruction				In-class ESL instruction				Bilingual education		
$s_{0|1}$	$n_{s_{0|1}}$	$n_{z=0,s_{0|1}}$	MMWS	$s_{1|1}$	$n_{s_{1|1}}$	$n_{z=1,s_{1|1}}$	MMWS	$s_{2|1}$	$n_{s_{2|1}}$	$n_{z=2,s_{2|1}}$	MMWS	$s_{3|1}$	$n_{s_{3|1}}$	$n_{z=3,s_{3|1}}$	MMWS
1	276	23	3.82	1	92	4	2.62	1	325	18	2.75	1	232	28	3.44
2	272	27	3.21	2	251	13	2.20	2	324	39	1.27	2	346	86	1.67
3	271	55	1.57	3	511	42	1.38	3	324	52	0.95	3	116	35	1.38
4	267	85	1.00	4	259	28	1.05	4	217	33	1.00	4	229	101	0.94
5	403	222	0.58	5	256	33	0.88	5	218	39	0.85	5	233	128	0.76
6	137	106	0.41	6	130	24	0.62	6	218	67	0.50	6	235	124	0.79
				7	127	41	0.35					7	235	173	0.56
Total	1626	518		Total	1626	185		Total	1626	248		Total	1626	675	

Adapted with permission from Hong (2012). © 2012 by the American Psychological Association. The use of this information does not imply endorsement by the publisher.

Table 7.2 Marginal mean weight through stratification for four treatment groups: Non-Hispanic

	No ELL services				Pull-out ESL instruction				In-class ESL instruction				Bilingual education		
$s_{0|0}$	$n_{s_{0|0}}$	$n_{z=0,s_{0|0}}$	MMWS	$s_{1|0}$	$n_{s_{1|0}}$	$n_{z=1,s_{1|0}}$	MMWS	$s_{2|0}$	$n_{s_{2|0}}$	$n_{z=2,s_{2|0}}$	MMWS	$s_{3|0}$	$n_{s_{3|0}}$	$n_{z=3,s_{3|0}}$	MMWS
1	226	41	2.96	1	225	16	2.87	1	222	8	3.78	1	188	11	2.10
2	226	91	1.33	2	221	23	1.96	2	225	15	2.05	2	185	9	2.53
3	224	130	0.92	3	225	43	1.07	3	224	24	1.27	3	186	10	2.29
4	225	153	0.79	4	224	52	0.88	4	226	41	0.75	4	188	24	0.96
5	221	187	0.63	5	170	62	0.56	5	225	65	0.47	5	188	26	0.89
				6	57	33	0.35					6	187	58	0.40
Total	1122	602		Total	1122	229		Total	1122	153		Total	1122	138	

Adapted with permission from Hong (2012). © 2012 by the American Psychological Association. The use of this information does not imply endorsement by the publisher.

Step 5. *Balance checking.* This step is crucial for detecting if any important pretreatment covariates have been omitted from the propensity score model or if the sample has been stratified inappropriately. Within each subpopulation in the weighted data, we examine whether each of the four logit scores has the same distribution across all four treatment groups. We test this null hypothesis through analyzing a weighted general linear model analogous to a one-way ANOVA. Similar hypothesis testing is conducted with all the observed pretreatment covariates. If more than a few observed covariates remain imbalanced in the weighted data and, in particular, if some strong confounders are imbalanced, it will become necessary to modify the propensity score model or restratify the sample.

Step 6. *Treatment effect estimation.* The causal question about whether the relative effectiveness of ELL services differs between Hispanic and non-Hispanic language minority students can be addressed by estimating the treatment effects for each subpopulation and comparing the effects between the two subpopulations. This is identical to the two-way ANOVA used in the randomized block design except that the observations are now weighted. Figure 7.1a shows the average outcome of the Hispanic students and that of the non-Hispanic students in each of the four treatment groups in the unweighted data. On average, Hispanic language minority students were rated lower in learning approaches by their teachers when compared with non-Hispanic students. Without adjustment for the differences in pretreatment composition between the four treatment groups in each subpopulation, results from a two-way ANOVA indicate no significant differences in the average outcome between the four treatment groups. This seems to be true for both Hispanic and non-Hispanic students. The mean comparisons of the outcome in the weighted data between the four treatment groups in each subpopulation are displayed in Figure 7.1b. The mean contrast between pull-out ESL and bilingual education for the non-Hispanic students seems to be particularly noteworthy; for Hispanic students, bilingual education appears to be associated with a lower average outcome in comparison with all three other treatment conditions. However, results from a weighted

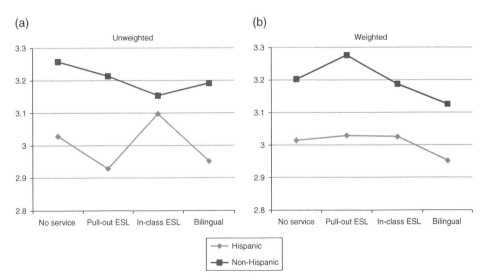

Figure 7.1   Mean of learning approaches for Hispanic and non-Hispanic language minority students by ELL services: (a) unweighted and (b) weighted

general linear model continue to suggest an insignificant main effect of ELL services and an insignificant interaction effect between ELL services and subpopulation membership. The statistical power for detecting a nonzero main effect and a nonzero interaction effect is likely constrained by the limited sample size and the relatively large unexplained variance in the outcome. To increase statistical power, we make additional covariance adjustment for child approaches to learning at kindergarten entry. This strategy reduces mean square error from 0.442 to 0.265 and increases R^2 from 0.028 to 0.418. We are then able to detect a significant main effect of ELL services, $F(3, 2728) = 3.033$, $p < 0.05$, while the interaction effect remains insignificant. Post hoc analysis of the weighted data within each subpopulation with adjustment for the type I error rate reveals that pull-out ESL benefitted non-Hispanic students especially when contrasted with bilingual education (coefficient $= 0.137$, SE $= 0.056$, $t = 2.453$, $p < 0.05$). The effect size is about one-fifth of a standard deviation of the outcome. For Hispanic students, bilingual education appears to be less effective when compared with the other three treatments in combination (coefficient $= -0.066$, SE $= 0.027$, $t = -2.471$, $p < 0.05$).

Step 7. *Sensitivity analysis.* The earlier results are causally valid if the identification assumptions hold. These assumptions take a similar form as (7.2) and (7.3) except that as many as four treatment conditions have been considered in the current study. It is assumed that, for Hispanic language minority students with the same observed pretreatment characteristics $\mathbf{X} = \mathbf{x}$, the subpopulation average potential outcome associated with treatment z can be estimated without bias by the mean observed outcome of Hispanic students with covariates $\mathbf{X} = \mathbf{x}$ who were actually assigned to treatment z. The same assumption is made for non-Hispanic students as well. The assumption applies to students who have a nonzero probability of being assigned to each of the four treatment conditions. The identification assumption can be summarized as

$$E[Y(z)|R=r,\mathbf{X}] = E[Y(z)|Z=z,R=r,\mathbf{X}] = E[Y|Z=z,R=r,\mathbf{X}];$$
$$0 < \mathrm{pr}(Z=z|R=r,\mathbf{X}) < 1$$

for $z = 0, 1, 2, 3$ and $r = 0, 1$. MMWS successfully removed or reduced the preexisting differences between the four treatment groups in each subpopulation in a large number of covariates. Yet one cannot rule out the possibility that the weighted estimates might nonetheless contain bias associated with some unmeasured confounders. For example, bilingual education tends to be offered in schools serving a large number of disadvantaged students under resource constraints. Hence, one may suspect that the aforementioned results questioning the benefits of bilingual education might be altered should additional adjustment for the school environment be made. Using school percentage of minority students and school percentage of free-lunch students as proxies for school environment, we find that, conditioning on the observed covariates in the propensity score model, neither of these two measures of school environment is associated with the outcome. Therefore, neither is likely to confound the existing results.

7.3   MMWS estimation of the joint effects of concurrent treatments

7.3.1   Application example

Earlier in this chapter, we described a case hypothesizing that the effect of within-class ability grouping in kindergarten literacy instruction depended on how much time the teacher allocated to such instruction (Hong and Hong, 2009). Specifically, constraint in literacy instruction time at most 1 hour/day was expected to offset the potential benefit of ability grouping, especially if grouping was employed with relatively high intensity taking up nearly half of literacy instruction time. Analyzing the ECLS-K data, the researchers evaluated the effects on kindergartners' literacy growth of high-intensity grouping and low-intensity grouping as opposed to no grouping under a high amount of time and a low amount of time.

Because US teachers tend to have considerable autonomy in allocating instructional time and determining whether and how often to use ability grouping (Firestone and Louis, 1999; Lortie, 1975), these decisions are typically made at the class level and would likely reflect class composition, teachers' professional orientation, and features of the school environment. The researchers employed MMWS to remove selection bias with the goal of approximating a factorial randomized experiment in which kindergarten classes would be assigned at random to either high time or low time and, within each level of instructional time, would be assigned at random to no grouping, low-intensity grouping, or high-intensity grouping. This would be equivalent to assigning classes at random to one of the six combinations of time and grouping.

7.3.2   Analytic procedure

As before, time and grouping are denoted by Z_1 and Z_2, respectively: Z_1 takes value 1 if a student received a high amount of literacy instruction time and 0 otherwise; Z_2 takes three different values 0, 1, and 2 representing no grouping, low-intensity grouping, and high-intensity grouping, respectively. The analysis followed a procedure that differs from the one described in the previous section primarily because no comparison is made between subpopulations. Steps 1 through 5 are carried out with no need to split the sample by subpopulation membership.

Step 1. *Propensity score estimation.* The researchers identified a total of 168 observed pretreatment covariates, created missing indicators to capture different missing patterns among categorical covariates, and imputed missing data in the continuous covariates and outcomes via maximum likelihood estimation. They then analyzed a multinomial logistic regression model at the class level, estimating six propensity scores per class corresponding to the six treatment conditions denoted by θ_{z_1,z_2} for $z_1 = 0, 1$ and $z_2 = 0, 1, 2$. Each propensity score is converted to a logit score $\eta_{z_1,z_2} = \theta_{z_1,z_2}/(1-\theta_{z_1,z_2})$ for subsequent analyses.

Step 2. *Common support identification.* Classes showing an almost zero probability of being assigned to one or more of the treatment conditions would not contribute to the causal inference and therefore were excluded from further analysis.

Step 3. *Propensity score stratification.* For each of the six treatment conditions, the analytic sample was divided into five or six strata on the basis of the corresponding logit score one at a time.

Step 4. *MMWS computation.* After the analytic sample has been stratified on the logit score η_{z_1,z_2}, for classes in the treatment condition jointly defined by z_1 and z_2 in stratum s_{z_1,z_2}, MMWS was computed as follows:

$$\text{MMWS} = \frac{\text{pr}(Z_1 = z_1, Z_2 = z_2)}{\text{pr}(Z_1 = z_1, Z_2 = z_2 | s_{z_1,z_2})}.$$

Here, the numerator is the proportion of classes in the analytic sample that adopted the treatment condition jointly defined by z_1 and z_2; the denominator is the proportion of such classes

Table 7.3   Computed MMWS for evaluating the joint effects of instructional time and grouping

	Low time			High time		
Stratum	No grouping	Low-intensity grouping	High-intensity grouping	No grouping	Low-intensity grouping	High-intensity grouping
1	3.04	3.31	3.05	3.38	4.57	3.50
2	1.63	1.23	1.07	1.75	2.08	1.39
3	0.92	0.78	0.72	0.78	1.31	0.55
4	0.40	0.41	0.30	0.53	0.83	0.32
5	0.23	0.27	0.23	0.44	0.63	0.18
6	—	—	—	0.28	0.45	—
Weighted n	296	290	240	308	518	206

in stratum s_{z_1, z_2}. Table 7.3 shows the computed MMWS by treatments and by stratum membership under each set of stratification. The weight has a mean of 1.0 and a standard deviation of 0.89.

Step 5. *Balance checking.* After weighting, all six treatment groups were expected to resemble the pretreatment composition of the entire analytic sample. The researchers analyzed a weighted general linear model analogous to a two-way ANOVA with a main effect of instructional time, a main effect of grouping, and a time-by-grouping interaction effect. All these effects were found indistinguishable from zero when each of the six logit scores was analyzed as the outcome. Additionally, about 95% of the 168 pretreatment covariates were balanced in distribution across the six treatment groups in the weighted data.

Step 6. *Treatment effect estimation.* Under the identification assumptions (7.4) and (7.5), once the pretreatment differences between the six treatment groups have been eliminated through weighting, the mean outcome of each of the six treatment groups is an unbiased estimate of the population average potential outcome associated with that treatment condition. The outcome model in its simplest form could be a weighted general linear model with a main effect of instructional time, a main effect of grouping, and a time-by-grouping interaction effect on the literacy outcome. Yet because the data contained repeated observations of students that have been vertically scaled to allow for an assessment of individual growth and because the students were nested in classes, the researchers analyzed a weighted three-level growth model with the repeated observations at level 1, students at level 2, and classes at level 3, applying the marginal mean weight to the kindergarten classes at level 3. The number of sampled kindergarten classes per school was too small to estimate the between-class variation in the literacy outcome within a school. The level 1 model for the literacy score of child i in class j at time t is

$$Y_{tij} = \pi_{0ij} + \pi_{1ij}(\text{Dur\_K})_{tij} + e_{tij}, e_{tij} \sim N\left(0, \sigma_{et}^2\right) \tag{7.7}$$

where (Dur_K) denotes the number of months between kindergarten entry and the assessment. The level 1 error variance σ_{et}^2 was considered to be known on the basis of the reliability of the assessment at time t denoted by λ_t. Specifically, $\sigma_{et}^2 = (1-\lambda_t)\sigma_{Yt}^2$ where σ_{Yt}^2 is the variance of the outcome Y at time t. The baseline score at kindergarten entry π_{0ij} and the monthly growth rate π_{1ij} varied randomly among students at level 2 and classes at level 3. The level 2 model was simply

$$\pi_{0ij} = \beta_{00j} + r_{0ij},$$

$$\pi_{1ij} = \beta_{10j} + r_{1ij};$$

$$\begin{pmatrix} r_{0ij} \\ r_{1ij} \end{pmatrix} \sim N \left[\begin{pmatrix} 0 \\ 0 \end{pmatrix}, \begin{pmatrix} \tau_{\pi00} & \tau_{\pi01} \\ \tau_{\pi10} & \tau_{\pi11} \end{pmatrix} \right]. \tag{7.8}$$

A saturated model was specified at level 3 for the class mean intercept β_{00j} and the class mean growth rate β_{10j}. As before, let D_1 and D_2 be dummy indicators for low-intensity grouping and high-intensity grouping, respectively. Each model takes a similar form as that in (7.6) except that there is no need to make covariance adjustment for all the pretreatment covariates:

$$\beta_{00j} = \gamma_{000} + \gamma_{001}D_{1j} + \gamma_{002}D_{2j} + \gamma_{003}Z_{1j} + \gamma_{004}Z_{1j}D_{1j} + \gamma_{005}Z_{1j}D_{2j} + u_{00j},$$

$$\beta_{10j} = \gamma_{100} + \gamma_{101}D_{1j} + \gamma_{102}D_{2j} + \gamma_{103}Z_{1j} + \gamma_{104}Z_{1j}D_{1j} + \gamma_{105}Z_{1j}D_{2j} + u_{10j};$$

$$\begin{pmatrix} u_{00j} \\ u_{10j} \end{pmatrix} \sim N \left[\begin{pmatrix} 0 \\ 0 \end{pmatrix}, \begin{pmatrix} \tau_{\beta00} & \tau_{\beta00.10} \\ \tau_{\beta10.00} & \tau_{\beta10} \end{pmatrix} \right]. \tag{7.9}$$

Hence, γ_{000} is the average baseline score of classes with low time and no grouping.

γ_{001} is the mean difference between low-intensity grouping and no grouping under low time.

γ_{002} is the mean difference between high-intensity grouping and no grouping under low time.

γ_{003} is the mean difference between high time and low time under no grouping.

γ_{004} will be nonzero if the mean difference between low-intensity grouping and no grouping under high time is different from the corresponding mean difference under low time;

γ_{005} will be nonzero if the mean difference between high-intensity grouping and no grouping under high time is different from the corresponding mean difference under low time.

We expect all these mean differences between treatment groups in the baseline score to be zero in the weighted data. When modeling the growth rate under weighting:

γ_{100} is the average monthly growth rate of classes with low time and no grouping.

γ_{101} is the effect on literacy growth of low-intensity grouping relative to no grouping under low time.

γ_{102} is the effect on literacy growth of high-intensity grouping relative to no grouping under low time.

γ_{103} is the effect of high time relative to low time under no grouping.

γ_{104} will be nonzero if the effect of low-intensity grouping relative to no grouping on literacy growth depends on instructional time.

γ_{105} will be nonzero if the effect of high-intensity grouping relative to no grouping on literacy growth depends on instructional time.

The estimation was conducted through pseudo-maximum likelihood that produced consistent estimates of the variance components and of the fixed effects (Hong and Raudenbush, 2008). Given the large sample of classes, hypothesis testing was based on robust standard errors clustered at the level of the classroom.

Hong and Hong (2009) reported that without the weighting adjustment, students attending classes with high time and high-intensity grouping had the lowest average baseline score, while those attending classes with low time and low-intensity grouping had the highest average baseline score. There was clearly a lack of equivalence between the treatment groups at kindergarten entry. These pretreatment differences disappeared in the weighted data. According to the weighted results, there was no benefit of either low-intensity grouping or high-intensity grouping relative to no grouping under low time. In contrast, when kindergartners received high literacy time, the monthly growth rate under low-intensity grouping and that under high-intensity grouping were both significantly higher than that under no grouping. These results are consistent with the initial hypothesis that the potential benefit of ability grouping in literacy instruction would become evident only if the teacher spent more than 1 hour/day teaching literacy. Converting the estimated treatment effects on monthly growth rate to the effects over a 9-month school year, Figure 7.2 illustrates the contrasts between the six treatment groups in the yearly growth. The moderating role of instructional time would have been disguised, however, if the causal effect of grouping had been examined in isolation. To examine the stability of these results under alternative model specifications, the researchers also obtained the weighted estimates of the treatment effects on the end-of-year reading achievement and arrived at the same conclusion.

Step 7. *Sensitivity analysis.* The weighted estimation of the treatment effects would be consistent under the assumption that the assignment to each treatment condition was independent of the unobserved confounders given the observed covariates. The richness of the pretreatment information in the ECLS-K data increased the plausibility of the strong ignorability assumption. Nonetheless, the researchers assessed the extent to which the causal conclusions would be altered by additional adjustment for potential unmeasured confounders. Among the observed pretreatment covariates, school safety rate was the strongest observed

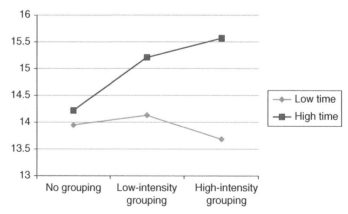

Figure 7.2   Weighted estimates of kindergartners' yearly growth rate in literacy by time and grouping

confounder of the treatment effects on literacy growth. Adjusting for either a positive or a negative hypothetical confounding effect of the same magnitude on the monthly growth rate, they found that the general conclusion with regard to the benefits of grouping under high time stayed the same. The last section in Chapter 3 provided detail on sensitivity analysis.

7.3.3   Joint treatment effects moderated by individual or contextual characteristics

When the effect of a treatment depends on a concurrent treatment, one may further investigate whether their joint effects may differ across different subpopulations or different contexts. Combining these two types of moderation questions enables one to gain even deeper understanding of the heterogeneity of treatment effects. Within-class homogeneous ability grouping again provides a classic example.

The greatest concern with ability grouping is that students with higher prior ability are encouraged and motivated to learn challenging academic content and thereby reach a higher level of achievement, while those with lower prior ability may suffer from lower teacher expectations and receive an overdose of rote drills that dull motivation and limit exposure to advanced content. The differentiated instruction under the ability grouping regime may generate a "Matthew effect" (Stanovich, 1986) that results in a widened gap between high and low achievers over time (Grant and Rothenberg, 1986; Oakes, 1995; Oakes, Gamoran, and Page, 1992; Rowan and Miracle, 1983; Trimble and Sinclair, 1987; Willms, 2002; Willms and Chen, 1989). Hence, it has been widely believed that individual prior ability moderates the effect of ability grouping.

Yet past attempts at comparing the relative effectiveness of grouping experienced by high-ability students with that experienced by low-ability students have often been misleading. The general observation that high-ability students placed in high reading groups learn faster than their low-ability peers in low reading groups, even after covariance adjustment for demographic characteristics and prior literacy skills, does not support the conclusion that low-achieving students, if placed in a high reading group instead, would have learned at the same rate as their high-achieving peers. Critics (Condron, 2008; Ferguson, 1998) pointed out that these are "apple–orange comparisons" due to the vast differences in prior academic skills between the high-ability students and the low-ability students. A sensible comparison is between low-ability students in grouped classes and their counterparts in nongrouped classes; and similarly, high-ability students in grouped classes are to be compared with their counterparts in nongrouped classes.

While the effects of ability grouping would likely depend on instructional time, researchers further hypothesized that the implications might be different for students at different prior ability levels (Hong *et al.*, 2012). Low-ability students tend to suffer most when instructional time is inadequate (Carroll, 1963) and would learn least when they are not directly supervised by the teacher under high-intensity grouping. In contrast, high-ability students tend to be self-regulated in learning and therefore are least vulnerable to suboptimal arrangements in instruction; medium-ability students are also expected to thrive when provided with ample time and substantial content for learning (Dreeben and Barr, 1988).

In a follow-up to the time-by-grouping study, the researchers (Hong *et al.*, 2012) identified high-ability, medium-ability, and low-ability students in grouped and nongrouped classes according to every student's relative standing in class. They then used the MMWS strategy to create a pseudo-subpopulation of students at each ability level. Specifically, the sample of students at a given ability level was stratified six times separately on the six estimated propensity scores corresponding to the six combinations of time and grouping. A weight was then computed for each child in each treatment group. The weighted sample of students at the same prior ability level therefore represented a pseudo-subpopulation within which kindergarten classes attended by these students could be viewed as if they had been assigned at random to different time-by-grouping treatments. Hence, the study approximated an aptitude–treatment interaction design (Cronbach and Snow, 1977; Snow, 1989; Snow, Federico, and Montague, 1980) in which two levels of instructional time and three levels of intensity of grouping constitute the experimental factors crossed by three ability levels. The joint effects of time and grouping could be estimated within each subpopulation, while differential treatment effects could be identified by comparing across the subpopulations.

A growth model was specified at level one that took the same form as that in (7.7). At level two (i.e., the student level), a dummy indicator for each of the three ability levels was entered in replacement of the single intercept:

$$\pi_{0ij} = \beta_{01j}(\text{high})_{ij} + \beta_{02j}(\text{medium})_{ij} + \beta_{03j}(\text{low})_{ij} + r_{0ij},$$
$$\pi_{1ij} = \beta_{11j}(\text{high})_{ij} + \beta_{12j}(\text{medium})_{ij} + \beta_{13j}(\text{low})_{ij} + r_{1ij}.$$

When implementing with multilevel modeling software such as HLM, one may leave the intercept β_{00j} and the slope β_{10j} in the level two model as placeholders for a level three (i.e., the class-level) random intercept u_{00j} and a level three random slope u_{10j} generic to all students in the same class regardless of their prior ability level (see Hong $et\ al.$ (2012) for details). Subsequently, at level three, a model comparing the baseline scores and another model comparing the monthly growth rates between the six treatment groups were specified for each ability subpopulation:

$$\beta_{pqj} = \gamma_{pq0} + \gamma_{pq1}D_{1j} + \gamma_{pq2}D_{2j} + \gamma_{pq3}Z_{1j} + \gamma_{pq4}Z_{1j}D_{1j} + \gamma_{pq5}Z_{1j}D_{2j}$$

for $p = 0, 1$ corresponding to the intercept and the slope and $q = 1, 2, 3$ corresponding to the three ability levels. To improve precision, the researchers entered additional pretreatment covariates at level two all centered at their respective subpopulation means.

The researchers examined an array of outcomes. They found no statistically significant effects of time and grouping on the average literacy growth of high-ability kindergartners; for medium-ability kindergartners, they reported an increase in literacy growth as literacy instruction time increased and when the intensity of grouping increased under high time; for low-ability kindergartners, there was clearly a detrimental effect of increasing the intensity of grouping under low time. The same patterns held across most literacy subdomains as well as in general learning behaviors for medium- and low-ability students. Interestingly, high-intensity grouping under low time showed a significant negative effect even for high-ability students learning vocabulary. The researchers reasoned that even though high-ability students were least vulnerable to ineffective instruction, in relatively advanced subdomains such as vocabulary, in which they were most in need of instructional guidance from the teacher, the detrimental effects of low instructional time with high-intensity grouping became apparent. Finally, in contradiction with the past belief that homogeneous grouping would harm low-ability students' self-esteem while improving high-ability students' self-esteem, the results showed that low-intensity grouping under high time apparently contributed to a reduction in internalizing problem behaviors (i.e., displays of sadness, loneliness, and self-withdrawal) among low-ability students. Grouping did not appear to affect medium- and high-ability kindergartners' social–emotional well-being regardless of time allocation. These findings provide useful evidence calling for a reevaluation of previous theories about the implications of homogeneous grouping for educational and social inequality.

In a related study, Hong and colleagues (Hong *et al.*, 2012) examined class manageability as a contextual factor that moderated the joint effects of time and grouping on kindergartners' externalizing problem behaviors (i.e., being disruptive, aggressive, and acting out) and literacy learning. Class manageability was measured on the basis of teacher reports of the frequency of disruptive behaviors in a class as a whole at the beginning of kindergarten. Good class management is presumably necessary for effective literacy instruction and may curtail disruptive behaviors. Analyzing the ECLS-K data, the researchers reported that the time-by-grouping effects depended on how difficult the class was to manage at the beginning of the year. In classes with frequent management problems, increasing time for grouped instruction reduced externalizing problem behaviors but had little impact on literacy learning. The instructional arrangements improved literacy learning only in well-managed classes and those with occasional disorder. Further analysis indicated that the impact of instruction was particularly evident for children with low or medium prior ability in well-managed classes.

Following an estimation of six propensity scores corresponding to the six time-by-grouping treatments for each class as described before, the analysis involved identifying the common support, stratifying on each of the propensity scores, computing MMWS, and checking balance within each level of class manageability. Specifically, the researchers constructed six weighted treatment groups within each of the three subsamples of classes defined by class manageability–classes with frequent disorder, occasional disorder, or well-managed. This strategy approximated a randomized block design where the blocks were defined by manageability. To quantify and test the time-by-grouping effects at each level of class manageability, the researchers specified a three-level model with students at level one, classes at level two, and schools at level three. The level one intercept representing the class mean outcome became an outcome at level two with three submodels, one for each level of class manageability. Each submodel estimated the effects of low-intensity grouping and high-intensity grouping relative to no grouping under low time or high time. As before, to avoid the inflated type I error in multiple comparisons, the researchers tested the omnibus null hypothesis regarding the effects of the six time-by-grouping treatments. They then tested whether a theory-based parsimonious model was as adequate as the full model. When this was true, they opted for the parsimonious model and tested the equivalence of the time-by-grouping effects across the three levels of manageability. Hypothesis testing was conducted through a series of likelihood ratio tests comparing nested models.

References

Carroll, J.B. (1963). A model of school learning. *Teachers College Record, 64*(8), 723–733.

Condron, D. (2008). Skill grouping and unequal reading gains in the elementary years. *Sociological Quarterly, 49*(2), 363–394.

Cronbach, L.J. and Snow, R.E. (1977). *Aptitudes and Instructional Methods: A Handbook for Research on Interactions*, Irvington, New York.

Dreeben, R. and Barr, R. (1988). Classroom composition and the design of instruction. *Sociology of Education*, *61*(3), 129–142.

Ferguson, R.F. (1998). Teachers' perceptions and expectations and the black-white test score gap, in *The Black-White Test Score Gap* (eds C. Jencks and M. Phillips), Brookings Institution Press, Washington, DC.

Firestone, W. and Louis, K. (1999). Schools as cultures, in *Handbook of Research on Educational Administration*, 2nd edn (eds J. Murphy and K. Louis), Jossey-Bass, San Francisco, CA.

Grant, L. and Rothenberg, J. (1986). The social enhancement of ability differences: teacher student interactions in first- and second-grade reading groups. *The Elementary School Journal*, *87*(1), 29–49.

Grodstein, F., Stampfer, M., Manson, J. *et al.* (1996). Postmenopausal estrogen and progestin use and the risk of cardiovascular disease. *New England Journal of Medicine*, *335*(7), 453–461.

Grodstein, F., Manson, J.E., Colditz, G.A. *et al.* (2000). A prospective, observational study of postmenopausal hormone therapy and primary prevention of cardiovascular disease. *Annals of Internal Medicine*, *133*(12), 933–941.

Grodstein, F., Manson, J.E. and Stampfer, M.J. (2006). Hormone therapy and coronary heart disease: the role of time since menopause and age at hormone initiation. *Journal of Women's Health*, *15*, 35–44.

Hernán, M.A., Alonso, A., Logan, R. *et al.* (2008). Observational studies analyzed like randomized experiments: an application to postmenopausal hormone therapy and coronary heart disease. *Epidemiology*, *19*(6), 766–779.

Hong, G. (2010). Marginal mean weighting through stratification: adjustment for selection bias in multilevel data. *Journal of Educational and Behavioral Statistics*, *35*(5), 499–531.

Hong, G. (2012). Marginal mean weighting through stratification: a generalized method for evaluating multi-valued and multiple treatments with non-experimental data. *Psychological Methods*, *17*(1), 44–60.

Hong, G. and Hong, Y. (2009). Reading instruction time and homogeneous grouping in kindergarten: an application of marginal mean weighting through stratification. *Educational Evaluation and Policy Analysis*, *31*(1), 54–81.

Hong, G. and Raudenbush, S.W. (2008). Causal inference for time-varying instructional treatments. *Journal of Educational and Behavioral Statistics*, *33*(3), 333–362.

Hong, G., Corter, C., Hong, Y. and Pelletier, J. (2012). Differential effects of literacy instruction time and homogeneous grouping in kindergarten: who will benefit? Who will suffer? *Educational Evaluation and Policy Analysis*, *34*(1), 69–88.

Hong, G., Pelletier, J., Hong, Y., and Corter, C. (2012). Does literacy instruction affect kindergartners' externalizing problem behaviors as well as their literacy learning? Taking class manageability into account (Working Paper). The University of Chicago and the Ontario Institute for Studies in Education of the University of Toronto.

Horvitz, D.G. and Thompson, D.J. (1952). A generalization of sampling without replacement from a finite universe. *Journal of the American Statistical Association*, *47*(260), 663–685.

Huang, I.-C., Frangakis, C., Dominici, F. *et al.* (2005). Approach for risk adjustment in profiling multiple physician groups on asthma care. *Health Services Research*, *40*, 253–278.

Kindler, A. (2002). Survey of the States' Limited English Proficient Students and Available Educational Programs and Services 2000–2001 Summary Report. Washington, DC: National Clearinghouse for English Language Acquisition.

Little, R.J.A. (1982). Models for nonresponse in sample surveys. *Journal of the American Statistical Association, 77*, 237–250.

Little, R.J.A. (1986). Survey nonresponse adjustments. *International Statistical Review, 54*, 139–157.

Little, R. and Pullum, T. (1979). *The General Linear Model and Direct Standardization: A Comparison.* International Statistical Institute, Voorburg. Occasional Papers No. 20.

Little, R.J.A. and Vartivarian, S. (2003). On weighting the rates in non-response weights. *Statistics in Medicine, 22*, 1589–1599.

Little, R.J.A. and Vartivarian, S. (2005). Does weighting for nonresponse increase the variance of survey means? *Survey Methodology, 31*, 161–168.

Lortie, D. (1975). *Schoolteacher: A Sociological Study*, The University of Chicago Press, Chicago, IL.

Manson, J.E. and Bassuk, S.S. (2007). Invited commentary: hormone therapy and risk of coronary heart disease why renew the focus on the early years of menopause? *American Journal of Epidemiology, 166*(5), 511–517.

Manson, J.E., Allison, M.A., Rossouw, J.E. *et al.* (2007). Estrogen therapy and coronary-artery calcification. *New England Journal of Medicine, 356*, 2591–2602.

Oakes, J. (1995). Two cities' tracking and within-school segregation. *Teachers College Record, 96*(4), 681–690.

Oakes, J. Gamoran, A. and Page, R. N. (1992). Curriculum differentiation: opportunities, outcomes, and meanings. In P. W. Jackson (ed), *Handbook of Research on Curriculum* (pp. 570–608). New York: Macmillan.

Oh, H.L. and Scheuren, F.S. (1983). Weighting adjustments for unit nonresponse, in *Incomplete Data in Sample Surveys*, vol. 2 (eds W.G. Madow, I. Olkin and D.B. Rubin), Academic Press, New York.

Reardon, S.F. and Galindo, C. (2009). The Hispanic–White achievement gap in math and reading in the elementary grades. *American Educational Research Journal, 46*, 853–891.

Robinson, J.P. (2008). Evidence of a differential effect of ability grouping on the reading achievement growth of language-minority Hispanics. *Educational Evaluation and Policy Analysis, 30*, 141–180.

Rosenbaum, P.R. and Rubin, D.B. (1983). The central role of the propensity score in observational studies for causal effects. *Biometrika, 70*, 41–55.

Rossouw, J.E., Prentice, R.L., Manson, J.E. *et al.* (2007). Postmenopausal hormone therapy and risk of cardiovascular disease by age and years since menopause. *The Journal of the American Medical Association, 297*(13), 1465–1477.

Rowan, B. and Miracle, A.W. (1983). Systems of ability grouping and the stratification of achievement in elementary schools. *Sociology of Education, 56*(3), 133–144.

Snow, R.E. (1989). Aptitude–treatment interaction as a framework for research on individual differences in learning, in *Learning and Individual Differences: Advances in Theory and Research* (eds P.L. Ackerman, R.J. Sternberg and R. Glaser), Freeman, New York.

Snow, R.E., Federico, P. and Montague, W. (1980). *Aptitude, learning, and instruction: cognitive process analyses*, Lawrence Erlbaum Associates, Hillsdale, NJ.

Stanovich, K.E. (1986). Matthew effects in reading: some consequences of individual differences in the acquisition of literacy. *Reading Research Quarterly, 21*, 36–407.

Taubes, G. (2007). Do We Really Know What Makes Us Healthy? New York Times Magazine, Sep 16, 2007.

Trimble, K.D. and Sinclair, R.L. (1987). On the wrong track: ability grouping and the threat to equity. *Equity and Excellence*, *23*, 15–21.

Willms, J.D. (2002). *Vulnerable Children: Findings From Canada's National Longitudinal Survey of Children and Youth*, University of Alberta Press, Edmonton, AB.

Willms, J.D. and Chen, M. (1989). The effects of ability grouping on the ethnic achievement gap in Israeli elementary schools. *American Journal of Education*, *97*(3), 237–257.

Writing Group for the Women's Health Initiative Investigators (2002). Risks and benefits of estrogen plus progestin in healthy postmenopausal women: principal results from the Women's Health Initiative randomized controlled trial. *The Journal of the American Medical Association*, *288*(3), 321–333.

Zanutto, E., Lu, B. and Hornik, R. (2005). Using propensity score subclassification for multiple treatment doses to evaluate a national antidrug media campaign. *Journal of Educational and Behavioral Statistics*, *30*(1), 59–73.

8

Cumulative effects of time-varying treatments

Human development is a longitudinal process in which the relationships between environmental inputs (considered as "treatments" here) and individual responses are of key interest. These relationships are highly dynamic partly because of the ongoing biological and psychological development in the respondent especially in periods of accelerated changes such as early childhood, adolescence, and the transition from middle age into old age. And it is partly because agents, in particular those who assume educators' or therapists' roles, often observe early responses and adapt subsequent treatments accordingly.

For example, it is now widely believed that increasing maternal verbal input increases a child's vocabulary (Hart and Risley, 1995). Yet it also seems that children who become better at utilizing the maternal input tend to engage their mothers in verbal exchanges more often and tend to elicit more verbal input from adults (Rowe, Pan, and Ayoub, 2005). Hence, a nonexperimental study may have difficulty separating a child's particular predisposition from the influence of maternal verbal input. This is because certain predispositions may manifest only under certain treatment conditions and because child predisposition and maternal verbal input may jointly contribute to a child's language development, adding to the confounding of genetic inheritance. Determining the impact of a series of maternal inputs in a longitudinal time frame—called "time-varying treatments"—is even more challenging due to the dynamic nature of the process. In particular, children who respond poorly to maternal input may subsequently receive reduced input or, if their mothers are on the alert, may instead be exposed to an enhanced dosage of verbal stimulation. The different ways in which agents adapt later treatments to prior individual responses may, therefore, result in different developmental trajectories.

This chapter starts with clarifying the causal questions of interest in studies of time-varying treatments and highlighting the vexing endogeneity problem of time-varying confounding. Time-varying confounders are intermediate outcomes of prior treatments that predict later treatment assignments and later outcomes. The chapter reviews the limitations of the standard methods including structural equation modeling (SEM) and fixed-effects econometric models and describes the existing experimental designs and causal inference approaches to time-varying confounding. The discussion then focuses on the rationale and the analytic procedure for using

Causality in a Social World: Moderation, Mediation and Spill-over, First Edition. Guanglei Hong.
© 2015 John Wiley & Sons, Ltd. Published 2015 by John Wiley & Sons, Ltd.

marginal mean weighting through stratification (MMWS) (Hong, 2010, 2012) to remove selection bias associated with both pretreatment and time-varying covariates. This is illustrated by a case study aimed at identifying an optimal length of English language learning (ELL) services for language minority students attending US elementary schools. Educators often decide whether a student should be assigned to an ELL program upon school entry and whether the treatment should continue every year on an individual basis. The impact of ELL services in a specific academic domain can be captured by a systematic shift in the individual growth trajectory. The same method applies to multiyear sequences of multivalued treatments.

8.1 Causal effects of treatment sequences

8.1.1 Application example

Language minority students, defined as those who speak a language other than English at home, are a fast-growing subpopulation in the United States constituting 21% of all school-aged children in 2009 (NCES, 2012). Many language minority students speak English with difficulty and are behind in their academic performance. The government allocates funds to support specialized services for English language learners (ELLs). ELLs are assessed annually for continued eligibility for services often on the basis of oral English proficiency, classroom performance, or grades. Yet states and districts vary in how they initially identify students as ELLs and how they determine whether a student is ready to exit an ELL program (Zehler *et al.*, 2003). The same student who might be considered not in need of ELL services initially or not in need of continuation of the services under one screening system might be provided with ELL services under a different system. The potential randomness in ELL services prevision may be attributed to teacher discretion, parental preference, shifting assessment criteria, measurement errors, or temporal changes in staffing and other local resources (Zehler *et al.*, 2008).

There have been intense debates about how many years of services are optimal for ELLs starting kindergarten in the United States. Researchers have argued that although it takes only several years of exposure to English for immigrant children to approach native-like fluency in conversational skills, it takes 4–9 years (Collier, 1987, 1989) or 5–7 years (Cummins, 1981) to become proficient in academic English essential for learning content areas (August and Hakuta, 1997). According to the nationally representative Early Childhood Longitudinal Study–Kindergarten (ECLS-K) cohort data, however, in practice, there has been vast variation across schools in the average number of years ELL services are provided, with 3 years being the mode. A policy-relevant question is: Does more than 3 years of ELL services benefit language minority students' academic learning more than three or fewer years of services? More specifically, one may ask about the contribution of every additional year of ELL services to academic learning.

Data for the application example are selected from the ECLS-K and are restricted to a sample of 2205 Spanish-speaking language minority students entering kindergarten in 1998. The students were observed in the fall of kindergarten, spring of kindergarten, spring of first grade, spring of third grade, and again in the spring of fifth grade. Among these students, 335 students never received ELL services throughout the elementary school years, while 440 students received more than 3 years of services starting from kindergarten. The outcome is math assessment scores vertically scaled over multiple waves of observations. The math assessment was administered in English if a student was proficient in English and was administered in Spanish if the student was proficient in Spanish but not in English.

8.1.2   Causal parameters

In the following, we introduce the general notation for time-varying treatments and time-varying potential outcomes. The notation will be used for defining the causal effects representing the relative effectiveness of different 2-year treatment sequences, which will then be extended to multiyear treatment sequences.

8.1.2.1   Time-varying treatments

We use z_t to denote the treatment received by a student in the year that ended at time t for $t = 0, 1, 3, 5$ representing the spring of kindergarten, spring of first grade, spring of third grade, and spring of fifth grade, respectively (no data were collected in second grade and fourth grade). Here, the value of t is chosen to correspond to the year. Let $z_t = 1$ if a student received ELL services in year t and 0 otherwise. In general, it is unlikely that a student would start ELL services in a later year. Nonetheless, a relatively small number of students received ELL services in first grade but not in kindergarten. To be complete, a 2-year treatment sequence is denoted by (z_0, z_1) that may take values $(1, 1)$, $(1, 0)$, $(0, 1)$, and $(0, 0)$ representing 2 years of treatment, treatment in kindergarten only, treatment in first grade only, and zero years of treatment, respectively. A student having ELL services in kindergarten, first grade, third grade, and fifth grade displays the treatment sequence $\bar{z}_5 = (z_0, z_1, z_3, z_5) = (1, 1, 1, 1)$. We use $\bar{z}_t$ to denote a treatment sequence from kindergarten up to time t. Examples of other dominant treatment sequences over the 6 years from kindergarten to fifth grade include $(1, 1, 1, 0)$, $(1, 1, 0, 0)$, $(1, 0, 0, 0)$, and $(0, 0, 0, 0)$. Among them, $(1, 1, 1, 1)$ and $(1, 1, 1, 0)$ are the treatment sequences indicating four or more years of ELL services, under the assumption that a student who received ELL services in first grade and third grade also received ELL services in second grade.

8.1.2.2   Time-varying potential outcomes

Let Y_t denote a student's math assessment score at time t for $t = -1, 0, 1, 3, 5$ where $t = -1$ corresponds to kindergarten entry. In general, a potential outcome at time t is a function of a treatment sequence that ends at time t. For simplicity, we invoke the stable unit treatment value assumption (SUTVA; Rubin, 1986), that is, the potential outcome associated with a given treatment or treatment sequence is stable regardless of how the treatment has been assigned and what treatments have been received by other individuals. The plausibility of SUTVA and possible ways of relaxing this assumption will be discussed in Chapters 14 and 15. Under SUTVA, the potential outcomes at the end of kindergarten are denoted by $Y_0(z_0)$, while those at the end of first grade are denoted by $Y_1(z_0, z_1)$ for $z_0 = 0, 1$ and for $z_1 = 0, 1$. In the current application, we also consider potential outcomes at the end of third grade and fifth grade denoted by $Y_3(\bar{z}_3)$ and $Y_5(\bar{z}_5)$, respectively.

8.1.2.3   Causal effects of 2-year treatment sequences

For a given student, the causal effect on the kindergarten outcome of having ELL services in kindergarten versus none is $Y_0(1) - Y_0(0)$. The causal effect on the first-grade outcome

Table 8.1   Glossary for the causal effects of 2-year treatment sequences

	Notation	Definition
Causal effect on the kindergarten outcome	$Y_0(1) - Y_0(0)$	Causal effect of being treated in kindergarten
Causal effects on the first grade outcome	$Y_1(1,0) - Y_1(0,0)$	Causal effect of being treated in kindergarten but not in first grade versus not having the treatment in either year
	$Y_1(0,1) - Y_1(0,0)$	Causal effect of being treated in first grade only versus not having the treatment in either year
	$Y_1(1,1) - Y_1(1,0)$	Causal effect of having a second year of treatment in first grade after having been treated in kindergarten
	$Y_1(1,1) - Y_1(0,0)$	Causal effect of having 2 years of treatment versus none

of having ELL services in kindergarten but not in first grade versus not having the treatment in either year is $Y_1(1,0) - Y_1(0,0)$, while the causal effect of having the treatment in first grade only versus none is $Y_1(0,1) - Y_1(0,0)$. If a student has already received ELL services in kindergarten, the causal effect of having a second year of treatment is defined by $Y_1(1,1) - Y_1(1,0)$. Finally, the causal effect of having 2 years of treatment versus none, $Y_1(1,1) - Y_1(0,0)$, can be represented as the sum of the effect of receiving ELL services in kindergarten only and the effect of receiving a second year of ELL services in first grade, that is, $[Y_1(1,0) - Y_1(0,0)] + [Y_1(1,1) - Y_1(1,0)]$. Table 8.1 provides a glossary for the causal effects of interest in an evaluation of 2-year treatment sequences.

The effect of having the treatment in kindergarten only, $Y_1(1,0) - Y_1(0,0)$, is not necessarily equal to the effect of having a second year of treatment, $Y_1(1,1) - Y_1(1,0)$; and neither of these two effects is necessarily equal to the effect of having the treatment in first grade only, $Y_1(0,1) - Y_1(0,0)$. One may reason that receiving ELL services in kindergarten is more crucial than receiving the treatment in later years for launching the academic growth of a language minority student; one may also reason that the effect of kindergarten ELL services would likely fade away a year later if not reinforced by a continuation of ELL services in first grade; one may further reason that because academic instruction tends to be much more intensive in first grade than in kindergarten, a delay of treatment that was much needed for a whole year might significantly undermine a student's prospect of catching up academically in first grade.

For a given student, the 2-year treatment sequence that leads to the maximal potential outcome value among $Y_1(1, 1)$, $Y_1(1, 0)$, $Y_1(0, 1)$, and $Y_1(0, 0)$ is considered to be optimal. Figure 8.1 illustrates, in a simplified scenario, four potential math growth trajectories of a hypothetical student. The student would display a linear growth throughout the kindergarten and first-grade years in the absence of ELL services, which is marked by the straight line on the bottom corresponding to the $(0, 0)$ sequence. The same student would experience a positive deflection from this trajectory in the amount of three points at the end of the kindergarten year if the student had received ELL services in kindergarten. Even if the ELL services did not

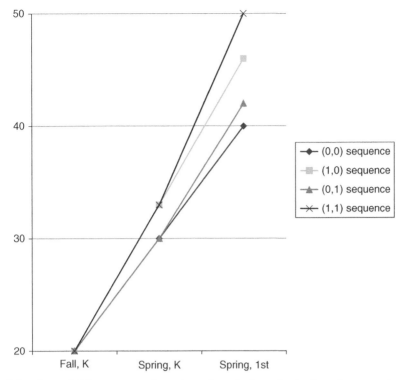

Figure 8.1   Potential 2-year growth trajectories of a hypothetical student associated with alternative treatment sequences

continue into first grade for this student, we postulate that the kindergarten treatment would nonetheless have a carryover effect that positively deflects the first-grade growth by another three points as marked by the trajectory that corresponds to the (1, 0) sequence. If the student received ELL services in first grade only, we suppose that the first-grade growth would be deflected by only two points as shown in the (0, 1) sequence. Finally, if the student received 2 years of ELL services, we suppose that there would be a three-point positive deflection to the kindergarten growth and another seven-point positive deflection to the first-grade growth represented by the (1, 1) sequence on the top.

We can observe only one of the four potential growth trajectories corresponding to the sequence of treatments that the student has actually received over the 2-year period. Hence, it is impossible to identify the optimal treatment sequence for a single student. Causal inferences are aimed at identifying the optimal treatment sequence for a given population of students by comparing the population average potential trajectories associated with different treatment sequences. We use $E[Y_0(0)]$ to denote the population average potential outcome at the end of kindergarten if, in a hypothetical world, no one in the population was assigned to ELL services in kindergarten; $E[Y_0(1)]$ represents the population average potential outcome at the end of kindergarten if, in a hypothetical world, the entire population was assigned to ELL services in kindergarten. Similarly, $E[Y_1(z_0, z_1)]$ represents the population average potential outcome at the end of first grade if the entire population was assigned to the treatment sequence (z_0, z_1) for $z_0 = 0, 1$ and $z_1 = 0, 1$.

8.1.2.4   Causal effects of multiyear treatment sequences

It is easy to show that the causal effect of having 4 years of treatment versus none is $Y_3(1,1,1) - Y_3(0,0,0)$; and the causal effect of having 6 years of treatment versus none is $Y_5(1,1,1,1) - Y_5(0,0,0,0)$. Relative to having the treatment for 2 years in kindergarten and first grade, the causal effect of having the treatment additionally in second and third grade is defined as $Y_3(1,1,1) - Y_3(1,1,0)$. If a child has already had 4 years of treatment from kindergarten to third grade, the causal effect of having the treatment additionally in fourth and fifth grade is $Y_5(1,1,1,1) - Y_5(1,1,1,0)$. An optimal 6-year treatment sequence leads to the maximal value among all the fifth-grade potential outcomes.

8.2   Existing strategies for evaluating time-varying treatments

8.2.1   The endogeneity problem in nonexperimental data

A major challenge to the evaluation of the causal effects of time-varying treatments was well documented in the epidemiology literature over 100 years ago as "the healthy worker effect," which consists of the "healthy worker hire effect" and the "healthy worker survivor effect" (Arrighi and Hertz-Picciotto, 1994). In the initial selection, healthy individuals are more likely to seek and secure employment than those who are less healthy; the health-based selection continues because workers who remain healthy are more likely to stay employed, while those whose health status deteriorates are more likely to leave employment. Therefore, comparing the health outcome of the employed with that of the unemployed or comparing the health outcome of those who have remained employed with the outcome of those who have left employment, one would likely underestimate the harm of exposure to occupational hazard to individual health.

A similar challenge arises in our evaluation of the causal effect of multiyear ELL services on academic learning. English proficiency is one of the most important predictors of a language minority student's academic outcome. At kindergarten entry, those who display a higher level of proficiency in English are less likely to be assigned to an ELL program; subsequently, at the end of each treatment year, those who have gained more proficiency in English are more likely to exit the ELL program. Therefore, comparing the average level of academic outcome of the treated students with that of the untreated or comparing the average outcome of those who have remained in an ELL program with those who have already exited, one would likely underestimate the potential benefit of ELL services.

The dynamic selection processes as described earlier result in the endogeneity problem in causal inference, which cannot be easily addressed by most methods for statistical adjustment. We use a 2-year treatment sequence to illustrate. As before, Z_0 and Z_1 are time-varying treatments in kindergarten and first grade, respectively, because a student may change treatment over the years. Let X_{-1} denote a student's English proficiency level at kindergarten entry; and let X_0 denote the student's English proficiency level at the end of the kindergarten year. Hence, X_{-1} is a baseline pretreatment covariate, while X_0 is a time-varying covariate, the value of which could be different from that at the baseline; Figure 8.2 highlights the key relationships among the covariates, treatments, and outcomes. Here, Y_{-1} is a baseline outcome (i.e., pretest math score) obtained at kindergarten entry; Y_0 and Y_1 are time-varying math outcomes

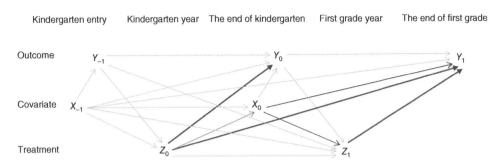

Figure 8.2   Endogeneity of time-varying treatments to time-varying covariates

obtained at the end of kindergarten and at the end of first grade, respectively. We can see that X_{-1} predicts both Z_0 and Y_0 and therefore confounds the effect of Z_0 on Y_0; X_{-1} also predicts both Z_0 and Y_1 and therefore confounds the effect of Z_0 on Y_1; X_{-1} additionally predicts both Z_1 and Y_1 and confounds the effect of Z_1 on Y_1; and finally, X_0 predicts both Z_1 and Y_1 and therefore confounds the effect of Z_1 on Y_1. Together, X_{-1} and X_0 confound the cumulative effect of Z_0 and Z_1 on Y_1.

In evaluating the causal effects of time-varying treatments, standard regression methods are suitable for adjusting for the confounding effects of baseline pretreatment covariates such as X_{-1} yet are unsuitable for adjusting for time-varying covariates such as X_0. This is because X_0 is an *intermediate outcome* of Z_0. As shown by Rosenbaum (1984), estimation of treatment effects is generally biased if one adjusts for a concomitant variable that has been affected by the treatment. In this case, adjustment for X_0 will likely bias the estimation of the effect of Z_0 on Y_0 and the effect of Z_0 on Y_1 and thereby derail our attempt to reveal how the treatment effects accumulate over the 2 years. A similar endogeneity problem applies to Y_0 that is an intermediate outcome of Z_0 and at the same time confounds the effect of Z_1 on Y_1.

The conventional approaches to evaluating time-varying treatments include SEM and fixed-effects econometric models. Neither provides a solution to the endogeneity problem. In recent advances in causal inference, researchers have conceptualized time-varying treatments under two alternative frameworks—namely, sequential randomization and dynamic treatment regimes (Lavori and Dawson, 2004; Murphy, 2003; Robins, 1986, 1987; Zajonc, 2012). These frameworks correspond to different experimental designs and require different analytic strategies to adjust for potential confounding in nonexperimental studies. In the following, we discuss their implications for the ELL study.

8.2.2   SEM

Applied researchers in psychology, sociology, and political science typically use SEM to address research questions that involve time-varying treatments and outcomes. In the example of 2-year treatment sequences, the analysis involves simultaneous modeling of Z_0, X_0, Y_0, Z_1, and Y_1 each as a function of its past treatments, covariates, and outcomes. These are written as a stack of equations. In particular, Y_1 is to be modeled as a function of Z_0, Z_1, X_{-1}, X_0, Y_{-1}, and Y_0, for example,

$$Y_1 = \alpha_0 + \beta_0 Z_0 + \beta_1 Z_1 + \beta_2 Z_0 Z_1 + \alpha_1 X_{-1} + \alpha_2 X_0 + \alpha_3 Y_{-1} + \alpha_4 Y_0 + e.$$

One may interpret β_0 as the effect on the first-grade outcome of having ELL services in kindergarten only, $E[Y_1(1,0) - Y_1(0,0)]$, and interpret β_1 as the effect on the first-grade outcome of having ELL services in first grade only, $E[Y_1(0,1) - Y_1(0,0)]$. Finally, β_2 may be interpreted as the effect on the first-grade outcome of having two consecutive years of ELL services above and beyond the sum of the effects of having ELL services in kindergarten only or in first grade only, which is equivalent to the difference between the effect of having a second year of ELL services in first grade after having already received ELL services in kindergarten and the effect of having ELL services only in first grade:

$$E[Y_1(1,1) - Y_1(0,0)] - E[Y_1(1,0) - Y_1(0,0)] - E[Y_1(0,1) - Y_1(0,0)]$$

$$= E[Y_1(1,1) - Y_1(1,0)] - E[Y_1(0,1) - Y_1(0,0)]. \tag{8.1}$$

However, as explained in the last section, because X_0 and Y_0 are measured at the end of kindergarten and are intermediate outcomes of the kindergarten treatment Z_0, when X_0 and Y_0 are controlled for in the aforementioned model for Y_1, β_0 would be biased for the effect on the first-grade outcome of having ELL services in kindergarten. Without controlling for X_0 and Y_0, however, β_1 would be biased for the effect of having ELL services in first grade. Furthermore, β_2 would be biased for the interaction effect one way or another.

8.2.3   Fixed-effects econometric models

Economists and public policy researchers have a tradition of employing fixed-effects models in analyzing time-varying data. When applied to the aforementioned example of 2-year treatment sequences, these models use person-specific fixed effects to eliminate selection bias associated with between-person variation in the covariates. Either time-varying assessment scores Y_{ti} or time-varying gain scores $\Delta Y_{ti} = Y_{ti} - Y_{t-1,i}$ for $t = 0,1$ for person i are modeled as functions of the individual's past treatments and covariates. Different versions of fixed-effects models impose alternative constraining assumptions such as either a complete decay or no decay with regard to the lagged treatment effects—that is, the effect of Z_0 on Y_1 is assumed to be either zero or equivalent to the effect of Z_0 on Y_0. Moreover, the effect of a treatment in a later year is always assumed to be independent of one's past treatment history. Finally, fixed-effects econometric models assume that there is no time-varying confounding and hence do not provide a solution to the endogeneity problem (Sobel, 2012).

8.2.4   Sequential randomization

To evaluate the causal effects of time-varying treatments, one may assign a population of language minority students at random to either an ELL program or a control condition at kindergarten entry. Subsequently, at the end of each year, students who have experienced the same treatment sequence in the past will again be assigned at random to either treatment or control. In its extreme version, the probability of treatment assignment at a later time is not informed by a student's response to the earlier treatments. Figure 8.3 illustrates these event

| Kindergarten entry | Kindergarten year | The end of kindergarten | First grade year | The end of first grade |

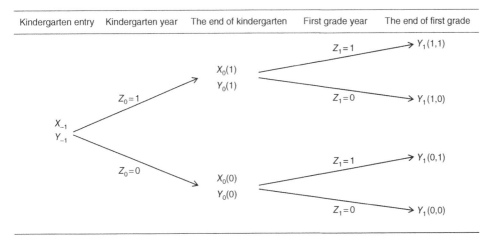

Figure 8.3   Sequential randomization of 2-year treatment sequences

trees in a relatively simple case of 2-year treatment sequences (Robins, 1986, 1987). The sequential randomization ensures the following:

1. The treatment group $Z_0 = 1$ and the control group $Z_0 = 0$ in kindergarten are comparable in their pretreatment composition $\mathbf{X}_{-1}$ and baseline outcome Y_{-1}. Here, $\mathbf{X}_{-1}$ denotes a vector of baseline covariates, observed or unobserved.

2. Within kindergarten treatment arm $Z_0 = z_0$ for $z_0 = 0, 1$, the treatment group and the control group in first grade are comparable at the end of kindergarten in the composition of pretreatment covariates $\mathbf{X}_{-1}$, baseline outcome Y_{-1}, time-varying covariates $\mathbf{X}_0(z_0)$, and time-varying outcomes $Y_0(z_0)$. Here, the time-varying covariates and the time-varying outcomes at the end of kindergarten are each represented as a function of the kindergarten treatment z_0.

This sequential randomization design is equivalent to assigning students at random to one of the four treatment sequences at the baseline. The average observed outcome of students assigned to a given treatment sequence (z_0, z_1) provides information on the average counterfactual outcome associated with that particular sequence for those assigned to each of the three other treatment sequences and therefore identifies the population average potential outcomes $E[Y_0(z_0)]$ and $E[Y_1(z_0, z_1)]$ should the entire population be hypothetically assigned to this same treatment sequence.

8.2.5   Dynamic treatment regimes

A treatment regime is a set of rules for choosing treatments for individuals given their prior status and ongoing progress. Some educational systems may prescribe a fixed number of years of ELL services as a mandatory sequence regardless of a student's initial or subsequent level of English proficiency. Other systems may require that educators screen their students periodically to determine whether a student should receive treatment in the next time period. This second type of treatment regime takes into account heterogeneity in the population, responds to each individual's ever-changing needs, and is therefore adaptive and dynamic in nature. For example, one dynamic treatment regime may specify that a language minority student who has become orally proficient in English should exit the ELL program, while another dynamic regime may require that a student falling behind in academic subjects continue the ELL program even if the student has achieved oral English proficiency.

Under a dynamic treatment regime with two time intervals, treatment assignment in kindergarten depends on pretreatment covariates $\mathbf{X}_{-1}$ and baseline outcome Y_{-1} following the rule $d_0(\mathbf{X}_{-1}, Y_{-1})$; treatment assignment in first grade, given the kindergarten treatment $Z_0 = z_0$, depends on pretreatment covariates $\mathbf{X}_{-1}$, baseline outcome Y_{-1}, time-varying covariates $\mathbf{X}_0(z_0)$, and time-varying outcomes $Y_0(z_0)$ and follows the rule $d_1(\mathbf{X}_{-1}, \mathbf{X}_0(z_0), Y_{-1}, Y_0(z_0), Z_0 = z_0)$. In general, Z_t is determined by $d_t(\bar{\mathbf{X}}_{t-1}, \bar{Y}_{t-1}, \bar{Z}_{t-1} = \bar{z}_{t-1})$ where $\bar{\mathbf{X}}_{t-1}$, $\bar{Y}_{t-1}$, and $\bar{Z}_{t-1}$ denote the past covariate history, past outcome history, and past treatment history up to time $t-1$. Should educators follow distinct treatment regimes in practice, contrasting the existing regimes would be of causal interest. A simple experiment would assign individual students (or, more realistically, individual schools or school districts) at random to different treatment regimes. The average observed outcome of those assigned to a treatment regime defined by the series of rules $\bar{d}_t$ is unbiased for the population average potential outcome should the entire population be hypothetically assigned to the same regime $E[Y_t(\bar{d}_t)]$.

In order to prospectively develop an optimal treatment regime, one may modify the sequential randomization design such that, at each decision point, individuals with identical covariate history, treatment history, and outcome history are randomized to the next set of treatments that represent different decision rules. One then identifies the set of dynamic rules that maximizes the desired outcome (Lavori and Dawson, 2004; Lei *et al.*, 2012; Murphy, 2003, 2005; Zajonc, 2012).

For determining the optimal length of ELL services, a sequential randomized experiment seems unrealistic because it would be against the law to deprive access to ELL services of students who are deemed eligible. Even though there are conceivably multiple treatment regimes in practice around the country, their decision rules are often unknown to the analyst. Hence, it would be difficult to analyze data as a quasiexperiment of ELL treatment regimes, that is, to view students as being assigned to different treatment regimes at the baseline with selection bias only associated with the baseline pretreatment covariates. Within a possible treatment regime, whether the decision rules are closely followed over time is another open question. As we mentioned earlier, teachers and parents may exercise their discretionary power, and all treatment assignments are constrained by the availability of resources. For example, some schools are initially unprepared for the arrival of new immigrants, while other schools with established ELL programs may suffer from unexpected teacher turnover.

8.2.6   Marginal structural models and structural nested models

In a nonexperimental study of time-varying treatments, researchers often employ marginal structural models through inverse-probability-of-treatment weighting (IPTW) (Hernán, Brumbeck, and Robins, 2000, 2001; Robins, 1999). This method transforms the data to approximate a sequential randomized design. By assigning to each individual observation a weight inverse to one's propensity of having the observed treatment sequence (i.e., the propensity score), the weighted mean observed outcome of a group experiencing a given treatment sequence consistently estimates the population mean associated with that treatment sequence. Hong and Raudenbush (2008) extended the IPTW method to multilevel data. Even though the same model applies within subpopulations defined by baseline characteristics, this method becomes limited if the optimal length of treatment varies across subpopulations of students according to their intermediate responses to the treatment sequences. Robins (1997) has also developed structural nested models for evaluating treatment regimes including dynamic regimes. These models allow one to test the null hypothesis that two alternative regimes generate the same outcome value given the observed history. When the null hypothesis is rejected, the same models are then employed to estimate, through a Monte Carlo repeated resampling procedure, the effect on the outcome of a final blip of treatment for those individuals with the same observed history. The models are fully parametric and require specifying the outcome distributions for constructing likelihood ratio tests.

However, estimation of causal effects through employing marginal structural models and structural nested models is known to be nonrobust. This is partly due to the frequent misspecification of statistical models for estimating the probability of treatment assignment at each time point (Hong, 2010; Kang and Schafer, 2007; Robins, 1997; Schafer and Kang, 2008; Waernbaum, 2012). Additionally, IPTW tends to be highly variable, leading to an increase in the variance of the causal effect estimator. Alternatively, Murphy (2003) proposed an analytic method for identifying optimal rules that utilizes a backward induction procedure and invokes smooth parametric assumptions on the relevant quantities. The applicability of this method is constrained in the current case because the decision rules are unknown to the analyst.

8.3 MMWS for evaluating 2-year treatment sequences

This chapter extends the MMWS method, a nonparametric strategy for implementing the marginal structural models, to evaluations of time-varying treatments. As shown in Chapter 3, the nonparametric procedure allows one to generate relatively robust estimates of the treatment effects despite possible misspecifications of the statistical models for predicting the probability of treatment assignment. Moreover, MMWS reduces the variation in the estimated weights and therefore improves efficiency by a noticeable amount. In this section, we discuss the identification assumptions, propensity score estimation, MMWS computation, and outcome model specification for estimating the causal effects of 2-year treatment sequences. By applying weight to the nonexperimental data, the goal is to approximate data from a hypothetical sequential randomized experiment. We illustrate with an evaluation of ELL services in kindergarten and first grade for Spanish-speaking language minority students (Hong, Gagne, and West, 2014; Gagne and Hong, 2015).

8.3.1 Sequential ignorability

Following the existing literature, we invoke the sequential ignorability assumption, that is, an individual's current treatment assignment is assumed independent of all the future potential outcomes given the past observed covariate history, treatment history, and outcome history. This assumption is essential for identifying the effects of time-varying treatments. In the ELL application with 2-year treatment sequences, personal and contextual characteristics measured at the beginning of kindergarten, denoted by $\mathbf{X}_{-1}$, include student demographic characteristics, family characteristics, preschool experience, baseline oral English proficiency, and baseline academic and social–emotional skills. Time-varying covariates measured at the end of kindergarten or the beginning of first grade, denoted by $\mathbf{X}_0$, include the evolving status of student oral English proficiency and academic and social–emotional skills. The sequential ignorability assumption states that among students who have the same observed baseline covariate values and baseline outcome values, whether a student receives ELL services in kindergarten is independent of the student's potential math outcomes at the end of kindergarten $Y_0(z_0)$ and the potential math outcomes at the end of first grade $Y_1(z_0, z_1)$:

$$Y_0(z_0), Y_1(z_0, z_1) \perp\!\!\!\perp Z_0 | Y_{-1}, \mathbf{X}_{-1}. \tag{8.2}$$

The sequential ignorability assumption additionally states that among students who have the same treatment assignment in kindergarten, the same observed end-of-kindergarten covariate values and outcome values, as well as the same observed baseline covariate values and baseline outcome values, whether a student receives ELL services in first grade is independent of the student's potential math outcomes at the end of first grade $Y_1(z_0, z_1)$:

$$Y_1(z_0, z_1) \perp\!\!\!\perp Z_1 | Z_0, Y_{-1}, Y_0, \mathbf{X}_{-1}, \mathbf{X}_0. \tag{8.3}$$

8.3.2   Propensity scores

Under the sequential ignorability assumption, a student's propensity of receiving ELL services in kindergarten is a function of y_{-1} and $\mathbf{x}_{-1}$:

$$\theta_0^{(Z_0 = 1)} = \theta_0^{(Z_0 = 1)}(y_{-1}, \mathbf{x}_{-1}) = \text{pr}(Z_0 = 1 | Y_{-1}, \mathbf{X}_{-1}). \tag{8.4}$$

One may estimate the kindergarten propensity score through analyzing a logistic regression model entering Y_{-1} and $\mathbf{X}_{-1}$ as predictors. As shown by Rosenbaum and Rubin (1983), the treated individuals and the control individuals with the same propensity score value are expected to have the same distribution of Y_{-1} and $\mathbf{X}_{-1}$.

Next, given a student's treatment history in kindergarten $Z_0 = z_0$ for $z_0 = 0$ or 1, the student's propensity of receiving ELL services in first grade is a function of y_{-1}, y_0, $\mathbf{x}_{-1}$, and $\mathbf{x}_0$:

$$\theta_1^{(Z_1 = 1 | Z_0 = z_0)} = \theta_1^{(Z_1 = 1 | Z_0 = z_0)}(y_{-1}, y_0, \mathbf{x}_{-1}, \mathbf{x}_0) = \text{pr}(Z_1 = 1 | Z_0 = z_0, Y_{-1}, Y_0, \mathbf{X}_{-1}, \mathbf{X}_0). \tag{8.5}$$

To estimate the first-grade propensity score, one may analyze a logistic regression model for a subset of students who received the same treatment in kindergarten, entering Y_{-1}, Y_0, $\mathbf{X}_{-1}$, and $\mathbf{X}_0$ as predictors. In other words, the first-grade propensity score $\theta_1^{(Z_1 = 1 | Z_0 = 0)}$ is estimated for those who did not receive ELL services in kindergarten, while $\theta_1^{(Z_1 = 1 | Z_0 = 1)}$ is estimated for those who were treated in kindergarten. In the subset of students who were not treated in kindergarten, those who started to receive treatment in first grade and those who continued to be untreated in first grade are expected to have the same baseline and time-varying characteristics if displaying the same value in the first-grade propensity score $\theta_1^{(Z_1 = 1|Z_0 = 0)}$. The same logic applies to the subset of students who had been treated in kindergarten. Those who continued to be treated in first grade and those whose treatment was discontinued in first grade, if displaying the same propensity score $\theta_1^{(Z_1 = 1|Z_0 = 1)}$, are expected to have the same distribution of baseline and time-varying characteristics.

In general, a student's propensity for receiving a 2-year treatment sequence (z_0, z_1) in kindergarten and first grade is a product of the kindergarten propensity score for receiving z_0 and the first-grade propensity score for receiving z_1 given that one has received z_0 in the previous year:

$$\theta^{(z_0, z_1)} = \theta_1^{(z_1|z_0)} \times \theta_0^{(z_0)} = \mathrm{pr}(Z_1 = z_1|Z_0 = z_0, Y_{-1}, Y_0, \mathbf{X}_{-1}, \mathbf{X}_0) \times \mathrm{pr}(Z_0 = z_0|Y_{-1}, \mathbf{X}_{-1}).$$

8.3.3 MMWS computation

In this evaluation of 2-year treatment sequences, we first apply weighting to the kindergarten data such that the treated group and the control group in kindergarten will have the same distribution of the kindergarten propensity score $\theta_0^{(Z_0 = 1)}$ and therefore the same pretreatment composition as that of the entire sample. Next, for a subsample of individuals who had not been treated in kindergarten, we apply weighting to the first-grade data such that the treated and the untreated in this subsample of first graders will have the same distribution of the first-grade propensity score $\theta_1^{(Z_1 = 1|Z_0 = 0)}$ and therefore the same distribution of pretreatment and time-varying characteristics as that of the entire subsample. Similarly, for a subsample of individuals who had already been treated in kindergarten, weighting is applied to their first-grade data to equate the distribution of the first-grade propensity score $\theta_1^{(Z_1 = 1|Z_0 = 1)}$ between those treated and those untreated in first grade. When the sequential ignorability assumption holds, the weighted data in kindergarten approximate a randomization of kindergartners to the treatment or the control condition; the weighted data in first grade approximate, within each subset of individuals who had the same treatment in kindergarten, a randomization of first graders to the treatment or the control condition.

8.3.3.1 MMWS adjustment for year 1 treatment selection

To remove selection bias associated with the baseline covariates and baseline outcomes, the IPTW for a student assigned to treatment z_0 in kindergarten is $\mathrm{pr}(Z_0 = z_0)/\theta_0^{(z_0)}$. The MMWS method reestimates $\theta_0^{(z_0)}$ nonparametrically by dividing the sample of kindergartners into a number of strata on the basis of the estimated kindergarten propensity score. The MMWS for a student in stratum $S_0^{(z_0)} = s_0$ who received ELL services in kindergarten is

$$\mathrm{MMWS}^{(z_0 = 1)} = \frac{\mathrm{pr}(Z_0 = 1)}{\mathrm{pr}\left(Z_0 = 1|S_0^{(z_0)} = s_0\right)}. \tag{8.6}$$

Here, the denominator is the proportion of students in stratum s_0 who received ELL services in kindergarten and is a nonparametric reestimate of the kindergarten propensity score for those in the given stratum; the numerator is the proportion of students in the entire sample who received ELL services in kindergarten. Similarly, the MMWS for a student in the same stratum who did not receive ELL services in kindergarten is

$$\text{MMWS}^{(z_0=0)} = \frac{\text{pr}(Z_0=0)}{\text{pr}\left(Z_0=0|S_0^{(z_0)}=s_0\right)}. \tag{8.7}$$

The denominator is the proportion of students in stratum s_0 who did not receive ELL services in kindergarten. The analyst must check the balance in the covariates under weighting in order to decide whether there is a need to modify the propensity score models or to modify the stratifications.

8.3.3.2   MMWS adjustment for two-year treatment sequence selection

The IPTW for a student assigned to the treatment sequence (z_0, z_1) is $\text{pr}(Z_0=z_0, Z_1=z_1)/\theta^{(z_0, z_1)}$, which is equivalent to

$$\frac{\text{pr}(Z_1=z_1|Z_0=z_0)}{\theta_1^{(z_1|z_0)}} \times \frac{\text{pr}(Z_0=z_0)}{\theta_0^{(z_0)}}.$$

The second quantity in the aforementioned product adjusts for the baseline difference between kindergarten treatment groups, while the first quantity adjusts for the time-varying difference prior to the beginning of the first-grade year between first-grade treatment groups who received the same kindergarten treatment. To nonparametrically estimate the first quantity, we may divide the subsample of students who received ELL services in kindergarten into five or six strata on the basis of the estimated first-grade propensity score $\widehat{\theta}_1^{(Z_1=1|Z_0=1)}$. Similarly, the subsample of students who did not receive ELL services in kindergarten will be stratified on the basis of $\widehat{\theta}_1^{(Z_1=1|Z_0=0)}$. For students who received treatment sequence (z_0, z_1) and who are in stratum $S_1^{(z_1|z_0)}=s_1$, the first quantity of the product is

$$\text{MMWS}_1^{(z_1|z_0)} = \frac{\text{pr}(Z_1=z_1|Z_0=z_0)}{\text{pr}\left(Z_1=z_1|Z_0=z_0, S_1^{(z_1|z_0)}=s_1\right)}. \tag{8.8}$$

Here, the denominator is the proportion of students with kindergarten treatment history z_0 in stratum s_1 who were assigned to first-grade treatment z_1; the numerator is the proportion of students in the subsample with kindergarten treatment history z_0 who were assigned to first-grade treatment z_1. Therefore, for a student whose treatment sequence consists of z_0 in kindergarten and z_1 in first grade, through the sequential stratification, the MMWS for the first-grade observation can be estimated nonparametrically as the product:

$$\text{MMWS}^{(z_0, z_1)} = \text{MMWS}^{(z_0)} \times \text{MMWS}_1^{(z_1|z_0)}. \tag{8.9}$$

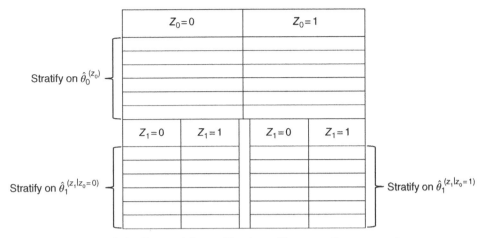

Figure 8.4   Sequential stratification of 2 years of treatment data

The sequential stratification of 2 years of treatment data is illustrated in Figure 8.4. After weighting, the treated group and the control group in kindergarten are expected to display the same pretreatment composition; and additionally, students in the treated group and the control group in first grade who shared the same past treatment history are expected to display the same composition of pretreatment and time-varying characteristics. If there is evidence suggesting that weighting has not adequately removed the preexisting differences in the baseline or time-varying characteristics between those treated and those untreated in first grade who had the same treatment in kindergarten, one may modify the first-grade propensity score model by explicitly including measures of such characteristics or restratify the subsample.

8.3.4   Two-year growth model specification

A growth model estimates an individual's growth trajectory as a function of time. Even in the absence of ELL services, a student may gain a certain amount of math knowledge and skills in kindergarten and again in first grade as shown in the trajectory corresponding to the 0-0 treatment sequence for the hypothetical student in Figure 8.1. The ELL services in kindergarten, if beneficial to the kindergartners' math learning, will positively deflect the hypothetical student's end-of-kindergarten-year math achievement by an amount equal to the kindergarten treatment effect. The treatment sequence in kindergarten and first grade together will deflect the hypothetical student's end-of-first-grade-year math achievement by an amount equal to the cumulative effect of the 2-year treatment sequence. A comparison of the cumulative treatment effect between the 1-0 sequence and the 0-0 sequence reveals the effect on the first-grade outcome of having ELL services in kindergarten only; a comparison between the 0-1 sequence and the 0-0 sequence shows the effect of having ELL services in first grade only; and a comparison between the 1-1 sequence and the 1-0 sequence indicates the effect of having a second year of ELL services. In the following, we explain the model specifications in detail.

8.3.4.1   Growth model in the absence of treatment

We first specify the model for a student's math growth in kindergarten and first grade in the absence of ELL services. This is a two-level model allowing every student to have an individual growth trajectory and can be generalized as

$$Y_{ti} = \gamma_{-1} + u_{-1i} + \sum_{h=0}^{t} (\gamma_h + u_{hi}) L_{hi} + \varepsilon_{ti}, \quad \varepsilon_{ti} \sim N(0, \sigma^2).$$

Here, Y_{ti} is student i's math score at time t for $t = -1, 0, 1$; γ_{-1} is the population average math score at the baseline; u_{-1i} is the student-specific increment at the baseline, representing the discrepancy between the student's baseline true score and the population average score; and ε_{-1i} is the measurement error at the baseline. L_{ti} is the lapse of time in months between the assessment at time $t-1$ and that at time t under the simplifying assumption that the assessment is always given at the end of a school year. Hence, γ_0 is the population average monthly growth rate in math over the kindergarten year; γ_1 is the population average monthly growth rate over the first-grade year; u_{0i} and u_{1i} are student-specific increments to the population average growth rates in kindergarten and first grade, respectively. Hence, if a student scored below the average at the baseline but displayed a kindergarten growth rate above the average and a first-grade growth rate at the average level, then u_{-1i} will be negative, u_{0i} will be positive, and u_{1i} will be equal to zero. The student-specific random effects u_{-1i}, u_{0i}, and u_{1i} are usually assumed to be multivariate normal with zero means. The parameters in the aforementioned model are identifiable when the level 1 variance $\sigma_{\varepsilon_t}^2$ is known. One can estimate the level 1 variance beforehand on the basis of the variance of each outcome measure $\sigma_{y_t}^2$ and its reliability λ_t, that is, $\sigma_{\varepsilon_t}^2 = (1 - \lambda_t) \sigma_{y_t}^2$. For completeness, one may also add class-level and school-level random effects.

8.3.4.2   Growth model in the absence of confounding

We then allow the individual growth trajectories to be deflected by time-varying treatments in the absence of confounding. To simplify the notation, let $\beta_{ti} = \gamma_t + u_{ti}$ for $t = -1, 0, 1$. Hence, β_{-1i}, β_{0i}, and β_{1i} represent student i's model-based intercept (i.e., baseline score), monthly growth rate in kindergarten, and monthly growth rate in first grade, respectively. The spring kindergarten math outcome is a function of the treatment in kindergarten:

$$Y_{0i} = \beta_{-1i} + \beta_{0i} L_{0i} + \delta_0^{(1)} I_i(Z_0 = 1) + \varepsilon_{0i}.$$

Here, $\delta_0^{(1)}$ identifies the kindergarten treatment effect $E[Y_0(1) - Y_0(0)]$. The spring first-grade math outcome is a function of the 2-year treatment sequence:

$$Y_{1i} = \beta_{-1i} + \beta_{0i} L_{0i} + \beta_{1i} L_{1i}$$
$$+ \delta_1^{(1,0)} I_i(Z_0 = 1) I_i(Z_1 = 0)$$
$$+ \delta_1^{(0,1)} I_i(Z_0 = 0) I_i(Z_1 = 1)$$
$$+ \delta_1^{(1,1)} I_i(Z_0 = 1) I_i(Z_1 = 1) + \varepsilon_{1i}.$$

Here, $I_i(z_0)$ and $I_i(z_1)$ are dummy indicators for the kindergarten treatment and the first-grade treatment, respectively. In general, $\delta_1^{(z_0,z_1)}$ identifies the cumulative effect of a 2-year treatment sequence (z_0, z_1) relative to the sequence of never being treated, that is, $E[Y_1(z_0,z_1) - Y_1(0,0)]$. To be specific, $\delta_1^{(1,0)}$ identifies $E[Y_1(1,0) - Y_1(0,0)]$, the effect of receiving ELL services in kindergarten only on the first-grade outcome; $\delta_1^{(0,1)}$ identifies $E[Y_1(0,1) - Y_1(0,0)]$, the effect of receiving ELL services in first grade only on the first-grade outcome; and $\delta_1^{(1,1)}$ identifies $E[Y_1(1,1) - Y_1(0,0)]$, the effect of receiving ELL services in both kindergarten and first grade. Note that $\delta_1^{(1,1)} - \delta_1^{(1,0)} - \delta_1^{(0,1)}$ identifies $E[Y_1(1,1) - Y_1(1,0)] - E[Y_1(0,1) - Y_1(0,0)]$, the difference in the first-grade treatment effect on the first-grade outcome between having received ELL services in kindergarten and having not. This last quantity is the interaction effect of kindergarten ELL services and first-grade ELL services. To estimate the parameters from sequential randomized experimental data, one may analyze a two-level model with repeated observations at level 1 and students at level 2. We use $D_{ti}(t)$ to be an indicator that takes value 1 if the observation was made in year t and zero otherwise. Similarly, let $D_{ti}(\geq t')$ take value 1 if the observation was made in year t' and later and let it be zero otherwise:

Level 1 $\quad Y_{ti} = \beta_{-1i} + \beta_{0i}L_{0i}D_{ti}(\geq 0) + \beta_{1i}L_{1i}D_{ti}(1)$

$$+ \delta_0^{(1)}I_i(Z_0 = 1)D_{ti}(0) + \left[\delta_1^{(1,0)}I_i(Z_0 = 1)I_i(Z_1 = 0) + \delta_1^{(0,1)}I_i(Z_0 = 0)I_i(Z_1 = 1)\right.$$

$$\left. + \delta_1^{(1,1)}I_i(Z_0 = 1)I_i(Z_1 = 1)\right]D_{ti}(1) + \varepsilon_{ti}, \quad \varepsilon_{ti} \sim N(0, \sigma^2).$$

Level 2 $\quad \beta_{-1i} = \gamma_{-1} + u_{-1i}$,

$\beta_{0i} = \gamma_0 + u_{0i}$,

$\beta_{1i} = \gamma_1 + u_{1i}$,

$$\begin{pmatrix} u_{-1i} \\ u_{0i} \\ u_{1i} \end{pmatrix} \sim N\left(\begin{pmatrix} 0 \\ 0 \\ 0 \end{pmatrix}, \begin{pmatrix} \tau_{\beta_{-1}} & \tau_{\beta_{-1}\cdot\beta_0} & \tau_{\beta_{-1}\cdot\beta_1} \\ \tau_{\beta_{-1}\cdot\beta_0} & \tau_{\beta_0} & \tau_{\beta_0\cdot\beta_1} \\ \tau_{\beta_{-1}\cdot\beta_1} & \tau_{\beta_0\cdot\beta_1} & \tau_{\beta_1} \end{pmatrix}\right). \tag{8.10}$$

Receiving 2 years of ELL services would be optimal if $\delta_0^{(1)}$ is positive and if $\delta_1^{(1,1)}$ is positive and greater than $\delta_1^{(1,0)}$ and $\delta_1^{(0,1)}$. Under the assumption that the data have been obtained from a sequential randomized design, the causal parameters obtained from the aforementioned model would explain whether the kindergarten and first-grade treatments lead to differences between the four potential math growth trajectories. Suppose that the hypothetical student whose potential growth trajectories have been illustrated in Figure 8.1 is a typical student in the population. We would then expect that $u_{-1i} = u_{0i} = u_{1i} = 0$. The population average causal effects are visualized in Figure 8.5.

Following is an alternative specification of the level 1 model analogous to that used by Hong and Raudenbush (2008) that can be viewed as a reparameterization of model (8.10):

$$Y_{ti} = \beta_{-1i} + \beta_{0i}L_{0i}D_{ti}(\geq 0) + \beta_{1i}L_{1i}D_{ti}(1) + \delta_0^{(1)}I_i(Z_0 = 1)D_{ti}(0)$$

$$+ \left[\delta_1^{(1,0)}I_i(Z_0 = 1) + \delta_1^{(0,1)}I_i(Z_1 = 1) + \delta_1^{(*)}I_i(Z_0 = 1)I_i(Z_1 = 1)\right]D_{ti}(1) + \varepsilon_{ti}, \tag{8.11}$$

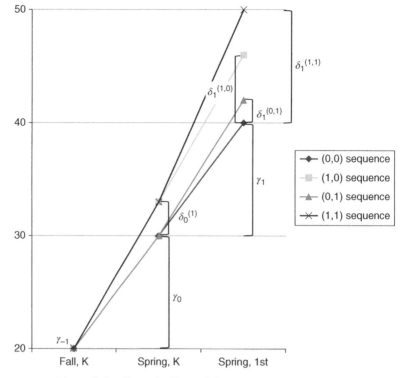

Figure 8.5   Causal effects of 2-year treatment sequences

where $\delta_1^{(*)} = \delta_1^{(1,1)} - \delta_1^{(1,0)} - \delta_1^{(0,1)}$ compares the effect of having a second year of treatment in first grade after already having been treated in kindergarten with the effect of being treated in first grade only. This corresponds to the interaction effect between kindergarten treatment and first-grade treatment derived in (8.1). A positive $\delta_1^{(*)}$ will indicate that being treated in kindergarten enhances the effect of the first-grade treatment. The level 2 model takes the same form as earlier.

8.3.4.3   Weighted 2-year growth model

To remove time-varying selection bias in nonexperimental data, we specify a two-level outcome model as shown previously and apply MMWS at level 1. Table 8.2 illustrates the level 1 data structure with four hypothetical students, each receiving one of the four treatment sequences. Student 1 never received ELL services in either kindergarten or first year; student 2 received ELL services in kindergarten only; student 3 received it in first grade only; and student 4 received the treatment in both years. The weight is 1.0 for all four students in fall kindergarten, is $\text{MMWS}^{(0)}$ for students 1 and 3 and $\text{MMWS}^{(1)}$ for students 2 and 4 in spring kindergarten, and is $\text{MMWS}^{(0.0)}$ for student 1, $\text{MMWS}^{(1.0)}$ for student 2, $\text{MMWS}^{(0.1)}$ for student 3, and $\text{MMWS}^{(1.1)}$ for student 4 in spring first grade. Suppose that the assessment in spring kindergarten was administered at 8 months from the beginning of the year for students

Table 8.2 Repeated observations at level 1 for a two-level model for evaluating 2-year treatment sequences

Student	Y_t	W_t	$L_0 I_t(t \geq 0)$	$L_1 I_t(t=1)$	$I(Z_0=1)I_t(t=0)$	$I(Z_0=1)I_t(t=1)$	$I(Z_1=1)I_t(t=1)$	$I(Z_0=1)I(Z_1=1)I_t(t=1)$
1	22	1	0	0	0	0	0	0
1	32	$MMWS^{(0)}$	8	0	0	0	0	0
1	42	$MMWS^{(0,0)}$	8	13	0	0	0	0
2	20	1	0	0	0	0	0	0
2	30	$MMWS^{(1)}$	8	0	1	1	0	0
2	45	$MMWS^{(1,0)}$	8	13	0	0	0	0
3	24	1	0	0	0	0	0	0
3	28	$MMWS^{(0)}$	9	0	0	0	0	0
3	48	$MMWS^{(0,1)}$	9	12	0	0	1	0
4	18	1	0	0	0	0	0	0
4	30	$MMWS^{(1)}$	9	0	1	1	0	0
4	40	$MMWS^{(1,1)}$	9	12	0	1	1	1

1 and 2 and was at 9 months from the beginning of the year for students 3 and 4. The assessment in spring first grade was administered at the same time for all four students, which was 13 months after the previous assessment for students 1 and 2 and was 12 months after the previous assessment for students 3 and 4. Under the assumption of sequential ignorability, $\delta_0^{(1)}$ is the average kindergarten treatment effect on the kindergarten outcome; $\delta_1^{(1,0)}$, $\delta_1^{(0,1)}$, and $\delta_1^{(1,1)}$ are the respective average effects on the first-grade outcome of having ELL services in kindergarten only, in first grade only, and in both kindergarten and first grade. Analyzing the Early Childhood Longitudinal Study Kindergarten cohort (ECLS-K) data, Gagne and Hong (2015) reported a major beneficial impact of receiving ELL services in kindergarten on Spanish-speaking students' math learning that was carried through the fifth grade. Continuing ELL services in first grade, however, showed no added value to these students' math learning. The authors cautioned that the conclusion should not be generalized to these students' learning of English language or other subject content.

8.4   MMWS for evaluating multiyear sequences of multivalued treatments

This section explains how to extend the MMWS method to multiyear sequences of binary treatments, to 2-year sequences of multivalued treatments, and, in the general case, to multiyear sequences of multivalued treatments.

8.4.1   Sequential ignorability of multiyear treatment sequences

To evaluate multiyear treatment sequences with nonexperimental data, the identification assumption states that given an individual's past observed treatment history, outcome history, and covariate history, the current treatment assignment is independent of all the future potential outcomes in a study of T time periods:

$$Y_t(\bar{z}_t), \ \ldots, Y_T(\bar{z}_T) \perp Z_t | \bar{Z}_{t-1}, \bar{Y}_{t-1}, \bar{X}_{t-1}. \tag{8.12}$$

Here, $\bar{Z}_{t-1}$ denotes the treatment history, $\bar{Y}_{t-1}$ denotes the outcome history, and $\bar{X}_{t-1}$ denotes the covariate history from the baseline up to time $t-1$.

8.4.2   Propensity scores for multiyear treatment sequences

In each time period, if the treatment is binary, we will estimate a single propensity score for being treated at time t for all those with the same treatment history $\bar{Z}_{t-1} = \bar{z}_{t-1}$ in the past:

$$\theta_t^{(Z_t = 1|\bar{z}_{t-1})} = \mathrm{pr}(Z_t = 1|\bar{Z}_{t-1} = \bar{z}_{t-1}, \bar{Y}_{t-1}, \bar{X}_{t-1}). \tag{8.13}$$

If, instead, there are K treatment categories in each time period, we will estimate, through analyzing a multinomial logistic regression model for those with the same treatment history, K propensity scores denoted by $\theta_t^{(z_t|\bar{z}_{t-1})}$ for $z_t = 1, \ldots, K$. An individual's propensity score for receiving a given treatment sequence from time 0 to t is the product of the propensity scores for the time-varying treatments up to time t.

8.4.3   MMWS computation

In a subsample of individuals who have shared the same treatment history $\bar{z}_{t-1}$ in the past, individuals might be assigned nonrandomly to different treatment groups at time t. If the treatment is binary, we will stratify such a subsample on the basis of the estimated propensity score $\theta_t^{(Z_t=1|\bar{z}_{t-1})}$ and compute the weight for the treated and control units under the same stratification. If the treatment has $K > 2$ categories, however, we will stratify the subsample on $\theta_t^{(z_t|\bar{z}_{t-1})}$ and compute the weight for those assigned to z_t for $z_t = 1, \ldots, K$. The same procedure is to be repeated for the individuals in each treatment category. In general, for individuals who have received treatment sequence $\bar{z}_{t-1}$ in the past and are in stratum $S_t^{(z_t|\bar{z}_{t-1})} = s_t$, the new element of MMWS for the observation at time t is

$$\text{MMWS}_t^{(z_t|\bar{z}_{t-1})} = \frac{\text{pr}(Z_t = z_t|\bar{Z}_{t-1} = \bar{z}_{t-1})}{\text{pr}\left(Z_t = z_t|\bar{Z}_{t-1} = \bar{z}_{t-1}, S_t^{(z_t|\bar{z}_{t-1})} = s_t\right)}. \tag{8.14}$$

Here, the denominator is the proportion of individuals with treatment history $\bar{z}_{t-1}$ in stratum s_t who are actually assigned to treatment z_t; the numerator is the proportion of individuals with treatment history $\bar{z}_{t-1}$ who are assigned to treatment z_t. Therefore, through the sequential stratification, the MMWS for the time t observation of an individual who has had a treatment history $\bar{z}_{t-1}$ is the product:

$$\text{MMWS}^{(\bar{z}_t)} = \text{MMWS}^{(z_0)} \times \text{MMWS}_1^{(z_1|z_0)} \times \cdots \times \text{MMWS}_t^{(z_t|\bar{z}_{t-1})}. \tag{8.15}$$

8.4.4   Weighted multiyear growth model

The effect of each treatment sequence is again represented by the deflection from the reference trajectory associated with the $\bar{Z}_t = \bar{0}$ treatment sequence where $\bar{0}$ represents a sequence of assignments to the control condition up to time t. For example, suppose that we intend to evaluate 2-year sequences of three-category ELL treatments—no ELL services, ESL instruction, and bilingual education—denoted by $z_0 = 0, 1, 2$, respectively, in kindergarten and by $z_1 = 0, 1, 2$, respectively, in first grade. The level 1 model takes the form as follows:

$$Y_{ti} = \beta_{-1i} + \beta_{0i} L_{0i} D_{ti}(\geq 0) + \beta_{1i} L_{1i} D_{ti}(1) + \sum_{z_0=1}^{2} \delta_0^{(z_0)} I_i(Z_0 = z_0) D_{ti}(0)$$

$$+ \left[\sum_{z_1=1}^{2} \delta_1^{(0,z_1)} I_i(Z_0 = 0) I_i(Z_1 = z_1) + \sum_{z_1=0}^{2} \sum_{z_0=1}^{2} \delta_t^{(z_0,z_1)} I_i(Z_0 = z_0) I_i(Z_1 = z_1)\right] D_{ti}(1) \tag{8.16}$$

$$+ \varepsilon_{ti}, \quad \varepsilon_{ti} \sim N(0, \sigma^2).$$

Here, $\delta_0^{(1)}$ identifies $E[Y_0(Z_0 = 1) - Y_0(Z_0 = 0)]$, that is, the effect of kindergarten ESL instruction relative to no ELL services on the kindergarten outcome, while $\delta_0^{(2)}$ identifies $E[Y_0(Z_0 = 2) - Y_0(Z_0 = 0)]$, the effect of kindergarten bilingual education relative to no ELL services on the kindergarten outcome. Additionally, $\delta_1^{(0,1)}$ and $\delta_1^{(0,2)}$ identify $E[Y_1(Z_0 = 0, Z_1 = 1) - Y_1(Z_0 = 0, Z_1 = 0)]$ and $E[Y_1(Z_0 = 0, Z_1 = 2) - Y_1(Z_0 = 0, Z_1 = 0)]$, respectively. The former is the effect on first-grade outcome of having first-grade ESL instruction only, while the latter the effect of having first-grade bilingual education only relative to receiving no ELL services for 2 years. Similarly, $\delta_1^{(1,0)}$, $\delta_1^{(1,1)}$, $\delta_1^{(1,2)}$, $\delta_1^{(2,0)}$, $\delta_1^{(2,1)}$, and $\delta_1^{(2,2)}$ identify the effect on first grade outcome of each of the other six treatment sequences relative to receiving no ELL services for 2 years.

Another example is an evaluation of 3-year sequences of binary treatments denoted by $z_t = 0,1$ for $t = 0,1,2$. The level 1 model can be written as follows:

$$Y_{ti} = \beta_{-1i} + \beta_{0i}L_{0i}D_{ti}(\geq 0) + \beta_{1i}L_{1i}D_{ti}(\geq 1) + \beta_{2i}L_{2i}D_{ti}(2)$$

$$+ \delta_0^{(1)}I_i(Z_0 = 1)D_{ti}(0) + \left[\delta_t^{(0,1)}I_i(Z_0 = 0)I_i(Z_1 = 1) + \sum_{z_1=0}^{1}\delta_t^{(1,z_1)}I_i(Z_0 = 1)I_i(Z_1 = z_1)\right]D_{ti}(1)$$

$$+ \left[\delta_t^{(0,0,1)}I_i(Z_0 = 0)I_i(Z_1 = 0)I_i(Z_2 = 1) + \sum_{z_2=0}^{1}\delta_t^{(0,1,z_2)}I_i(Z_0 = 0)I_i(Z_1 = 1)I_i(Z_2 = z_2)\right.$$

$$\left.+ \sum_{z_1=0}^{1}\sum_{z_2=0}^{1}\delta_t^{(1,z_1,z_2)}I_i(Z_0 = 1)I_i(Z_1 = z_1)I_i(Z_2 = z_2)\right]D_{ti}(2) + \varepsilon_{ti}, \ \varepsilon_{ti} \sim N(0,\sigma^2). \quad (8.17)$$

Appendix 8.A shows a growth model in the general case of T time periods with K treatment categories in each time period. To reduce the number of causal parameters, the analyst may impose a parsimonious structure on the basis of theoretical assumptions with regard to how the treatment effects accumulate. One assumption is that treatments received in different time periods have exchangeable impacts (Hernán, Brumbeck, and Robins, 2000). For example, if a student is to receive 2 years of ELL services, the impact on the student's achievement is assumed to be the same regardless of when the outcome is observed and regardless of whether the student receives ELL services in kindergarten and first grade or in first grade and second grade. Applying this assumption to model (8.17), we will have that

$$\delta_0^{(1)} = \delta_1^{(0,1)} = \delta_1^{(1,0)} = \delta_2^{(0,0,1)} = \delta_2^{(0,1,0)} = \delta_2^{(1,0,0)}, \ \delta_1^{(1,1)} = \delta_2^{(0,1,1)} = \delta_2^{(1,1,0)} = \delta_2^{(1,0,1)}.$$

We may use $\delta^{\{c\}}$ to denote the causal effect of having c years of ELL services for $0 < c \leq T$. In evaluating a binary treatment over three time periods, we simplify the level 1 model in (8.17) as follows:

$$Y_{ti} = \beta_{-1i} + \beta_{0i}L_{0i}D_{ti}(\geq 0) + \beta_{1i}L_{1i}D_{1i}(\geq 1) + \beta_{2i}L_{2i}D_{2i}(2)$$

$$+ \delta^{\{1\}}[I_i(Z_0 = 1)D_{ti}(0) + I_i(Z_0 + Z_1 = 1)D_{ti}(1) + I_i(Z_0 + Z_1 + Z_2 = 1)D_{ti}(2)]$$

$$+ \delta^{\{2\}}[I_i(Z_0 + Z_1 = 2)D_{ti}(1) + I_i(Z_0 + Z_1 + Z_2 = 2)D_{ti}(2)]$$

$$+ \delta^{\{3\}}[I_i(Z_0 + Z_1 + Z_2 = 3)D_{ti}(2)] + \varepsilon_{ti}, \ \varepsilon_{ti} \sim N(0,\sigma^2). \quad (8.18)$$

8.4.5   Issues of sample size

Following the tradition in psychological research, in a sequential randomized design, at least 30 individuals should be assigned to a treatment group in any time period and to any given treatment sequence. A larger sample would be needed for detecting treatment effects of a smaller effect size. In analyses of nonexperimental data, the smallest sample size in a treatment sequence constrains the number of predictors one may enter in a propensity score model without overfitting the model. A rule of thumb is that a propensity score model with 10 predictors would need at least 100 units in each treatment sequence.

8.5   Conclusion

This chapter has pointed out a number of major challenges in evaluations of time-varying treatments: sequential randomization is rare; treatments are often dynamically adapted to an individual's evolving status under unspecified treatment regimes; and the existing methods including SEM and fixed-effects econometric models are unsuitable for adjusting for time-varying confounding in the presence of the endogeneity problem. Despite its unique strength in handling time-varying confounding, IPTW is known for results that are not robust and low in efficiency. The nonparametric MMWS method, when extended to studies of time-varying treatments, holds the promise of generating robust and relatively more efficient results and can accommodate multiyear sequences of multivalued treatments.

In analyses of nonexperimental data, the causal validity of the MMWS-adjusted treatment effect estimates nonetheless depends on the strong assumption of sequential ignorability. Moreover, as the number of time periods and the number of treatment categories at each time increase, it becomes necessary to invoke theory-based assumptions with regard to how the treatment effects accumulate. The analysis will then focus on a reduced number of causal parameters. Finally, subpopulation-specific optimal treatment sequences are often particularly relevant to theory building as well as to decision-making in practice. Future research may extend the MMWS procedure for evaluating time-varying treatments within each subpopulation. This is sometimes feasible when the sample size of each subpopulation is sufficiently large. One would expect that the amount of selection bias is already considerably reduced within a relatively homogeneous subpopulation, the variance of the weight would be reduced as well, and therefore the treatment effect estimation within each subpopulation would perhaps remain precise.

Appendix 8.A: A saturated model for evaluating multivalued treatments over multiple time periods

The number of causal parameters in a saturated model increases quickly as the number of time periods increases and as the number of treatment categories increases. In general, with K treatment categories in each time period, we include in the level 1 model for the time t observation dummy indicators for $K-1$ treatment categories at time t conditioning on the past treatment history. This specification applies to observations from $t = 0, \ldots, T$ where T denotes the last time period with $\bar{0} = (0, \ldots, 0)$ being the reference sequence. We apply MMWS at level 1 for each observation of each individual:

$$Y_{ti} = \beta_{-1i} + \sum_{h=0}^{t} \beta_{hi} L_{hi} D_{ti} (\geq h)$$

$$+ \sum_{z_0=1}^{K-1} \delta_0^{(z_0)} I_i(Z_0 = z_0) D_{ti}(0) + \cdots + \left\{ \sum_{z_T=1}^{K-1} \left[\delta_T^{(0,\cdots,0,z_T)} \left(\prod_{h=0}^{T-1} I_i(Z_h = 0) \right) I_i(Z_T = z_T) \right] \right.$$

$$+ \sum_{z_{T-1}=1}^{K-1} \sum_{z_T=0}^{K-1} \left[\delta_T^{(0,\cdots,z_{T-1},z_T)} \left(\prod_{h=0}^{T-2} I_i(Z_h = 0) \right) \prod_{h=T-1}^{T} I_i(Z_h = z_h) \right] + \cdots$$

$$\left. + \sum_{z_0=1}^{K-1} \cdots \sum_{z_{T-1}=0}^{K-1} \sum_{z_T=0}^{K-1} \left[\delta_T^{(z_0,\cdots,z_{T-1},z_T)} \prod_{h=0}^{T} I_i(Z_h = z_h) \right] \right\} D_{ti}(T) + \varepsilon_{ti}, \; \varepsilon_{ti} \sim N(0, \sigma^2).$$

The level 2 model has $T + 2$ equations:

$$\beta_{-1i} = \gamma_{-1} + u_{-1i},$$

$$\beta_{hi} = \gamma_h + u_{hi}, \text{ for } h = 0, \ldots, T,$$

$$\begin{pmatrix} u_{-1i} \\ \vdots \\ u_{Ti} \end{pmatrix} \sim N \left(\begin{pmatrix} 0 \\ \vdots \\ 0 \end{pmatrix}, \begin{pmatrix} \tau_{\beta_{-1}} & \cdots & \tau_{\beta_{-1} \cdot \beta_T} \\ \vdots & \ddots & \vdots \\ \tau_{\beta_{-1} \cdot \beta_T} & \cdots & \tau_{\beta_T} \end{pmatrix} \right).$$

References

Arrighi, H.M. and Hertz-Picciotto, I. (1994). The evolving concept of the healthy worker survivor effect. *Epidemiology*, 5(2), 189–196.

August, D. and Hakuta, K. (1997). *Improving Schooling for Language-Minority Children: A Research Agenda*, National Academy Press, Washington, DC.

Collier, V.P. (1987). Age and rate of acquisition of second language for academic purposes. *TESOL Quarterly*, 21, 617–641.

Collier, V.P. (1989). How long? A synthesis of research on academic achievement in a second language. *TESOL Quarterly*, 23, 509–531.

Cummins, J. (1981). Age on arrival and immigrant second language learning in Canada: a reassessment. *Applied Linguistics*, 2, 132–149.

Gagne, J., and Hong, G. (2015). How crucial is ELL support in kindergarten and first grade for Spanish-speaking students' math learning? Paper presented at the 2015 annual meeting of the, American Educational Research Association, Chicago, IL.

Hart, B. and Risley, T.R. (1995). *Meaningful Differences in the Everyday Experience of Young American Children*, Paul H. Brookes Publishing, Baltimore, MD.

Hernán, M.A., Brumback, B. and Robins, J.M. (2000). Marginal structural models to estimate the causal effect of zidovudine on the survival of HIV-positive men. *Epidemiology*, 11(5), 561–570.

Hernán, M.A., Brumback, B. and Robins, J.M. (2001). Marginal structural models to estimate the joint causal effect of nonrandomized treatments. *Journal of American Statistical Association*, *96*(454), 440–448.

Hong, G. (2010). Marginal mean weighting through stratification: adjustment for selection bias in multilevel data. *Journal of Educational and Behavioral Statistics*, *35*(5), 499–531.

Hong, G. (2012). Marginal mean weighting through stratification: a generalized method for evaluating multi-valued and multiple treatments with non-experimental data. *Psychological Methods*, *17*(1), 44–60.

Hong, G., Gagne, J., and West, A. (2014). Optimal sequence of ELL services in kindergarten and first grade for Spanish-speaking students. Paper presented at the Society for Research on Educational Effectiveness Spring Conference, Washington, DC.

Hong, G. and Raudenbush, S.W. (2008). Causal inference for time-varying instructional treatments. *Journal of Educational and Behavioral Statistics*, *33*(3), 333–362.

Kang, J.D. and Schafer, J.L. (2007). Demystifying double robustness: a comparison of alternative strategies for estimating a population mean from incomplete data. *Statistical Science*, *22*(4), 523–539.

Lavori, P.W. and Dawson, R. (2004). Dynamic treatment regimes: practical design considerations. *Clinical Trials*, *1*, 9–20.

Lavori, P.W. and Dawson, R. (2008). Adaptive treatment strategies in chronic disease. *Annual Review of Medicine*, *59*, 443–453.

Lei, H., Nahum-Shani, I., Lynch, K. *et al.* (2012). A "SMART" design for building individualized treatment sequences. *Annual Review of Clinical Psychology*, *8*, 21–48.

Murphy, S.A. (2003). Optimal dynamic treatment regimes. *Journal of the Royal Statistical Society, Series B (with discussion)*, *65*(2), 331–366.

Murphy, S.A. (2005). An experimental design for the development of adaptive treatment strategies. *Statistics in Medicine*, *24*(10), 1455–1481.

National Center for Education Statistics (2012). *The Condition of Education 2011 (NCES 2011-045)*, U.S. Department of Education, Washington, DC.

Robins, J. (1997). Causal inference from complex longitudinal data. In M. Berkane (ed), *Latent Variable Modeling and Applications to Causality*. New York: Springer Verlag (pp. 69–117). Lecture Notes in Statistics, v. *120*.

Robins, J. (1999). Marginal structural models versus structural nested models as tools for causal inference. In M. E. Halloran and D. Berry (eds), *Statistical Models in Epidemiology, the Environment, and Clinical Trials* (pp. 95–134). New York: Springer.

Rosenbaum, P.R. (1984). The consequences of adjustment for a concomitant variable that has been affected by the treatment. *Journal of the Royal Statistical Society, Series A (General)*, *147*(5), 656–666.

Rosenbaum, P.R. and Rubin, D.B. (1983). The central role of the propensity score in observational studies for causal effects. *Biometrika*, *70*, 41–55.

Rowe, M.L., Pan, B.A. and Ayoub, C. (2005). Predictors of variation in maternal talk to children: a longitudinal study of low-income families. *Parenting: Science and Practice*, *5*(3), 259–283.

Rubin, D.B. (1986). Comment: which Ifs have causal answers. *Journal of the American Statistical Association*, *81*, 961–962.

Schafer, J.L. and Kang, J. (2008). Average causal effects from nonrandomized studies: a practical guide and simulated example. *Psychological Methods*, *13*(4), 279–313.

Sobel, M.E. (2012). Does marriage boost men's wages? identification of treatment effects in fixed effects regression models for panel data. *Journal of the American Statistical Association*, *107*(498), 521–529.

Waernbaum, I. (2012). Model misspecification and robustness in causal inference: comparing matching with doubly robust estimation. *Statistics in Medicine*, *31*, 1572–1581.

Zajonc, T. (2012). Bayesian inference for dynamic treatment regimes: mobility, equity, and efficiency in student tracking. *Journal of the American Statistical Association, 107*(497), 80–92.

Zehler, A.M., Fleischman, H.L., Hopstock, P.J. *et al.* (2003). *Descriptive Study of Services to LEP Students and LEP Students with Disabilities. Volume 1: Research Report.* Report submitted to the U.S. Department of Education, Office of English Language Acquisition, Development Associates, Inc., Arlington, VA.

Zehler, A.M., Adger, C., Coburn, C. *et al.* (2008). *Preparing to Serve English Language Learner Students: School Districts with Emerging English Language Learner Communities*, U.S. Department of Education, Institute of Education Sciences National Center for Education Evaluation and Regional Assistance, Washington, DC.

Part III

MEDIATION

9

Concepts of mediated treatment effects and experimental designs for investigating causal mechanisms

Much of the literature on quantitative research methods in social sciences has been devoted to a conceptual clarification of the distinction between moderation and mediation (Baron and Kenny, 1986; Bauer, Preacher, and Gil, 2006; Fairchild and MacKinnon, 2009; Frazier, Tix, and Baron, 2004; Hayes, 2013; Hoyle and Robinson, 2003; James and Brett, 1984; Jose, 2013; Judd and Kenny, 1981; Kraemer *et al.*, 2001, 2002; Muller, Judd, and Yzerbyt, 2005; Preacher, Rucker, and Hayes, 2007; Rose *et al.*, 2004; Wu and Zumbo, 2008). Typical moderation questions include for whom a treatment works and under what conditions it works; typical mediation questions have to do with the process through which a treatment produces its effect. Yet in applied research, the confusion centers primarily around the concept of mediation.

This chapter reviews the concepts of the indirect effect and the direct effect initially defined in the framework of path analysis. These definitions have dominated discussions of mediation for decades. The indirect effect and the direct effect defined in terms of path coefficients are meaningful, however, only when the functional forms of the path models are correctly specified. This framework typically overlooks the fact that a treatment may generate an impact on the outcome through not only changing the mediator value but also changing the mediator–outcome relationship (Judd and Kenny, 1981). The potential outcomes framework provides a useful alternative because it enables one to define mediated and unmediated causal effects of a treatment without making assumptions about the unknown functional forms. Having defined the causal parameters in terms of potential outcomes, naturally, one would search for an experimental design for studying causal mediation mechanisms that invokes perhaps the weakest identifying assumptions. The last part of this chapter reviews and compares a variety of experimental designs for this purpose.

Causality in a Social World: Moderation, Meditation and Spill-over, First Edition. Guanglei Hong.
© 2015 John Wiley & Sons, Ltd. Published 2015 by John Wiley & Sons, Ltd.

The section on mediation includes four more chapters. Chapter 10 reviews and compares the existing analytic methods for investigating causal mediation. Chapter 11 introduces the ratio-of-mediator-probability weighting (RMPW) strategy for decomposing a total treatment effect into a direct effect and an indirect effect transmitted through a focal mediator. The RMPW strategy is particularly flexible to use in the presence of treatment-by-mediator interactions without having to rely on the functional form of the outcome model. Chapter 12 presents extensions of the RMPW strategy to multivalued mediators and binary outcomes as well as to multilevel experimental designs including cluster randomized trials and multisite randomized trials. Chapter 13 discusses RMPW extensions when the research interest lies in mediation effects moderated by pretreatment individual or contextual characteristics. The chapter also extends RMPW to investigations of complex mediation mechanisms that involve concurrent or consecutive mediators.

9.1   Introduction

Why does a treatment work? Or why does it fail to work as intended? In research on human development, almost every treatment is built on specific theories with regard to how an external intervention would bring about changes in an individual's experience and imminent behavior. Such behavioral changes often reflect treatment-induced changes in cognitive or social–emotional process, which would subsequently transform the individual's developmental outcomes. A mediator, according to Baron and Kenny (1986), "represents the generative mechanism through which the focal independent variable is able to influence the dependent variable of interest" (p. 1173). The "focal independent variable" corresponds to the "treatment," while the "dependent variable of interest" corresponds to the "outcome" in our discussion. For example, the well-known and often controversial Kumon method, used for after-school supplementary learning of reading and math, emphasizes the value of incremental development of academic skills, self-learning habits, and confidence through the daily routine of practice at home (Ukai, 1994). Let a student's sign-up in a Kumon program be the treatment of interest. Daily practice at home, viewed as the key vehicle that moves a student steadily toward a higher level of academic accomplishment, is a hypothesized mediator that would transmit the impact, if there is any, of a Kumon program on a student's educational attainment.

A theoretical understanding of the process or mechanism through which a treatment is expected to work is essential not only for advancing scientific knowledge but also for improving treatment effectiveness (Judd and Kenny, 1981). As some researchers have argued or illustrated, mediation could occur even when the total effect of the treatment on the outcome is zero (Collins, Graham, and Flaherty, 1998; Imai, Keele, and Tingley, 2010; MacKinnon, Krull, and Lockwood, 2000; Preacher and Hayes, 2008; Sheets and Braver, 1999; Shrout and Bolger, 2002). If a treatment fails to improve the targeted outcomes, one possibility is that flaws in implementation weakened its impact on the targeted intermediate experiences. A solution might be to allocate more resources to enhance implementation. For example, Kumon educators would argue that simply signing up for a Kumon program would unlikely bring a student measurable success unless the student completes his/her assignments and makes all the corrections on a daily basis, which would require a considerable amount of parental involvement especially for young children. A second possibility is that the intervention did change the intermediate experiences as expected, but no change in the targeted outcomes resulted. This would call for a reexamination of the hypothesized causal mechanisms.

The theory could be flawed because the hypothesized relationships between the intermediate experiences and the targeted outcomes do not hold or because the potential benefits were offset by the worsening of some other intermediate experiences not targeted by the intervention. A treatment may turn out to be ineffective due to one or more issues mentioned earlier. Other possibilities include that the treatment is beneficial to one subpopulation while being detrimental to another or that the success of an intervention requires a change in some concurrent conditions such as a reorganization of the home environment to create a peaceful time and space for study. A comprehensive evaluation would distinguish among these explanations. Even if a treatment does improve the targeted outcomes, there is still a possibility that the treatment theory might have failed to fully represent the actual causal mechanisms (Hong, 2012).

Causal mediation analysis generates and examines evidence and thereby evaluates the theory of causal mechanisms. Once a theory has been well specified, statistical concepts come to aid by translating the theory into mathematical quantities that one may identify from observable data (Heckman, 2010). In multiple social science disciplines, "the direct effect" and "the indirect effect" are terms conventionally used to represent a way of decomposing the total effect of a treatment on the outcome. When there is a single mediator of interest, the indirect effect of the treatment, channeled through the hypothesized mediator, quantifies the contributing role of the mediator, while the direct effect of the treatment works possibly through other unspecified mechanisms (Alwin and Hauser, 1975). Most researchers agree that the indirect effect involves at least two causal chains: the treatment must have a causal impact on the mediator; and the treatment-induced change in the mediator must subsequently have a causal impact on the outcome (Shadish and Sweeney, 1991).

9.2   Path coefficients

Researchers in social sciences have found a convenient use of path diagrams to represent mediational relationships. Figure 9.1 has been most commonly used to depict the basic causal chains involved when M mediates the effect of treatment Z on outcome Y. Here a single-headed arrow represents a causal relationship (Baron and Kenny, 1986; Preacher and Hayes, 2004). This path diagram corresponds to two regression models, one for the mediator and the other for the outcome, representing the analysts' belief with regard to how the data are generated:

$$M = \gamma_0 + aZ + \varepsilon_M; \tag{9.1}$$

$$Y = \beta_0 + bM + cZ + \varepsilon_Y. \tag{9.2}$$

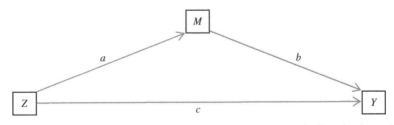

Figure 9.1   Path diagram representing the effect of Z on Y indirectly through M

Here, γ_0 and ε_M are the intercept and the error term in the mediator model; and β_0 and ε_Y are the intercept and the error term in the outcome model. Following Judd and Kenny (1981) and Baron and Kenny (1986), applied researchers have routinely regarded c as the direct effect and ab the indirect effect. Alternatively, one may specify a third model regressing the outcome on the treatment:

$$Y = \beta_0' + c'Z + \varepsilon_Y'. \tag{9.3}$$

Considering $c' = ab + c$ as the total effect of the treatment on the outcome, one may then represent the indirect effect as $c'-c$ (Duncan, 1966; Finney, 1972; Preacher and Hayes, 2004).

However, the path coefficients represent the causal effects of interest only when the functional form of each of these models is correctly specified. The direct effect is no longer c and the indirect effect no longer ab if Z and M, Z and Y, and M and Y are not linearly associated and if Z and M do not additively influence Y. Every theory of causal mediation represents a novel attempt at revealing the structure; yet it is rare that a given theory fully specifies the structural relationships in mathematical forms. In contrast to the path models, defining the causal effects in terms of potential outcomes does not involve the specification of any functional form.

9.3   Potential outcomes and potential mediators

Holland (1988) made an initial attempt at extending to mediation problems the Neyman–Rubin causal model, discussed in detail in Chapter 2 (Holland, 1986; Neyman, 1935; Rubin, 1978). He drew his example from an encouragement design in which students are assigned at random either to an experimental condition that encourages study for the test or to a control condition. Holland argued that the amount a student studies is a response to the encouragement condition and is a self-imposed treatment that may have an effect on test performance. Hence, one may view the amount of study as a mediator, denoted by M, of the effect of encouragement Z on test performance Y. Holland then defined the potential outcomes for student i being exposed to either the experimental condition $(Z_i = 1)$ or the control condition $(Z_i = 0)$:

$M_i(1)$ is the amount that student i studies if encouraged to study.

$M_i(0)$ is the amount that student i studies if not encouraged to study.

$Y_i(1, M_i(1))$ is the student's test score under encouragement and can be simplified as $Y_i(1)$.

$Y_i(0, M_i(0))$ is the test score if the student is not encouraged to study and can be simplified as $Y_i(0)$.

$Y_i(1, m)$ is the test score if the student is encouraged to study and studies for m hours.

$Y_i(0, m)$ is the test score if the student is not encouraged to study and studies for m hours.

A contrast of the two potential mediators $M_i(1) - M_i(0)$ defines the causal effect of encouragement on student i's amount of study. A contrast of the potential outcome when treated and the potential outcome when untreated $Y_i(1, M_i(1)) - Y_i(0, M_i(0))$, which is equivalent to the familiar form $Y_i(1) - Y_i(0)$, defines the causal effect of encouragement on student i's test score.

9.3.1 Controlled direct effects

Holland (1988) defined two additional causal effects of interest:

$Y_i(1,m) - Y_i(0,m)$ is the effect of encouragement on test scores when student i studies m hours regardless of being encouraged or not.

$Y_i(z,m) - Y_i(z,m')$ is the effect of studying m hours relative to studying m' hours on student i's test score when the student is under treatment condition z, which can be written as $Y_i(1,m) - Y_i(1,m')$ if the student is encouraged to study and as $Y_i(0,m) - Y_i(0,m')$ if the student is not encouraged to study.

The aforementioned causal effects defined in terms of the potential outcomes do not invoke any assumptions with regard to the functional form relating the treatment to the mediator or the outcome or relating the treatment and the mediator to the outcome. Yet the definitions are based on the principle of *manipulable treatment* and *manipulable mediator*. A student may have exposure to encouragement or may have no such exposure; the amount that the student studies may be fixed at a certain value (e.g., m hours) or may be changed to a certain alternative value (e.g., m' hours) under either treatment condition. Given Holland's position on "no causation without manipulation" (1986), he argued that, in principle, the mediator as well as the treatment can conceivably be manipulated by an experimenter or by other external forces. In the current example, the student may study more or less or not at all, which allows the amount of study to be the second cause in the causal chains.

Pearl (2001) pointed out that experimentally controlling the mediator at a fixed value such as forcing all students to study m hours reflects a prescriptive conceptualization. Hence, in Pearl's vocabulary, $Y_i(1,m) - Y_i(0,m)$ is the *controlled direct effect* of encouragement on test scores when student i studies m hours regardless of being encouraged or not; and $Y_i(z,m) - Y_i(z,m')$ is the *controlled direct effect* of studying m hours relative to studying m' hours on student i's test score when the student is under treatment condition z. Table 9.1 provides a glossary of the causal effects related to causal mechanisms.

9.3.2 Controlled treatment-by-mediator interaction effect

A controlled treatment-by-mediator interaction effect arises if the controlled direct effect of the treatment depends on the mediator value or, analogously, if the controlled direct effect of the mediator depends on the treatment. Suppose that a student, if assigned to the experimental condition, would become more attentive in study than he would under the control condition. If the student were not given any time to study, the assignment to the experimental condition rather than the control condition perhaps would make no difference to his test performance, that is, $Y_i(1,0) - Y_i(0,0) = 0$. However, if the student were given 4 hours/day to study, studying more attentively under the experimental condition than under the control condition for 4 hours/day would perhaps make a noticeable difference in his test score. In other words, one would expect that $Y_i(1,4) - Y_i(0,4) > 0$. The controlled direct effect of the treatment in this case would depend on the amount of study. On the basis of the same reasoning, studying 4 hours/day rather than none under the experimental condition is expected to produce a larger gain in the student's test score when compared with studying 4 hours/day rather than none under the control condition, that is, $Y_i(1,4) - Y_i(1,0) > Y_i(0,4) - Y_i(0,0)$. When we control

Table 9.1 Glossary of causal effects in mediation analysis

Label	Notation	Definition
Treatment effect on the mediator	$M(1)-M(0)$	The effect of being assigned to the experimental condition versus the control condition on an individual's mediator
Treatment effect on the outcome	$Y(1)-Y(0)$	The effect of being assigned to the experimental condition versus the control condition on an individual's outcome
Controlled direct effect of the treatment	$Y(1, m)-Y(0, m)$	The effect of the treatment on an individual's outcome when the mediator is held at a fixed value m
Controlled direct effect of the mediator	$Y(z, m)-Y(z, m')$	The effect of the mediator on an individual's outcome under a fixed treatment condition z
Controlled treatment-by-mediator interaction effect	$[Y(1, m)-Y(1, m')]-[Y(0, m)-Y(0, m')]$	The difference in the mediator effect on an individual's outcome between the experimental condition and the control condition
Natural direct effect of the treatment on the outcome	$Y(1, M(0))-Y(0, M(0))$	The effect of the treatment on an individual's outcome should the individual's mediator value remain unchanged by the treatment
Natural indirect effect of the treatment on the outcome	$Y(1, M(1))-Y(1, M(0))$	The effect of the treatment-induced change in an individual's mediator value on the outcome under the experimental condition
Pure indirect effect of the treatment on the outcome	$Y(0, M(1))-Y(0, M(0))$	The effect of the treatment-induced change in an individual's mediator value on the outcome under the control condition
Natural treatment-by-mediator interaction effect	$[Y(1, M(1))-Y(1, M(0))]-[Y(0, M(1))-Y(0, M(0))]$	The difference between the experimental condition and the control condition in how the treatment-induced change in an individual's mediator value affects the outcome

the mediator value at m as opposed to m' through experimental manipulation, the controlled treatment-by-mediator interaction effect can be denoted by

$$[Y_i(1,m) - Y_i(0,m)] - [Y_i(1,m') - Y_i(0,m')],$$

representing the difference in the controlled direct effect of the treatment between two different mediator values specified by the experimenter. This is equivalent to

$$[Y_i(1,m) - Y_i(1,m')] - [Y_i(0,m) - Y_i(0,m')], \tag{9.4}$$

representing the difference in the controlled direct effect of the mediator between the experimental condition and the control condition.

9.4 Causal effects with counterfactual mediators

In the aforementioned definitions, an individual's mediator under each treatment condition can conceivably be fixed at one value or another through experimental manipulation. Yet the amount of study set by the experimenter might be different from the amount of study a student would actually have in response to the treatment that he has been assigned to. Moreover, the potential mediators $M_i(1)$ and $M_i(0)$ and their difference $M_i(1) - M_i(0)$ may take values that vary randomly among individuals. To many, an important goal in mediation research is to decompose the total treatment effect $Y_i(1, M_i(1)) - Y_i(0, M_i(0))$ so that it can be determined the extent to which the total effect is transmitted through the treatment effect on the mediator. The controlled direct effects, however, do not generally add up to the total effect.

Pearl (2001) made a distinction between prescriptive and descriptive definitions of direct effects. The controlled direct effects are prescriptive because the amount that a student studies is strictly under the experimenter's control and is not affected by the initial treatment assignment. In contrast, the descriptive conceptualization maintains the natural relationship between the treatment and the mediator. This conceptual distinction was similarly laid out by Robins and Greenland (1992). Most importantly, as we will explain in the following, when the interest is in decomposing the total effect into a *natural direct effect* and a *natural indirect effect*, the definitions of these natural effects involve potential outcomes of counterfactual mediators.

9.4.1 Natural direct effect

Suppose that the assignment to encouragement introduces no change in the natural amount of study of any student, one may anticipate no change in the students' test scores. That is, the natural direct effect of encouragement is anticipated to be zero. If the natural direct effect is in fact nonzero, this would indicate a possible existence of other unspecified pathways. For example, along with the encouragement to study, students in the experimental group may receive study materials that resemble the test items. As a result, *without increasing the amount of study*, a student in the experimental group may nonetheless score higher than he would have had he instead been assigned to the control condition. Pearl formally defines the natural direct effect for student i as

$$Y_i(1, M_i(0)) - Y_i(0, M_i(0)). \tag{9.5}$$

Here, $Y_i(1, M_i(0))$ is the test score of student i if the student is encouraged to study yet counterfactually studies only as much as he would under the control condition.

9.4.2   Natural indirect effect

We now ask whether an increase in the amount of study mediates the effect of encouragement on a student's test score. Understandably, there is no mediation if the encouragement does not influence the amount that the student studies or if, despite a change in the amount of study, the student scores the same as he would under the counterfactual control condition. The natural indirect effect is defined as the difference between a student's test score if encouraged to study and studying the amount as he would when encouraged, denoted by $Y_i(1, M_i(1))$, and the student's test score if encouraged to study yet counterfactually studying the amount as he would if not encouraged, denoted by $Y_i(1, M_i(0))$. This difference is solely attributable to the change in his amount of study induced by encouragement, that is, the change of study hours from $M_i(0)$ to $M_i(1)$:

$$Y_i(1, M_i(1)) - Y_i(1, M_i(0)). \tag{9.6}$$

The sum of the natural direct effect and the natural indirect effect is the total effect of the treatment on the outcome: $Y_i(1, M_i(1)) - Y_i(0, M_i(0))$.

9.4.3   Natural treatment-by-mediator interaction effect

As Judd and Kenny (1981) pointed out, a treatment may produce its effects not only through changing the mediator value but also in part by altering the mediational process that normally produces the outcome. Hence, they emphasized that investigating treatment-by-mediator interactions should be an important component of mediation analysis, a point echoed in the more recent discussions (Kraemer *et al.*, 2002). In the current example, suppose that when a student is assigned to the experimental condition, as a result of receiving encouragement, the student will not only spend more time to study for the test but also study more attentively than he would under the control condition. Due to the improvement in attentiveness, increasing the amount of study from $M_i(0)$ to $M_i(1)$ under the experimental condition may produce a greater amount of gain in test performance than under the control condition. This extra amount of gain is to be attributed to an intervention-induced change in the relationship between the mediator and the outcome. This is a case in which the intervention exerts its impact on test performance partly through increasing the number of study hours and partly through increasing the amount of learning produced by every additional hour of study. The natural treatment-by-mediator interaction effect for student i, representing how the treatment-induced change in the mediator affects the outcome differently between the experimental condition and the control condition, can be defined as

$$[Y_i(1, M_i(1)) - Y_i(1, M_i(0))] - [Y_i(0, M_i(1)) - Y_i(0, M_i(0))]. \tag{9.7}$$

Here, the first quantity represents, under the experimental condition, the impact of the encouragement-induced change in the amount of study on the student's test score. In parallel, the second quantity represents, under the control condition, the impact of the encouragement-induced change in the amount of study on the student's test score. Robins and Greenland (1992) called the former "the total indirect effect" and the latter "the pure indirect effect."

The pure indirect effect and the natural treatment-by-mediator interaction effect add up to the natural indirect effect:

$$Y_i(1,M_i(1)) - Y_i(1,M_i(0))$$
$$= [Y_i(0,M_i(1)) - Y_i(0,M_i(0))]$$
$$+ \{[Y_i(1,M_i(1)) - Y_i(1,M_i(0))] - [Y_i(0,M_i(1)) - Y_i(0,M_i(0))]\}.$$

Both components are functions of the treatment effect on the mediator and would both be 0 in the case that $M_i(1) - M_i(0) = 0$. When the treatment changes the mediator value without changing the mediator-outcome relationship, the natural indirect effect will be simply equal to the pure indirect effect.

Treatment-by-mediator interactions may sometimes provide an explanation for why an intervention fails to produce its intended effect on the outcome. For example, an encouragement that comes with an undue amount of pressure may increase study hours yet at the same time may reduce the amount of learning produced per hour. Even though an increase in the amount of study is expected to increase learning, with a fixed amount of study, the student would learn less under the experimental condition than under the control condition, leading to a possible null effect or even a negative effect of the encouragement treatment.

The controlled treatment-by-mediator interaction effect defined in (9.4) is distinct from the natural treatment-by-mediator interaction effect defined in (9.7). This is because the mediator values m and m' to be set by the experimenter for evaluating the controlled treatment-by-mediator interaction effect may not correspond to an individual's amount of study under either treatment condition. For example, the treatment may have no effect on a student's amount of study, that is, $M_i(1) - M_i(0) = 0$, and therefore the natural treatment-by-mediator interaction effect is zero. Yet the experimenter might be able to change the student's amount of study from m' to m through other ways of manipulation such that the controlled treatment-by-mediator interaction effect could possibly be nonzero.

9.4.4   Unstable unit treatment value

Rubin (1986) emphasized the key role of the stable unit treatment value assumption (SUTVA) in defining causal effects. When the treatment effect on the mediator is defined as $M_i(1) - M_i(0)$, each potential outcome $M_i(z)$ for $z = 0, 1$ is a function of who the student is and to which treatment the student is assigned. According to Rubin, SUTVA requires that the amount that student i studies if encouraged to study will be the same "no matter what mechanism is used to assign" the treatment and "no matter what treatments the other units receive" (1986, p. 961). The first part of SUTVA is sometimes named the "consistency assumption," while the second part is known as the "no interference assumption." SUTVA further implies that no other random events beyond the experimenter's control could possibly alter the value of $M_i(z)$. Such events may include, for example, that the student falls sick and is unable to study as many hours as he has planned to.

The definitions of the natural effects, however, involve the counterfactual outcomes $Y_i(1, M_i(0))$ and $Y_i(0, M_i(1))$ for student i. Here, $M_i(0)$ is counterfactual when the student is treated, while $M_i(1)$ is counterfactual when the student is untreated. The conception of a potential outcome under one treatment condition as a function of a "counterfactual mediator" associated with an alternative treatment condition is seemingly in contradiction with SUTVA. Suppose that student i would study 4 hours/day if encouraged and would not study at all if not encouraged. Under SUTVA, the student would never study 4 hours/day when he is not encouraged and would never study zero hours if he is encouraged. Therefore, neither $Y_i(1, M_i(0))$ nor $Y_i(0, M_i(1))$ is conceivable. These counterfactual outcomes are conceivable only if the mediator value under a given treatment condition is "unstable." We may conceive $M_i(1)$ and $M_i(0)$ to each have its distribution reflecting possible influences of random events in a student's life. For example, student i may study anywhere between 0 and 7 hours/day if encouraged and may study 0–5 hours if not encouraged.

9.5   Population causal parameters

Social scientists are often interested in theories about causal mechanisms that can be generalized to a well-defined population. Important variations in causal mechanisms across subpopulations will be discussed in Chapter 13. For each of the individual-specific causal effects defined previously, one may define the population average causal effect accordingly. As before, we use $E[\cdot]$ to denote the population mean of a random variable.

Table 9.2 illustrates with a population of six individuals in the simplified case of a binary treatment and a binary mediator. For example, let $Z_i = 1$ if individual i is randomized to receive encouragement and let $Z_i = 0$ if the individual does not receive encouragement; let $M_i = 1$ if the individual studies and let $M_i = 0$ if the individual does not study. Three of these individuals are assigned to the experimental group and three to the control group. For each individual, we list two potential mediator values corresponding to the two possible treatment conditions and four potential outcomes. For the first three individuals, the only observables are $M(1)$ and $Y(1, M(1))$, while for the next three, the only observables are $M(0)$ and $Y(0, M(0))$.

For example, the first individual would study when encouraged $(M_1(1) = 1)$ and would also study if counterfactually not encouraged $(M_1(0) = 1)$. The treatment effect on this individual's mediator values is therefore zero $(M_1(1) - M_1(0) = 0)$. Because encouragement fails to change this individual's study behavior, the natural indirect effect of encouragement on this student's test score when encouraged, transmitted through the encouragement-induced change in study behavior, $Y_1(1, M_1(1)) - Y_1(1, M_1(0))$, is zero given that $M_1(1) = M_1(0)$, while the natural direct effect $Y_1(1, M_1(0)) - Y_1(0, M_1(0))$ equals the total effect for this student $Y_1(1, M_1(1)) - Y_1(0, M_1(0))$. Similarly, the pure indirect effect of encouragement on this student's test score when not encouraged, transmitted through the encouragement-induced change in study behavior, $Y_1(0, M_1(1)) - Y_1(0, M_1(0))$, is also zero. Hence, the natural treatment-by-mediator interaction effect, defined as the difference between the natural indirect effect and the pure indirect effect, is zero.

In contrast, the second individual would study when encouraged $(M_2(1) = 1)$ but would not study if counterfactually not encouraged $(M_2(0) = 0)$. Hence, the treatment has a nonzero effect on this individual's mediator values $(M_2(1) - M_2(0) = 1)$. In this case, the natural indirect effect of encouragement on this individual's test scores when encouraged, transmitted through the encouragement-induced change in study behavior, $Y_2(1, M_2(1)) - Y_2(1, M_2(0))$, is possibly nonzero. Similarly, the pure indirect effect of encouragement on this individual's test scores when not encouraged, transmitted through the encouragement-induced change in study behavior, $Y_2(0, M_2(1)) - Y_2(0, M_2(0))$, may also be nonzero. Yet the natural indirect

Table 9.2 Potential mediators and potential outcomes

Individual	Treatment Z	Potential mediators		Potential outcomes			
		M(1)	M(0)	Y(1, M(1))	Y(1, M(0))	Y(0, M(1))	Y(0, M(0))
1	1	1	1	Y(1,1)	Y(1,1)	Y(0,1)	Y(0,1)
2	1	1	0	Y(1,1)	Y(1,0)	Y(0,1)	Y(0,0)
3	1	0	0	Y(1,0)	Y(1,0)	Y(0,0)	Y(0,0)
4	0	1	1	Y(1,1)	Y(1,1)	Y(0,1)	Y(0,1)
5	0	1	0	Y(1,1)	Y(1,0)	Y(0,1)	Y(0,0)
6	0	0	0	Y(1,0)	Y(1,0)	Y(0,0)	Y(0,0)
Population average		E[M(1)]	E[M(0)]	E[Y(1, M(1))]	E[Y(1, M(0))]	E[Y(0, M(1))]	E[Y(0, M(0))]

effect does not have to equal the pure indirect effect. For example, if the student studies more effectively when encouraged than when not encouraged, the natural treatment-by-mediator interaction effect is possibly nonzero.

Taking an average of each of the potential mediators and potential outcomes over all the individuals in the population, we obtain the population average potential mediators $E[M(1)]$ and $E[M(0)]$ and the population average potential outcomes $E[Y(1, M(1))]$, $E[Y(1, M(0))]$, $E[Y(0, M(1))]$, and $E[Y(0, M(0))]$. Below, we define the population average causal effects:

$E[M(1)-M(0)]$ is the population average causal effect of the treatment on the mediator, that is, the population average causal effect of encouragement on the amount students actually study, and is equivalent to $E[M(1)-M(0)]$.

$E[Y(1,M(1))-Y(0,M(0))]$ is the population average causal effect of the treatment on the outcome, that is, the population average causal effect of encouragement on test scores, and can be simplified in notation as $E[Y(1)-Y(0)]$.

$E[Y(1,m)-Y(0,m)]$ is the population average controlled direct effect of the treatment, that is, the population average causal effect of encouragement on test scores when all students study m hours regardless of being encouraged or not.

$E[Y(z,m)-Y(z,m')]$ is the population average controlled direct effect of the mediator, that is, the average causal effect on test scores when all students in the population study m hours relative to studying m' hours under treatment condition z.

$E[\{Y(1,m)-Y(1,m')\}-\{Y(0,m)-Y(0,m')\}]$ is the population average controlled treatment-by-mediator interaction effect, that is, the difference between the experimental condition and the control condition in the controlled direct effect of the mediator.

$E[Y(1,M(0))-Y(0,M(0))]$ is the population average natural direct effect, that is, the population average causal effect of encouragement on test scores if encouragement, perhaps counterfactually, fails to change any student's amount of study.

$E[Y(1,M(1))-Y(1,M(0))]$ is the population average natural indirect effect, that is, when all the students are encouraged to study, the population average amount of change in test scores solely attributable to the encouragement-induced change in the amount of study.

When the goal is to decompose the total effect of the treatment on the outcome, the population average total effect $E[Y(1)-Y(0)]$ can be represented as a sum of the population average natural direct effect $E[Y(1,M(0))-Y(0,M(0))]$ and the population average natural indirect effect $E[Y(1,M(1))-Y(1,M(0))]$, and the latter can be further decomposed into the population average pure direct effect $E[Y(0,M(1))-Y(0,M(0))]$ and the population average natural treatment-by-mediator interaction effect $E[\{Y(1,M(1))-Y(1,M(0))\}-\{Y(0,M(1))-Y(0,M(0))\}]$.

9.5.1   Population average natural direct effect

Should the controlled treatment-by-mediator interaction effect be zero, the population average natural direct effect $E[Y(1,M(0))-Y(0,M(0))]$ would be simply equal to the population average controlled direct effect of the treatment $E[Y(1,m)-Y(0,m)]$ because the latter would be a constant irrespective of the mediator value m. In the more general case, however, the natural direct effect may differ across subpopulations of students who would study different amounts under the control condition. We may define a relatively homogeneous subpopulation of students who would study the same amount under the control condition, $M(0) = m$. The population average natural direct effect is an average of the subpopulation-specific natural direct effect $E[Y(1, M(0)) - Y(0, M(0))|M(0) = m]$ weighted by the proportion of students in each subpopulation $pr(M(0) = m)$ for all possible values of m.

9.5.2   Population average natural indirect effect

In general, the population average natural indirect effect $E[Y(1, M(1)) - Y(1, M(0))]$ cannot be equated with the population average controlled direct effect of the mediator under the experimental condition $E[Y(1, m) - Y(1, m')]$ unless $M(1) = m$ and $M(0) = m'$ are constants for all individuals. We may represent the natural indirect effect as the difference between the two population means $E[Y(1, M(1))]$ and $E[Y(1, M(0))]$. The first population mean can be written as an average of subpopulation means over the distribution of $M(1)$, where each subpopulation consists of students who would study the same amount when encouraged; the second population mean is an average of subpopulation means over the distribution of $M(0)$, where the students in each subpopulation would study the same amount when not encouraged. When the treatment has a nonzero effect on the mediator on average, the distribution of $M(1)$ is expected to be different from the distribution of $M(0)$

9.6   Experimental designs for studying causal mediation

As Holland (1986) made clear, a major challenge in causal inference is that the potential outcomes associated with counterfactual treatment conditions are unobserved. Identification relates the causal parameters to the observable quantities under necessary assumptions with regard to the unobservable quantities. Which identification assumptions need to be invoked and how plausible they are depend first of all on the research design. There is an increasing awareness among applied researchers that, even when the treatment is randomized, individuals who have assigned to the same treatment condition may differ in their mediator values, which often reflects individual differences in personal traits or environmental resources that may predict the outcome. For example, among those assigned at random to receive encouragement to study for a test, a student who studies 4 hours/day is perhaps more motivated than another student who does not study at all. A third student who studies 8 hours/day is perhaps less prepared for and more anxious about the test than the first student. Scholarly discussions about how to design an experiment for investigating causal mediation mechanisms have started to emerge only recently in the statistics and methodology literature, all aimed at eliminating the confounding of mediator–outcome relationships through manipulating the mediator value assignment. In the following, we briefly review eight experimental designs that have been proposed. In each case, we clarify the causal effects to be identified and the key assumptions required for the identification. These are summarized in Table 9.3.

Table 9.3   A comparison of experimental designs for studying causal mediation

Type of designs	Causal effects	Identification assumptions
Sequentially randomized designs	$E[M(1)-M(0)]$; $E[Y(z,m)-Y(z,m')]$; $E[Y(1,m)-Y(1,m')]-[Y(0,m)-Y(0,m')]$; $E[Y(1,M(0))-Y(0,M(0))]$; $E[Y(1,M(1))-Y(1,M(0))]$; $E[\{Y(1,M(1))-Y(1,M(0))\}-\{Y(0,M(1))-Y(0,M(0))\}]$.	Ignorable treatment assignment; ignorable mediator value assignment informed by pretreatment covariates under each treatment condition
Two-phase experimental designs	$E[M(1)-M(0)]$; $E[Y(1)-Y(0)]$; $E[Y(z,m)-Y(z,m')]$; $E[Y(1,m)-Y(1,m')]-[Y(0,m)-Y(0,m')]$; $E[Y(1,M(0))-Y(0,M(0))]$; $E[Y(1,M(1))-Y(1,M(0))]$; $E[\{Y(1,M(1))-Y(1,M(0))\}-\{Y(0,M(1))-Y(0,M(0))\}]$.	Same as above
Three- and four-arm designs	$E[Y(1,M(0))-Y(0,M(0))]$; $E[Y(1,M(1))-Y(1,M(0))]$; $E[\{Y(1,M(1))-Y(1,M(0))\}-\{Y(0,M(1))-Y(0,M(0))\}]$.	Ignorable assignment to treatment-by-mediator combinations
Causal-chain designs	$E[M(1)-M(0)]$; $E[Y(1)-Y(0)]$; $E[Y(z,m)-Y(z,m')]$; $E[Y(1,m)-Y(1,m')]-[Y(0,m)-Y(0,m')]$.	Ignorable treatment assignment; ignorable mediator value assignment under each treatment condition
Moderation-of-process designs	$E[Y(1,m)-Y(0,m)]$; $E[Y(1,m)-Y(0,m')]$.	Ignorable treatment assignment; ignorable mediator value assignment; no concurrent mediators

Design	Estimand(s)	Assumptions
Augmented encouragement designs	$E[Y(1,m) - Y(0,m)|M(z=1,g) = M(z=0,g)]$.	Ignorable treatment assignment; ignorable encouragement assignment; exclusion restriction; monotonicity
Parallel experimental designs	$E[M(1) - M(0)]$; $E[Y(1) - Y(0)]$; $E[Y(1,M(1)) - Y(1,M(0))]$; $E[Y(1,M(0)) - Y(0,M(0))]$.	Ignorable treatment assignment; ignorable assignment to treatment-by-mediator combinations; no treatment-by-mediator interaction
Parallel encouragement designs	$E[Y(1,M(1)) - Y(1,M(0))|M(z,g=1) > M(z,g=0)]$.	Ignorable treatment assignment; ignorable encouragement assignment; exclusion restriction; monotonicity
Crossover experimental designs	$E[M(1) - M(0)]$; $E[Y(1) - Y(0)]$; $E[Y(1,M(0)) - Y(0,M(0))]$; $E[Y(1,M(1)) - Y(0,M(1))]$; $E[Y(1,M(1)) - Y(1,M(0))]$; $E[\{Y(1,M(1)) - Y(1,M(0))\} - \{Y(0,M(1)) - Y(0,M(0))\}]$.	Ignorable treatment assignment; ignorable mediator value assignment under each treatment condition; no carry-over effects
Crossover encouragement designs	$E[M(1) - M(0)]$; $E[Y(1) - Y(0)]$; $E[Y(1,M(0)) - Y(0,M(0))|M(z,g=1) = M(z',g=1),M(z,g=0) \neq M(z',g=0)]$; $E[Y(1,M(1)) - Y(0,M(1))|M(z,g=1) = M(z',g=1),M(z,g=0) \neq M(z',g=0)]$; $E[Y(1,M(1)) - Y(1,M(0))|M(z,g=1) = M(z',g=1),M(z,g=0) \neq M(z',g=0)]$.	Ignorable treatment assignment; ignorable encouragement assignment; exclusion restriction; monotonicity; no carry-over effects

9.6.1   Sequentially randomized designs

This design was initially proposed for investigating the mediator–outcome relationship under the experimental condition in randomized clinical trials. Rubin (1991) described a hypothetical dose–response experiment aimed at disclosing the causal relationship between dose (M) and response (Y). In the first phase of this hypothetical experiment, patients would be assigned at random to either the experimental condition or the control condition. In phase two, patients in the control group would have no access to the prescription $(M(0)=0)$. Instead of allowing patients in the experimental group to comply at their own will, every treated patient would be randomly assigned a certain dose $(M(1)=m)$, and 100% compliance would be enforced. Probabilities of assignment to a possible dose would be either a stochastic function or determined by a vector of pretreatment covariates $\mathbf{X}$. Because complete randomization of dosage levels might not be realistic, Rubin suggested that the experimenter model the dosage assignment mechanism as a function of a set of pretreatment covariates $\mathbf{x}$. The predicted dosage as a function of $\mathbf{x}$ is then used to create subclasses. The probability of random assignment to a certain dosage is the same within a subclass but differs between subclasses. Then the dose–response curve can be estimated within each subclass and be averaged over all subclasses, leading to an estimate of the average controlled direct effect of the mediator on the outcome under the experimental condition, $E[Y(1,m)-Y(1,m')]$.

The aforementioned design can be extended to a placebo-controlled clinical trial. Efron and Feldman (1991) noted self-selection of dosage level in the placebo control group as well as in the experimental group. To eliminate the selection bias, subsequent to the initial treatment randomization, the experimenter might conduct a random assignment of dosage of the active drug in the experimental group and of the placebo in the control group. The treatment assignment and the dosage assignment under each treatment condition are both independent of the potential outcomes and therefore satisfy the *ignorability* assumption. Under such a design, the experimenter would obtain an unbiased estimate of the average treatment effect on the mediator $E[M(1)-M(0)]$ and, under each treatment condition, an unbiased estimate of the controlled direct effect of the mediator on the outcome $E[Y(1,m)-Y(1,m')]$ and $E[Y(0,m)-Y(0,m')]$. Such a design also has the potential of decomposing the total effect into a natural direct effect and a natural indirect effect.

9.6.2   Two-phase experimental designs

Extending Rubin's sequentially randomization designs, the two-phase experimental designs are more explicit about how to systematically acquire an empirical basis for determining the probability of dosage assignment as a function of pretreatment covariates. Hong (2006) argued that, without taking into account the natural variation among individuals in their predisposition of implementing the treatment assigned, even if the randomization of dosage intake is 100% enforced, the results will have limited external validity in naturalistic settings. Hong (2006, 2013) further pointed out that, despite an individual's latent predisposition of implementation, the observed implementation behavior of the person (i.e., the observed mediator value) might be a result of exogenous random factors in the natural world such as unanticipated sickness despite one's intention to study. For example, if encouraged to study, a student may have a 0.8 probability of studying at least 6 hours/day and, if not encouraged, may have only a 0.3

probability of studying at least 6 hours/day. An individual's latent predisposition of implementation under a given treatment condition, as in the current example, may be represented by the person's conditional probability of studying at least 6 hours/day as a function of pretreatment characteristics. The goal of the phase-one experiment is to find out each individual's conditional probability of implementation under each treatment condition. These individual-specific probabilities are then used for mediator value assignment in the phase-two experiment.

According to Hong (2006), the phase-one experiment would draw a random sample of individuals and record a comprehensive set of covariates X prior to the random treatment assignment. The experimenter would then observe the implementation behavior of each individual subsequent to the treatment assignment. Most importantly, the observed data need to be comprehensive enough such that implementation under each treatment condition is independent of the potential outcomes within each level of X. The experimenter would then analyze the conditional probability of implementation under each treatment condition as a function of the observed x. The empirical information about which types of individuals are more likely to implement the treatment assigned would be summarized in a prediction function, one for each treatment condition.

The phase-two experiment would apply a sequentially randomized block design as proposed by Rubin (1991) to a new sample of individuals. These could be individuals who have been assigned at random to a waiting list at the onset of the phase-one study. A comprehensive set of pretreatment covariates X again needs to be recorded for each individual. After random treatment assignment, the experimenter would apply the prediction functions specified in phase one to predict, in the phase-two experiment, each individual's conditional probability of implementation under the treatment that one has been assigned to. Those with the same conditional probability of implementation under each treatment condition would constitute blocks. The predicted conditional probabilities would be used for the subsequent within-block random assignment of the mediator value under each treatment condition. Hong (2013) argued that, analogous to a two-stage adaptive design in clinical trials (Bauer and Kieser, 1999; Liu, Proschan, and Pledger, 2002), the covariate-informed randomization of mediator values should have a higher compliance rate than simply assigning the mediator values at random. Moreover, by comparing the average treatment effects on the outcome across the phase-one and phase-two experiments, one may partially test the consistency assumption–i.e., the assumption that the potential outcome associated with a given treatment condition and a given mediator value does not depend on whether the mediator value has been self-selected or randomly assigned. The phase-two experiment additionally allows one to test the no treatment-by-mediator interaction assumption.

In addition, Hong (2006) defined subpopulations of individuals on the basis of their predicted implementation behaviors and considered the identification of subpopulation-specific causal effects. These include the controlled direct effect of the treatment, the controlled direct effect of the mediator, and the controlled treatment-by-mediator interaction effect. Chapter 13 will provide an in-depth discussion of mediation mechanisms moderated by individual characteristics. Hong (2013) further delineated a ratio-of-mediator-probability weighting (RMPW) strategy applicable to data from a two-phase experimental design for estimating the natural direct effect, the natural indirect effect, and the natural treatment-by-mediator interaction effect. The same strategy can be applied to data from sequentially randomized designs. The rationale and the analytic procedure of RMPW are presented in Chapter 11.

9.6.3   Three- and four-treatment arm designs

To decompose the total effect into a natural direct effect and a natural indirect effect, Sobel and Stuart (2012) proposed a three-treatment arm experimental design for causal mediation analysis. They discussed a template of this experiment in the context of a study by Hong and Nomi (2012) that evaluated a policy requiring algebra for all ninth graders. Many schools implemented the policy not only by eliminating remedial math classes for low-ability students but also by mixing low- and high-ability students in algebra classes. Hong and Nomi investigated the unintended indirect effect of the policy on low-ability students' math achievement mediated by the rise in the ability levels of their class peers due to mixing. Would the low-ability students be better off if the policy were implemented without creating mixed-ability classes? Sobel and Stuart suggested that, in an ideal world, one would investigate the mechanism with a three-arm experimental design assigning schools at random to:

1. The status quo in which many low-ability students take remedial math, while high-ability students take algebra
2. Algebra course requirement with a mixing of low- and high-ability students in the same class
3. Algebra course requirement without mixing low- and high-ability students in the same class

Because the assignment to the treatment-by-mediator combinations would be ignorable, the average observed outcomes of the three treatment groups would be unbiased estimates of the respective population average potential outcomes $E[Y(0, M(0))]$, $E[Y(1, M(1))]$, and $E[Y(1, M(0))]$. The mean difference in the observed outcome between groups (2) and (1) estimates the total effect. The mean difference between groups (2) and (3) estimates the natural indirect effect, while that between groups (3) and (1) estimates the natural direct effect. A negative natural indirect effect would indicate that, indeed, the low-ability students would be better off if the policy were implemented without creating mixed-ability classes.

To further decompose the natural indirect effect into a pure indirect effect and a natural treatment-by-mediator interaction effect, in theory, one may add a fourth treatment arm:

4. Many low-ability students taking remedial math with a mixing of low- and high-ability students in the same class

This fourth arm is apparently unfeasible in the current example because few high-ability students would take remedial math together with low-ability students. Nonetheless, in this thought experiment, the mean difference in the observed outcome between groups (4) and (1) would estimate the pure indirect effect $E[Y(0, M(1)) - Y(0, M(0))]$, that is, the expected change in a low-ability student's test score when the student attends mixed-ability classes as opposed to low-ability classes when algebra is not required. The natural treatment-by-mediator interaction effect can be estimated as the difference between the (2)–(3) contrast and the (4)–(1) contrast. This design, however, does not estimate the treatment effect on the mediator. Moreover, the number of treatment arms may increase dramatically for multivalued mediators such as dosage.

9.6.4   Experimental causal-chain designs

Spencer and colleagues (2005) proposed an experimental design that involves (i) manipulating the treatment and observing the mediator as a response to the treatment and (ii) manipulating the mediator and observing the outcome as a response to the mediator. Unlike the sequentially randomized designs, these are conducted as two consecutive but separate studies. The first experiment tests the causal relationship between the treatment and the mediator, that is, $E[M(1)-M(0)]$, while the second one tests the causal relationship between the mediator and the outcome under each treatment condition $E[Y(z,m)-Y(z,m')]$. The authors argued that a series of experiments of this sort would establish the causal chain of events and would thereby provide strong evidence for the theoretically hypothesized mediational mechanism even without testing for mediation statistically.

One major difficulty in linking the two studies, as the authors pointed out, is to make a convincing case that the intermediate process as a response to the treatment in the first study is the same as the intermediate process manipulated in the second study. This has been highlighted as *the consistency assumption* in the causal inference literature and is encompassed by SUTVA. In our earlier example, this assumption would imply that if a student is forced to study 6 hours/day, the student would perform the same on the test as he would if he instead made up his own mind to study 6 hours/day.

An additional difficulty is that the design does not facilitate an easy analysis of how much of the treatment effect on the outcome is transmitted through the mediator. This is because, under a certain treatment condition, the mediator value that a participant is randomly assigned to could be quite different from the mediator value that the participant would display as a natural response to the given treatment. In their criticism of the experimental causal-chain designs, Imai and colleagues (2011) presented a numerical example showing that, even when the average treatment effect on the mediator and the average mediator effect on the outcome are both positive, the average natural indirect effect can possibly be negative. In general, one should not simply equate the product of the average treatment effect on the mediator and the average mediator effect on the outcome with the average natural indirect effect.

9.6.5   Moderation-of-process designs

Spencer and colleagues (2005) proposed an alternative type of designs that involves, in a single study, a manipulation of the treatment followed by a manipulation of the mediator. The latter is meant to either amplify or suppress the treatment–mediator relationship. For example, to suppress the relationship between encouragement and the amount of study, the experimenter may randomly assign students, regardless of their initial treatment assignment, to a condition that requires participation in full-time activities that are unrelated to the test such that no one will have time to study. The relationship between encouragement and the amount of study could be amplified if only the control students are assigned at random to the condition that leaves no time for study. The authors reasoned that if the treatment shows a greater effect on the outcome when the treatment–mediator relationship is amplified and a smaller effect on the outcome when the treatment–mediator relationship is suppressed, such findings would provide evidence that the treatment causes a change in the outcome through changing the mediator. A potential drawback of these designs, according to the authors, is that the instrument used to amplify or suppress the treatment–mediator relationship may affect some other intermediate process such as a student's level of anxiety about the forthcoming test.

For example, students who were encouraged to study and were then prevented from studying might become more anxious and might subsequently perform not better but rather worse than the control students. If a nonfocal mediator such as anxiety is changed along with the focal mediator by the instrument, the inference about the causal mediation mechanism operating through the focal mediator will likely be incorrect. Another drawback is that these designs again do not quantify the treatment effect on the outcome transmitted through the mediator.

9.6.6   Augmented encouragement designs

Mattei and Mealli (2011) presented an experimental design in which the treatment Z (e.g., assignment to a therapy) is randomized, while the mediator M (e.g., whether one complies with the treatment assigned) can be influenced by a second randomization of an encouragement G. It is assumed that the encouragement G affects the outcome only through its effect on the mediator M (i.e., the exclusion restriction). The randomized encouragement G therefore serves as an instrumental variable that allows the experimenter to at least partially manipulate the mediator without bringing additional variation to the outcome. The authors showed how the exogenous variation in the mediator M induced by the randomized encouragement G might lead to a finer partitioning of the population into subpopulations, called "principal strata" (Frangakis and Rubin, 2002; see a review of "principal stratification" in Chapters 6 and 10). Each subpopulation consists of individuals who would display the same set of mediator values $M(z=1,g=1), M(z=1,g=0), M(z=0,g=1)$, and $M(z=0,g=0)$ in response to all possible combinations of treatment assignment and encouragement assignment. Under this design, the causal parameter of interest is confined to the controlled direct effect of the treatment when the mediator is held at a fixed value $E[Y(1,m)-Y(0,m)]$ for a unique subpopulation of individuals. Namely, these are individuals who, under the same encouragement assignment, would display a constant mediator value regardless of the initial treatment assignment—that is, $M(z=1,g=1)=M(z=0,g=1)$ if encouraged and $M(z=1,g=0)=M(z=0,g=0)$ if not encouraged. In other words, these individuals' mediator values would be manipulated uniformly by the randomized encouragement G. The counterfactual mediator values are inferred under the monotonicity assumption–that is, the encouragement would not reduce therapy use. However, the subpopulation-specific controlled direct effect is likely different from the population average controlled direct effect. Because the controlled direct effects remain unknown for other subpopulations, this analytic strategy does not decompose the total treatment effect into a population average natural direct effect and a population average natural indirect effect.

9.6.7   Parallel experimental designs and parallel encouragement designs

Imai, Tingley, and Yamamoto (2013) proposed several experimental designs that might help identify natural indirect effects. A *parallel experimental design* would involve two parallel experiments to which individual participants would be assigned at random initially. In one experiment, only the treatment variable would be randomized, whereas in the other, the treatment and the mediator would be simultaneously randomized. The average natural indirect effect $E[Y(1,M(1))-Y(1,M(0))]$ is to be computed as the difference between the average total treatment effect on the outcome $E[Y(1,M(1))-Y(0,M(0))]$ estimated in the first experiment and the natural direct effect $E[Y(1,M(0))-Y(0,M(0))]$ estimated in the second experiment. The latter is computed as the controlled direct effect of the treatment on the outcome averaged over the distribution of the mediator values $\sum_m E[Y(1,m)-Y(0,m)]\times \mathrm{pr}(M=m|Z=0)$. To identify the average natural direct effect in this way, however, the authors must invoke an assumption unjustified by the aforementioned design, that is, the assumption of no treatment-by-mediator interaction for every individual unit. This assumption can be partially tested by examining, with data from the second experiment, whether the average controlled treatment-by-mediator interaction effect is zero. The authors further suggested collecting pretreatment characteristics known to be related to the magnitude of the interaction effects and implementing the parallel design within blocks defined by these pretreatment variables.

However, as the authors acknowledged, perfectly manipulating the mediator is often difficult. Moreover, the consistency assumption may not hold. That is, the mediator–outcome relationship when the mediator is manipulated may not be equivalent to the mediator–outcome relationship when the mediator is a natural response to the treatment. To address these limitations, a *parallel encouragement design* permits the use of indirect and subtle manipulation. Individuals assigned to the second experiment in a parallel design would be randomly encouraged to take certain values of the mediator after having been randomized to a treatment. The parallel encouragement design requires that the effect of the encouragement assignment G on the outcome should be transmitted entirely through the mediator and that the encouragement assignment G should not change the mediator in an unintended direction. Under these assumptions, one might make informative inferences about "the average complier indirect effect," that is, the natural indirect effect $E[Y(1,M(1))-Y(1,M(0))]$ for the "compliers" whose mediator value would be affected by either positive or negative encouragement in the intended direction, that is, $M(z,g=1)>M(z,g=0)$ with positive encouragement and $M(z,g=1)<M(z,g=0)$ with negative encouragement.

9.6.8 Crossover experimental designs and crossover encouragement designs

In the same paper, Imai, Tingley, and Yamamoto (2013) proposed an alternative set of experimental designs that would sequentially expose every participant to both treatment and control conditions. In the first experimental period, the assignment to the treatment and control conditions is conducted randomly. The experimenter observes every participant's mediator and outcome values. Subsequently, the second treatment assignment is opposite to a participant's first treatment condition, while the mediator value is manipulated to be the same as that observed in the first period. In other words, across the two experimental periods, the mediator takes the same value for an individual despite the change in the treatment. The goal is to directly estimate the natural direct effects $E[Y(1,M(0))-Y(0,M(0))]$ and $E[Y(1,M(1))-Y(0,M(1))]$. The difference between these two effects equals the natural treatment-by-mediator interaction effect $E[\{Y(1,M(1))-Y(1,M(0))\}-\{Y(0,M(1))-Y(0,M(0))\}]$. Subtracting the natural direct effect estimate from the total treatment effect estimate obtained in the first period, one would also obtain an estimate of the natural indirect effect $E[Y(1,M(1))-Y(1,M(0))]$. This design heavily relies on the key assumption that there are no carry-over effects, that is, the treatment that one has received in the first experimental period does not affect one's outcome in the second period. This assumption cannot be justified by the design and is often violated in human subject research. For example, if a student has received encouragement to study in the first period, the benefit of the first-period study will likely show in the test scores at the end of the second period.

A crossover encouragement design instead employs randomized encouragement in the second experimental period. After assigning every participant to the treatment condition opposite to the first-period treatment, the experimenter selects a random subset of participants and encourages them to take mediator values equal to their observed mediator values in the first period. Comparing the encouragement group $(g = 1)$ with the group receiving no encouragement $(g = 0)$, one may estimate the proportion of participants who are *pliable*, that is, whose mediator values in the second period would be the same as their observed mediator values in the first period only if they are encouraged. The subpopulation of pliable participants can be defined by $M(1, g = 1) = M(0, g = 1)$ and $M(1, g = 0) \neq M(0, g = 0)$. Key assumptions include no carry-over effects, consistency, and that the encouragement would affect the outcome only through the mediator and would not change the mediator in an unintended direction. When these assumptions hold, one may identify the average natural indirect effect of the treatment transmitted through the mediator for the pliable participants as the difference between the average treatment effect on the outcome and the controlled direct effect of the treatment on the outcome averaged over all mediator values for this subpopulation.

9.6.9   Summary

As shown in Table 9.3, these various experimental designs do not always estimate the same causal effects. For example, the causal-chain designs and the moderation-of-process designs do not estimate the natural direct effect and the natural indirect effect without further assumptions. The three encouragement designs each estimate one or more causal effects for a uniquely defined subpopulation only.

Each of the eight experimental designs invokes a series of identification assumptions. The ignorable treatment assignment assumption is assured, in theory, by all these designs; the assumption of ignorable mediator value assignment is invoked by most of the designs except for the encouragement designs. The latter has replaced this second ignorability with a set of assumptions including ignorable encouragement assignment, exclusion restriction, and monotonicity. The assumption of no carry-over effects required by the crossover designs is particularly strong and often implausible.

Bullock and colleagues (Bullock and Green, 2013; Bullock, Green, and Ha, 2010) pointed out three potential problems that can distort causal mechanisms when the experimenter attempts to manipulate the mediator values. They argued that it is often difficult in social science research to manipulate a particular mediator without inadvertently affecting other potential mediators. For example, the algebra-for-all requirement may change teacher recruitment and teaching resource allocation within a school in addition to the change in how students are assigned to classes. This criticism is applicable to most experimental designs reviewed previously and highlights the importance of theorizing and empirically investigating the relationship of a focal mediator with other concurrent or temporally adjacent mediators. Experimental designs will need to be modified accordingly either to block the confounding of other mediators or to explicitly test theories with regard to how multiple mediators contribute jointly to the causal mechanisms. Chapter 13 will discuss the extension of causal mediation analysis to multiple mediators.

A second concern is possible noncompliance with the mediator value assignment. The two-phase experimental designs informed by pretreatment covariates are deliberately configured to circumvent this problem. The goal of encouragement designs is to manipulate individual mediator values on a limited yet relatively realistic scale. Viewing noncompliance as a reflection of heterogeneity in the population, principal stratification (to be discussed further in the next chapter) may provide a potentially useful framework for analyzing subpopulation-specific causal mechanisms. Yet, as suggested by Berzuini (2013), the complier (or noncomplier) status might be "a random variable, causing individuals to move in and out of the group of compliers in an unpredictable way" (p. 33), rather than a fixed and time-invariant attribute of an individual. If a "complier" could possibly be a "noncomplier" in an unpredictable way, a causal effect for the "compliers" would then become ill-defined.

Finally, the consistency assumption is required by all the experimental designs reviewed previously. The causal mechanism would be misunderstood if individual responses when the mediator values are experimentally manipulated differ from the natural responses of the same individuals when the mediator values are self-selected. These last two concerns, however, are not unique to experiments for studying mediation.

References

Alwin, D.F. and Hauser, R.M. (1975). The decomposition of effects in path analysis. *American Sociological Review*, 40, 37–47.

Baron, R.M. and Kenny, D.A. (1986). The moderator-mediator variable distinction in social psychological research: conceptual, strategic, and statistical considerations. *Journal of Personality and Social Psychology*, 51, 1173–1182.

Bauer, P. and Kieser, M. (1999). Combining different phases in the development of medical treatments within a single trial. *Statistics in Medicine*, 18, 1833–1848.

Bauer, D.J., Preacher, K.J. and Gil, K.M. (2006). Conceptualizing and testing random indirect effects and moderated mediation in multilevel models: new procedures and recommendations. *Psychological Methods*, 11, 142–163.

Berzuini, C. (2013). Discussion on "Experimental designs for identifying causal mechanisms" by K. Imai, D. Tingley, and T. Yamamoto. *Journal of the Royal Statistical Society*, 176, 33–34.

Bullock, J.G. and Green, D.P. (2013). Discussion on "Experimental designs for identifying causal mechanisms" by K. Imai, D. Tingley, and T. Yamamoto. *Journal of the Royal Statistical Society, Serial A*, 176, 38–39.

Bullock, J.G., Green, D.P. and Ha, S.E. (2010). Yes, but what's the mechanism? (don't expect an easy answer). *Journal of Personality and Social Psychology*, 98, 550–558.

Collins, L., Graham, J. and Flaherty, B. (1998). An alternative framework for defining mediation. *Multivariate Behavioral Research*, 33, 295–312.

Duncan, O.D. (1966). Path analysis: sociological examples. *The American Journal of Sociology*, 72(1), 1–16.

Efron, B. and Feldman, D. (1991). Compliance as an explanatory variable in clinical trials. *Journal of the American Statistical Association*, 86(413), 9–17.

Fairchild, A.J. and MacKinnon, D.P. (2009). A general model for testing mediation and moderation effects. *Prevention Science*, 10, 87–99.

Finney, J. (1972). Indirect effects in path analysis. *Sociological Methods and Research*, 1(2), 175–187.

Frangakis, C.E. and Rubin, D.B. (2002). Principal stratification in causal inference. *Biometrics*, *58*, 21–29.

Frazier, P.A., Tix, A.P. and Baron, K.E. (2004). Testing moderator and mediator effects in counseling psychology. *Journal of Counseling Psychology*, *51*, 115–134.

Hayes, A.F. (2013). *Introduction to Mediation, Moderation, and Conditional Process Analysis: A Regression-Based Approach*, The Guilford Press, New York.

Heckman, J.J. (2010). Building bridges between structural and program evaluation approaches to evaluating policy. *Journal of Economic Literature*, *48*(2), 356–398.

Holland, P.W. (1986). Statistics and causal inference. *Journal of the American Statistical Association*, *81*(396), 945–960.

Holland, P.W. (1988). Causal interference, path analysis, and recursive structural equations models. *Sociological Methodology*, *18*, 449–484.

Hong, G. (2006). *Multi-level experimental designs and quasi-experimental approximations for studying intervention implementation as a mediator*. Paper presented at the Annual Meeting of the American Educational Research Association, San Francisco, CA.

Hong, G. (2012). Editorial comments. *Journal of Research on Educational Effectiveness,* special issue on the statistical approaches to studying mediator effects in education research, *5*(3), 213–214.

Hong, G. (2013). Covariate-informed parallel design: discussion on "Experimental designs for identifying causal mechanisms" by K. Imai, D. Tingley, and T. Yamamoto. *Journal of the Royal Statistical Society, Serial A, 176*, 35.

Hong, G. and Nomi, T. (2012). Weighting methods for assessing policy effects mediated by peer change. *Journal of Research on Educational Effectiveness,* special issue on the statistical approaches to studying mediator effects in education research, *5*(3), 261–289.

Hoyle, R.H. and Robinson, J.I. (2003). Mediated and moderated effects in social psychological research: measurement, design, analysis issues, in *Handbook of Methods in Social Psychology* (eds C. Sansone, C. Morf and A.T. Panter), Sage, Thousand Oaks, CA.

Imai, K., Keele, L. and Tingly, D. (2010). A general approach to causal mediation analysis. *Psychological Methods*, *15*(4), 309–334.

Imai, K., Keele, L., Tingley, D. and Yamamoto, T. (2011). Unpacking the black box of causality: learning about causal mechanisms from experimental and observational studies. *American Political Science Review*, *105*(4), 765–789.

Imai, K., Tingley, D. and Yamamoto, T. (2013). Experimental designs for identifying causal mechanisms. *Journal of the Royal Statistical Society*, *176*(1), 5–32.

James, L.R. and Brett, J.M. (1984). Mediators, moderators, and test for mediation. *Journal of Applied Psychology*, *69*, 307–321.

Jose, P.E. (2013). *Doing Statistical Mediation and Moderation*, The Guilford Press, New York.

Judd, C.M. and Kenny, D.A. (1981). Process analysis: estimating mediation in treatment evaluation. *Evaluation Review*, *5*, 602–619.

Kraemer, H.C., Stice, E., Kazdin, A. *et al.* (2001). How do risk factors work together? Mediators, moderators, and independent, overlapping, and proxy risk factors. *American Journal of Psychiatry*, *158*, 848–856.

Kraemer, H.C., Wilson, G.T., Fairburn, C.G. and Agras, W.S. (2002). Mediators and moderators of treatment effects in randomized clinical trials. *Archives of General Psychiatry*, *59*, 877–883.

Liu, Q., Proschan, M.A. and Pledger, G.W. (2002). A unified theory of two-stage adaptive designs. *Journal of the American Statistical Association*, *97*, 1034–1041.

MacKinnon, D.P., Krull, J.L. and Lockwood, C.M. (2000). Equivalence of the mediation, confounding and suppression effect. *Prevention Science*, *1*(4), 173–181.

Mattei, A. and Mealli, F. (2011). Augmented designs to assess principal strata direct effects. *Journal of the Royal Statistical Society, 73*(5), 729–752.

Muller, D., Judd, C.M. and Yzerbyt, V.Y. (2005). When moderation is mediated and mediation is moderated. *Journal of Personality and Social Psychology, 89*, 852–863.

Neyman, J. (1935). On the problem of confidence intervals. *The Annals of Mathematical Statistics, 6*(3), 111–169.

Pearl, J. (2001). Direct and Indirect Effects. Proceedings of the Seventeenth Conference on Uncertainty in Artificial Intelligence. Morgan Kaufmann, San Francisco, CA, pp. 411–420.

Preacher, K.J. and Hayes, A.F. (2004). SPSS and SAS procedures for estimating indirect effects in simple mediation models. *Behavior Research Methods, Instruments, & Computers, 36*(4), 717–731.

Preacher, K.J. and Hayes, A.F. (2008). Asymptotic and resampling strategies for assessing and comparing indirect effects in multiple mediator models. *Behavior Research Methods, 40*(3), 879–891.

Preacher, K.J., Rucker, D.D. and Hayes, A.F. (2007). Addressing moderated mediation hypotheses: theory, methods, and prescriptions. *Multivariate Behavioral Research, 42*, 185–227.

Robins, J.M. and Greenland, S. (1992). Identifiability and exchangeability for direct and indirect effects. *Epidemiology, 3*(2), 143–155.

Rose, B.M., Holmbeck, G.N., Coakley, R.M. and Franks, E.A. (2004). Mediator and moderator effects in developmental and behavioral pediatric research. *Developmental and Behavioral Pediatrics, 25*, 58–67.

Rubin, D.B. (1978). Bayesian inference for causal effects: the role of randomization. *The Annals of Statistics, 6*(1), 34–58.

Rubin, D.B. (1986). Statistics and causal inference: comment: which ifs have causal answers. *Journal of the American Statistical Association, 81*(396), 961–962.

Rubin, D.B. (1991). Practical implications of modes of statistical inference for causal effects and the critical role of the assignment mechanism. *Biometrics, 47*, 1213–1234.

Shadish, W.R. and Sweeney, R.B. (1991). Mediators and moderators in meta-analysis: there's a reason we don't let dodo birds tell us which psychotherapies should have prizes. *Journal of Consulting and Clinical Psychology, 59*, 883–893.

Sheets, V. and Braver, S.L. (1999). Organizational status and perceived sexual harassment: detecting the mediators of a null effect. *Personality and Social Psychology Bulletin, 25*, 1159–1171.

Shrout, P.E. and Bolger, N. (2002). Mediation in experimental and nonexperimental studies: new procedures and recommendations. *Psychological Methods, 4*, 422–445.

Sobel, M.E. and Stuart, E.A. (2012). Comments. *Journal of Research on Educational Effectiveness, special issue on the statistical approaches to studying mediator effects in education research, 5*(3), 290–293.

Spencer, S.J., Zanna, M.P. and Fong, G.T. (2005). Establishing a causal chain: why experiments are often more effective than mediational analysis in examining psychological processes. *Journal of Personality and Social Psychology, 89*, 845–851.

Ukai, N. (1994). The Kumon approach to teaching and learning. *Journal of Japanese Studies, 20*, 87–113.

Wu, A.D. and Zumbo, B.D. (2008). Understanding and using mediators and moderators. *Social Indicator Research, 87*, 367–392.

10

Existing analytic methods for investigating causal mediation mechanisms

This chapter reviews the existing methods for investigating causal mechanisms with experimental or quasiexperimental data. These include path analysis and structural equation modeling (SEM), modified regression approaches, marginal structural models, conditional structural models, the resampling approach, alternative weighting methods, the instrumental variable (IV) method, and the principal stratification approach. Some of these methods are well known, whereas others have been proposed only recently. To ease the comparison across these various analytic methods, this review is restricted to the case of a single mediator. Discussions of methods for investigating multiple mediators are left to Chapter 13. The causal effects of key interest are the natural direct effect and the natural indirect effect, which decompose the total effect as defined in Chapter 9. These definitions assume no interference between individuals, that is, one individual's potential outcomes cannot be influenced by other individuals' treatment assignment, which is known as the "stable unit treatment value assumption" (SUTVA; see Rubin, 1986). Chapter 15 discusses spill-over between units as a mediation mechanism when SUTVA is relaxed.

In applications, experimental data typically involve the randomization of treatment only; quasiexperimental data do not involve experimental manipulation of either the treatment or the mediator. Hence, all the analytic methods must address the issue of confounding. The methods reviewed here differ not only in statistical modeling and estimation but, more importantly, in the identification assumptions that they invoke. Some methods require relatively intensive computation, and some others rely on a larger number of and often relatively strong identification assumptions and therefore are at a higher risk of possibly violating the assumptions. Section 10.8 reviews several ways of assessing the sensitivity of estimation results to potential violations of a key assumption shared by most of the analytic methods. A comparison of identification and model-based assumptions across all these methods is summarized in Tables 10.1 and 10.2.

Table 10.1   Identification assumptions for estimating the natural direct and indirect effects

(I)	The treatment assignment is independent of the potential mediators
(II)	The treatment assignment is independent of the potential outcomes
(III.a)	The mediator value assignment under each treatment is independent of the potential outcomes under the same treatment
(III.b)	The mediator value assignment under each treatment is independent of the potential outcomes under the same treatment within levels of pretreatment covariates and posttreatment covariates
(III.c)	The mediator value assignment under one treatment is independent of the potential outcome under the alternative treatment
(IV.a)	There is no treatment-by-mediator interaction
(IV.b)	The direct effect assumption: the average controlled direct effect of the treatment at any fixed mediator value does not differ across subgroups who may take different mediator values under a given treatment
(V.a)	Each causal effect is constant in the population
(V.b)	The treatment effect on the mediator is independent of the mediator effect on the outcome
(V.c)	The monotonicity assumption: the treatment does not have opposite impacts on different individuals at the intermediate stage
(VI)	The exclusion restriction: the direct effect of the treatment on the outcome must be zero for all units in the population
(VII)	The treatment has a nonzero effect on the mediator
(VIII)	Assumptions with regard to the functional form of the outcome model
(IX)	Assumptions with regard to the functional form of the mediator model

In general, an assumption may apply within levels of pretreatment covariates.

10.1   Path analysis and SEM

10.1.1   Analytic procedure for continuous outcomes

Over the past decades, path analysis (Alwin and Hauser, 1975; Duncan, 1966; Finney, 1972; Wright, 1934) and SEM (Bollen, 1987; Jo, 2008; Jöreskog, 1970; MacKinnon, 2008) have been the most commonly used techniques in social science research for analyzing mediation. A path model represents the supposed structural relationships between an outcome and one or more predictors and uses a single measure for each variable; in contrast, SEM can incorporate measurement models when the structural models involve latent constructs each measured by multiple items. Path analysis typically employs ordinary least squares for estimation, while SEM analyzes multiple models simultaneously through maximum likelihood. To focus on key issues in mediation analysis, in the following text, we assume no measurement error in the observed variables.

Let Y be the outcome, Z be the treatment, and M be the mediator. Following the example in Chapter 9, we may consider encouragement to study to be the treatment of interest, the actual hours a student studies for the test to be the mediator, and the student's test score to be the outcome. In a linear additive framework, the analytic procedure (Baron and Kenny, 1986; Judd and Kenny, 1981; MacKinnon and Dwyer, 1993) involves analyzing regression models (10.1) and (10.2)

Table 10.2 Comparisons of identification assumptions across the alternative analytic methods

Analytic methods	Sequential ignorability					Interaction		Homo-/heterogeneity			Modeling			
	(I)	(II)	(III.a)	(III.b)	(III.c)	(IV.a)	(IV.b)	(V.a)	(V.b)	(V.c)	(VI)	(VII)	(VIII)	(IX)
Path analysis and SEM	√	√	√			√		√/	√/				√	√
Modified regression approaches	√	√	√		√		√						√	√
Marginal structural models	√	√		√		√		√/	√/				√	
Conditional structural models	√	√		√	√			√/						√
Alternative weighting methods	√	√	√		√								√	
Resampling approach	√	√			√	√							√	√
Instrumental variable method	√	√				√		√/		√/		√	√	√
Principal stratification	√	√								√/			√	√

√, an identification assumption required by the analytic method. √/, an identification assumption replaceable by an alternative assumption with regard to homogeneous or heterogeneous causal effects.

$$M = \gamma_0 + aZ + \varepsilon_M; \tag{10.1}$$

$$Y = \beta_0 + bM + cZ + \varepsilon_Y. \tag{10.2}$$

Here, a is interpreted as the causal effect of the treatment on the mediator (e.g., the effect of encouragement on how many hours one studies for the test); b is interpreted as the mediator effect on the outcome (e.g., the effect of study on test score) under each treatment condition; and c is interpreted as the direct effect of the treatment on the outcome when the mediator is fixed at a certain value (e.g., the effect of encouragement on test score when the number of hours one studies is held constant).

Replacing M in (10.2) with the right-hand side of the equation in (10.1), we obtain a reduced-form model as follows:

$$
\begin{aligned}
Y &= \beta_0 + b(\gamma_0 + aZ + \varepsilon_M) + cZ + \varepsilon_Y \\
&= (\beta_0 + b\gamma_0) + (ab + c)Z + (b\varepsilon_M + \varepsilon_Y) \\
&= \beta_0' + c'Z + \varepsilon_Y'.
\end{aligned}
\tag{10.3}
$$

Here, $c' = ab + c$ is interpreted as the total effect of the treatment on the outcome (e.g., the effect of encouragement on test score). Hence, through analyzing regression models (10.2) and (10.3), one may obtain a sample estimate of $c' - c$ (Alwin and Hauser, 1975; Judd and Kenny, 1981), which has been interpreted as the indirect effect. The validity of these interpretations depends on a series of assumptions to be explicated in the following text. When these assumptions hold, c identifies the natural direct effect $E[Y(1, M(0))] - E[Y(0, M(0))]$. The natural indirect effect $E[Y(1, M(1))] - E[Y(1, M(0))]$ is identified either as the product of a and b or as the difference between the total effect and the natural direct effect $c' - c$. (Note that in past literature, c is often used to denote the total effect and c' for the direct effect.)

As MacKinnon and colleagues (2002) pointed out, the aforementioned multistep approaches do not provide a joint test of a, b, and c', nor do they provide a direct estimate of the natural indirect effect and its standard error for constructing a confidence interval. Obtaining correct standard errors for the sample estimates is crucial for hypothesis testing. To obtain the standard error for a sample estimate of ab or of $c' - c$ requires additional computation. A Sobel test (1982) has been widely used for testing $ab = 0$. The large sample test statistic is

$$z \approx \frac{\hat{a}\hat{b}}{\sqrt{\hat{a}^2 \mathrm{Var}(\hat{b}) + \hat{b}^2 \mathrm{Var}(\hat{a})}}.$$

Here, $\hat{a}$ and $\hat{b}$ denote the sample estimates of a and b, respectively; $\mathrm{Var}(\hat{a})$ and $\mathrm{Var}(\hat{b})$ represent the sampling variability of $\hat{a}$ and $\hat{b}$, respectively. The denominator, derived through the multivariate delta method based on a first-order Taylor series approximation, estimates the standard error of $\hat{a}\hat{b}$ under the assumption that $\hat{a}$ and $\hat{b}$ are multivariate normal. However, MacKinnon, Lockwood, and Hoffman (1998) showed that the sampling distribution of $\hat{a}\hat{b}$ tends to be asymmetric with high kurtosis. Hence, the standard error estimate, though accurate according to simulation results, cannot be used for constructing a confidence interval. McGuigan and Langholtz (1988) derived the standard error for sample estimates of $c' - c$.

When there is no covariance adjustment for any pretreatment covariates in the structural models, the statistic for testing $c'-c=0$ follows a t distribution with $N-2$ degrees of freedom where N is the total sample size:

$$t = \frac{\hat{c}'-\hat{c}}{\sqrt{\mathrm{Var}(\hat{c}')+\mathrm{Var}(\hat{c})-2\mathrm{Cov}(\hat{c}',\hat{c})}}.$$

Here, $\hat{c}'$ and $\hat{c}$ are the sample estimates of c' and c, respectively; $\mathrm{Cov}(\hat{c}', \hat{c})$ represents the covariance between $\hat{c}'$ and $\hat{c}$. Freedman and Schatzkin (1992) proposed an alternative testing procedure. MacKinnon and colleagues conducted a series of simulations to compare different procedures for statistical inference (MacKinnon and Dwyer, 1993; MacKinnon et al., 2002).

10.1.2   Identification assumptions

Path analysis and SEM require a series of strong assumptions. A 1986 article by Baron and Kenny, serving as guidelines for mediation analysis in social sciences, stated only two assumptions: one is that there is no measurement error in the mediator; and the other is that the outcome does not cause the mediator. Although neither of these assumptions are trivial as measurement error in the mediator may attenuate the estimated mediator–outcome relationship, while a mediator caused by the outcome would suggest a completely different structural model, these assumptions are far from adequate for identifying the causal effects in mediation analysis.

Holland (1988) attempted to explicate the identification assumptions that are required for using path analysis or SEM to estimate the average treatment effect on the mediator, the average treatment effect on the outcome, the average direct effect of the treatment on the outcome when the mediator is fixed at a particular value, and the average effect of the mediator on the outcome under a given treatment condition. These assumptions, revisited by Sobel (2008) and Jo (2008), become clear when individual-specific treatment effects are defined in terms of potential outcomes. Specifically, we let $A_i = M_i(1)-M_i(0)$ denote the treatment effect on the mediator for individual i; let $\Delta_i = Y_i(1)-Y_i(0)$ denote the treatment effect on this individual's outcome; let $C_i = Y_i(1,m)-Y_i(0,m)$ denote the treatment effect on the individual's outcome when the mediator is set at value m; and let $B_i = Y_i(z,m)-Y_i(z,m')$ denote the effect on the individual's outcome when the mediator value is changed by one unit from m' to m under treatment condition z for $z=0, 1$. According to Holland, under the following identification assumptions, the path coefficients can be equated with the population average causal effects, that is, $a=E(A_i)$, $b=E(B_i)$, $c=E(C_i)$, and $c'=E(\Delta_i)$. The numbering system for the assumptions below will be used consistently throughout the chapter:

(I) The treatment assignment is independent of the potential mediators.

(II) The treatment assignment is independent of the potential outcomes.

(III.a) The mediator value assignment under each treatment is independent of the potential outcomes under the same treatment.

(IV.a) There is no treatment-by-mediator interaction (i.e., the mediator–outcome relationship is the same across the treatment conditions).

(V.a) Each causal effect is constant in the population.

(VIII) The treatment and the mediator are linearly associated with the outcome.

(IX) The treatment is linearly associated with the mediator.

In the current example, assumption (I) states that whether or not one is encouraged to study for the test is a treatment decision made regardless of how many hours one will study. The assumption is violated, for example, if the encouragement is given to students who will tend to increase their hours of study with the encouragement and is withheld from students who will tend not to increase their hours of study even with the encouragement.

Assumption (II) states that a student is assigned to receive encouragement regardless of how he or she will perform on the test. This assumption is violated if the students who actually receive the encouragement are those who will tend to perform poorly without the encouragement; the assumption is also violated if the students who actually receive the encouragement are those who will tend to perform well when encouraged.

Assumptions (I) and (II), together named the assumption of *ignorable treatment assignment*, rule out possible pretreatment differences between the treated group and the control group associated with how many hours one would study and how well one would perform on the test if both groups were encouraged or if neither group was encouraged. If the treated students were initially better prepared for the test than the control students, for example, such pretreatment differences between the treated group and the control group will bias the estimation of the treatment effect on the mediator and of the treatment effect on the outcome.

Assumption (III.a), also named the assumption of *within-treatment ignorable mediator value assignment*, states that, under each treatment condition, how much one would study does not depend on how well one was to perform on the test. This assumption rules out possible pretreatment and posttreatment differences associated with future test performance between those who study relatively more hours and those who study relatively fewer hours under the same treatment. The assumption is violated, for example, if students who study 8 hours would perform better (or worse) than those who study 4 hours when both groups were to study the same number of hours under encouragement.

The assumption of *no treatment-by-mediator interaction* (IV.a), also called the *additivity* assumption, states that additional hours of study will have the same impact on test performance regardless of whether one is encouraged or not. Under this assumption, the controlled direct effect of the mediator on the outcome under the experimental condition $Y_i(1,m) - Y_i(1,m')$, defined in Chapter 9, is assumed equal to that under the control condition $Y_i(0,m) - Y_i(0,m')$ for individual i. Under this same assumption, the natural direct effect is equal to the controlled direct effect of the treatment on the outcome $Y_i(1,m) - Y_i(0,m)$ because the latter is assumed to be independent of any particular value m that the mediator is set at. This assumption will be violated if a student studies more attentively and effectively when encouraged than when he or she is not encouraged such that an additional hour of study will produce a greater improvement in test score under encouragement than under the control condition.

In the presence of a treatment-by-mediator interaction, the path coefficients c and ab are biased for the natural direct effect and the natural indirect effect, respectively. As shown in Appendix 10.A, in the relatively simple case of a binary treatment and a binary mediator, for the natural direct effect, the bias is a product of three elements: the controlled treatment-by-mediator interaction effect on the outcome, the treatment effect on the mediator, and the proportion of units assigned to the control group. For the natural indirect effect, the bias will have the same magnitude but take the opposite sign.

The *additivity* assumption (IV.a) and the *linearity* assumptions (VIII) and (IX) together constitute the model-based assumptions that are required, along with the other identification assumptions, for path coefficients c and ab to be causally meaningful.

According to the *constant treatment effect* assumption (V.a), there is no heterogeneity in any of the causal effects among individuals in a population. This assumption seems highly implausible in social science research. We have used random variable A to denote the treatment effect on the mediator and random variable B for the mediator effect on the outcome under a given treatment. Without the current assumption, A and B may each take different values across individuals in the population. The corresponding population average causal effects are $E(A) = a$ and $E(B) = b$. If all the other assumptions hold, the individual-specific natural indirect effect will be equal to AB. However, as pointed out by Bullock and colleagues (Bullock, Green, and Ha, 2010), the population average natural indirect effect $E(AB)$ is equal to $ab + \text{Cov}(A,B)$ where $\text{Cov}(A, B)$ denotes the covariance between A and B. This is because

$$E(AB) = E(A) \times E(B) + \text{Cov}(A,B) = ab + \text{Cov}(A,B),$$

which is unequal to ab except when $\text{Cov}(A,B) = 0$.

Therefore, in studies of heterogeneous populations with the path analysis procedure, the constant treatment effect assumption (V.a) is to be replaced by the *no covariance* assumption (V.b) stated as follows:

(V.b) The treatment effect on the mediator is independent of the mediator effect on the outcome, that is, $\text{Cov}(A,B) = 0$.

The *no covariance* assumption can be alternatively written as $E(B|A) = E(B)$, that is, in terms of potential outcomes, $E[Y(z,m) - Y(z,m')|M(1) - M(0)] = E[Y(z,m) - Y(z,m')]$. This assumption will be violated if the students who are already well prepared for the test will unlikely increase their study hours in response to the encouragement because, for these students, there is a diminishing return from any additional hour of study. In contrast, the students who are underprepared for the test will likely increase their study hours due to the encouragement. These are the same students whose additional hours of study will likely produce a large gain in test score.

In contrast to the approach of using the product ab to identify the natural indirect effect, $c' - c$ is unbiased under the ignorability assumptions (I), (II), and (III.a) and the model-based assumptions (IV.a), (VIII), and (IX). This is because $E(C' - C) = E(C') - E(C) = c' - c$ where C' denotes a random variable for the individual-specific total treatment effect on the outcome and C a random variable for the individual-specific direct effect of the treatment on the outcome.

Treatment randomization will guarantee the *ignorable treatment assignment* assumptions (I) and (II); and mediator randomization within each treatment will guarantee the *within-treatment ignorable mediator value assignment* assumption (III.a). Yet experimental designs, including the ones discussed in the previous chapter, do not warrant the *constant treatment effect* assumption (V.a). Nor do they warrant the model-based assumptions of *linearity* and *additivity*. In most social science research, only the treatment is randomized, or neither the treatment nor the mediator is randomized. In the best possible scenario, a study may have collected comprehensive pretreatment information denoted by a vector of covariates $\mathbf{X}$ such that the ignorability assumptions (I), (II), and (III.a) hold within levels of $\mathbf{X} = \mathbf{x}$. Here, $\mathbf{X}$ denotes a vector containing multiple elements $X_1, X_2, \ldots$. In other words, within each subpopulation of individuals sharing the same pretreatment characteristics $\mathbf{X} = \mathbf{x}$, it may become plausible that the treatment–mediator relationship, the treatment–outcome relationship, and the mediator–outcome relationship are not confounded by unobserved covariates. Researchers would modify the path analysis models typically by making covariance adjustment for the observed pretreatment covariates:

$$M = \gamma_0 + aZ + \gamma_1 \mathbf{X} + \varepsilon_M; \tag{10.4}$$

$$Y = \beta_0 + bM + cZ + \beta_1 \mathbf{X} + \varepsilon_Y; \tag{10.5}$$

$$Y = \beta_0' + c'Z + \beta_1' \mathbf{X} + \varepsilon_Y'. \tag{10.6}$$

Models (10.4)–(10.6) assume that each covariate is linearly associated with the mediator or the outcome and that there are no interactions among the covariates. Additionally, model (10.4) assumes that the treatment effect on the mediator does not depend on any of the covariate values; model (10.5) assumes that the direct effect of the treatment on the outcome and the mediator effect on the outcome within each treatment condition do not depend on the covariate values; and model (10.6) similarly assumes that the total treatment effect on the outcome does not depend on the covariate values. If X is a confounder in any of these models and additionally if there is a nonzero X-by-Z interaction or X-by-M interaction in the same model, omitting the interaction would bias the corresponding causal effect estimates. Hence, in general, path analysis and SEM identify the natural direct and indirect effects under the crucial assumptions that the mediator model and the outcome model are correctly specified in their functional forms.

Textbooks often combine the ignorability assumptions with the model-based assumptions and state them as the assumption of zero covariance between the error term and the predictors in each model and that of zero covariance between the error terms from simultaneous models. Specifically, applied researchers have been typically taught to assume that

$$\mathrm{Cov}(Z, \varepsilon_M) = \mathrm{Cov}(Z, \varepsilon_Y) = \mathrm{Cov}(M, \varepsilon_Y) = \mathrm{Cov}(\varepsilon_M, \varepsilon_Y) = 0$$

or that

$$\mathrm{Cov}(Z, \varepsilon_Y') = \mathrm{Cov}(Z, \varepsilon_Y) = \mathrm{Cov}(M, \varepsilon_Y) = \mathrm{Cov}(\varepsilon_Y', \varepsilon_Y) = 0.$$

The impacts of omitting a confounder in a linear or nonlinear form, of omitting its possibly nonzero interaction with the treatment or the mediator, and of omitting a possibly nonzero treatment-by-mediator interaction would be lumped together in the error terms. Such assumptions are rather obscure and have not enabled applied researchers to examine their plausibility on scientific grounds.

10.1.3   Analytic procedure for discrete outcomes

The path analysis and SEM procedures have been modified for applications in which the outcomes are measured on a categorical scale. For example, for a binary outcome, models (10.2) and (10.3) may be replaced by logistic regression or probit regression. Model (10.1) may be replaced by logistic regression or probit regression accordingly for a binary mediator.

However, as MacKinnon and colleagues (MacKinnon, 2008; MacKinnon and Dwyer, 1993; MacKinnon *et al.*, 2007) pointed out, when a logistic regression model or a probit model is applied to a binary outcome, the product of coefficients approach and the difference of coefficients approach typically give different estimates of the natural indirect effect defined in Chapter 9, and neither stays unbiased. Imai, Keele, and Tingly (2010) have shown through simulations that the bias can be substantial when the product of coefficients approach is employed. The only exceptions found by VanderWeele and Vansteelandt (2010) are when the outcome is rare. Corrections have been proposed through either standardizing the path coefficients (MacKinnon, 2008; MacKinnon and Dwyer, 1993; Winship and Mare, 1983) or specifying threshold models for the latent continuous constructs (Ditlevsen *et al.*, 2005). These developments are of particular interest to applied researchers. Yet without clarifying and assessing identification assumptions, these approaches generate results that do not necessarily have an unambiguous causal interpretation (Kaufman *et al.*, 2005b).

10.2   Modified regression approaches

To identify the causal effects in mediation analysis while allowing for treatment-by-mediator interaction, some recent modifications of the regression approach resort to specifying an outcome model as a function of the treatment, the mediator, the observed pretreatment covariates, and their interactions (Pearl, 2010; Petersen, Sinisi, and van der Laan, 2006; Preacher, Rucker, and Hayes, 2007; Valeri and VanderWeele, 2013; VanderWeele and Vansteelandt, 2009). For example, within the path analysis framework, Preacher and colleagues (2007) changed model (10.2) to the following:

$$Y = \beta_0 + b_1 M + cZ + b_2 MZ + \varepsilon_Y.$$

They then used $a(b_1 + b_2 z)$ to identify the conditional indirect effect, the value of which depends on the treatment condition z. This quantity or its expectation over the distribution of Z, however, does not correspond to the average natural indirect effect defined in Chapter 9 without further assumptions. In the meantime, other regression-based analytic methods have been developed under the potential outcomes framework and are reviewed in the following text.

10.2.1   Analytic procedure for continuous outcomes

Petersen, Sinisi, and van der Laan (2006) proposed to estimate the natural direct effect through an analysis that involves three major steps. *Step 1* is to analyze a multiple regression model regressing the outcome Y on the treatment Z, the mediator M, pretreatment covariates $\mathbf{X}$, and their interactions. They gave an example of a linear regression model predicting the outcome:

$$\hat{Y} = \beta_0 + \beta_1 m + \beta_2 z + \beta_3 zm + \boldsymbol{\beta}_4 \mathbf{x} + \boldsymbol{\beta}_5 z\mathbf{x} + \boldsymbol{\beta}_6 m\mathbf{x} + \boldsymbol{\beta}_7 zm\mathbf{x}. \tag{10.7}$$

Because in a nonexperimental study, treatment assignment and mediator value assignment may be associated with a large number of selection factors, $\mathbf{X}$ is typically multidimensional. Therefore, $\boldsymbol{\beta}_4$, $\boldsymbol{\beta}_5$, $\boldsymbol{\beta}_6$, and $\boldsymbol{\beta}_7$ are each a vector that may contain as many coefficients as the number of covariates in the $\mathbf{X}$ vector. Suppose that the treatment is binary and takes values $z = 0, 1$. When the mediator is held at a given value $M = m$ and within levels of the pretreatment covariates $X = x$, the controlled direct effect of the treatment on the outcome $E[Y(1,m) - Y(0,m)|X = x]$ can be estimated as $\beta_2 + \beta_3 m + \beta_5 x + \beta_7 m x$. To estimate the average natural direct effect, one needs to take an average of the aforementioned controlled direct effect over the distribution of M under the control condition and over the distribution of X and obtain $\beta_2 + \beta_3 E[M(0)] + \beta_5 E[X] + \beta_7 E[M(0)X]$. The average natural direct effect is zero if $\beta_2 = \beta_3 = \beta_5 = \beta_7 = 0$, a null hypothesis that can be tested. Upon rejection of this null hypothesis, estimates of $E[M(0)]$, $E[X]$, and $E[M(0)X]$ are obtained in the next step.

Step 2 is to analyze a multiple regression model and predict the mediator M as a function of the treatment z and pretreatment covariates x:

$$\hat{M} = \gamma_0 + \gamma_1 z + \gamma_2 x + \gamma_3 z x. \tag{10.8}$$

Under the control condition $Z = 0$ and within levels of the pretreatment covariates $X = x$, the expected mediator value $E[M|Z = 0, X = x]$ can be estimated as $\gamma_0 + \gamma_2 x$. Having estimated $E[X]$ as the sample mean of each covariate and $E[X^2]$ as the sample mean of each covariate squared, one then obtains the sample estimates of $E[M(0)]$ and $E[M(0)X]$ as follows, both of which involve taking an average over the distribution of X:

$$\hat{E}[M(0)] = \gamma_0 + \gamma_2 \hat{E}[X].$$
$$\hat{E}[M(0)X] = \gamma_0 \hat{E}[X] + \gamma_2 \hat{E}[X^2].$$

In *Step 3*, one obtains a point estimate of the average natural direct effect $\beta_2 + \beta_3 \hat{E}[M(0)] + \beta_5 \hat{E}[X] + \beta_7 \hat{E}[M(0)X]$, which can be simplified as

$$\beta_2 + \beta_3 \gamma_0 + (\beta_3 \gamma_2 + \beta_5)\hat{E}[X] + \beta_7 \left(\gamma_0 \hat{E}[X] + \gamma_2 \hat{E}[X^2] \right).$$

The standard error may be estimated by the multivariate delta method; or one may obtain the confidence interval through bootstrapping. Assuming that the model specifications in (10.7) and (10.8) could be simplified by omitting the covariate interactions with the treatment and the mediator, that is, assuming that $\beta_5 = \beta_6 = \beta_7 = \gamma_3 = 0$, VanderWeele and Vansteelandt (2009) derived the standard error for the average natural direct effect and the average natural indirect effect. Valeri and VanderWeele (2013) showed how the analysis with the simplified model specifications could be implemented with SAS and SPSS macros. The derivation would be considerably more complex and the computation considerably more intensive should the effects of treatment-by-covariate interactions, mediator-by-covariate interactions, or treatment-by-mediator-by-covariate interactions be nonzero.

10.2.2   Identification assumptions

The modified regression approach still requires that the *ignorability* assumptions (I), (II), and (III.a) hold within levels of $X = x$. The *additivity* assumption (IV.a) is replaced by the *direct effect* assumption:

(IV.b) The average controlled direct effect of the treatment at any fixed mediator value is a constant across subgroups who may differ in mediator values under a given treatment.

That is,

$$E[Y(1,m)-Y(0,m)|Z=z,\mathbf{X}=\mathbf{x}] = E[Y(1,m)-Y(0,m)|Z=z,M(z)=m,\mathbf{X}=\mathbf{x}].$$

For example, among students who have the same pretreatment characteristics $\mathbf{X}=\mathbf{x}$, some may study 8 hours (i.e., $M(1)=8$), and others may study 4 hours (i.e., $M(1)=4$) when encouraged. This assumption states that if they all study the same number of hours when encouraged (i.e., $M(1)=m$), the impact of encouragement on test performance denoted by $Y(1,m)-Y(0,m)$ will be the same for these students on average. Unlike the *additivity* assumption (IV.a), the *direct effect* assumption (IV.b) can be satisfied by design when the mediator value assignment is randomized within levels of $\mathbf{x}$ in each treatment group. The *constant treatment effect* assumption (V.a) is no longer required. Nor is the *no covariance* assumption (V.b) when the procedure applies only to the estimation of the natural direct effect. Most importantly, by relaxing the assumption of no treatment-by-mediator interaction, this approach requires relatively weaker assumptions in comparison with the conventional regression approach such as path analysis and SEM.

However, similar to path analysis and SEM, the causal effect estimates obtained from the modified regression approach are still prone to bias when the mediator model or the outcome model is not correctly specified. Model specification is especially challenging when $\mathbf{X}$ is high dimensional. These regression-based approaches are additionally constrained by the requirement that the mediator–outcome relationship be linear.

10.2.3   Analytic procedure for binary outcomes

Causal mediation analyses with the existing methods tend to be challenging when the outcome is binary. In the earlier example of an encouragement study, a binary outcome would be whether a student fails the test. Adopting the potential outcomes framework, Huang and colleagues (2004) proposed a "logistic mediational model" that defines the total effect of the treatment assignment as the difference between the logit of $\mathrm{pr}(Y(1)=1)$ and the logit of $\mathrm{pr}(Y(0)=1)$, the natural direct effect as the difference between the logit of $\mathrm{pr}(Y(1,M(0))=1)$ and the logit of $\mathrm{pr}(Y(0,M(0))=1)$, and the natural indirect effect as the difference between the logit of $\mathrm{pr}(Y(1,M(1))=1)$ and the logit of $\mathrm{pr}(Y(1,M(0))=1)$. Each causal effect after exponential transformation corresponds to an odds ratio. The odds ratio for the total effect decomposes into a product of the odds ratio for the natural direct effect and that for the natural indirect effect. In the case in which the treatment, the mediator, and the outcome are all binary, the estimation involves analyzing a logistic regression model regressing Y on Z and M and a second logistic regression model regressing Z on M. One then computes the following marginal probabilities and their logit values:

$$\mathrm{pr}(Y(1,M(1))=1) = \mathrm{pr}(Y=1|Z=1,M=1)\times\mathrm{pr}(M=1|Z=1)$$
$$+\mathrm{pr}(Y=1|Z=1,M=0)\times\mathrm{pr}(M=0|Z=1);$$

$$\mathrm{pr}(Y(0,M(0))=1) = \mathrm{pr}(Y=1|Z=0,M=1)\times\mathrm{pr}(M=1|Z=0)$$
$$+\mathrm{pr}(Y=1|Z=0,M=0)\times\mathrm{pr}(M=0|Z=0);$$

$$\mathrm{pr}(Y(1,M(0))=1) = \mathrm{pr}(Y=1|Z=1,M=1)\times\mathrm{pr}(M=1|Z=0)$$
$$+\mathrm{pr}(Y=1|Z=1,M=0)\times\mathrm{pr}(M=0|Z=0).$$

Standard errors or confidence intervals are obtained through a multivariate delta method, bootstrapping, or Bayesian inference. However, the above estimation results hold only when data are generated from a sequential randomized design such that there is no need to make statistical adjustment for covariates $\mathbf{X}$. This approach becomes inapplicable when the treatment or the mediator is not randomized. This is because the conditional odds ratio for $Y = 1$ given z, m, and $\mathbf{x}$ averaged over the distribution of $\mathbf{X}$ is unequal to the conditional odds ratio for $Y = 1$ given z and m. In general, statistical adjustment for the covariates in a logistic regression model changes the treatment effect estimate and even the parameter. This paradoxical behavior of odds ratios called "noncollapasibility" has been well documented in the epidemiology literature (Bishop, Fienberg, and Holland, 1975; Breslow, 1981; Fienberg, 2007; Greenland, 1987; Greenland, Robins, and Pearl, 1999; Miettinen and Cook, 1981).

VanderWeele and Vansteelandt (2010) similarly defined the natural direct effect and the natural indirect effect on the odds ratio scale yet conditioning on observed pretreatment covariates. They developed a technique combining a logistic regression for the outcome and a linear regression for the mediator when the outcome is of low prevalence (i.e., a rare event) and when the mediator is continuous:

$$\text{logit}(\text{pr}(Y = 1|z, m, \mathbf{x})) = \beta_0 + \beta_1 m + \beta_2 z + \beta_3 zm + \boldsymbol{\beta}_4 \mathbf{x};$$

$$E(M|z, \mathbf{x}) = \gamma_0 + \gamma_1 z + \boldsymbol{\gamma}_2 \mathbf{x}.$$

Unlike the modified regression models proposed by Petersen, Sinisi, and van der Laan (2006), the above two models impose much stronger model-based assumptions as they do not include two-way or three-way interactions that involve $\mathbf{X}$. The natural indirect effect odds ratio is *approximately* equal to $\exp[(\beta_1 + \beta_3)\gamma_1]$. This result is constrained only to outcomes of low prevalence. The authors suggested using the multivariate delta method to obtain the standard error and using bootstrapping to obtain the confidence interval. According to the authors, the expression for the natural direct effect is more complicated and differs for units with different covariate values $\mathbf{x}$.

The two methods described earlier for binary outcomes relax the *additivity* assumption (IV.a). Yet the method proposed by Huang and colleagues (2004) strictly requires, without adjustment for pretreatment covariates, the *ignorable treatment assignment* assumptions (I) and (II), the assumption of *within-treatment ignorable mediator value assignment* (III.a) and additionally the assumption of *cross-treatment ignorable mediator value assignment*. This last assumption can be stated as follows:

(III.c) The mediator value assignment under one treatment is independent of the potential outcome under the alternative treatment.

According to assumption (III.c), students who study 8 hours and those who study 4 hours when encouraged should have no difference in their test performance if they were counterfactually not encouraged. The aforementioned four ignorability assumptions are warranted when the treatment assignment and the mediator value assignment are both randomized. VanderWeele and Vansteelandt's (2010) strategy invokes these four ignorability assumptions within levels of observed pretreatment covariates $\mathbf{x}$. Hence, these assumptions are met when the treatment assignment and the mediator value assignment are both randomized in subpopulations defined by $\mathbf{x}$. When all four ignorability assumptions hold, the *no covariance* assumption (V.b) becomes unnecessary. Nonetheless, the two methods generate causally valid estimates only when the mediator model and the outcome model are both correctly specified.

10.3   Marginal structural models

10.3.1   Analytic procedure

Marginal structural models, well known to epidemiologists when implemented with inverse-probability-of-treatment weighting (IPTW) (Coffman and Zhong, 2012; Robins, 2003; Robins, Hernan, and Brumback, 2000), can be applied to estimating the controlled direct effect of the treatment and the controlled direct effect of the mediator. As shown by Vander-Weele (2009), in correspondence with a marginal structural model for the population average potential outcome associated with treatment z and mediator value m for all possible values of z and m,

$$E[Y(z,m)] = \beta_0 + \beta_1 m + \beta_2 z + \beta_3 zm,$$

one may specify and analyze a weighted model regressing the observed outcome Y on the treatment Z, the mediator M, and their interaction ZM. The weighted sample approximates data from a sequential randomized experiment in which units are first assigned at random to the treatment and then, within each treatment group, are assigned at random to different mediator values.

Unlike the regression-based approaches, a major strength of marginal structural models is that they enable the researcher to make adjustment for posttreatment covariates that may confound the mediator–outcome relationship under a given treatment condition. For example, after receiving the encouragement to study, some students may become much more anxious about the test than they had been before the treatment assignment. Here, a student's anxiety level measured after the treatment assignment is a posttreatment variable because it could have been affected by the treatment assignment. An appropriate amount of anxiety may increase one's study hours and may also improve one's test performance; an excessive amount of anxiety, however, may impede one's ability to study and to perform on the test. Hence, a spurious association between the mediator and the outcome in the treated group may be explained by the confounding role of posttreatment anxiety.

For a unit with treatment z and mediator value m and with pretreatment covariate values $\mathbf{x}$ and posttreatment covariate values $\mathbf{l}$, the weight is $\text{IPTW}^{(z)} \times \text{IPTW}^{(m|z)}$ where

$$\text{IPTW}^{(z)} = \frac{\text{pr}(Z = z)}{\text{pr}(Z = z | \mathbf{X} = \mathbf{x})} \tag{10.9}$$

and

$$\text{IPTW}^{(m|z)} = \frac{\text{pr}(M = m | Z = z)}{\text{pr}(M = m | Z = z, \mathbf{X} = \mathbf{x}, \mathbf{L} = \mathbf{l})} \tag{10.10}$$

Importantly, although the covariates that condition the probability of treatment assignment in $\text{pr}(Z = z | \mathbf{X} = \mathbf{x})$ include only pretreatment measures, those that condition the probability of mediator value assignment in $\text{pr}(M = m | Z = z, \mathbf{X} = \mathbf{x}, \mathbf{L} = \mathbf{l})$ may also include posttreatment measures. This is because the treatment assignment and the subsequent mediator value assignment within each treatment are analogous to sequential assignments of time-varying treatments. Hence, one may view $\mathbf{L}$ as time-varying covariates that may have been caused by the earlier treatment and may confound the relationship between the later treatment (in this case the mediator) and the outcome. IPTW adjustment for time-varying covariates does not introduce bias, whereas regression adjustment for time-varying covariates does. See Chapter 8 for related discussions.

Having obtained consistent estimates of the coefficients β_0, β_1, β_2, and β_3 through analyzing the weighted regression model, one may then estimate (i) the controlled direct effect of the treatment on the outcome when the mediator is held at a fixed value m, that is, $E[Y(1,m) - Y(0,m)] = \beta_2 + \beta_3 m$; (ii) the mediator effect on the outcome under the control condition $E[Y(0,m) - Y(0,m')] = \beta_1(m - m')$; and (iii) the mediator effect on the outcome under the experimental condition $E[Y(1,m) - Y(1,m')] = (\beta_1 + \beta_3)(m - m')$. When the mediator is binary, these three causal effects are simply represented by β_1, β_2, and $\beta_2 + \beta_3$, respectively.

Under a different set of identification assumptions to be explicated later, marginal structural models may be applied to the estimation of the natural direct and indirect effects as well. The procedure typically involves analyzing two weighted regression models taking the same form as that in path analysis models (10.1) and (10.2). The product of $\text{IPTW}^{(z)}$ and $\text{IPTW}^{(m|z)}$ is applied to the outcome model, while $\text{IPTW}^{(z)}$ is applied to the mediator model. Similar to path analysis and SEM, the natural direct effect is estimated by the coefficient for the treatment in the weighted model (10.2), while the natural indirect effect is estimated by the product of the coefficient for the treatment in the weighted model (10.1) and that for the mediator in the weighted model (10.2).

Using IPTW to adjust for the confounding effects of the covariates has some important advantages. One major advantage is that it becomes possible to adjust for the posttreatment covariates that confound the mediator–outcome relationships. In contrast, entering the posttreatment covariates as predictors in regression models for covariance adjustment would bias the estimation of the controlled direct effect of the treatment and of the natural direct effect. Another advantage is that by omitting all the covariates from the structural models, the data analyst is no longer at risk of mis-specifying the covariate–mediator relationships and the covariate–outcome relationships. Nonetheless, it is now well known that IPTW suffers from its lack of robustness when individuals without counterfactual information in the data receive a nonzero weight (Frölich, 2004; Khan and Tamer, 2010). In addition, mis-specifying the functional form of the propensity score model for the treatment may lead to severe bias in the causal effect estimation (Hong, 2010a; Kang and Schafer, 2007; Schafer and Kang, 2008; Waernbaum, 2012). The same problem occurs when the functional form of the propensity score model for the mediator is mis-specified. The marginal mean weighting through stratification (MMWS) approach (Hong, 2010a, 2012) introduced in Chapter 4 of this book may be employed in replacement of IPTW because the former has been found to be relatively robust despite functional form mis-specifications.

10.3.2   Identification assumptions

IPTW estimation of the controlled direct effects invokes three identification assumptions. The reader may use Table 10.1 to compare the identification assumptions across different analytic methods. As before, assumptions (I) and (II) state that the treatment assignment is independent of the potential mediators and the potential outcomes within levels of pretreatment covariates $\mathbf{X} = \mathbf{x}$. Yet instead of invoking the *within-treatment ignorable mediator value assignment* assumption (III.a), IPTW requires the following assumption:

> (III.b) The mediator value assignment under each treatment is independent of the potential outcomes not only within levels of pretreatment covariates $\mathbf{X} = \mathbf{x}$ but also within levels of posttreatment covariates $\mathbf{L} = \mathbf{l}$.

Assumption (III.b) is more plausible than assumption (III.a) because the latter only conditions on the pretreatment covariates. Under these identification assumptions, the weighted sample approximates data from a sequential randomized experimental design. Hence, the weighted subsample of units actually assigned to treatment z and mediator value m is representative of a pseudopopulation in which all units are assigned to treatment z and mediator value m. The weighted mean of the observed outcome of this subsample therefore estimates the population average potential outcome $E[Y(z, m)]$.

When applying marginal structural models to the estimation of the natural direct effect and the natural indirect effect, one must additionally invoke the *additivity* assumption (IV.a) and the *no covariance* assumption (V.b). As Coffman and Zhong (2012) acknowledged, without assuming additivity, marginal structural models cannot be used to obtain an estimate of the natural indirect effect. Therefore, assumption (IV.a) limits applications of this particular method in the same way as it limits applications of path analysis and SEM. Additional constraints arise when the mediator–outcome relationship is nonlinear.

10.4   Conditional structural models

10.4.1   Analytic procedure

When the goal is to estimate the average natural direct and indirect effects, a major constraint of the marginal structural models is that they cannot accommodate treatment-by-mediator interactions. To overcome this limitation, VanderWeele (2009) proposed conditional structural models as an alternative. This approach requires the specification of a conditional potential outcome model and a conditional potential mediator model, conditioning on a subset of observed pretreatment covariates $\mathbf{X}^- = \mathbf{x}^-$. The subset of covariates is chosen to define subpopulations within which the mediator–outcome relationships are *not* confounded by posttreatment covariates. Within subpopulations defined by $\mathbf{x}^-$, the potential outcome model represents the conditional average potential outcome $E[Y(z, m) | \mathbf{X}^- = \mathbf{x}^-]$ associated with treatment z and mediator value m and allows for treatment-by-mediator interaction. The structural relationships between the treatment, the mediator, and the potential outcome are assumed to be constant across all levels of $\mathbf{x}^-$. In particular, the mediator–outcome relationship under each treatment condition is assumed to be linear. The potential mediator model represents the conditional average potential mediator $E[M(z) | \mathbf{X}^- = \mathbf{x}^-]$ associated with treatment z conditioning on $\mathbf{x}^-$ and assumes that the same relationship between the treatment and the potential mediator holds across all levels of $\mathbf{x}^-$. In the case of a binary treatment, the mediator model generates, for individual units with pretreatment covariates $\mathbf{x}^-$, the potential mediator value when untreated $E[M(0)|\mathbf{X}^- = \mathbf{x}^-]$ and the potential mediator value when treated $E[M(1)|\mathbf{X}^- = \mathbf{x}^-]$. Replacing the mediator in the outcome model with these predicted potential mediator values, one can represent the subpopulation average potential outcomes $E[Y(0,M(0))|\mathbf{X}^- = \mathbf{x}^-]$, $E[Y(1,M(1))|\mathbf{X}^- = \mathbf{x}^-]$, and $E[Y(1,M(0))|\mathbf{X}^- = \mathbf{x}^-]$ each as a function of $\mathbf{x}^-$. An expression for the average natural direct effect and one for the average natural indirect effect can be derived after taking an average over the distribution of $\mathbf{x}^-$.

To estimate the coefficients in the potential outcome model and those in the potential mediator model, VanderWeele (2009) suggested analyzing two corresponding regression models, the first regressing Y on M, Z, ZM, and $\mathbf{X}^-$ and the second regressing M on Z and $\mathbf{X}^-$. IPTW is applied to these regression models to remove confounding due to selective treatment assignment and mediator value assignment. Let $\mathbf{X}$ denote the entire set of pretreatment covariates. For a unit assigned to treatment z and mediator value m and with covariate values $\mathbf{x}$, the outcome is weighted by $\mathrm{IPTW}^{(z|\mathbf{x}^-)} \times \mathrm{IPTW}^{(m|z,\mathbf{x}^-)}$ and the mediator is weighted by $\mathrm{IPTW}^{(z|\mathbf{x}^-)}$:

$$\mathrm{IPTW}^{(z|\mathbf{x}^-)} = \frac{\mathrm{pr}(Z=z|\mathbf{X}^- = \mathbf{x}^-)}{\mathrm{pr}(Z=z|\mathbf{X}=\mathbf{x})};$$

$$\mathrm{IPTW}^{(m|z,\mathbf{x}^-)} = \frac{\mathrm{pr}(M=m|Z=z,\mathbf{X}^- = \mathbf{x}^-)}{\mathrm{pr}(M=m|Z=z,\mathbf{X}=\mathbf{x})}.$$

This weighting procedure creates a pseudo-subpopulation of units sharing pretreatment characteristics $\mathbf{x}^-$. The subpopulation average potential outcome associated with treatment z and mediator value m is estimated by the weighted mean observed outcome of units with covariate value $\mathbf{x}^-$ who were actually assigned to treatment z and mediator value m.

10.4.2 Identification assumptions

Clearly, the conditional structural models are a combination of the marginal structural models and the modified regression approaches. This combined approach requires the assumptions (I) and (II) of *ignorable treatment assignment* within levels of the pretreatment covariates $\mathbf{X} = \mathbf{x}$ and the assumption (III.b) of *within-treatment ignorable mediator value assignment* within levels of the pretreatment covariates $\mathbf{X} = \mathbf{x}$ and the posttreatment covariates $\mathbf{L} = \mathbf{l}$. The conditional structural models additionally require the following assumption within levels of $\mathbf{x}^-$:

(III.c) The mediator value assignment under one treatment is independent of the potential outcome under the alternative treatment.

When this assumption of *cross-treatment ignorable mediator value assignment* is met, this approach can relax the *additivity* assumption (IV.a) required by the marginal structural models approach. However, as Robins (2003) pointed out, assumption (III.c) is hard to satisfy unless the mediator value is as if randomized under each treatment condition in subpopulations defined by $\mathbf{x}^-$. In such a case, the mediator value assignment would become independent of other pretreatment covariates and posttreatment covariates conditional on $\mathbf{x}^-$, and hence, $\mathrm{IPTW}^{(m|z,\mathbf{x}^-)}$ would become 1.0.

One may argue that a potential advantage of the conditional structural models approach, when compared with the modified regression approaches, is that the IPTW strategy enables adjustment for a relatively large number of covariates without having to enter all these covariates in the conditional structural models. However, as the set of pretreatment covariates $\mathbf{X}^-$ required for satisfying assumption (III.c) increases, this potential advantage vanishes. The conditional structural models approach and the modified regression approaches share a number of other limitations. Both procedures apply primarily to continuous outcomes that have a linear relationship with the mediator; both strategies require combining parameters estimated from the outcome model and the mediator model to compute the point estimates of the natural direct effect or the natural indirect effect. Extra computationally intensive steps are involved for estimating the standard errors. Similar to the modified regression approaches, the conditional structural models are prone to specification errors in the outcome model and the mediator model. In the meantime, the IPTW adjustment cannot be relied upon when the functional forms of the propensity score models are mis-specified.

10.5   Alternative weighting methods

Alternative weighting methods have been proposed in the recent years for decomposing the total treatment effect into a natural direct effect and a natural indirect effect. A common rationale is to identify the population average potential outcomes $E[Y(1, M(1))]$, $E[Y(1, M(0))]$, $E[Y(0, M(1))]$, and $E[Y(0, M(0))]$ from observable outcome data one at a time. This is done through weighting such that the mediator distributions become identical across the treatment groups. Here we briefly discuss two alternative strategies. One strategy, known as inverse probabiity weighting (IPW), is to assign weights to observations by their inverse conditional propensities to be in a particular treatment condition as a function of the observed mediator and pretreatment covariates (Huber, 2014). The other strategy, called inverse odds ratio-weighted estimation (IORW), assigns a weight that is the inverse of an odds ratio function relating the treatment and the mediator within levels of pretreatment covariates (Tchetgen Tchetgen, 2013). It is easy to show that these two types of weights are mathematically identical to ratio-of-mediator-probability weighting (RMPW) (Hong, 2010; Hong, Deutsch, & Hill, 2011, in press; Hong & Nomi, 2012; Tchetgen Tchetgen & Shpitser, 2012) to be presented in detail in the next three chapters.

10.5.1   Analytic procedure

Huber (2014) derived a weighted estimator of the population average potential outcome associated with the experimental condition $E[Y(1, M(1))]$, where the weight for each individual in the treated group is inverse to the individual's propensity of being treated as a function of pretreatment covariates $\mathbf{x}$:

$$\frac{1}{\mathrm{pr}(Z = 1 | \mathbf{X})}.$$

Similarly, to estimate the population average potential outcome associated with the control condition, the weight for each individual in the control group is inverse to the individual's propensity of being untreated:

$$\frac{1}{\text{pr}(Z=0|\mathbf{X})}.$$

When the treatment is randomized, the weight for those in the treated group becomes a constant. So does the weight for those in the control group.

To estimate the population average counterfactual outcome $E[Y(1, M(0))]$, the weight for a treated individual is instead equal to the following:

$$\frac{1}{\text{pr}(Z=1|\mathbf{X})} \times \frac{\text{pr}(Z=0|M,\mathbf{X})}{\text{pr}(Z=0|\mathbf{X})}.$$

And finally, to estimate the population average counterfactual outcome $E[Y(0, M(1))]$, the weight for an untreated individual is

$$\frac{1}{\text{pr}(Z=0|\mathbf{X})} \times \frac{\text{pr}(Z=1|M,\mathbf{X})}{\text{pr}(Z=1|\mathbf{X})}.$$

The analysis involves specifying a propensity score model for the treatment as a function of pretreatment covariates $\mathbf{x}$ and a second propensity score model for the treatment as a function of $\mathbf{x}$ and mediator value m. It should be noted that, because the mediator is expected to be caused by the treatment, the observed mediator M is equal to $M(1)$ for the treated and is equal to $M(0)$ for the untreated. The causal effects of interest are then estimated through pairwise comparisons of the estimated population average potential outcome.

According to Huber (2014), the weighting is aimed at balancing the distributions of the mediator and $\mathbf{X}$ between the experimental group and the control group. In particular, after weighting, the mediator distribution in the experimental group will resemble the distribution of the corresponding potential mediator $M(1)$ in the entire population; in parallel, the mediator distribution in the control group will resemble the population distribution of $M(0)$. However, as Hong, Deutsch, and Hill (in press) pointed out, modeling the treatment as a function of the mediator and covariates does not have immediate substantive interpretations given that the treatment causally precedes rather than succeeds the mediator.

The inverse odds ratio-weighted (IORW) approach was proposed by Tchetgen Tchetgen (2013) for decomposing the total treatment effect into a natural direct effect and a natural indirect effect conditional on pretreatment covariates in generalized linear models with a nonlinear link function. The same approach applies to the Cox proportional hazards model for a possibly right-censored survival outcome.

To estimate the natural direct effect, the first step is to fit a logistic regression model for the treatment Z as a function of the mediator value m and the pretreatment covariates $\mathbf{x}$ and obtain the coefficient estimates. In the second step, one analyzes a weighted outcome model by regressing Y on Z and $\mathbf{X}$ and obtains an estimate of the natural direct effect (i.e., the coefficient for Z) conditioning on $\mathbf{X}$ where the weight is

$$\frac{\text{pr}(Z=0|M=m,\mathbf{X})}{\text{pr}(Z=1|M=m,\mathbf{X})} \times \frac{\text{pr}(Z=1|M=0,\mathbf{X})}{\text{pr}(Z=0|M=0,\mathbf{X})}$$

for an individual whose observed mediator value is $M = m$ with $M = 0$ being the reference value. In the earlier example of randomizing students to receive encouragement to study for a test, this reference value would represent not studying before the test. This weight is the ratio of one's odds of being treated when not studying to the odds of being treated when studying an amount as observed. Each of the four conditional probabilities is estimated through applying the coefficient estimates obtained in the first step. An estimate of the natural indirect effect is then obtained by subtracting the natural direct effect estimate from the total effect estimate.

10.5.2   Identification assumptions

The aforementioned two additional weighting approaches are closely related and require the same set of identification assumptions. Most importantly, the causal effects are consistently estimated under the *ignorable treatment assignment* assumptions (I) and (II) and the assumptions of *ignorable mediator value assignment within each treatment condition* (IIIa) and across treatment conditions (IIIc) within levels of pretreatment covariates. In addition, Tchetgen Tchetgen (2013) emphasized that each individual must have a nonzero probability of displaying each of the possible mediator values under each treatment condition. For both approaches, due to the lack of robustness of parametric weighting, it is crucial that the propensity score model for the treatment be correctly specified in its functional form. An important strength of these approaches, however, is that they both allow for treatment-by-mediator interaction.

10.6   Resampling approach

10.6.1   Analytic procedure

Imai and colleagues (Imai, Keele, and Tingley, 2010; Imai, Keele, and Yamamoto, 2010) have developed two computationally intensive Monte Carlo simulation algorithms for generating the distribution of the counterfactual outcome $E[Y(1, M(0))]$ and that of $E[Y(0, M(1))]$. These algorithms require fitting a mediator model and an outcome model followed by repeatedly simulating the potential values of the mediator and the potential outcome values given the simulated values of the mediator. One then obtains the summary statistics for the causal effects of interest and their estimation uncertainty. The algorithm for parametric inference involves simulating model parameter values from their sampling distributions and requires resampling at least 1000 times at each step. The algorithm for nonparametric inference allows the analyst to specify nonparametric or semiparametric regression models (e.g., a spline regression) on the basis of relatively weaker functional form assumptions yet requires a much larger number of bootstrap resamples. The implementation requires extensive programming and specialized software in R that the authors have freely provided online.

The resampling algorithms are applicable to any statistical models. Hence, the authors claimed that the estimation procedure "can accommodate linear and nonlinear relationships, parametric and nonparametric models, continuous and discrete mediators, and various types of outcome variables." In particular, the resampling approach enables researchers to estimate not only the average causal effects but also quantile causal effects of interest such as the median natural indirect effect and thus can be useful for detecting heterogeneous causal effects.

10.6.2   Identification assumptions

Imai and colleagues used the term "sequential ignorability"—that is, the *ignorability of treatment assignment* and subsequently the *ignorability of mediator value assignment* given the treatment assigned—to summarize identification assumptions (I), (II), (III.a), and (III.c) conditional on the observed pretreatment covariates. When the sequential ignorability holds, the treatment effect on the mediator is independent of the mediator effect on the outcome. Therefore, assumption (IV.b) is already satisfied. Importantly, this method allows for treatment-by-mediator interactions. Even though the authors claimed that their identification assumptions do not require reference to specific statistical models, correct specifications of the outcome model involving multiway interactions among the treatment, the mediator, and the covariates as well as of the mediator model are nonetheless crucial for generating unbiased estimates of the causal effects in mediation analyses.

10.7   IV method

10.7.1   Rationale and analytic procedure

The IV method, widely employed by economists for bias removal, can be viewed as a special class of linear structural equation models (White, 1982; Wooldridge, 2002). In considering the impact of mediator M on outcome Y when the mediator values are not randomized, the IV method takes advantage of the fact that the treatment Z is often randomized by the experimenter or otherwise imposed on the individual participants by stochastic events in the world. Z may serve as an IV if Z predicts M in the mediator model but not Y in the outcome model conditioning on M:

$$M = \gamma_0 + aZ + \varepsilon_M;$$
(10.11)

$$Y = \beta_0 + bM + \varepsilon_Y.$$
(10.12)

The reduced-form outcome model is

$$Y = \beta'_0 + c'Z + \varepsilon'_Y.$$

Models (10.11) and (10.12) imply that the effect of Z on Y is completely mediated by M and hence the natural indirect effect is equal to the total effect, that is, $ab = c'$. When Z is randomized, the treatment effect on the mediator a and the treatment effect on the outcome c' can be estimated without bias. Hence, the mediator effect on the outcome b can be estimated as the ratio of $\hat{c}'$ to $\hat{a}$ as long as $a \neq 0$. The IV method can be implemented through two-stage least squares estimation or maximum likelihood estimation. In cases in which the treatment is not randomized, Altonji, Elder, and Taber (2005) derived the bias in the IV estimate. Tan (2006) proposed two possible remedies: one is to make covariance adjustment for observed pretreatment covariates, and the other is to use IPTW to equate the pretreatment composition between treatment groups.

10.7.2   Identification assumptions

In addition to the assumptions of ignorable treatment assignment (I) and (II), the key assumptions invoked by the IV method include the following (Angrist, Imbens, and Rubin, 1996; Sobel, 2008):

(VI) The *exclusion restriction*, that is, the direct effect of the treatment on the outcome must be zero for all units in the population.

The exclusion restriction assumption (VI) is especially restrictive because it makes the IV method inapplicable when the direct effect, assumed to be zero here, is a key parameter of interest. This is often the case when the intervention of interest is a multielement package, while the focal mediator under consideration measures the implementation of only one of the elements. One may find it tempting to modify model (10.12) by regressing Y on M and Z for testing whether the direct effect c is zero. However, just like most other identification assumptions except for assumption (VII) stated below, the exclusion restriction cannot be empirically verified unless the mediator values have been randomized under each treatment condition or is constant across the treatment conditions (Altonji, Elder, and Taber, 2005). The exclusion restriction clearly implies assumption (IV.a) that there is no treatment-by-mediator interaction.

(VII) The instrument variable has a nonzero average causal effect on the mediator (i.e., $a \neq 0$). Or in other words, the treatment must induce a change in at least one individual's mediator value.

In practice, however, a weak instrument that explains little of the variation in the mediator can lead to not only low precision but also large inconsistency and bias in the IV estimate (Bound, Jaeger, and Baker, 1995).

Additionally, analysts using the IV method often implicitly invoke the *constant causal effect* assumption (V.a). In the case of a binary treatment and a binary mediator, Imbens and Angrist (1994) and Angrist, Imbens, and Rubin (1996) relaxed this assumption by allowing for heterogeneity in individual responses to the treatment. They nonetheless replaced assumption (V.a) with assumption (V.c) summarized as the *monotonicity* assumption when the mediator is binary:

(V.c). The treatment does not have opposite impacts on different individuals at the intermediate stage.

In an example in which the mediator of interest is program participation, this assumption states that there is not a single defier in the population who would participate only if assigned to the control condition and would instead not participate if assigned to the experimental condition. When the previous identification assumptions hold, the IV method estimates the "local average treatment effect," that is, the average treatment effect for the compliers defined as those whose program participation corresponds to the treatment assignment.

When the mediator is multivalued, however, the *no covariance* assumption (V.b) is required in replacement of the *monotonicity* assumption. Let A, B, and C' denote random variables representing individual-specific causal effects. Specifying the structural models at the individual level, Raudenbush, Reardon, and Nomi (2012) derived that

$$E(C') = c' = E(AB) = E(A) \times E(B) + \text{Cov}(A,B) = ab + \text{Cov}(A,B).$$

The IV method requires that $\text{Cov}(A,B) = 0$, that is, treatment assignment does not induce a greater level of participation among individuals who are more likely to benefit from the program. The same assumption is required by path analysis and SEM as pointed out by Bullock, Green, and Ha (2010). However, this assumption is in contradiction with Heckman's (1997) argument that in a heterogeneous population, individuals may possess and act on private information about potential gains from the program in making decisions to participate and that such information may not be fully predicted by pretreatment covariates. Without measurement and adjustment for such private information, the *no covariance* assumption (V.b) may not be highly plausible.

Another strand of research uses the interaction between a baseline covariate X and a random treatment assignment Z as the IV to remove unmeasured confounding (Small, 2012). The causal parameters of interest are (i) the controlled direct effect of the mediator on the outcome under a given treatment $\delta_M = E[Y(z,m) - Y(z,m')]$ and (ii) the controlled direct effect of the treatment on the outcome at a given mediator value $\delta_Z = E[Y(1,m) - Y(0,m)]$. A two-stage least squares procedure regresses mediator M on Z, X, and ZX at the first stage and obtains the predicted mediator value for each individual. At the second stage, Y is regressed on Z, X, and the predicted mediator. In the second-stage results, the coefficient for M estimates δ_M; and the coefficient for Z estimates δ_Z. More generally, a vector of baseline covariates X can be interacted with treatment assignment to generate multiple IVs.

A major advantage of this approach, in comparison with the standard IV method, is that it does not require the exclusion restriction. Yet in addition to assuming ignorable treatment assignment (I) and (II), which is warranted by treatment randomization, this approach also shares with the standard IV method the assumption of *a nonzero IV effect on the mediator* (VII) and the *additivity* assumption (IV.a). The *constant causal effect* assumption (V.a) invoked by Ten Have and colleagues (2007) was relaxed by Small (2012) who allowed for heterogeneous causal effects among individuals within the same subclass defined by the baseline covariates $X = x$. Nonetheless, Small (2012) emphasized that the subclass-specific causal effects must be constant across all values of X. In other words, the baseline covariates cannot be moderators of the causal effects δ_M and δ_Z. This assumption remains to be a strong one especially in light of the discussion on moderation earlier in this book. Moreover, this approach requires that mediator M be independent of δ_M given the treatment and the baseline covariates, which is nearly as strong as the assumption of *ignorable mediator value assignment* within each treatment and within levels of pretreatment covariates (III.a). Finally, the approach cannot avoid making model-based assumptions with regard to the linear covariate–mediator associations at the first stage and the covariate–outcome associations at the second stage of the ordinary least squares analysis.

10.8 Principal stratification

10.8.1 Rationale and analytic procedure

The principal stratification framework, formalized by Frangakis and Rubin (2002), has been exploited by some researchers for investigating causal mechanisms across subpopulations (Barnard *et al.*, 2003; Elliott, Raghunathan, & Li, 2010; Gallop *et al.*, 2009; Jin and Rubin,

2008, 2009; Jo, 2008; Jo *et al.*, 2011; Page, 2012). Principal strata are conceptually defined by the potential mediator values under alternative treatment conditions. Individual units within a principal stratum are homogeneous in how they would respond to each of the treatment conditions at the intermediate stage and hence constitute a meaningful subpopulation.

For example, Chapter 6 reviewed a study by Barnard and colleagues (2003) who examined the impact of offering a voucher for attending a private school on student outcomes. Whether a student actually switched to a private school from a public school can be viewed as a mediator in this context. Each principal stratum is defined by a pair of potential mediator values $M(1)$ and $M(0)$. For *always-takers* who would always switch to a private school ($M(1) = M(0) = 1$) and for *never-takers* who would never do so ($M(1) = M(0) = 0$) regardless of the voucher offer, the total treatment effect is equal to the controlled direct effect of the treatment. This is because the treatment fails to induce a change in the mediator value in these two subpopulations. For the always-takers, the ITT effect

$$E[Y(1,M(1)) - Y(0,M(0)|M(1) = M(0) = 1]$$

becomes the controlled direct effect of the treatment when a student would always switch school:

$$E[Y(1,1) - Y(0,1)|M(1) = M(0) = 1].$$

And for the never-takers, the ITT effect

$$E[Y(1,M(1)) - Y(0,M(0))|M(1) = M(0) = 0]$$

becomes the controlled direct effect of the treatment when a student would never switch school:

$$E[Y(1,0) - Y(0,0)|M(1) = M(0) = 0].$$

Due to the perfect correspondence between the treatment values and the mediator values for the compliers ($M(1) = 1, M(0) = 0$), the ITT effect

$$E[Y(1,M(1) - Y(0,M(0))|M(1) = 1, M(0) = 0]$$

is equivalent to the mediator effect on the outcome of these students. The aforementioned categorization has, quite reasonably in this particular case, ruled out the existence of *defiers* who would not switch to a private school if offered a voucher and would switch only when not offered a voucher. Even though the analyst can never empirically identify the always-takers in the experimental group, the never-takers in the control group, and the compliers, the average causal effect defined for each subpopulation can be estimated through Bayesian inference.

10.8.2 Identification assumptions

Similar to most other analytic methods reviewed in this chapter, the principal stratification approach typically assumes (I) and (II) that the treatment assignment is independent of the potential mediators and the potential outcomes. It is also necessary for the purpose of identification to assume (V.c) monotonicity. For multivalued mediators (e.g., Page, 2012), the monotonicity assumption excludes the possibility of having subpopulations of individuals whose mediator values would be changed by the treatment in an unintended direction. However, unlike the IV method, the principal stratification approach does not require (VI) the exclusion restriction. Rather, this approach explicitly estimates the direct effect of the treatment on the outcome but only for subpopulations of individuals whose mediator values are not influenced by the treatment assignment (i.e., the "always-takers" and the "never-takers"). Assumption (IV.a) that there is no treatment-by-mediator interaction is easy to relax by allowing the outcome model specification to be different between the experimental group and the control group. Finally, even though the principal stratification approach does not require assumption (III.a) or its alternatives (III.b) and (III.c), in other words, the mediator value assignment is not required to be ignorable within levels of pretreatment covariates, the importance of pretreatment covariates cannot be understated in practice. In particular, as Hill (2012) pointed out, when the parametric model specification assumptions fail to distinguish between the principal strata, the burden of identification falls on strong predictive covariates.

VanderWeele (2011, 2012) emphasized that the principal stratification approach does not decompose the total effect into a natural direct effect and a natural indirect effect for the compliers unless one is willing to make the overly strong assumption that the average direct effect is the same across all principal strata. Under this additional assumption that cannot be verified, one would subtract the average natural direct effect estimate for always-takers and never-takers from the ITT effect estimate for compliers and claim the difference to be an estimate of the natural indirect effect for compliers.

10.9   Sensitivity analysis

The goal of a sensitivity analysis is to determine whether a plausible violation of an identification assumption could easily reverse the empirical conclusion. Most of the analytic methods reviewed earlier invoke the assumption that the mediator–outcome relationship under each treatment condition is not confounded within levels of the observed pretreatment covariates, an assumption that has been the focus of sensitivity analysis in the literature on causal mediation analysis. Although this assumption can never be empirically verified, analytic conclusions that are harder to alter by possible violations of this assumption are expected to add a higher value to scientific knowledge. This section reviews two alternative ways of conducting sensitivity analyses after estimating the natural direct and indirect effects. The two strategies differ in how they represent the unadjusted confounding associated with an omitted pretreatment covariate. The first strategy represents the amount of unadjusted confounding effect as a product of two hypothetical regression coefficients specifying the association between the omitted covariate and the treatment at a fixed mediator level and its association with the outcome when the treatment and the mediator level are fixed; while the second uses a hypothetical correlation coefficient between the error in the mediator model and that in the outcome model to represent the severity of confounding.

10.9.1   Unadjusted confounding as a product of hypothetical regression coefficients

Under the framework of linear structural equation models that allows for treatment-by-mediator interaction, VanderWeele (2010) proposed a procedure for assessing, in estimating the natural direct and indirect effects, the amount of confounding associated with an omitted pretreatment covariate U that confounds the mediator–outcome relationships within and across treatment conditions. Suppose that U is binary and that the effect of U on Y, denoted by λ_Y, is constant across levels of the treatment $Z = z$, the mediator $M = m$, and the observed pretreatment covariates $\mathbf{X} = \mathbf{x}$. Additionally, suppose that the difference in the prevalence of U between the experimental and the control units with mediator level m, denoted by δ_m, is constant across levels of $\mathbf{x}$. Under these simplifying assumptions, the bias formula for the natural direct effect $E[Y(1, M(0)) - Y(0, M(0))]$ is given by

$$Bias(\text{NDE}) = \lambda_Y \sum_{\mathbf{x}} \sum_{m} \delta_m \mathrm{pr}(M(0) = m | Z = 0, \mathbf{X} = \mathbf{x}) \mathrm{pr}(\mathbf{X} = \mathbf{x}). \qquad (10.13)$$

If U has no association with the mediator under each treatment condition, δ_m will be zero and so will $Bias(\text{NDE})$. The bias formula for the natural indirect effect $E[Y(1, M(1)) - Y(1, M(0))]$ is simply

$$Bias(\text{NIE}) = -Bias(\text{NDE}).$$

However, the above formula is hard to implement when $\mathbf{X}$ is high dimensional or when $\mathbf{X}$ and M are multivalued. If it is further simplified such that the difference in the prevalence of U between the experimental and the control units is a constant δ across levels of m, then $Bias(\text{NDE}) = \delta\lambda_Y$. The empirical conclusion is considered to be sensitive to the potential omission of U if the conclusion can be easily altered once the hypothetical amount of bias associated with U has been removed. For example, the confidence interval for the natural direct effect or the natural indirect effect, initially away from zero, may encompass zero once a hypothetical U is taken into account.

Importantly, if the observed pretreatment covariates $\mathbf{X}$ have explained a great majority of the variation in the outcome, the remaining variance in Y to be explained by U will be relatively small, which limits the magnitude of λ_Y as well as the potential bias that could be introduced by U. For data analysts, measuring and adjusting for pretreatment covariates that account for as much variation in the outcome as possible seems to be an effective way of reducing the sensitivity of their results.

10.9.2   Unadjusted confounding reflected in a hypothetical correlation coefficient

Imai and colleagues (Imai, Keele, and Tingley, 2010; Imai, Keele, and Yamamoto, 2010) developed an alternative set of sensitivity analyses for assessing the robustness of conclusions when the mediator–outcome relationships are potentially confounded by omitted covariates. They presented the basic form of the sensitivity analyses in the context of the linear SEM framework as follows that allows for treatment-by-mediator interaction:

$$Y = \beta_0' + \beta_2' Z + \boldsymbol{\beta}_4' \mathbf{X} + e_Y', \qquad (10.14)$$

$$M = \gamma_0 + \gamma_1 Z + \boldsymbol{\gamma}_2 \mathbf{X} + e_M, \qquad (10.15)$$

$$Y = \beta_0 + \beta_1 M + \beta_2 Z + \beta_3 ZM + \boldsymbol{\beta}_4 \mathbf{X} + e_Y. \qquad (10.16)$$

The sensitivity parameter is the correlation ρ between the error in the mediator model e_M and that in the outcome model e_Y, assumed to be constant across the treatment groups. The magnitude of this correlation corresponds to the amount of potential confounding of the mediator–outcome relationship. The empirical results are considered to be sensitive if a relatively small ρ would lead to an alternative conclusion.

The researchers derived the natural indirect effect $E[Y(1,M(1))-Y(1,M(0))]$ as a function of ρ and other parameters that can be consistently estimated from the sample residuals when the treatment is randomized:

$$\gamma_1 \times \frac{\sigma_{e'_Y}(1)}{\sigma_{e_M}(1)} \times \left[\widetilde{\rho}(1)-\rho\times\sqrt{\frac{1-\widetilde{\rho}^2(1)}{(1-\rho)}}\right].$$

Here, $\sigma_{e'_Y}(1)$ is the standard deviation of e'_Y in the experimental group; $\sigma_{e_M}(1)$ is the standard deviation of e_M in the same group; and $\widetilde{\rho}(1)$ is the correlation between e'_Y and e_M in the experimental group. When the treatment assignment is ignorable, Z is independent of e'_Y in model (10.14) and is also independent of e_M in model (10.15). Hence, the sample standard deviation of the residual obtained from analyzing model (10.14) estimates $\sigma_{e'_Y}(1)$, while the sample standard deviation of the residual obtained from analyzing model (10.15) estimates $\sigma_{e_M}(1)$. One may then compare the new estimate of the natural indirect effect with the original result. An iterative procedure can be used to obtain the confidence intervals for different values of ρ for determining how likely a relatively small value of ρ would lead to a change in the conclusion.

Similarly, the pure indirect effect $E[Y(0,M(1))-Y(0,M(0))]$ as defined in Chapter 9 is equal to

$$\gamma_1 \times \frac{\sigma_{e'_Y}(0)}{\sigma_{e_M}(0)} \times \left[\widetilde{\rho}(0)-\rho\times\sqrt{\frac{1-\widetilde{\rho}^2(0)}{(1-\rho)}}\right].$$

Here, $\sigma_{e'_Y}(0)$, $\sigma_{e_M}(0)$, and $\widetilde{\rho}(0)$ are the standard deviation of e'_Y, the standard deviation of e_M, and their correlation, respectively, in the control group.

To ease the interpretation of ρ in relation to researchers' substantive knowledge, Imai, Keele, and Yamamoto (2010) proposed an alternative formulation by representing each error term as a function of an omitted covariate (or a linear combination of omitted covariates) U:

$$e_M = \lambda_M U + \varepsilon_M,$$

$$e_Y = \lambda_Y U + \varepsilon_Y.$$

After adjusting for the observed pretreatment covariates, let $\widetilde{R}^2_M = [\mathrm{Var}(e_M)-\mathrm{Var}(\varepsilon_M)]/\mathrm{Var}(M)$ represent the proportion of variance in M explained by U; and let $\widetilde{R}^2_Y = [\mathrm{Var}(e_Y)-\mathrm{Var}(\varepsilon_Y)]/\mathrm{Var}(Y)$ represent the proportion of variance in Y explained by U. Hence,

$$\rho = \operatorname{sgn}(\lambda_M \lambda_Y) \times \frac{\widetilde{R}_M \widetilde{R}_Y}{\sqrt{(1-R_M^2)(1-R_Y^2)}},$$

where $\operatorname{sgn}(\lambda_M \lambda_Y)$ is the sign of $\lambda_M \lambda_Y$ and R_M^2 and R_Y^2 are the respective proportions of variance in the mediator and the outcome actually explained by the observed pretreatment covariates $\mathbf{X}$. Under this formulation, researchers can speculate a specific omitted covariate U and its plausible direction of the associations (i.e., positive or negative) with M and Y indicated by $\operatorname{sgn}(\lambda_M \lambda_Y)$ and the magnitude of the associations with M and Y indicated by $\widetilde{R}_M$ and $\widetilde{R}_Y$, respectively.

10.9.3   Limitations when the selection mechanism differs by treatment

It is important to note that the aforementioned strategies show limitations when the omitted covariate U is associated with M differently across the treatment conditions. In many applications, the treatment may change the relationship between U and M. For example, when encouraged to study for a test, students who are quick at following directions (denoted by $U = 1$) will likely spend more hours studying than those who have a tendency of ignoring directions (denoted by $U = 0$). In the absence of the encouragement, however, these two types of students may spend a similar amount of time studying. Therefore, a student's tendency of following directions is associated with hours of study under the experimental condition but not under the control condition. It seems reasonable to expect that, in general, students who follow directions more closely tend to perform better in the test than their peers who pay less attention to directions, even if these two types of students study the same amount of time. In this scenario, U confounds the mediator–outcome relationship under the experimental condition but not under the control condition.

The simplified version of the first strategy for sensitivity analysis would not apply because it requires that between the treated units and the control units with the same mediator value m, the difference in the prevalence of U is constant across levels of m, that is, $\delta_m = \delta$ conditioning on the observed pretreatment covariates. Apparently, when the encouragement to study mobilizes only those students who are quick at following directions, such students would have a larger representation under the experimental condition than under the control condition among those who study relatively more hours and would have a smaller representation under the experimental condition than under the control condition among those who study relatively few hours. In other words, δ_m is expected to increase as the mediator value m increases. In this case, the analyst must return to (10.13) which, as noted earlier, is hard to implement. The second strategy similarly requires that the relationship between U and M represented by λ_M does not depend on the treatment condition. In conclusion, in causal mediation analysis, a more general strategy for sensitivity analysis is needed that accommodates differing selection mechanisms across treatment conditions and summarizes a potential bias as a function of sensitivity parameters that are relatively easy to obtain. Finally, many may find it potentially disconcerting if a sensitivity analysis must rely on simplifying mathematical assumptions that are hard to parse on scientific grounds or are substantively unconvincing in most applications.

Cox and colleagues (2014) reviewed the aforementioned two methods and contrasted them with a third one adapted from Mauro (1990) that used correlations of a potential confounder with the mediator and the outcome to assess "the left-out variable error." As discussed in Chapter 3 of this book, Ken Frank (2000) adopted a similar approach in assessing the impact of omitting a confounder of the treatment–outcome relationship in treatment effect evaluation. Cox and colleagues extended this approach to the evaluation of the controlled direct effect of the treatment and the controlled direct effect of the mediator when the outcome is regressed on the treatment and the mediator without treatment-by-mediator interaction.

10.9.4   Other sensitivity analyses

Other sensitivity analyses may be employed for determining the possible changes in the causal effect estimates in anticipation of possible violations of other identification assumptions or model-based assumptions. For example, if the exclusion restriction is violated and if the hypothetical direct effect of the treatment on the outcome is c, a ratio of the ITT effect of the treatment on the outcome to the ITT effect of the treatment on the mediator would be $b+c/a$, which would suggest that the estimated mediator effect on the outcome obtained through an IV analysis contains a possible bias equal to c/a. The sensitivity to the exclusion restriction, therefore, depends on the magnitude of c relative to a. One may use scientific reasoning to determine a plausible range of values of c. The empirical result may be considered sensitive if a relative small value of c leads to an alteration of the conclusion. Sensitivity of results to model-based assumptions is typically assessed through modifying the functional forms of the mediator and the outcome models.

10.10   Conclusion

Among the eight approaches to causal mediation analysis reviewed in this chapter, path analysis/SEM and the IV method are the most well known and are relatively easy to implement. Yet these two methods are most stringent in their respective sets of identification assumptions. The alternative approaches that have emerged in the recent years represent attempts to relax some of the key assumptions yet often involve intensive computation in estimating the causal parameters and their standard errors or confidence intervals, which make them less accessible than the conventional methods. The following discussions compare the identification assumptions and the model-based assumptions across these eight methods.

10.10.1   The essentiality of sequential ignorability

The sequential ignorability assumption states that the treatment assignment and the subsequent mediator value assignment under each treatment are ignorable—that is, are as if randomized—within levels of the observed pretreatment covariates. As Imai and colleagues (Imai, Keele, and Tingley, 2010; Imai, Keele, and Yamamoto, 2010) argued, this assumption encompasses assumptions (I) and (II) with regard to the treatment assignment and assumptions (III.a) and (III.c) with regard to the mediator value assignment. Assumptions (III.b), (IV.b), and (V.b) are simultaneously satisfied when the sequential ignorability holds. The sequential ignorability assumption is clearly essential for estimating the natural direct effect and the natural indirect effect. Among the eight approaches reviewed here, only the IV method and the principal stratification approach do not invoke the ignorable mediator value assignment assumption. Yet neither the IV method nor the principal stratification approach decomposes the total effect of a treatment into a direct effect and an indirect effect and hence provides relatively limited understanding of the causal mediation mechanism. Most applications of the conventional approaches such as path analysis and SEM tend to take the sequential ignorability for granted especially when the treatment is randomized. It is common to see in the published scientific and methodological literature mediator models and outcome models with no adjustment for potential confounding covariates. This assumption has received great emphasis in the causal inference literature on mediation analysis. Alternative strategies for confounder adjustment and for sensitivity analysis have been proposed in the past years as reviewed earlier in this chapter.

10.10.2   Treatment-by-mediator interactions

As discussed in Chapter 9, a mediator may transmit the effect of the treatment on the outcome not only through a treatment-induced change in the mediator value but also through a treatment-induced change in the mediator–outcome relationship. The latter is viewed as a change in the mediational process that is often crucial for understanding the causal mechanism (Judd and Kenny, 1981). Yet path analysis/SEM, the IV method, and the marginal structural models are unsuitable for empirical studies in which treatment-by-mediator interactions are likely. The modified regression approach, the conditional structural models, the alternative weighting methods, the resampling approach, and the principal stratification approach all relax the assumption of no treatment-by-mediator interaction and therefore have greater applicability. However, none of these methods explicitly generate an estimate of the treatment-by-mediator interaction effect as a component of the total effect. Due to this limitation, the extent to which the effect of the treatment is transmitted through a change in the mediational process remains unknown.

10.10.3   Homogeneous versus heterogeneous causal effects

Path analysis/SEM, the IV method, the marginal structural models, and the conditional structural models all involve analyzing simultaneous structural equations and all represent the natural indirect effect as a product of the treatment effect on the mediator and the mediator effect on the outcome. This common framework requires either the assumption of homogeneous causal effects or, when heterogeneity is anticipated, the assumption of no covariance between the treatment effect on the mediator and the mediator effect on the outcome or the assumption that pretreatment covariates do not moderate the mediator effect on the outcome. The assumption of monotonicity is a special case of the assumption of no covariance and is often invoked in applications of the IV method and the principal stratification approach. For example, when the treatment and the mediator are both binary, the no covariance assumption holds for always-takers and never-takers as the treatment effect on the mediator is zero in these two principal strata; the no covariance assumption is unnecessary for the compliers as only the ITT effect is estimated in this third principal stratum. Yet these labels may misrepresent the reality because random events beyond the control of the experimenter may conceivably turn an always-taker or a never-taker into a complier.

10.10.4   Model-based assumptions

Correct specifications of the functional forms of the outcome model and of the mediator model are required by nearly all the analytic methods. Most of the methods reviewed here, with the exception of the marginal structural models and the alternative weighting methods, involve entering covariates in the outcome model when the mediator value assignment is not random and entering covariates in both the mediator model and the outcome model when the treatment assignment is not random. It is well known that when the covariates confound the treatment–outcome relationship or the mediator–outcome relationship, mis-specifying the covariate–outcome relationship in the outcome model may introduce bias (Drake, 1993). Similarly, when the covariates confound the treatment–mediator relationship, mis-specifying the covariate–mediator relationship in the mediator model will be costly. With the exception of the resampling approach and the alternative weighting methods, all other methods that involve continuous mediators require that the mediator–outcome relationship be linear. In other words, the outcome model cannot include polynomial terms (e.g., M^2, M^3,...) of the mediator (VanderWeele, 2009). Finally, as discussed in Chapter 3, covariance adjustment for pretreatment covariates often buys efficiency. SEM gains even more efficiency by invoking the assumption that the conditional joint distribution of the mediator and outcome is multivariate normal, yet with a potential trade-off for bias when the distribution assumption does not hold.

Most of these methods except for SEM require that the variables are free of measurement errors. It is well known that measurement errors in the treatment and the mediator will attenuate the estimated causal relationships, while measurement errors in the pretreatment covariates will compromise the effort of bias removal. When multiple items are used to measure a latent construct, SEM allows one to analyze the measurement models and the structural models simultaneously. Alternatively, one may conduct a separate factor analysis or employ item response theory to obtain the scale score for a latent construct and then use it in causal mediation analysis (Heckman, Pinto, and Savelyev, 2013).

The challenges faced by applied researchers in causal mediation analysis have triggered a great amount of interest among statisticians. Many novel ideas are emerging quickly and cannot be fully presented here due to the space constraint. For example, posttreatment confounders of the mediator–outcome relationship under a given treatment condition often exist even after statistical adjustment for the pretreatment covariates. Marginal structural models have made it possible to adjust for post-treatment covariates but only in the absence of treatment-by-mediator interaction. We have omitted discussions of analytic strategies for dealing with this issue by deriving upper and lower bounds on the direct and indirect effects (Cai *et al.*, 2008; Kaufman *et al.*, 2005a; Sjölander, 2009) on the basis of partial identification (Manski, 1989).

Chapter 11 will introduce the ratio-of-mediator-probability weighting (RMPW) method (Hong, 2010b; Hong and Nomi, 2012; Hong, Deutsch, and Hill, 2011, in press; Tchetgen Tchetgen & Shpitser, 2012) that minimizes model-based assumptions and limits identification assumptions to the sequential ignorability only. Its theoretical rationale and analytic strategy will be illustrated through an application.

Appendix 10.A: Bias in path analysis estimation due to the omission of treatment-by-mediator interaction

For simplicity, suppose that the treatment and the mediator are both binary. Also, suppose that the treatment assignment and the mediator value assignment under each treatment are both randomized. We will show that when the mediator–outcome relationship depends on the treatment, the bias in the path analysis estimate of the direct effect is a product of three elements:

the controlled treatment-by-mediator interaction effect, the treatment effect on the mediator, and the proportion of units assigned to the control group. The bias in the indirect effect estimate takes the opposite sign.

Let $Z = 1$ if a unit is treated and 0 if the unit is assigned to the control condition. As a mediator, $M(z)$ is a function of treatment assignment z for $z = 0$, 1 and can be generated by $M(z) = \beta_0 + \beta_1 z + \varepsilon_m$ where ε_m is a random error. In other words, we have that $\text{pr}(M(0) = 1) = \beta_0$ and that $\text{pr}(M(1) = 1) = \beta_0 + \beta_1$. We denote the potential outcome by Y (z, m) if a unit is assigned to treatment z and displays mediator value m. Suppose that the data generation function for the potential outcomes is $Y(z, m) = \theta_0 + \theta_1 z + \theta_2 m + \theta_3 zm + \varepsilon_Y$. Hence, the total effect is $\theta_1 + \theta_3 \beta_0 + (\theta_2 + \theta_3)\beta_1$, the natural direct effect is $\theta_1 + \theta_3 \beta_0$, and the natural indirect effect is $(\theta_2 + \theta_3)\beta_1$. Path analysis invokes the assumption of linearity and additivity (Holland, 1988) and specifies the observed outcome model as $Y = \gamma_0 + \gamma_1 Z + \gamma_2 M + e$. We can show that $\gamma_2 = \theta_2 + \theta_3 \times \text{pr}(Z = 1)$. The indirect effect estimate obtained from path analysis is $\gamma_2 \beta_1 = \theta_2 \beta_1 + \theta_3 \beta_1 \times \text{pr}(Z = 1) = (\theta_2 + \theta_3)\beta_1 - \theta_3 \beta_1 \times \text{pr}(Z = 0)$. Hence, the bias in the indirect effect estimate is $-\theta_3 \beta_1 \times \text{pr}(Z = 0)$, which is equivalent to

$$-E\{(Y(1,1) - Y(0,1)) - (Y(1,0) - Y(0,0))\} \times \{\text{pr}(M(1) = 1) - \text{pr}(M(0) = 1)\} \times \text{pr}(Z = 0).$$

References

Altonji, J.G., Elder, T.E. and Taber, C.R. (2005). An evaluation of instrumental variable strategies for estimating the effects of Catholic schooling. *Journal of Human Resources*, *XL*(4), 791–821.

Alwin, D.F. and Hauser, R.M. (1975). The decomposition of effects in path analysis. *American Sociological Review*, 40, 37–47.

Angrist, J.D., Imbens, G.W. and Rubin, D.B. (1996). Identification of causal effects using instrumental variables. *Journal of the American Statistical Association*, *91*(434), 444–472.

Barnard, J., Frangakis, C., Hill, J. *et al.* (2003). Principal stratification approach to broken randomized experiments: a case study of school choice vouchers in New York City (with discussion). *Journal of the American Statistical Association*, 98, 299–323.

Baron, R.M. and Kenny, D.A. (1986). The moderator-mediator variable distinction in social psychological research: conceptual, strategic, and statistical considerations. *Journal of Personality and Social Psychology*, *51*, 1173–1182.

Bishop, Y.M.M., Fienberg, S.E. and Holland, P.W. (1975). *Discrete Multivariate Analysis*, MIT Press, Cambridge, MA.

Bollen, K.A. (1987). Total, direct, and indirect effects in structural equation models, in *Sociological Methodology* (ed C. Clogg), American Sociological Association, Washington, DC.

Bound, J., Jaeger, D.A. and Baker, R.M. (1995). Problems with instrumental variables estimation when the correlation between the instruments and the endogenous explanatory variable is weak. *Journal of the American Statistical Association*, *90*(430), 443–450.

Breslow, N. (1981). Odds ratio estimators when the data are sparse. *Biometrika*, *68*(1), 73–84.

Bullock, J.G., Green, D.P. and Ha, S.E. (2010). Yes, but what's the mechanism? (don't expect an easy answer). *Journal of Personality and Social Psychology*, 98, 550–558.

Cai, Z., Kuroki, M., Pearl, J. *et al.* (2008). Bounds on direct effects in the presence of confounded intermediate variables. *Biometrics*, 64, 695–701.

Coffman, D.L. and Zhong, W. (2012). Assessing mediation using marginal structural models in the presence of confounding and moderation. *Psychological Methods, 17*(4), 642–664.

Cox, M.G., Kisbu-Sakarya, Y., Miočević, M. *et al.* (2014). Sensitivity plots for confounder bias in the single mediator model. *Evaluation Review, 37,* 405–431.

Ditlevsen, S., Christensen, U., Lynch, J. *et al.* (2005). The mediation proportion: a structural equation approach for estimating the proportion of exposure effect on outcome explained by an intermediate variable. *Epidemiology, 16,* 114–120.

Drake, C. (1993). Effects of misspecification of the propensity score on estimators of treatments effects. *Biometrics, 49,* 1231–1236.

Duncan, O.D. (1966). Path analysis: sociological examples. *The American Journal of Sociology, 72*(1), 1–16.

Elliott, M.R., Raghunathan, T.E., and Li, Y. (2010). Bayesian inference for causal mediation effects using principal stratification with dichotomous mediators and outcomes. *Biostatistics, 11*(2), 353–372.

Fienberg, S.E. (2007). *The Analysis of Cross-Classified Categorical Data,* Springer, New York.

Finney, J. (1972). Indirect effects in path analysis. *Sociological Methods and Research, 1*(2), 175–187.

Frangakis, C.E. and Rubin, D.B. (2002). Principal stratification in causal inference. *Biometrics, 58,* 21–29.

Frank, K.A. (2000). Impact of a confounding variable on a regression coefficient. *Sociological Methods and Research, 29*(2), 147–194.

Freedman, L.S. and Schatzkin, A. (1992). Sample size for studying intermediate endpoints within intervention trials of observational studies. *American Journal of Epidemiology, 136,* 1148–1159.

Frölich, M. (2004). Finite sample properties of propensity-score matching and weighting estimators. *The Review of Economics and Statistics, 86,* 77–90.

Gallop, R., Small, D.S., Lin, J.Y. *et al.* (2009). Mediation analysis with principal stratification. *Statistics in Medicine, 28,* 1108–1130.

Greenland, S. (1987). Interpretation and choice of effect measures in epidemiologic analyses. *American Journal of Epidemiology, 125*(5), 761–768.

Greenland, S., Robins, J.M. and Pearl, J. (1999). Confounding and collapsibility in causal inference. *Statistical Science, 14*(1), 29–46.

Heckman, J.J. (1997). Instrumental variables: a study of implicit behavioral assumptions used in making program evaluations. *The Journal of Human Resources, 32*(3), 441–462.

Heckman, J., Pinto, R. and Savelyev, P. (2013). Understanding the mechanisms through which an influential early childhood program boosted adult outcomes. *American Economic Review, 103*(6), 2052–2086.

Hill, J. (2012). Discussion of 'principal stratification as a framework for investigating mediational processes in experimental settings' by Lindsay Page. *Journal of Research on Educational Effectiveness, 5*(3), 254–257.

Holland, P.W. (1988). Causal interference, path analysis, and recursive structural equations models. *Sociological Methodology, 18,* 449–484.

Hong, G. (2010a). Marginal mean weighting through stratification: adjustment for selection bias in multilevel data. *Journal of Educational and Behavioral Statistics, 35*(5), 499–531.

Hong, G. (2010b). *Ratio of Mediator Probability Weighting for Estimating Natural Direct and Indirect Effects.* JSM Proceedings, Biometrics Section. American Statistical Association, Alexandria, VA, pp. 2401–2415.

Hong, G. (2012). Marginal mean weighting through stratification: a generalized method for evaluating multi-valued and multiple treatments with non-experimental data. *Psychological Methods, 17*(1), 44–60.

Hong, G. and Nomi, T. (2012). Weighting methods for assessing policy effects mediated by peer change. *Journal of Research on Educational Effectiveness* special issue on the statistical approaches to studying mediator effects in education research, *5*(3), 261–289.

Hong, G., Deutsch, J., and Hill, H. (2011). *Parametric and Non-Parametric Weighting Methods for Estimating Mediation Effects: An Application to the National Evaluation of Welfare-to-Work Strategies.* JSM Proceedings, Social Statistics Section. American Statistical Association, Alexandria, VA, pp. 3215–3229.

Hong, G., Deutsch, J., and Hill, H.D. (in press). Ratio-of-mediator-probability weighting for causal mediation analysis in the presence of treatment-by-mediator interaction. *Journal of Educational and Behavioral Statistics.*

Huang, B., Sivaganesan, S., Succop, P. *et al.* (2004). Statistical assessment of mediational effects for logistic mediational models. *Statistics in Medicine, 23,* 2713–2728.

Huber, M. (2014). Identifying causal mechanisms (primarily) based on inverse probability weighting. *Journal of Applied Econometrics, 29*(6), 920–943.

Imai, K., Keele, L. and Tingly, D. (2010a). A general approach to causal mediation analysis. *Psychological Methods, 15*(4), 309–334.

Imai, K., Keele, L. and Yamamoto, T. (2010b). Identification, inference and sensitivity analysis for causal mediation effects. *Statistical Science, 25*(1), 51–71.

Imbens, G.W. and Angrist, J. (1994). Identification and estimation of local average treatment effects. *Econometrica, 62,* 467–476.

Jin, H. and Rubin, D.B. (2008). Principal stratification for causal inference with extended partial compliance. *Journal of the American Statistical Association, 103,* 101–111.

Jin, H. and Rubin, D.B. (2009). Public schools versus private schools: causal inference with partial compliance. *Journal of Educational and Behavioral Statistics, 34,* 24–45.

Jo, B. (2008). Causal inference in randomized experiments with mediational processes. *Psychological Methods, 13*(4), 314–336.

Jo, B., Stuart, E.A., MacKinnon, D.P. *et al.* (2011). The use of propensity scores in mediation analysis. *Multivariate Behavioral Research, 46*(3), 425–452.

Jöreskog, K.G. (1970). A general method for analysis of covariance structures. *Biometrika, 57,* 239–251.

Judd, C.M. and Kenny, D.A. (1981). Process analysis: estimating mediation in treatment evaluation. *Evaluation Review, 5,* 602–619.

Kang, J.D. and Schafer, J.L. (2007). Demystifying double robustness: a comparison of alternative strategies for estimating a population mean from incomplete data. *Statistical Science, 22*(4), 523–539.

Kaufman, S., Kaufman, J., MacLenose, R. *et al.* (2005a). Improved estimation of controlled direct effects in the presence of unmeasured confounding of intermediate variables. *Statistics in Medicine, 24,* 1683–1702.

Kaufman, J.S., MacLehose, R.F., Kaufman, S. *et al.* (2005b). The mediation proportion. *Epidemiology, 16*(5), 710.

Khan, S. and Tamer, E. (2010). Irregular identification, support conditions, and inverse weight estimation. *Econometrica, 78,* 2021–2042.

MacKinnon, D.P. (2008). *Introduction to Statistical Mediation Analysis,* Erlbaum, Mahwah, NJ.

MacKinnon, D.P. and Dwyer, J.H. (1993). Estimating mediated effects in prevention studies. *Evaluation Review, 17*(2), 144–158.

MacKinnon D.P., Lockwood C.M., and Hoffman J.M. (1998). A new method to test for mediation. Paper presented at the Annual Meeting of the Society for Prevention Research. Park City, UT.

MacKinnon, D.P., Lockwood, C.M., Hoffman, J.M. *et al.* (2002). A comparison of methods to test mediation and other intervening variable effects. *Psychological Methods, 7,* 83–104.

MacKinnon, D.P., Lockwood, C.M., Brown, C.H. *et al.* (2007). The intermediate endpoint effect in logistic and probit regression. *Clinical Trials, 4*, 499–513.

Manski, C.F. (1989). Anatomy of the selection problem. *The Journal of Human Resources, 24*, 343–360.

Mauro, R. (1990). Understanding L.O.V.E. (left out variables error): a method for estimating the effects of omitted variables. *Psychological Bulletin, 108*, 314–329.

McGuigan, K., and Langholtz, B. (1988). A note on testing mediation paths using ordinary least-squares regression. Unpublished note. According to MacKinnon and Dwyer (1993), McGuigan and Langholtz (1988) derived the standard error for sample estimates of $c' - c$.

Miettinen, O.S. and Cook, E.F. (1981). Confounding: essence and detection. *American Journal of Epidemiology, 114*(4), 595–603.

Page, L.C. (2012). Principal stratification as a framework for investigating mediational processes in experimental settings. *Journal of Research on Educational Effectiveness* special issue on the statistical approaches to studying mediator effects in education research, *5*(3), 215–244.

Pearl, J. (2010). An introduction to causal inference. *The International Journal of Biostatistics, 6*(2), 1–59.

Petersen, M.L., Sinisi, S.E. and van der Laan, M.J. (2006). Estimation of direct causal effects. *Epidemiology, 17*(3), 276–284.

Preacher, K.J., Rucker, D.D. and Hayes, A.F. (2007). Addressing moderated mediation hypotheses: theory, methods, and prescriptions. *Multivariate Behavioral Research, 42*, 185–227.

Raudenbush, S.W., Reardon, S.F. and Nomi, T. (2012). Statistical analysis for multisite trials using instrumental variables with random coefficients. *Journal of Research on Educational Effectiveness* special issue on the statistical approaches to studying mediator effects in education research, *5*(3), 303–332.

Robins, J.M. (2003). Semantics of causal DAG models and the identification of direct and indirect effects, in *Highly Structured Stochastic Systems* (eds P. Green, N. Hjort and S. Richardson), Oxford University Press, Oxford.

Robins, J.M., Hernan, M.A. and Brumback, B. (2000). Marginal structural models and causal inference in epidemiology. *Epidemiology, 11*(5), 550–560.

Rubin, D.B. (1986). Comment: which Ifs have causal answers. *Journal of the American Statistical Association, 81*, 961–962.

Schafer, J.L. and Kang, J. (2008). Average causal effects from nonrandomized studies: a practical guide and simulated example. *Psychological Methods, 13*(4), 279–313.

Sjölander, A. (2009). Bounds on natural direct effects in the presence of confounded intermediate variables. *Statistics in Medicine, 28*, 558–571.

Small, D. (2012). Mediation analysis without sequential ignorability: using baseline covariates interacted with random assignment as instrumental variables. *Journal of Statistical Research, 46*, 91–103.

Sobel, M.E. (1982). Asymptotic confidence intervals for indirect effects in structural models, in *Sociological Methodology* (ed S. Leinhardt), Jossey-Bass, San Francisco, CA, pp. 290–312.

Sobel, M.E. (2008). Identification of causal parameters in randomized studies with mediating variables. *Journal of Educational and Behavioral Statistics, 33*(2), 230–251.

Tan, Z. (2006). Regression and weighting methods for causal inference using instrumental variables. *Journal of the American Statistical Association, 101*(476), 1607–1618.

Tchetgen Tchetgen, E.J. (2013). Inverse odds ratio-weighted estimation for causal mediation analysis. *Statistics in Medicine, 32*(26), 4567–4580.

Tchetgen Tchetgen, E.J., and Shpitser, I. (2012). Semiparametric theory for causal mediation analysis: Efficiency bounds, multiple robustness and sensitivity analysis. *The Annals of Statistics, 40*(3), 1816–1845.

Ten Have, T.R., Joffe, M.M., Lynch, K.G. *et al.* (2007). Causal mediation analysis with rank preserving models. *Biometrics*, *63*, 926–934.

Valeri, L. and VanderWeele, T.J. (2013). Mediation analysis allowing for exposure-mediator interactions and causal interpretation: theoretical assumptions and implementation with SAS and SPSS macros. *Psychological Methods*, *18*, 137–150.

VanderWeele, T.J. (2009). Marginal structural models for the estimation of direct and indirect effects. *Epidemiology*, *20*, 18–26.

VanderWeele, T.J. (2010). Bias formulas for sensitivity analysis for direct and indirect effects. *Epidemiology*, *21*(4), 540–551.

VanderWeele, T.J. (2011). Principal stratification: uses and limitations. *International Journal of Biostatistics*, *7*, 1–14.

VanderWeele, T.J. (2012). Comments: should principal stratification be used to study mediational processes? *Journal of Research on Educational Effectiveness* special issue on the statistical approaches to studying mediator effects in education research, *5*(3), 245–249.

VanderWeele, T.J. and Vansteelandt, S. (2009). Conceptual issues concerning mediation, interventions and composition. *Statistics and Its Interface*, *2*(4), 457–468.

VanderWeele, T. and Vansteelandt, S. (2010). Odds ratios for mediation analysis for a dichotomous outcome. *American Journal of Epidemiology*, *171*(12), 1339–1348.

Waernbaum, I. (2012). Model misspecification and robustness in causal inference: comparing matching with doubly robust estimation. *Statistics in Medicine*, *31*, 1572–1581.

White, H. (1982). Instrumental variables regression with independent observations. *Econometrica*, *50*, 483–499.

Winship, C. and Mare, R.D. (1983). Structural equations and path analysis for discrete data. *American Journal of Sociology*, *89*, 54–110.

Wooldridge, J. (2002). *Econometric Analysis of Cross Section and Panel Data*, MIT Press, Cambridge, MA.

Wright, S. (1934). The method of path coefficients. *Annals of Mathematical Statistics*, *5*(3), 161–215.

11

Investigations of a simple mediation mechanism

As reviewed in Chapter 10, most conventional methods are not suitable for decomposing the total treatment effect into a natural direct effect and a natural indirect effect in the presence of a treatment-by-mediator interaction. This chapter introduces the ratio-of-mediator-probability weighting (RMPW) method for decomposing the total effect in the presence of such an interaction (Hong, 2010b; Hong & Nomi, 2012; Tchetgen Tchetgen & Shpitser, 2012). The natural indirect effect can be further decomposed into a pure indirect effect and a natural treatment-by-mediator interaction effect. The latter captures the treatment effect transmitted through a change in the mediational process. The RMPW strategy is illustrated with an analysis of the impact of a welfare-to-work program on maternal depression mediated by employment experience when there is evidence that employment (the mediator) affects depressive symptoms (the outcome) differently under alternative policy conditions (the treatment) (Hong, Deutsch, and Hill, 2011, in press). For simplicity, we focus on a binary treatment and a binary mediator in the current chapter and discuss RMPW extensions in the next two chapters.

After introducing the application example, this chapter explains the RMPW rationale first in an ideal sequentially randomized design and next in a hypothetical sequentially randomized block design. Under these hypothetical designs, the causal effects can be identified under minimal assumptions. We then show the extension to a standard experiment in which only the treatment is randomized and clarify the identification assumptions. The chapter then explains a parametric procedure and a nonparametric alternative for implementing the RMPW method. Simulation results demonstrate satisfactory performance of both procedures when the functional form of the propensity score model is correctly specified and reveal the relative robustness of the nonparametric results when this is not true. The last section assesses the strengths and limitations of the RMPW strategy. In comparison with some other techniques for mediation analysis, RMPW requires relatively few assumptions about the distribution of the outcome, the distribution of the mediator, and the functional form of the outcome model.

The parametric and nonparametric RMPW procedures can be carried out in any standard statistical software. However, a stand-alone RMPW program greatly eases the computation especially when the number of pretreatment covariates becomes daunting. The RMPW

Causality in a Social World: Moderation, Mediation and Spill-over, First Edition. Guanglei Hong.
© 2015 John Wiley & Sons, Ltd. Published 2015 by John Wiley & Sons, Ltd.

program and the SAS, Stata, and R syntax files along with a data example are all freely available online at the publisher's web site: http://www.wiley.com/go/social_world.

11.1   Application example: national evaluation of welfare-to-work strategies

11.1.1   Historical context

In the late 1990s, the US government's six decade-long welfare cash assistance program (i.e., Aid to Families with Dependent Children, AFDC) was replaced nationwide by a new program (i.e., Temporary Assistance for Needy Families, TANF). The old program entitled low-income single parents with dependent children to collect cash benefits without working. In contrast, the new program, which was designed to move low-income mothers from welfare to work, put time limits on the receipt of cash assistance and required parents to seek and secure employment as a condition for receiving assistance and for avoiding financial penalties. This change in federal policy was heavily influenced by the National Evaluation of Welfare-to-Work Strategies (NEWWS) as well as other experiments conducted earlier in the decade. The experimental results showed increased employment and earnings for welfare recipients as a result of employment-focused incentives and services (Michalopoulos, Schwartz, and Adams-Ciardullo, 2001). The increase in employment was particularly evident in Riverside, California, in a comparison between participants who were assigned at random to the labor force attachment program (henceforth LFA) and their counterparts assigned to the control condition. The LFA program implemented key elements of TANF, while the control group members continued to receive public assistance from AFDC. None of the participants in either group were working full time initially (defined as 30 or more hours per week).

The political debate surrounding welfare reform focused on the potential for employment to benefit welfare recipients psychologically as well as financially. Low-income single mothers with young children experience disproportionately high rates of depressive symptoms and clinical depression (Coiro, 2001). Despite rhetoric and past evidence suggesting that welfare-to-work programs would benefit or harm the psychological well-being of welfare recipients (Cheng, 2007; Jagannathan, Camasso, and Sambamoorthi, 2010; Knab, McLanahan, and Garfnikel, 2008; Morris, 2008), LFA in Riverside did not show a statistically significant total effect on maternal depression (Hamilton *et al.*, 2002). However, the null total effect does not rule out possible mediation. There are at least two distinct scenarios in which the null total effect of the program on maternal depression would mask mediated effects. In both cases, the direct and indirect effects of the program on an individual could offset one another.

First, program-induced employment might benefit a participant's mental health by boosting self-efficacy and reducing depressive symptoms (a beneficial indirect effect due to a change in the mediator value, i.e., employment), while other aspects of the program, such as the threat of sanctions, might be stressful and adversely affect the participant's mental health (a detrimental direct effect). If similar in size, these countervailing effects could result in a null total effect.

Second, program expectations with regard to employment and the threat of sanctions could alter the relationship between employment and depression such that employment would be more beneficial, and lack of employment more detrimental, to psychological well-being

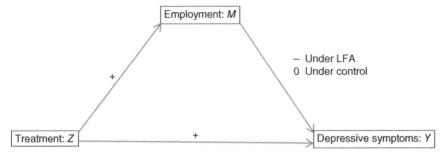

Figure 11.1   Hypothesized causal relationships between the treatment (Z), mediator (M), and outcome (Y)

if a mother was assigned to LFA than if she was assigned to the control condition. This second scenario, a classic case of treatment-by-mediator interaction, highlights an indirect effect due to a change in the mediational process, which again could possibly be offset by a direct effect.

Figure 11.1 illustrates the hypothesized causal relationships. These include (i) an expected increase in employment if one is assigned to LFA rather than to the control condition; (ii) an expected increase in depressive symptoms if one is assigned to LFA rather than to the control condition yet the treatment, perhaps counterfactually, fails to increase one's employment; (iii) an expected decrease in depressive symptoms under LFA should the treatment induce an increase in employment; and (iv) a possible lack of change in depressive symptoms under the control condition despite a similar amount of increase in employment.

11.1.2   Research questions

To empirically investigate the hypothesized causal mechanisms discussed in Section 11.1.1, Hong, Deutsch, and Hill (2011, in press) asked, for a low-income mother with preschool-aged children, whether the new welfare-to-work policy as represented by the LFA program would have an impact on her depressive symptoms through changing her employment experience and what the policy impact on her depressive symptoms would be if the policy, perhaps coun-terfactually, failed to change her employment experience. The first question concerns the person-specific indirect effect of the treatment on the outcome transmitted through the focal mediator, while the second question is about the direct effect of the treatment on the outcome should there be no change in the mediator value. Moreover, they asked whether a change in her employment in the amount that the policy was expected to induce would affect her depres-sive symptoms differently under the LFA program than under the control condition. The answer would be yes in the presence of a treatment-by-mediator interaction effect.

11.1.3   Causal parameters

We use Z to denote treatment assignment, M for employment experience during the 2 years after randomization, and Y for depressive symptoms 2 years after randomization. A participant's employment information during the 2 years after randomization was extracted from quarterly administrative data maintained by the State of California. A self-administered questionnaire at the 2-year follow-up included 12 items from the Center for Epidemiology

Studies—Depression Scale (CES-D; Radloff, 1977) measuring depressive symptoms during the past week (e.g., I could not get going).

Let $Z_i = 1$ if individual i was assigned to the LFA program and $Z_i = 0$ if assigned to the control condition. The corresponding *potential mediators* are denoted by $M_i(1)$ and $M_i(0)$, representing the individual's employment status under LFA and under the control condition, respectively. Let $M_i(1) = 1$ if the individual was assigned to LFA and was ever employed during the 2 years after randomization; let $M_i(1) = 0$ if she was assigned to LFA yet was never employed during the 2-year period. Let $M_i(0) = 1$ if the same individual was assigned to the control condition and was ever employed and let $M_i(0) = 0$ if she was assigned to the control condition and was never employed during the same time period. Hence, the treatment-induced change in her employment experience is $M_i(1)-M_i(0)$. The definition of the potential mediators assumes (i) that one's employment is affected by one's own treatment assignment and is not affected by other individuals' treatment assignment and (ii) that one's employment associated with a given treatment does not depend on whether the individual selected the treatment on her own or was assigned at random to the treatment (Rubin, 1986).

The individual's *potential outcome* under LFA is typically denoted by $Y_i(1)$; and that under the control condition is denoted by $Y_i(0)$. Because each potential outcome in this case is also a function of the potential employment experience corresponding to the given treatment assignment, we may write $Y_i(1)$ as $Y_i(1, M_i(1))$ and write $Y_i(0)$ as $Y_i(0, M_i(0))$. Additionally, we use $Y_i(1, M_i(0))$ to denote the mother's counterfactual outcome if assigned to the LFA program yet experiencing employment as she would have had under the control condition, and use $Y_i(0, M_i(1))$ to denote her counterfactual outcome if assigned to the control condition yet experiencing employment as she would have had under LFA. These counterfactual outcomes are conceivable because, despite one's actual treatment assignment and actual employment experience associated with the treatment, there is often a nonzero probability that uncertainties in the job market or in one's personal environment—in other words, *pure luck*—might have altered one's employment opportunities unexpectedly. For example, a participant assigned to the LFA program who otherwise would have become employed might remain unemployed due to an economic downturn or an unexpected health problem of a family member. Therefore, we allow an individual's potential mediator value under a given treatment to be shifted possibly by random events often beyond the control of the experimenter.

The *natural direct effect* of the policy on depression, defined by $Y_i(1, M_i(0))-Y_i(0, M_i(0))$, represents the effect of the policy on maternal depression if the policy, perhaps counterfactually, failed to change one's employment experience. The direct effect is to be attributed to other unspecified mediational processes that are unrelated to employment. For example, the threat of sanctions under LFA might heighten depression, while interactions with caseworkers in the LFA program might lead to improved access to community mental health services.

The *natural indirect effect* of the policy on depression mediated by employment, $Y_i(1, M_i(1))-Y_i(1, M_i(0))$, represents the change in a mother's depressive symptoms under LFA solely attributable to the policy-induced change in her employment experience (i.e., a change from $M_i(0)$ to $M_i(1)$). According to the earlier reasoning, the LFA program relative to the control condition may affect maternal depression partly through increasing employment and partly through altering the mediational process such that employment would be more beneficial under the LFA program than under the control condition. In such cases, the natural indirect effect may be further decomposed into two elements. The first element $Y_i(0, M_i(1))$ $-Y_i(0, M_i(0))$ is the change in the mother's depressive symptoms under the control condition should her employment increase by an amount that could be induced by a hypothetical assignment to the LFA program rather than the control condition. Robins and Greenland (1992) called this "the pure indirect effect." According to the earlier hypothesis, should the same increase in employment occur under LFA, there might be a greater change in the mother's depressive symptoms. Hence, the second element is the difference between the natural indirect effect and the pure indirect effect and is called the natural treatment-by-mediator interaction effect, $[Y_i(1, M_i(1))-Y_i(1, M_i(0))]-[Y_i(0, M_i(1))-Y_i(0, M_i(0))]$. It reflects how the policy-induced change in employment would affect the mother's depression differently between the experimental condition and the control condition.

The total effect of the treatment on the outcome is the sum of the natural direct effect and the natural indirect effect, and the latter is the sum of the pure indirect effect and the natural treatment-by-mediator interaction effect.

Of course, individuals in the population may respond to the treatment differently due to differences in past experiences and current personal circumstances. Questions of immediate relevance to policy-making have to do with the population average natural direct effect $E[Y(1, M(0))-Y(0, M(0))]$, the population average natural indirect effect $E[Y(1, M(1))-Y(1, M(0))]$, and the population average natural treatment-by-mediator interaction effect $E\{[Y(1, M(1))-Y(1, M(0))]-[Y(0, M(1))-Y(0, M(0))]\}$.

11.1.4   NEWWS Riverside data

The Riverside sample included 208 LFA group members and 486 control group members with at least one child aged 3–5 years. In addition to the employment record and the measurement of depressive symptoms at the 2-year follow-up, the baseline survey contains rich information about participant characteristics shown previously to be important predictors of both employment and depressive symptoms. These include measures of (i) maternal psychological well-being; (ii) history of employment and welfare use, employment status, earnings, and income in the quarter prior to randomization; (iii) human capital (i.e., educational credentials and literacy and math skills); (iv) personal attitudes toward employment, including the preference to work, willingness to accept a low-wage job, and shame to be on welfare; (v) perceived social support and barriers to work; (vi) practical support and barriers to work such as childcare arrangement and extra family burden; (vii) household composition, including number and age of children and marital status; (viii) teen parenthood; (ix) public housing residence and residential mobility; and (x) demographic features including age and race/ethnicity.

11.2   RMPW rationale

Causal mediation analysis in the current application faces two major methodological challenges. The first is that the treatment might have changed not only the mediator value but also the mediator–outcome relationship. When this is the case, the treatment-by-mediator interaction becomes an important component of the causal mediation mechanism. The second is that although the random treatment assignment enables one to estimate without bias the treatment effect on the mediator and the treatment effect on the outcome, such a design does not generate an unbiased estimate of the mediator effect on the outcome under each treatment condition. For example, participants with a higher level of human capital, a longer employment history, a shorter duration of past dependence on welfare, and more social support and fewer barriers to work are expected to have a higher employment rate on average; the same participants are also expected to have a lower level of depressive symptoms on average under each treatment condition.

The RMPW strategy addresses both challenges. This section first explains how RMPW can be employed to decompose the total effect in a hypothetical sequentially randomized experiment. Subsequently, RMPW is extended to a hypothetical sequentially randomized block design where levels of observed pretreatment covariates define blocks. Finally, identification assumptions are derived for applying the RMPW strategy to studies in which only the treatment is randomized.

11.2.1   RMPW in a sequentially randomized design

As discussed in Chapter 9, a sequentially randomized design enables one to estimate without bias the treatment effect on the mediator and the mediator effect on the outcome under each treatment condition. Such an experiment would, in the first step, assign welfare applicants at random to either the LFA program or the control condition. In the second step, applicants within each treatment group would be assigned at random to employment. Suppose that according to earlier research, the employment rate was 40% in the control group and 70% in the LFA group, the second step randomization would assign the control units to employment at random with a probability of 0.4 and would assign the LFA units to employment with a probability of 0.7. The mean observed outcome obtained from each of the four treatment-by-employment combinations would provide an unbiased estimate of the corresponding population average potential outcome:

$E[Y(1,1)]$ if the entire population was assigned to LFA and employed

$E[Y(1,0)]$ if the entire population was assigned to LFA and unemployed

$E[Y(0,1)]$ if the entire population was assigned to the control condition and employed

$E[Y(0,0)]$ if the entire population was assigned to the control condition and unemployed

One would thereby obtain an unbiased estimate of the employment effect on depression under LFA $E[Y(1,1)-Y(1,0)]$ and that under the control condition $E[Y(0,1)-Y(0,0)]$ and test whether the employment effect on depression depends on the treatment condition. However, such an analysis does not immediately allow one to decompose the total effect into a natural direct effect and a natural indirect effect without further assumptions such as the assumption of no treatment-by-mediator interaction.

To decompose the total treatment effect, it is essential to obtain a consistent estimate of each of the following four potential outcomes:

(1) The population average potential outcome should all the units be assigned to the control condition $E[Y(0, M(0))]$

(2) The population average potential outcome should all the units be assigned to LFA $E[Y(1, M(1))]$

(3) The population average potential outcome should all the units be assigned to LFA, yet the mediator values would counterfactually remain the same as that under the control condition $E[Y(1, M(0))]$

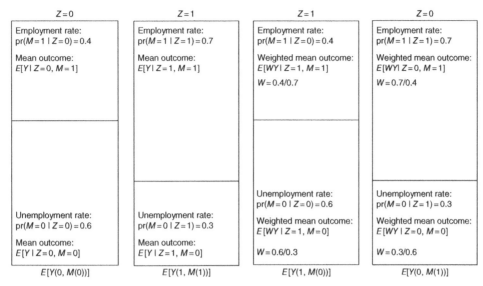

Figure 11.2   Average potential outcomes under a hypothetical sequentially randomized experiment

(4) The population average potential outcome should all the units be assigned to the control condition, yet the mediator would counterfactually take the same values as those under LFA $E[Y(0, M(1))]$

$(2) - (1)$ gives the population average total effect of the treatment on the outcome; $(3) - (1)$ gives the population average natural direct effect of the treatment on the outcome; $(2) - (3)$ gives the population average natural indirect effect; and $(4) - (1)$ gives the population average pure indirect effect. Finally, $[(2) - (3)] - [(4) - (1)]$ gives the population average natural treatment-by-mediator interaction effect.

From a sequentially randomized design, the observable mean outcome of the control group and that of the LFA group are unbiased for (1) and (2), respectively. These are shown in the first two vertical panels of Figure 11.2. Each vertical panel is divided into the upper section and the lower section representing the relative proportions of individuals employed and unemployed, respectively. The third panel shows that through applying RMPW to the LFA group, one obtains a weighted observable outcome that is unbiased for (3). In the fourth panel, applying RMPW to the control group generates a weighted observable outcome that is unbiased for (4). In the following, we show that the first two population average potential outcomes can be computed each as an average over its corresponding mediator distribution without weighting. The next two population average counterfactual outcomes can be computed each as a weighted average over the mediator distribution.

11.2.1.1 $E[Y(0, M(0))]$

The average potential outcome associated with the control condition $E[Y(0, M(0))]$ is the average of the potential outcome of employment under the control condition $E[Y(0,1)]$ and that of unemployment under the control condition $E[Y(0,0)]$ proportionally weighted by the employment rate and the unemployment rate, respectively, under the control condition:

$$E[Y(0,M(0))] = E[Y(0,1)] \times \mathrm{pr}(M(0)=1) + E[Y(0,0)] \times \mathrm{pr}(M(0)=0).$$

Here, $\mathrm{pr}(M(0)=1)$ is the employment rate and $\mathrm{pr}(M(0)=0)$ the unemployment rate if the entire population would be assigned to the control condition. Suppose that in a hypothetical sequentially randomized experiment that involves the entire population, the probability of employment is 0.4 and that of unemployment is 0.6 under the control condition. One would expect to observe that 40% of the control units are employed and 60% unemployed, that is, $\mathrm{pr}(M=1|Z=0)=0.4$ and $\mathrm{pr}(M=0|Z=0)=0.6$. Even if the probabilities of employment assignment are unknown to the data analyst, because of the sequentially randomized design, the former is unbiased for $\mathrm{pr}(M(0)=1)$, while the latter is unbiased for $\mathrm{pr}(M(0)=0)$. Additionally, due to sequential randomization, the mean observable outcome of the employed control units $E[Y|Z=0, M=1]$ is unbiased for $E[Y(0,1)]$; similarly, the mean observable outcome of the unemployed control units $E[Y|Z=0, M=0]$ is unbiased for $E[Y(0,0)]$. Hence, as shown in the first panel in Figure 11.2, $E[Y(0, M(0))]$ is equal to

$$E[Y|Z=0,M=1] \times \mathrm{pr}(M=1\,|Z=0) + E[Y|Z=0,M=0] \times \mathrm{pr}(M=0\,|Z=0).$$

11.2.1.2 $E[Y(1, M(1))]$

Similarly, the average potential outcome associated with LFA $E[Y(1, M(1))]$ is the average of the potential outcome of employment under LFA $E[Y(1,1)]$ and that of unemployment under LFA $E[Y(1,0)]$ proportionally weighted by the employment rate and the unemployment rate, respectively, under LFA:

$$E[Y(1,M(1))] = E[Y(1,1)] \times \mathrm{pr}(M(1)=1) + E[Y(1,0)] \times \mathrm{pr}(M(1)=0).$$

Suppose that the probability of employment is 0.7 and that of unemployment is 0.3 under LFA. As shown in the second panel of Figure 11.2, one would observe that 70% of the LFA units are employed and 30% unemployed, that is, $\mathrm{pr}(M=1|Z=1)=0.7$ and $\mathrm{pr}(M=0|Z=1)=0.3$. The mean observable outcome of the employed LFA units $E[Y|Z=1, M=1]$ is unbiased for $E[Y(1,1)]$; while the mean observable outcome of the unemployed LFA units $E[Y|Z=1, M=0]$ is unbiased for $E[Y(1,0)]$. Hence, $E[Y(1, M(1))]$ is equal to

$$E[Y|Z=1,M=1] \times \mathrm{pr}(M=1|Z=1) + E[Y|Z=1,M=0] \times \mathrm{pr}(M=0\,|Z=1).$$

11.2.1.3 $E[Y(1, M(0))]$

If the entire population would be assigned to LFA yet each individual's employment status would counterfactually remain the same as that under the control condition, the employment rate would be 0.4 rather than 0.7, while the unemployment rate would be 0.6 rather than 0.3 as shown in the third panel of Figure 11.2. Hence, the average potential outcome $E[Y(1, M(0))]$ is the average of the potential outcome of employment under LFA and that of unemployment under LFA proportionally weighted by the employment rate and the unemployment rate, respectively, under the control condition:

$$E[Y(1,M(0))] = E[Y(1,1)] \times \mathrm{pr}(M(0)=1) + E[Y(1,0)] \times \mathrm{pr}(M(0)=0).$$

We may simply transform the employment rate in the LFA group to resemble that in the control group by down-weighting the employed LFA units and up-weighting the unemployed LFA units. The transformation can be done through weighting because the above is equal to

$$E\left[\frac{\mathrm{pr}(M(0)=1)}{\mathrm{pr}(M(1)=1)} \times Y(1,1)\right] \times \mathrm{pr}(M(1)=1) + E\left[\frac{\mathrm{pr}(M(0)=0)}{\mathrm{pr}(M(1)=0)} \times Y(1,0)\right] \times \mathrm{pr}(M(1)=0).$$

Here, the potential outcome of employment under LFA $Y(1,1)$ is weighted by the ratio of the probability of employment under the control condition to that under LFA:

$$\frac{\mathrm{pr}(M(0)=1)}{\mathrm{pr}(M(1)=1)}.$$

In parallel, the potential outcome of unemployment under LFA $Y(1,0)$ is weighted by the ratio of the probability of unemployment under the control condition to that under LFA:

$$\frac{\mathrm{pr}(M(0)=0)}{\mathrm{pr}(M(1)=0)}.$$

In a sequentially randomized design, $\mathrm{pr}(M=1|Z=0)/\mathrm{pr}(M=1|Z=1) = 0.4/0.7 = 4/7$ is the RMPW for the employed LFA units, while $\mathrm{pr}(M=0|Z=0)/\mathrm{pr}(M=0|Z=1) = 0.6/0.3 = 2$ is the RMPW for the unemployed LFA units.

11.2.1.4 $E[Y(0, M(1))]$

If, on the contrary, the entire population would be assigned to the control condition yet each individual's employment status would counterfactually be the same as that under LFA, the employment rate would be 0.7 rather than 0.4, while the unemployment rate would be 0.3 rather than 0.6 as shown in the fourth panel of Figure 11.2. The average potential outcome $E[Y(0, M(1))]$ is the average of the potential outcome of employment and that of unemployment under the control condition proportionally weighted by the employment rate and the unemployment rate, respectively, under LFA:

$$E[Y(0,M(1))] = E[Y(0,1)] \times \mathrm{pr}(M(1)=1) + E[Y(0,0)] \times \mathrm{pr}(M(1)=0).$$

A transformation of the employment rate in the control group to resemble that in the LFA group can be done through up-weighting the employed control units and down-weighting the unemployed control units. This is because the above is equal to

$$E\left[\frac{\mathrm{pr}(M(1)=1)}{\mathrm{pr}(M(0)=1)} \times Y(0,1)\right] \times \mathrm{pr}(M(0)=1) + E\left[\frac{\mathrm{pr}(M(1)=0)}{\mathrm{pr}(M(0)=0)} \times Y(0,0)\right] \times \mathrm{pr}(M(0)=0).$$

In this sequentially randomized design, $\mathrm{pr}(M=1|Z=1) \,/\, \mathrm{pr}(M=1|Z=0) = 0.7/0.4 = 7/4$ is the RMPW for the employed control units, while $\mathrm{pr}(M=0|Z=1)/\mathrm{pr}(M=0|Z=0) = 0.3/0.6 = 1/2$ is the RMPW for the unemployed control units.

11.2.1.5 Nonparametric outcome model

The estimation of the average natural direct effect $E[Y(1, M(0))-Y(0, M(0))]$ simply involves a mean contrast between the weighted LFA group and the unweighted control group, while the estimation of the average natural indirect effect $E[Y(1, M(1))-Y(1, M(0))]$ involves a mean contrast between the unweighted LFA group and the weighted LFA group. Therefore, the outcome model can be viewed as nonparametric. To implement, we may combine the control group and the LFA group with a duplicate set of the LFA group. Let $D1$ be a dummy indicator that takes value 1 for the duplicate LFA units and 0 otherwise. We assign the weight as follows:

For the control units (i.e., $Z=0$, $D1=0$), the weight is 1.0.

For the employed LFA units (i.e., $Z=1$, $D1=0$, $M=1$), the weight is $\mathrm{pr}(M=1|Z=0)$ $/\mathrm{pr}(M=1|Z=1)$.

For the unemployed LFA units (i.e., $Z=1$, $D1=0$, $M=0$), the weight is $\mathrm{pr}(M=0|Z=0)/$ $\mathrm{pr}(M=0|Z=1)$.

For the duplicate LFA units (i.e., $Z=1$, $D1=1$), the weight is 1.0.

We then regress Y on the treatment indicator Z and the indicator for LFA duplicate $D1$ in a weighted model:

$$Y=\gamma^{(0)}+\gamma^{(DE.0)}Z+\gamma^{(IE.1)}D1+e. \tag{11.1}$$

Here, $\gamma^{(0)}$, the mean observed outcome of the control group, is unbiased for $E[Y(0, M(0))]$; $\gamma^{(0)}+\gamma^{(DE.0)}$, the weighted mean observed outcome of the LFA group, is unbiased for $E[Y(1, M(0))]$; $\gamma^{(0)}+\gamma^{(DE)}+\gamma^{(IE.1)}$, the mean observed outcome of the LFA group, is unbiased for $E[Y(1, M(1))]$.

Hence, $\gamma^{(DE.0)}$ is unbiased for the population average natural direct effect; and $\gamma^{(IE.1)}$ is unbiased for the population average natural indirect effect. To account for the duplication of every LFA unit, one may identify individual units as clusters and obtain cluster-robust standard errors when the weight is known rather than estimated. In general, in computing the asymptotic standard errors for $\hat{\gamma}^{(DE.0)}$ and $\hat{\gamma}^{(IE.1)}$, the analyst needs to consider the statistical uncertainty in the two-step estimation—i.e., estimating the conditional probabilities to construct the weight followed by estimating the causal effects (Bein et al, 2015; Cameron & Trivedi, 2005; Newey, 1984; Wooldridge, 2012). This is implemented in the stand-alone free RMPW software program and can also be carried out with the generalized method of moments (GMM) procedure in Stata. Others have alternatively computed the standard error and obtained the confidence interval for each causal parameter through bootstrapping (e.g., Huber, 2014). This involves constructing the sampling distribution of a sample estimate through random resampling from the observed data with replacement (Efron, 1975, 1988).

In the current application, the research interest also lies in estimating the pure indirect effect and the natural treatment-by-mediator interaction effect. We may additionally create a duplicate set of the control group. Let $D0$ be a dummy indicator that takes value 1 for the duplicate control units and 0 otherwise. We assign additional weight as follows:

For the duplicate set of the employed control units (i.e., $Z=0$, $D0=1$, $M=1$), the weight is $\mathrm{pr}(M=1|Z=1)/\mathrm{pr}(M=1|Z=0)$.

For the duplicate set of the unemployed control units (i.e., $Z=0$, $D0=1$, $M=0$), the weight is $\text{pr}(M=0|Z=1)/\text{pr}(M=0|Z=0)$.

The weighting scheme is summarized in Table 11.1. A mean contrast between the weighted control group and the unweighted control group estimates the average pure indirect effect. Finally, the difference between the estimated average natural indirect effect and the estimated pure indirect effect is an estimate of the average natural treatment-by-mediator interaction effect. One may conduct a weighted analysis regressing the outcome Y on the treatment indicator Z, the indicator for LFA duplicates $D1$, and the indicator for control duplicates $D0$:

$$Y = \gamma^{(0)} + \gamma^{(DE.0)}Z + \gamma^{(IE.1)}D1 + \gamma^{(IE.0)}D0 + e. \tag{11.2}$$

Here, $\gamma^{(IE.0)}$ is unbiased for the population average pure indirect effect, and hence, $\gamma^{(IE.1)} - \gamma^{(IE.0)}$ is unbiased for the population average natural treatment-by-mediator interaction effect. Its standard error is $\text{Var}\left(\hat{\gamma}^{(IE.1)}\right) + \text{Var}\left(\hat{\gamma}^{(IE.0)}\right) - 2\text{Cov}\left(\hat{\gamma}^{(IE.1)}, \hat{\gamma}^{(IE.0)}\right)$. Appendix 11.A shows the computation of the cluster-robust standard errors when the weight is known rather than estimated.

11.2.2 RMPW in a sequentially randomized block design

In a hypothetical sequentially randomized block design, individuals with homogeneous pretreatment characteristics $\mathbf{X} = \mathbf{x}$ constitute blocks. Those in the same block are randomized first to LFA or the control condition and subsequently to employment or unemployment. While the probability of treatment assignment may be constant for all units, the probability of employment assignment under each treatment may vary across blocks according to the participants' likelihood of being employed, reflecting the current knowledge of how individual background predicts employment prospects in a particular job market. For example, a block of participants with some post-secondary education and a solid record of employment in the past may be assigned at random to employment with a probability of 0.85 if under LFA and 0.75 if under the control condition; in contrast, another block of participants with no high school diploma and a long history of welfare dependence may be assigned at random to employment with a probability of 0.5 if under LFA and 0.2 if under the control condition.

To investigate the mediation mechanism that may transmit the treatment effect on maternal depression through changing one's employment experience, the analyst may apply RMPW to data within each block and then summarize the results over all blocks. In a block defined by $\mathbf{X} = \mathbf{x}$, the block-specific probability of employment assignment is predetermined by the experimenter to be $\text{pr}(M(1) = 1|\mathbf{X} = \mathbf{x})$ under LFA and $\text{pr}(M(0) = 1|\mathbf{X} = \mathbf{x})$ under the control condition. Even if these probabilities of employment assignment are unknown to the data analyst, because of the sequential randomization within each block, the proportion of LFA units who are employed in this block $\text{pr}(M = 1|Z = 1, \mathbf{X} = \mathbf{x})$ is unbiased for $\text{pr}(M(1) = 1|\mathbf{X} = \mathbf{x})$; similarly, the proportion of control units who are employed in this block $\text{pr}(M = 1| Z = 0, \mathbf{X} = \mathbf{x})$ is unbiased for $\text{pr}(M(0) = 1|\mathbf{X} = \mathbf{x})$. For estimating the average counterfactual outcome in the given block $E[Y(1, M(0))|\mathbf{X} = \mathbf{x}]$, the observed outcomes of the employed LFA units would be weighted by

$$\frac{\text{pr}(M = 1|Z = 0, \mathbf{X} = \mathbf{x})}{\text{pr}(M = 1|Z = 1, \mathbf{X} = \mathbf{x})},$$

while those of the unemployed LFA units would be weighted by

Table 11.1 RMPW applied to data from a sequentially randomized design

	E[Y(0, M(0))]	E[Y(1, M(1))]	E[Y(1, M(0))]	E[Y(0, M(1))]
Z	0	1	1	0
D1	0	1	0	0
D0	0	0	0	1
M	0 or 1	0 or 1	0	0
			1	1
W	1.0	1.0	$\dfrac{\mathrm{pr}(M=0\mid Z=0)}{\mathrm{pr}(M=0\mid Z=1)}$	$\dfrac{\mathrm{pr}(M=0\mid Z=1)}{\mathrm{pr}(M=0\mid Z=0)}$
			$\dfrac{\mathrm{pr}(M=1\mid Z=0)}{\mathrm{pr}(M=1\mid Z=1)}$	$\dfrac{\mathrm{pr}(M=1\mid Z=1)}{\mathrm{pr}(M=1\mid Z=0)}$

Table 11.2 RMPW applied to data from a sequentially randomized block design

	E[Y(0, M(0))]	E[Y(1, M(1))]	E[Y(1, M(0))]	E[Y(0, M(1))]
Z	0	1	1	0
D1	0	1	0	0
D0	0	0	0	1
M	0 or 1	0 or 1	0	0
			1	1
W	1.0	1.0	$\dfrac{\mathrm{pr}(M=0\mid Z=0,\mathbf{X}=\mathbf{x})}{\mathrm{pr}(M=0\mid Z=1,\mathbf{X}=\mathbf{x})}$	$\dfrac{\mathrm{pr}(M=0\mid Z=1,\mathbf{X}=\mathbf{x})}{\mathrm{pr}(M=0\mid Z=0,\mathbf{X}=\mathbf{x})}$
			$\dfrac{\mathrm{pr}(M=1\mid Z=0,\mathbf{X}=\mathbf{x})}{\mathrm{pr}(M=1\mid Z=1,\mathbf{X}=\mathbf{x})}$	$\dfrac{\mathrm{pr}(M=1\mid Z=1,\mathbf{X}=\mathbf{x})}{\mathrm{pr}(M=1\mid Z=0,\mathbf{X}=\mathbf{x})}$

$$\frac{\text{pr}(M=0|Z=0,\mathbf{X}=\mathbf{x})}{\text{pr}(M=0|Z=1,\mathbf{X}=\mathbf{x})}.$$

For estimating the average counterfactual outcome $E[Y(0, M(1))|\mathbf{X}=\mathbf{x}]$, the observed outcomes of the employed control units would be weighted by

$$\frac{\text{pr}(M=1|Z=1,\mathbf{X}=\mathbf{x})}{\text{pr}(M=1|Z=0,\mathbf{X}=\mathbf{x})},$$

while those of the unemployed control units would be weighted by

$$\frac{\text{pr}(M=0|Z=1,\mathbf{X}=\mathbf{x})}{\text{pr}(M=0|Z=0,\mathbf{X}=\mathbf{x})}.$$

Subsequently, following the same procedure as that delineated in Section 11.2.1.5, a weighted analysis of model (2) generates estimates of the population average natural direct effect, natural indirect effect, pure indirect effect, and natural treatment-by-mediator interaction effect. See Table 11.2 for a summary of the weighting scheme.

11.2.3   RMPW in a standard randomized experiment

Welfare agencies in the United States are not generally in a position to offer jobs to applicants or assign them at random to employment. While it might be conceivable to randomize employment in "New Deal"-type public jobs programs, the sequential experimental designs with or without blocking were impractical in the context of the welfare-to-work programs in the 1990s. The NEWWS data are representative of many applications in which only the treatment is randomized. Within each treatment group, some individuals might have a higher likelihood of employment than others due to their prior education and training, personal predispositions, past employment experience, and family situations. Failure to account for such individual variation in employment selection would inevitably lead to bias in mediation analyses.

Suppose that an individual's probability of employment under a given treatment is a function of the observed pretreatment characteristics $\mathbf{x}$. We may envision that the data approximate a sequentially randomized block design in which individuals with homogeneous pretreatment characteristics $\mathbf{X}=\mathbf{x}$ constitute blocks. Among those who have been assigned at random to the same treatment, subsequent hypothetical randomization to employment within each block could be a result of unpredictable events that occur at random in nature.

Imagine that Mary and Cathy are identical twins whose past life history and current family situation have been all identical; both applied for welfare and both were assigned at random to LFA; both were eager to find a job and applied for the same position—the only opening available in a stagnant local economy. Unfortunately, Mary missed her job interview because her 3-year-old son happened to be sick and needed urgent care on that particular day. The fact that Cathy became employed while Mary remained unemployed can be viewed as a result of "pure luck." Therefore, arguably Cathy's depressive symptoms at the 2-year follow-up may provide the counterfactual outcome for Mary should Mary have been employed instead under LFA. Viewing one's depressive symptoms in the past week as a random variable the value of which could be swayed by a variety of random factors, formally, we may represent the aforementioned reasoning as $E\left[Y_{\text{Mary}}\left(1,M_{\text{Mary}}(1)\right)\right]=E\left[Y_{\text{Cathy}}\left(1,M_{\text{Cathy}}(1)\right)\right]=E\left[Y_{\text{Cathy}}|Z_{\text{Cathy}}=1,\right.$

$M_{Cathy}(1) = 1$]. Or in words, Mary's counterfactual outcome if employed under LFA is expected to be equal to Cathy's potential outcome if employed under LFA; and the latter is expected to be equal to Cathy's observed outcome given that she was actually employed under LFA.

We may also imagine that while Mary was assigned at random to LFA, her twin sister Cathy might have been assigned to the control condition. In that case, Cathy's employment experience under the control condition may serve as the counterfactual mediator for Mary if Mary had instead been assigned to the control condition. This reasoning is formally represented as $E[M_{Mary}(0)] = E[M_{Cathy}(0)] = E[M_{Cathy}|Z_{Cathy} = 0]$. In words, Mary's counterfactual mediator if assigned to the control condition is expected to be equal to Cathy's potential mediator under the control condition; and the latter is expected to be equal to Cathy's observed mediator given that she was actually assigned to the control condition.

This discussion should make clear that when only the treatment has been randomized by the experimenter, the success of a causal mediation analysis depends on whether the data analyst is able to empirically identify homogeneous blocks of individuals. Those in the same block, given their pretreatment background, are equally likely to be employed under a given treatment condition and are equally likely to be depressed should they share the same employment experience under the same treatment. One may apply RMPW to the data that can be viewed as approximating a sequentially randomized block design. In the following, we formally state the identification assumptions. Sections 11.3 and 11.4 give details of the parametric and nonparametric RMPW procedure when only the treatment is randomized.

11.2.4   Identification assumptions

Similar to most existing methods for causal mediation analysis, the RMPW approach identifies the causal effects of interest if the data resemble what one would obtain from a sequentially randomized block design satisfying the "sequential ignorability" assumption (Imai, Keele, and Yamamoto, 2010; Imai, Keele, and Tingley, 2010). That is, the treatment assignment and the mediator value assignment under each treatment can be viewed as randomized within levels of the observed pretreatment covariates. In the causal inference literature, other authors (e.g., Pearl, 2001; Robins, 2003; VanderWeele, 2009) have spelled out the components of the "sequential ignorability" including assumptions (I) and (II) of *ignorable treatment assignment* and assumptions (III.a) and (III.c) of *ignorable mediator value assignment under each treatment condition*. The numbering of these assumptions here is consistent with the numbering system used in Chapter 10.

In the NEWWS Riverside data, because of the randomized treatment assignment, the assumptions of ignorable treatment assignment are satisfied. Hence, the mean observed outcome of the control units provides an unbiased estimate of $E[Y(0, M(0))]$, while the mean observed outcome of the LFA units provides an unbiased estimate of $E[Y(1, M(1))]$. In a nonexperimental study, these two assumptions may hold within levels of the observed pretreatment characteristics.

The assumptions of ignorable mediator value assignment under each treatment are applied within levels of the observed pretreatment covariates. Under these assumptions, every unit has a nonzero probability of being assigned to each mediator value under each treatment condition. In the NEWWS context, given that the job market was at least partially governed by uncertainty, many of those who were unemployed under the LFA condition may have had a nonzero probability of being employed; similarly, many of those who were employed under the control condition may have had a nonzero probability of becoming unemployed. Also under these assumptions, mediator value assignment under a given treatment condition is independent of the potential outcomes associated with the same treatment condition and those associated with an alternative treatment condition. In other words, it is assumed that the observed pretreatment covariates adequately account for all the potential confounding of the mediator–outcome relationships within and across the treatment conditions. These assumptions would be violated if the mediator–outcome relationship is confounded by an unobserved pretreatment covariate or by an observed or unobserved posttreatment covariate.

Under the sequential ignorability, the RMPW-adjusted mean observed outcome of the LFA units in the NEWWS data provides an unbiased estimate of $E[Y(1, M(0))]$, while the RMPW-adjusted mean observed outcome of the control units provides an unbiased estimate of $E[Y(0, M(1))]$. Most importantly, unlike many of the existing methods, the RMPW strategy does not require the no treatment-by-mediator interaction assumption. In other words, it allows employment to affect depression differently under LFA than under the control condition.

Appendix 11.B provides the theorems and a formal proof of the consistency of RMPW estimation under the sequential ignorability. The theoretical results apply to multivalued ediators as well as to quasiexperimental data in which neither the treatment nor the mediator under each treatment is randomized. These extensions will be discussed in Chapter 12.

11.3   Parametric RMPW procedure

Applying the parametric approach, one computes RMPW as a ratio of the estimated conditional probability of mediator value assignment under the control condition to that under LFA. The former is the estimated propensity score for employment under the control condition; and the latter is the estimated propensity score for employment under LFA. The analytic procedure involves seven steps.

Step 1: Select and prepare pretreatment covariates. According to Hong, Deutsch, and Hill (2011, in press), in the NEWWS Riverside data, 86 pretreatment covariates were considered to be theoretically associated with maternal depression or with employment. After creating a missing category for each categorical covariate with missing information, they generated five impute data sets in which missing data in the outcome and in the continuous covariates were all imputed (Little and Rubin, 2002). Steps 2 through 7 were then applied to each imputed data set one at a time. At the end, the estimated causal effects were combined over the five imputed data sets. For simplicity, here, we discuss the analytic procedure with one imputed data set. If the treatment assignment has not been randomized or has suffered from nonrandom attrition, IPTW can be employed to equate the pretreatment composition of the experimental group and the control group. Chapter 4 described this procedure in Section 4.2.4.

Step 2: Specify the propensity score model for the mediator under each treatment condition. After stepwise selection of the outcome predictors or taking alternative procedures for variable selection, one may fit a logistic regression model to each treatment group. Analyzing data from the LFA group, one may predict an LFA unit's propensity of employment under LFA, denoted by

$$\theta_{M(1)} = \theta_{M(1)}(\mathbf{x}) = \mathrm{pr}(M(1) = 1 | \mathbf{X} = \mathbf{x}) = \mathrm{pr}(M = 1 | Z = 1, \mathbf{X} = \mathbf{x}),$$

as a function of the unit's observed pretreatment characteristics $\mathbf{x}$. Similarly, analyzing data from the control group, one may predict a control unit's propensity of employment under the control condition, denoted by

$$\theta_{M(0)} = \theta_{M(0)}(\mathbf{x}) = \mathrm{pr}(M(0) = 1 | \mathbf{X} = \mathbf{x}) = \mathrm{pr}(M = 1 | Z = 0, \mathbf{X} = \mathbf{x}).$$

By virtue of the random treatment assignment, the propensity score model specified under the control condition would apply to the LFA units had they been counterfactually assigned to the control condition instead; the same is true vice versa. Hence, applying the coefficient estimates obtained from the second propensity score model, one may predict each LFA unit's $\theta_{M(0)}$, that is, the unit's propensity score for employment under the counterfactual control condition. Similarly, applying the coefficient estimates obtained from the LFA group, one may predict each control unit's counterfactual propensity score for employment under LFA $\theta_{M(1)}$. However, the problem of overfitting (Hawkins, 2004) is a potential concern when the sample size is small relative to the number of parameters in the prediction model. This is the well-known problem of tailoring model specification only to a particular sample, which may inflate the prediction error when the model is applied to cases not in the current sample. Alternatives to stepwise variable selection for avoiding model over-fitting include cross-validation, leave-one-out bootstrapping, Ridge regression, and LASSO (Hastie et al, 2009).

Step 3: Identify the common support for mediation analysis in each treatment group. Among those who display the same propensity score for employment under a given treatment condition, the employed units are expected to have their unemployed counterparts and vice versa. Units who do not have counterparts are excluded from the subsequent mediation analysis due to their lack of counterfactual information. To implement, one may compare the distribution of the logit of $\theta_{M(1)}$ and that of $\theta_{M(0)}$ across the employed LFA units, the unemployed LFA units, the employed control units, and the unemployed control units and identify cases in which the distribution of either propensity score does not overlap across all four groups. One may add 20% of a standard deviation of the logit of each propensity score at each end to expand the range of the common support (Austin, 2011). If the treatment has a particularly strong effect on the mediator—in the extreme case, if nearly all the LFA participants become employed and nearly all the control group members become unemployed—this step will reveal the lack of common support. In such a case, the RMPW method and other alternative methods that rely on the sequential ignorability assumptions will become unsuitable, whereas the instrumental variable (IV) method may offer a convenient solution.

Step 4: Check balance in covariate distribution across the treatment-by-mediator combinations. Even though the identification assumptions cannot be empirically verified, if a considerable portion of the observed pretreatment covariates remains predictive of the mediator after propensity score adjustment, we view this as evidence that the adjustment fails to approximate data from a sequentially randomized block design. One may assess the adequacy of propensity score adjustment by applying inverse-probability-of-treatment weighting (IPTW) (Robins, 1999), assigning the weight $\mathrm{pr}(M = 1 | Z = 1)/\theta_{M(1)}$ to the employed LFA units, $\mathrm{pr}(M = 0 | Z = 1)/(1 - \theta_{M(1)})$ to the unemployed LFA units, $\mathrm{pr}(M = 1 | Z = 0)/\theta_{M(0)}$ to the employed control units, and $\mathrm{pr}(M = 0 | Z = 0)/(1 - \theta_{M(0)})$ to the unemployed control units.

When the adjustment for observed confounding is adequate, approximately 95% of the time, a categorical covariate is expected to show equal proportion distribution, and a continuous covariate (including the logit of each propensity score) is expected to show equal mean and variance across these four groups according to the results of a weighted Chi-square test and a weighted F test, respectively. One may improve the balance through modifying the propensity score models.

Step 5: Estimate the mediator effect on the outcome under each treatment condition. Contrasting the estimated population average potential outcome under each treatment when all units are employed with that when all units are unemployed, this step produces useful evidence with regard to whether the mediator–outcome relationship differs by treatment. Applying IPTW to the data, one may simply regress the outcome on the binary treatment, the binary mediator, and their interaction:

$$Y = \beta_0 + \beta_1 Z + \beta_2 M + \beta_3 ZM + e. \tag{11.3}$$

Here, $\widehat{\beta}_2$ estimates the mediator effect on the outcome under the control condition $E[Y(0,1) - Y(0,0)]$, $\widehat{\beta}_2 + \widehat{\beta}_3$ estimates the mediator effect on the outcome under LFA $E[Y(1,1) - Y(1,0)]$, and hence, $\widehat{\beta}_3$ estimates the controlled treatment-by-mediator interaction effect, that is, $E\{[Y(1,1) - Y(1,0)] - [Y(0,1) - Y(0,0)]\}$.

Step 6: Create a duplicate and compute the parametric RMPW. The data within the common support identified in Step 3 are then reconstructed to include a duplicate for each control unit and one for each LFA unit. Let $D1$ be a dummy indicator that takes value 1 for an LFA duplicate and 0 otherwise; let $D0$ be a dummy indicator that takes value 1 for a control duplicate and 0 otherwise. Most importantly, to estimate $E[Y(1, M(0))]$, the weight for the employed LFA units ($Z = 1$, $M = 1$, $D1 = 0$, $D0 = 0$) is $\theta_{M(0)}/\theta_{M(1)}$ and that for the unemployed LFA units ($Z = 1$, $M = 0$, $D1 = 0$, $D0 = 0$) is $(1 - \theta_{M(0)})/(1 - \theta_{M(1)})$; to estimate $E[Y(0, M(1))]$, the weight for the employed control units ($Z = 0$, $M = 1$, $D1 = 0$, $D0 = 1$) is $\theta_{M(1)}/\theta_{M(0)}$ and that for the unemployed control units ($Z = 0$, $M = 0$, $D1 = 0$, $D0 = 1$) is $(1 - \theta_{M(1)})/(1 - \theta_{M(0)})$. The weight is 1.0 for the duplicate LFA units and the original control units. Table 11.3 summarizes the weighting scheme.

Step 7: Estimate the causal effects. Finally, conducting a weighted analysis of model (11.1), one obtains estimates of the natural direct effect and the natural indirect effect. Analyzing the weighted model (11.2), one additionally obtains estimates of the pure indirect effect

Table 11.3   Parametric RMPW applied to data from a standard randomized experiment

	$E[Y(0, M(0))]$	$E[Y(1, M(1))]$	$E[Y(1, M(0))]$		$E[Y(0, M(1))]$	
Z	0	1	1		0	
$D1$	0	1	0		0	
$D0$	0	0	0		1	
M	0 or 1	0 or 1	0	1	0	1
W	1.0	1.0	$\dfrac{1 - \theta_{M(0)}}{1 - \theta_{M(1)}}$	$\dfrac{\theta_{M(0)}}{\theta_{M(1)}}$	$\dfrac{1 - \theta_{M(1)}}{1 - \theta_{M(0)}}$	$\dfrac{\theta_{M(1)}}{\theta_{M(0)}}$

and the natural treatment-by-mediator interaction effect. One may improve precision by making additional covariance adjustment for strong predictors of the continuous outcome.

Analyzing the NEWWS data, Hong, Deutsch, and Hill (in press) reported the estimates of the treatment effect on the outcome and of the treatment effect on the mediator. The summary score of depressive symptoms at the 2-year follow-up ranged from 0 to 34 with a mean equal to 7.49 and a standard deviation equal to 7.74. An intention-to-treat (ITT) effect analysis shows that the average treatment effect on depressive symptoms cannot be statistically distinguished from zero (coefficient = 0.11, SE = 0.64, $t = 0.18$, $p = 0.86$). Another ITT effect analysis shows that assignment to LFA increased the employment rate from 39.5 to 65.4%.

According to the results from Step 5, the employment effect on depressive symptoms differed by treatment. Specifically, having all participants employed as opposed to having none employed would reduce depressive symptoms under LFA (coefficient = −2.49, SE = 1.20, $t = -2.07$, $p < 0.05$) but not under the control condition (coefficient = 0.74, SE = 0.76, $t = 0.97$, $p = 0.33$). The treatment-by-mediator interaction is statistically significant (coefficient = 3.23, SE = 1.42, $t = 2.27$, $p < 0.05$). According to these results, had all welfare mothers continued to be covered by the old policy, even a landscape change in employment would not have affected maternal depression by a significant amount. Once employment became one of the primary qualifications for welfare receipt, full-scale success in employment apparently would lead to a reduction in depressive symptoms. However, the analysis in Step 5 does not decompose the total effect to reveal the mediation mechanism.

The employment rate in the RMPW-adjusted LFA group (38.1%) approximates that in the control group (39.5%), while that in the RMPW-adjusted control group (66.9%) approximates that in the LFA group (65.4%). Step 7 then decomposes the total treatment effect. The estimated natural direct effect is 1.29 (SE = 0.87, $t = 1.48$, $p = 0.14$), about 17% of a standard deviation of the outcome; the estimated natural indirect effect is −0.87 (SE = 0.47, $t = -1.87$, $p = 0.06$). The natural direct effect estimate indicates that if the assignment to LFA rather than to the control condition had counterfactually generated no impact on employment (i.e., if the employment rate had remained at 39.5% rather than increasing to 65.4%), maternal depression would not have increased by a statistically significant amount, on average. According to the natural indirect effect estimate, if all individuals were hypothetically assigned to LFA, the LFA-induced change in employment (i.e., the increase in employment rate from 39.5 to 65.4%) almost reached the point of producing a statistically significant reduction in maternal depression on average. The natural indirect effect is then decomposed into a "pure indirect effect" and a "natural treatment-by-mediator interaction effect." The results indicate that if all individuals were hypothetically assigned to the control condition instead, the same amount of change in employment as reported earlier would not have a statistically significant impact on the average level of depression (coefficient = 0.32, SE = 0.27, $t = 1.48$, $p = 0.14$). The estimated natural treatment-by-mediator interaction effect is −1.19 (SE = 0.53, $t = -2.26$, $p < 0.05$), providing evidence that the LFA-induced increase in employment reduced depression under the LFA condition in a way that did not happen under the control condition.

11.4   Nonparametric RMPW procedure

In general, nonparametric analyses are more robust than their parametric counterparts because the former are less reliant on model-based assumptions. For example, past research has shown that in evaluating the relative effectiveness of different treatments, parametric IPTW often generates biased results especially when the propensity score models are misspecified in their functional forms (Hong, 2010a; Schafer and Kang, 2008). In contrast, nonparametric weighting methods such as marginal mean weighting through stratification (MMWS) produce robust results despite the misspecification of the propensity score models (Hong, 2010a, 2012). IPTW and MMWS, however, are not suitable for decomposing the total effect in the presence of treatment-by-mediator interaction. This section introduces a nonparametric RMPW procedure for mediation analysis. Its performance is compared with that of the parametric RMPW procedure through simulations in the next section.

In essence, the nonparametric RMPW procedure recomputes the conditional probability of employment under each treatment condition on the basis of propensity score stratification. It differs from the parametric RMPW procedure only in Steps 4, 5, and 6.

Step 4: Check balance in covariate distribution across the treatment-by-mediator combinations. Instead of checking the balance in the data adjusted by IPTW, one may adapt the nonparametric MMWS procedure to adjust for employment selection associated with the observed pretreatment covariates. For example, one may first rank the sampled units by the logit of $\theta_{M(1)}$ and divide the sample into three even portions. Within each of these three subclasses, one then ranks and subdivides the units again by the logit of $\theta_{M(0)}$. Let $s = 1, \ldots, 9$ denote the nine strata. With a relatively large sample size, one may increase the number of strata to 4×3 or 4×4. Within stratum s, one then computes MMWS as follows:

$\text{MMWS} = \text{pr}(M = 1|Z = 1)/\text{pr}(M = 1|Z = 1, S = s)$ for the employed LFA units.

$\text{MMWS} = \text{pr}(M = 0|Z = 1)/\text{pr}(M = 0|Z = 1, S = s)$ for the unemployed LFA units.

$\text{MMWS} = \text{pr}(M = 1|Z = 0)/\text{pr}(M = 1|Z = 0, S = s)$ for the employed control units.

$\text{MMWS} = \text{pr}(M = 0|Z = 0)/\text{pr}(M = 0|Z = 0, S = s)$ for the unemployed control units.

Here, $\text{pr}(M = 1|Z = 1, S = s)$ is the proportion of LFA units in stratum s who were employed; $\text{pr}(M = 0|Z = 1, S = s)$ is the proportion of LFA units in stratum s who were unemployed; $\text{pr}(M = 1|Z = 0, S = s)$ and $\text{pr}(M = 0|Z = 0, S = s)$ represent, in stratum s, the respective proportions of control units who were employed and unemployed. These four conditional probabilities are nonparametric estimates of the corresponding propensity scores. One then examines covariate balance across the four treatment-by-mediator categories in the MMWS-adjusted sample.

Step 5: Estimate the mediator effect on the outcome under each treatment condition. Applying MMWS to the data, one may analyze the weighted regression model specified in (11.3) and test whether the mediator–outcome relationship depends on the treatment condition.

Step 6: Create a duplicate and compute the nonparametric RMPW. Under the same stratification as described in Step 4, RMPW can now be computed nonparametrically. To estimate $E[Y(1, M(0))]$, the nonparametric weight for the employed LFA units is

$$\text{RMPW} = \frac{\text{pr}(M = 1|Z = 0, S = s)}{\text{pr}(M = 1|Z = 1, S = s)}.$$

The weight for the unemployed LFA units is

$$\text{RMPW} = \frac{\text{pr}(M=0|Z=0,S=s)}{\text{pr}(M=0|Z=1,S=s)}.$$

To estimate $E[Y(0, M(1))]$, the nonparametric weight for the employed control units is

$$\text{RMPW} = \frac{\text{pr}(M=1|Z=1,S=s)}{\text{pr}(M=1|Z=0,S=s)}.$$

The weight for the unemployed control units is

$$\text{RMPW} = \frac{\text{pr}(M=0|Z=1,S=s)}{\text{pr}(M=0|Z=0,S=s)}.$$

Table 11.4 summarizes the weighting scheme.

The nonparametric RMPW is then applied to the outcome models specified in (11.1) and (11.2). Under a four-by-four stratification, the natural direct effect estimate is 1.34 (SE = 0.79, $t = 1.70$, $p = 0.09$), and the natural indirect effect estimate is −0.93 (SE = 0.38, $t = -2.43$, $p < 0.05$). The natural indirect effect is then decomposed into the pure indirect effect (coefficient = 0.45, SE = 0.30, $t = 1.50$, $p = 0.13$) and the natural treatment-by-mediator interaction effect (coefficient = −1.38, SE = 0.49, $t = -2.85$, $p < 0.01$). Table 11.5 compares the parametric and nonparametric results side by side. The table shows that the point estimates obtained under the four-by-four stratification are converging to the parametric weighting results. Yet the estimation with nonparametric weighting appears to be relatively more efficient. Hence, it becomes possible to detect a statistically significant negative natural indirect effect of the treatment. According to these results, the LFA-induced increase in employment rate would reduce maternal depression should the entire population be assigned to LFA. Additionally, there is clear evidence that the treatment changed the mediational process: the LFA-induced increase in employment would become beneficial to participants' mental health only under the LFA program while showing no benefit under the control condition.

11.5   Simulation results

Hong, Deutsch, and Hill (2011, in press) reported results from a series of Monte Carlo simulations assessing the performance of the nonparametric RMPW procedure relative to the parametric RMPW procedure in estimating the direct and indirect effects in the case of a binary randomized treatment, a binary mediator, and a continuous outcome. When the nonparametric RMPW procedure was applied, they compared 3×3 with 4×4 strata. Additionally, they assessed the robustness of the estimation results between the parametric and the nonparametric procedures when the propensity score models are misspecified in functional forms. Finally, they compared through simulations the RMPW results with the corresponding results from path analysis and the IV method.

11.5.1   Correctly specified propensity score models

When the propensity score models are correctly specified, the parametric RMPW and the nonparametric RMPW procedures both perform generally well. The parametric procedure

Table 11.4 Nonparametric RMPW applied to data from a standard randomized experiment

	$E[Y(0, M(0))]$	$E[Y(1, M(1))]$	$E[Y(1, M(0))]$	$E[Y(0, M(1))]$
Z	0	1	1	0
$D1$	0	1	0	0
$D0$	0	1	0	1
M	0 or 1	0 or 1	0 1	0 1
W	1.0	1.0	$\dfrac{\mathrm{pr}(M=1\mid Z=0,S=s)}{\mathrm{pr}(M=1\mid Z=1,S=s)}$ $\dfrac{\mathrm{pr}(M=0\mid Z=0,S=s)}{\mathrm{pr}(M=0\mid Z=1,S=s)}$	$\dfrac{\mathrm{pr}(M=0\mid Z=1,S=s)}{\mathrm{pr}(M=0\mid Z=0,S=s)}$ $\dfrac{\mathrm{pr}(M=1\mid Z=1,S=s)}{\mathrm{pr}(M=1\mid Z=0,S=s)}$

Table 11.5 Parametric and nonparametric RMPW estimation of the causal effects

	Parametric RMPW				4×4 Nonparametric RMPW			
	Coefficient	SE	t	p	Coefficient	SE	t	p
Natural direct effect ($\gamma^{(DE.0)}$)	1.29	0.87	1.48	0.14	1.34	0.79	1.70	0.09
Natural indirect effect ($\gamma^{(IE.1)}$)	−0.87	0.47	−1.87	0.06	−0.93	0.38	−2.43	<0.05
Pure indirect effect ($\gamma^{(IE.0)}$)	0.32	0.27	1.48	0.14	0.45	0.30	1.50	0.13
Natural treatment-by-mediator interaction effect ($\gamma^{(IE.1)} - \gamma^{(IE.0)}$)	−1.19	0.53	−2.26	<0.05	−1.38	0.49	−2.85	<0.01

removes nearly 100% of the bias; the nonparametric procedure with 3×3 strata removes 85% or more of the initial bias while that with 4×4 strata removes 90% or more of the bias when the sample size is relatively large ($N = 5000$). However, in a relatively small sample ($N = 800$), increasing strata does not necessarily remove a higher proportion of bias when the average value of the mediator under each treatment condition is shifting away from 0.5. With a relatively large sample size, the nonparametric estimates often show a higher efficiency and a smaller MSE when compared with the parametric estimates. However, with a relatively small sample size, an increase in the number of strata seems to result in a loss of efficiency. Finally, the cluster-robust standard error estimates are compared with the corresponding sampling standard deviations approximated on the basis of 1000 random samples. The average discrepancy is close to zero across the simulated cases and never exceeds 0.047 standard deviations of a potential outcome in any single case. However, further simulation results have shown that the cluster-robust standard error estimates underperform the bootstrapped standard error estimates and the asymptotic standard error estimates that account for the uncertainty in the estimated propensity scores (Bein et al, 2005).

11.5.2   Misspecified propensity score models

Another series of simulations compared the parametric and the nonparametric RMPW results when nonlinear, nonadditive propensity score models are misspecified as linear additive. The parametric RMPW procedure generates estimates that are increasingly biased as the degree of nonlinearity or nonadditivity increases regardless of sample size. In contrast, the nonparametric RMPW results remain robust in all cases when the same size is sufficiently large. There are also nonparametric approaches to propensity score estimation, including generalized boosted models, which reduce model misspecification errors (McCaffrey, Ridgeway, & Morral, 2004). Future research may investigate the application of these approaches to RMPW analysis.

11.5.3   Comparisons with path analysis and IV results

To evaluate the performance of RMPW relative to path analysis and the IV method, Hong, Deutsch, and Hill (2011, in press) analyzed naïve path analysis models with no statistical adjustment for any covariates and with no treatment-by-mediator interaction. The naïve path analysis results then served as the basis for comparing across different strategies for removing selection bias. Adjusted path analysis makes linear covariance adjustment for the observed pretreatment covariates yet excludes the treatment-by-mediator interaction. They concluded that, when there is no treatment-by-mediator interaction in the simulated data, the RMPW results replicate the results from path analysis. When there is a zero direct effect in the simulated data, the RMPW results replicate the IV results. As expected, RMPW outperforms these conventional methods especially when the assumption of no treatment-by-mediator interaction does not hold or when the exclusion restriction does not hold.

11.6   Discussion

11.6.1   Advantages of the RMPW strategy

The RMPW strategy shows its strengths in comparison with some other existing methods discussed in Chapter 10 because it relies on relatively fewer identification assumptions and model-based assumptions and because it can be implemented fairly easily with standard statistical software. Table 10.2 in Chapter 10 compared the identification assumptions across the existing methods. Similar to the alternative weighting methods and the resampling approach, the RMPW strategy invokes the sequential ignorability as its key identification assumptions while avoiding the additivity assumption, that is, the assumption that there is no treatment-by-mediator interaction. The additivity assumption, clearly implausible in the NEWWS application example and in many other applications, is required by the conventional path analysis/SEM approach and the marginal structural models. The modified regression approach, the conditional structural models, and the resampling approach accommodate treatment-by-mediator interactions by resorting to model-based assumptions with regard to how the treatment, the mediator, and the covariates interact in the outcome model. It is well-known that misspecifications of the outcome model would often bias causal effect estimation (Drake, 1993). In contrast, the RMPW strategy, similar to the alternative weighting methods (e.g., Huber, 2014), not only relaxes the no treatment-by-mediator interaction assumption but also greatly simplifies the specification of the outcome model. By directly estimating each causal parameter of interest, the RMPW strategy does not have to represent the natural indirect effect as a function of multiple parameters (such as representing the natural indirect effect as the product of the treatment effect on the mediator and the mediator effect on the outcome under each treatment condition). This is a major advantage because it then avoids the constant treatment effect assumption that is clearly implausible and also simplifies hypothesis testing.

The RMPW strategy has broad applications regardless of the distribution of the outcome, the distribution of the mediator, or the functional relationship between the outcome and the mediator. This method can be readily implemented with standard software such as Stata, SPSS, SAS, and R. Syntax files for implementing RMPW analysis with the standard software are available online. In addition, a stand-alone RMPW software, also available online, eases the programming burden on the user especially when there is a need to impute missing data and when the number of covariates is relatively large.

11.6.2   Limitations of the RMPW strategy

RMPW identifies the causal effects of interest under the untestable assumption of sequential ignorability. Even though the ignorability of treatment assignment can be warranted by treatment randomization, mediator value assignment is typically not randomized. Similar to most existing methods, RMPW removes selection bias associated with the observed pretreatment covariates. The result will be biased if, for those who share the same observed pretreatment characteristics, the mediator–outcome relationship is confounded by omitted pretreatment or posttreatment covariates. Posttreatment covariates can be viewed as potential mediators that are not independent of the focal mediator. In the NEWWS Riverside example, immediately after the randomization of treatment assignment, suppose that some participants' depressive symptoms would be heightened if assigned to LFA but not if assigned to the control condition. The postrandomization depressive symptoms at a heightened level under LFA would likely impede one's ability to secure employment and might also independently predict depression at the 2-year follow-up. In causal mediation analyses that allow for treatment-by-mediator interactions, the potential confounding effect of posttreatment covariates cannot be adjusted for directly (e.g., Imai *et al.*, 2011) but only indirectly through the adjustment for the related pretreatment covariates such as baseline depressive symptoms in the current example. Some attempts have been made to analyze posttreatment confounders as additional mediators (Huber, 2014; Imai and Yamamoto, 2013), a topic that we will leave to Chapter 13. The analyst may also employ sensitivity analysis to assess the consequence of a possible omission (Imai, Keele, and Tingley, 2010; Imai, Keele, and Yamamoto, 2010; VanderWeele, 2010). However, in the case that the observed pretreatment covariates have explained nearly all the systematic variation in the outcome, the remaining potential bias associated with the omitted pretreatment and posttreatment covariates may become negligible.

Appendix 11.A: Causal effect estimation through the RMPW procedure

Suppose that an analytic sample consists of n individuals, which includes n_1 LFA individuals and n_0 control individuals. As shown in Section 11.2.1, after applying RMPW to the LFA observations and merging these with the unweighted LFA observations and the unweighted control observations, the researcher analyzes a weighted outcome model specified in (11.2). Here, we represent the outcome model in matrix form for observation i of individual j:

$$Y_{ij} = \mathbf{A}_{ij}^{\mathrm{T}}\boldsymbol{\gamma} + e_{ij}$$

where $\mathbf{A}_{ij} = \begin{pmatrix} 1 \\ Z_j \\ D1_{ij} \\ D0_{ij} \end{pmatrix}$ and $\boldsymbol{\gamma} = \begin{pmatrix} \gamma^{(0)} \\ \gamma^{(DE.0)} \\ \gamma^{(IE.1)} \\ \gamma^{(IE.0)} \end{pmatrix}$. We use W_{ij} to denote the weight for observation i of individual j. One obtains estimates of the regression coefficients as follows:

$$\widehat{\boldsymbol{\gamma}} = \left(\sum_{j=1}^{n}\sum_{i=1}^{2} W_{ij}\mathbf{A}_{ij}\mathbf{A}_{ij}^{\mathrm{T}}\right)^{-1} \sum_{j=1}^{n}\sum_{i=1}^{2} W_{ij}\mathbf{A}_{ij}Y_{ij}.$$

When the weight is known, one may compute cluster-robust standard errors which are square roots of the diagonal elements in the following matrix:

$$\mathrm{Var}(\widehat{\boldsymbol{\gamma}}) = \left(\sum_{j=1}^{n}\sum_{i=1}^{2} W_{ij}\mathbf{A}_{ij}\mathbf{A}_{ij}^{\mathrm{T}}\right)^{-1} \sum_{j=1}^{n}\mathbf{U}_j\mathbf{U}_j^{\mathrm{T}} \left(\sum_{j=1}^{n}\sum_{i=1}^{2} W_{ij}\mathbf{A}_{ij}\mathbf{A}_{ij}^{\mathrm{T}}\right)^{-1},$$

where for observation i of individual j,

$$\mathbf{U}_j = \sum_{i=1}^{2} W_{ij}\left(Y_{ij} - \mathbf{A}_{ij}^{\mathrm{T}}\hat{\boldsymbol{\gamma}}\right)\mathbf{A}_{ij}.$$

Appendix 11.B: Proof of the consistency of RMPW estimation

This appendix presents the theoretical results clarifying the identification assumptions under which RMPW removes selection bias in estimating $E[Y(z, M(z))]$ and $E[Y(z, M(z'))]$. Following van der Laan and Petersen (2008), Hong (2010b) represented the joint distribution of the observed data $O = (\mathbf{X}, Z, M(z), Y(z, M(z)))$ in general as follows:

$$f^{(z,m)}(Y(z,m)|Z=z, M(z)=m, \mathbf{X}) \times p^{(z)}(M(z)=m|Z=z, \mathbf{X}) \times q(Z=z|X) \times h(\mathbf{X}),$$

where $\mathbf{X}$ denotes a vector of observed pretreatment covariates and $f^{(z,m)}(\cdot)$, $p^{(z)}(\cdot)$, $q(\cdot)$, and $h(\cdot)$ are probability or density functions. For simplicity, we use $f(\cdot)$ to represent $f^{(z,m)}(\cdot)$ henceforth. Theorem 1 summarizes the results from past research (Robins, 1999; Rosenbaum, 1987) showing that $E[Y(z, M(z))]$ is identified through IPTW.

Theorem 1. $E(Y^*|Z=z) \equiv E\left(W^{(z,M(z))}Y|Z=z\right)$ is an observed data estimand for $E[Y(z, M(z))]$, where

$$W^{(z,M(z))} = \frac{q(Z=z)}{q(Z=z|\mathbf{X})}$$

for all possible values of z under assumption (II) that the treatment assignment is independent of the potential outcomes within levels of the pretreatment covariates.

Theorem 2 summarizes the results for identifying $E[Y(z, m)]$ through IPTW (VanderWeele, 2009).

Theorem 2. $E(Y^*|Z=z) \equiv E\left(W^{(z,m)}Y|Z=z, M(z)=m\right)$ is an observed data estimand for $E[Y(z, m)]$, where

$$W^{(z,m)} = \frac{p^{(z)}(M(z)=m|Z=z)}{p^{(z)}(M(z)=m|Z=z, \mathbf{X})} \times \frac{q(Z=z)}{q(Z=z|\mathbf{X})}$$

for all possible values of z and m under assumption (II) and assumption (IIIa) that the mediator value assignment within each treatment is independent of the potential outcomes given the pretreatment covariates.

Theorem 3 summarizes the results for identifying $E[Y(z, M(z'))]$ through RMPW (Hong, 2010b; Hong and Nomi, 2012; Hong, Deutsch, and Hill, 2011, in press).

Theorem 3. $E(Y^*|Z=z) \equiv E\left(W^{(z,M(z'))}Y|Z=z\right)$ is an observed data estimand for $E[Y(z, M(z'))]$, where

$$W^{(z,M(z'))} = \frac{p^{(z')}(M(z')=m|Z=z', \mathbf{X})}{p^{(z)}(M(z)=m|Z=z, \mathbf{X})} \times \frac{q(Z=z)}{q(Z=z|\mathbf{X})}$$

for all possible values of z and m under the sequential ignorability. This encompasses, in addition to assumptions (II) and (IIIa), assumption (I) that the treatment assignment is independent of the potential mediators and assumption (IIIc) that the mediator value assignment under one treatment is independent of the potential outcome under an alternative treatment given the pretreatment covariates.

Below is a proof of Theorem 3:

$$E[Y(z,M(z'))] \equiv E\{E[Y(z,M(z'))|\mathbf{X}]\}.$$

By assumption (II), the above is equal to

$$E\{E[Y(z,M(z'))|Z=z,\mathbf{X}]\}$$
$$\equiv \int_x \int_m \int_y y \times f(Y(z,m)=y|Z=z,M(z')=m,\mathbf{X}=\mathbf{x})$$
$$\times p^{(z')}(M(z')=m|Z=z,\mathbf{X}=\mathbf{x}) \times h(\mathbf{X}=\mathbf{x})dydmd\mathbf{x},$$

which, by assumptions (I), (IIIa), and (IIIc), is equal to

$$\int_x \int_m \int_y y \times f(Y(z,m)=y|Z=z,M(z)=m,\mathbf{X}=\mathbf{x}) \times p^{(z')}(M(z')=m|Z=z',\mathbf{X}=\mathbf{x}) \times h(\mathbf{X}=\mathbf{x})dydmd\mathbf{x},$$

which, by Bayes theorem and by assumption (II), is equal to

$$\int_x \int_m \int_y y \times f(Y(z,m)=y|Z=z,M(z)=m,\mathbf{X}=\mathbf{x}) \times p^{(z)}(M(z)=m|Z=z,\mathbf{X}=\mathbf{x}) \times h(\mathbf{X}=\mathbf{x}|Z=z)$$

$$\times \frac{p^{(z')}(M(z')=m|Z=z',\mathbf{X}=\mathbf{x})}{p^{(z)}(M(z)=m|Z=z,\mathbf{X}=\mathbf{x})} \times \frac{q(Z=z)}{q(Z=z|\mathbf{X}=\mathbf{x})}dydmd\mathbf{x}$$
$$= E(Y^*|Z=z),$$

where $Y^* = W^{(z,M(z'))}Y$ and $W^{(z,M(z'))} = \dfrac{p^{(z')}(M(z')=m|Z=z',\mathbf{X})}{p^{(z)}(M(z)=m|Z=z,\mathbf{X})} \times \dfrac{q(Z=z)}{q(Z=z|\mathbf{X})}.$

Here, we view the mediator as continuous. For a discrete mediator, the proof can be modified accordingly by taking a summation over the discrete mediator values rather than taking the integral over its distribution. Similarly, for a discrete outcome, the integral over its distribution can be replaced by a summation over the discrete values.

References

Austin, P.C. (2011). Optimal caliper widths for propensity-score matching when estimating differences in means and differences in proportions in observational studies. *Pharmaceutical Statistics*, 10(2), 150–161.

Bein, E., Deutsch, J., Porter, K., Qin, X., Yang, C., and Hong, G. (2015). *Technical report on two-step estimation in RMPW analysis*. NY: Manpower Development Research Corporation.

Cheng, T. (2007). Impact of work requirements on the psychological well-being of TANF recipients. *Health & Social Work*, *32*(1), 41–48.

Coiro, M.J. (2001). Depressive symptoms among women receiving welfare. *Women & Health*, *32*, 1–23.

Drake, C. (1993). Effects of misspecification of the propensity score on estimators of treatments effects. *Biometrics*, *49*, 1231–1236.

Efron, B. (1975). Bootstrap methods: another look at the jackknife. *Annals of Statistics*, *7*(1), 1–26.

Efron, B. (1988). Bootstrap confidence intervals: good or bad? *Psychological Bulletin*, *104*(2), 293–296.

Hamilton, G., United States. Administration for Children and Families *et al.* (2002). *How Effective Are Different Welfare-to-Work Approaches: Five-Year Adult and Child Impacts for Eleven Programs*, U.S. Department of Health and Human Services, and U.S. Department of Education, Washington, DC.

Hastie, T., Tibshirani, R., and Friedman, J. (2009). *The Elements of Statistical Learning: Data Mining, Inference, and Prediction* (2nd ed.). NY: Springer.

Hawkins, D.M. (2004). The problem of overfitting. *Journal of Chemical Information and Computer Sciences*, *44*(1), 1–12.

Heckman, J.J. (1997). Instrumental variables: a study of implicit behavioral assumptions used in making program evaluations. *The Journal of Human Resources*, *32*(3), 441–462.

Hong, G. (2010a). Marginal mean weighting through stratification: adjustment for selection bias in multilevel data. *Journal of Educational and Behavioral Statistics*, *35*(5), 499–531.

Hong, G. (2010b). *Ratio of Mediator Probability Weighting for Estimating Natural Direct and Indirect Effects*. JSM Proceedings, Biometrics Section. American Statistical Association, Alexandria, VA, pp. 2401–2415.

Hong, G. (2012). Marginal mean weighting through stratification: a generalized method for evaluating multi-valued and multiple treatments with non-experimental data. *Psychological Methods*, *17*(1), 44–60.

Hong, G. and Nomi, T. (2012). Weighting methods for assessing policy effects mediated by peer change. *Journal of Research on Educational Effectiveness* special issue on the statistical approaches to studying mediator effects in education research, *5*(3), 261–289.

Hong, G., Deutsch, J., and Hill, H. (2011). *Parametric and Non-Parametric Weighting Methods for Estimating Mediation Effects: An Application to the National Evaluation of Welfare-to-Work Strategies*. JSM Proceedings, Social Statistics Section. American Statistical Association, Alexandria, VA, pp. 3215–3229.

Hong, G., Deutsch, J., and Hill, H. D. (in press). Ratio-of-mediator-probability weighting for causal mediation analysis in the presence of treatment-by-mediator interaction. *Journal of Educational and Behavioral Statistics*.

Huber, M. (2014). Identifying causal mechanisms (primarily) based on inverse probability weighting. *Journal of Applied Econometrics*, *29*(6), 920–943.

Imai, K. and Yamamoto, T. (2013). Identification and sensitivity analysis for multiple causal mechanisms: revisiting evidence from framing experiments. *Political Analysis*, *21*(2), 141–171.

Imai, K., Keele, L. and Tingly, D. (2010a). A general approach to causal mediation analysis. *Psychological Methods*, *15*(4), 309–334.

Imai, K., Keele, L. and Yamamoto, T. (2010b). Identification, inference and sensitivity analysis for causal mediation effects. *Statistical Science*, *25*(1), 51–71.

Imai, K., Keele, L., Tingley, D. and Yamamoto, T. (2011). Unpacking the black box of causality: learning about causal mechanisms from experimental and observational studies. *American Political Science Review*, *105*(4), 765–789.

Jagannathan, R., Camasso, M.J. and Sambamoorthi, U. (2010). Experimental evidence of welfare reform impact on clinical anxiety and depression levels among poor women. *Social Science and Medicine*, *71*(1), 152–160.

Knab, J.T., Garfinkel, I. and McLanahan, S. (2008). The effects of welfare and child support policies on maternal health and wellbeing, in *Making Americans Healthier: Social and Economic Policy as Health Policy* (eds J. House, R. Schoeni, H. Pollack and G. Kaplan), Russell Sage, New York.

Little, R.J. and Rubin, D.B. (2002). *Statistical Analysis With Missing Data*, 2nd edn, Wiley, New York.

McCaffrey, D. F., Ridgeway, G., and Morral, A. R. (2004). Propensity score estimation with boosted regression for evaluating causal effects in observational studies. *Psychological Methods*, 9(4), 403–425.

Michalopoulos, C., Schwartz, C. and Adams-Ciardullo, D. (2001). *What Works Best for Whom: Impacts of 20 Welfare-to-Work Programs by Subgroup*. Manpower Demonstration Research Corporation. Washington, DC: US Department of Health and Human Services, Administration for Children and Families and Office of the Assistant Secretary for Planning and Evaluation and US Department of Education.

Morris, P.A. (2008). Welfare program implementation and parents' depression. *Social Service Review*, 82(4), 579–614.

Newey, W. K. (1984). A method of moments interpretation of sequential estimators. *Economics Letters*, 14, 201–206.

Pearl, J. (2001). *Direct and Indirect Effects*. Proceedings of the Seventeenth Conference on Uncertainty in Artificial Intelligence. Morgan Kaufmann, San Francisco, CA, pp. 411–420.

Radloff, L.S. (1977). The CES-D scale: a self-report depression scale for research in the general population. *Applied Psychological Measurement*, 1(3), 385–401.

Robins, J. (1999). Marginal structural models versus structural nested models as tools for causal inference, in *Statistical Models in Epidemiology, the Environment, and Clinical Trials* (eds M.E. Halloran and D. Berry), Springer, New York, pp. 95–134.

Robins, J.M. (2003). Semantics of causal DAG models and the identification of direct and indirect effects, in *Highly Structured Stochastic Systems* (eds P. Green, N. Hjort and S. Richardson), Oxford University Press, Oxford.

Robins, J.M. and Greenland, S. (1992). Identifiability and exchangeability for direct and indirect effects. *Epidemiology*, 3(2), 143–155.

Rosenbaum, P.R. (1987). Model based direct adjustment. *Journal of the American Statistical Association*, 82(398), 387–394.

Rubin, D.B. (1986). Statistics and causal inference: comment: which ifs have causal answers. *Journal of the American Statistical Association*, 81(396), 961–962.

Schafer, J.L. and Kang, J. (2008). Average causal effects from nonrandomized studies: a practical guide and simulated example. *Psychological Methods*, 13(4), 279–313.

Tchetgen Tchetgen, E. J., and Shpitser, I. (2012). Semiparametric theory for causal mediation analysis: Efficiency bounds, multiple robustness and sensitivity analysis. *The Annals of Statistics*, 40(3), 1816–1845.

van der Laan, M.J. and Peterson, M.L. (2008). Direct effect models. *The International Journal of Biostatistics*, 4(1), 1–27.

VanderWeele, T.J. (2009). Mediation and mechanism. *European Journal of Epidemiology*, 24(5), 217–24.

VanderWeele, T.J. (2010). Bias formulas for sensitivity analysis for direct and indirect effects. *Epidemiology*, 21(4), 540–551.

Wooldridge, J. M. (2012). *Introductory econometrics: A modern approach* (5th ed). Mason, OH: Cengage Learning.

12

RMPW extensions to alternative designs and measurement

For simplicity, Chapter 11 has restricted the discussion to the case of a binary treatment, a binary mediator, and a continuous outcome. The first part of this chapter extends the RMPW strategy to mediators and outcomes having alternative distributions. These include categorical and continuous mediators and binary outcomes. The second part of this chapter discusses RMPW extensions to quasiexperimental data and then considers additional research questions and methodological challenges that arise in causal mediation analysis with multilevel data including data from cluster randomized trials and multisite randomized trials. In each case, the RMPW strategy is contrasted with other existing strategies. The last section of this chapter describes an alternative set of decomposition of the total treatment effect that may become relevant in some applications.

12.1 RMPW extensions to mediators and outcomes of alternative distributions

As reviewed in Chapter 10, the existing methods for causal mediation analysis are mostly constrained by the requirement of having a linear mediator–outcome relationship; some also require the relationship to be additive. Among these methods, the resampling method (Imai, Keele, and Tingly, 2010; Imai, Keele, and Yamamoto, 2010) and the alternative weighting methods (Huber, 2014; Tchetgen Tchetgen, 2013) are the most flexible with regard to the distribution of the mediator and that of the outcome. The resampling method nonetheless requires explicit specification of the functional form of the outcome model. The alternative weighting methods, on the other hand, require that the propensity score model for the treatment as a function of the mediator and the covariates be correctly specified in its functional form. By providing a nonparametric alternative to these other weighting methods, RMPW promises a relatively higher level of robustness in estimation. We will show in this chapter that the RMPW method can be extended to multicategory and continuous mediators and binary outcomes with no need to specify the functional relationship between the mediator

Causality in a Social World: Moderation, Mediation and Spill-over, First Edition. Guanglei Hong.
© 2015 John Wiley & Sons, Ltd. Published 2015 by John Wiley & Sons, Ltd.

and the outcome. In addition, the computation involved in the RMPW method is considerably simpler than that involved in the resampling method.

12.1.1   Extensions to a multicategory mediator

Many applications involve a mediator measured on a multicategory scale. In evaluating the effect of the welfare-to-work program on maternal depression mediated by a program-induced increase in employment, the analysis described in Chapter 11 made a binary divide between those who were ever employed and those who were never employed during the two-year postrandomization period. Yet it may seem more informative to distinguish among participants who were employed to varying degrees. For example, one may examine a three-category mediator: never employed ($M = 0$), low employment ($M = 1$) (i.e., employed for no more than 50% of the 2-year period), and high employment ($M = 2$) (i.e., employed for more than 50% of the 2-year period). The natural direct effect and the natural indirect effect of treatment assignment on maternal depression are defined the same as before.

Most existing methods for causal mediation analysis, with the resampling method and the alternative weighting methods being the exceptions, cannot handle mediators measured on a multicategory scale. The discussions in the following show how the parametric and the nonparametric RMPW procedures can be extended for analyzing the three-category mediator, highlighting modifications in Steps 2–6. The same procedure applies when the mediator contains more than three categories.

12.1.1.1   Parametric RMPW procedure

Step 1. Select and prepare pretreatment covariates.

Step 2. Specify the propensity score model for the mediator under each treatment condition. In the NEWWS Riverside experiment, every participant now has six propensity scores, three under the labor force attachment (LFA) program and three under the control condition. Suppose that an individual displays a vector of pretreatment covariates $\mathbf{X} = \mathbf{x}$. Let $\theta_{M(z)=m} = \theta_{M(z)=m}(\mathbf{x}) = \mathrm{pr}(M(z) = m|Z = z, \mathbf{X} = \mathbf{x})$ denote the individual's conditional probability of having employment level m under treatment condition z for $z = 0, 1$ and $m = 0, 1, 2$. Therefore, $\theta_{M(1)=0}$, $\theta_{M(1)=1}$, and $\theta_{M(1)=2}$ are the propensity scores for having no employment, low employment, and high employment, respectively, under LFA; $\theta_{M(0)=0}$, $\theta_{M(0)=1}$, and $\theta_{M(0)=2}$ are the propensity scores for having these three levels of employment under the control condition. A comparison between a multinomial logistic regression model and an ordinal model shows that, in this case, the latter fits the data as adequately as the former. Let $\mathbf{X}$ be a vector of pretreatment covariates with C elements, that is, $(X_1 \ldots X_C)^{\mathrm{T}}$. The propensity score model for the three-category measure of employment under LFA and that under the control condition each fits an ordinal logistic regression specified as follows:

$$\ln\left(\frac{\mathrm{pr}(M > m|Z = 1, \mathbf{X} = \mathbf{x})}{\mathrm{pr}(M \leq m|Z = 1, \mathbf{X} = \mathbf{x})}\right) = \alpha_m^{(Z=1)} + \beta_1^{(Z=1)} X_1 + \cdots + \beta_C^{(Z=1)} X_C;$$

$$\ln\left(\frac{\mathrm{pr}(M > m|Z = 0, \mathbf{X} = \mathbf{x})}{\mathrm{pr}(M \leq m|Z = 0, \mathbf{X} = \mathbf{x})}\right) = \alpha_m^{(Z=0)} + \beta_1^{(Z=0)} X_1 + \cdots + \beta_C^{(Z=0)} X_C$$

for $m = 0, 1$. To simplify the notation, let $\varphi^{(M(1))} = \varphi^{(M(1))}(\mathbf{x}) = \beta_1^{(Z=1)}X_1 + \cdots + \beta_C^{(Z=1)}X_C$ and let $\varphi^{(M(0))} = \varphi^{(M(0))}(\mathbf{x}) = \beta_1^{(Z=0)}X_1 + \cdots + \beta_C^{(Z=0)}X_C$. These ordinal models assume that, under each treatment, the partial association between a pretreatment covariate and the odds of being assigned to a certain category of employment or below is the same across all employment categories. Under this assumption, $\varphi^{(M(1))}$ is a single balancing score for all three employment categories under LFA; and similarly, $\varphi^{(M(0))}$ is a single balancing score under the control condition.

Step 3. Identify the common support for mediation analysis in each treatment group. The common support comprises the range of the logit of $\theta_{M(1)=m}$ and that of $\theta_{M(0)=m}$ for $m = 0, 1, 2$ within which the distributions of the three groups under LFA and the three groups under the control condition overlap. In other words, within the common support, every participant with a given mediator value under a given treatment has counterparts with each of the two other mediator values under the same treatment as well as counterparts with each of the three mediator values under the alternative treatment.

Step 4. Check balance in covariate distribution across the treatment-by-mediator combinations. With adjustment for the propensity scores, inverse-probability-of-treatment weighting (IPTW) is again employed for checking balance in the distribution of each observed pretreatment covariate across the six treatment-by-mediator categories. The IPTW is $\mathrm{pr}(M = m|Z = 1)/\theta_{M(1)=m}$ for the LFA units displaying employment level m and is $\mathrm{pr}(M = m|Z = 0)/\theta_{M(0)=m}$ for the control units displaying employment level m for $m = 0, 1, 2$.

Step 5. Estimate the mediator effect on the outcome under each treatment condition. Let $M^{\#}$ be a dummy indicator for $M = 1$ and $M^{\#\#\#}$ a dummy indicator for $M = 2$. Applying IPTW to the data, one may estimate the mediator effect on the outcome under each treatment condition by regressing the outcome Y on the binary treatment Z, the dummy indicators $M^{\#}$ and $M^{\#\#\#}$, and the interactions $ZM^{\#}$ and $ZM^{\#\#\#}$:

$$Y = \beta_0 + \beta_1 Z + \beta_2 M^{\#} + \beta_3 M^{\#\#\#} + \beta_4 ZM^{\#} + \beta_5 ZM^{\#\#\#} + e. \tag{12.1}$$

The result will indicate whether changes from unemployment to low employment, from unemployment to high employment, and from low employment to high employment under the control condition would lead to a reduction or inflation in depressive symptoms and whether such changes under LFA would have a similar impact on depressive symptoms.

Step 6. Create a duplicate and compute the parametric RMPW. To decompose the total effect, the data are reconstructed to include a duplicate set. The weight is 1.0 for the original control units and the duplicate LFA units. The parametric RMPW is $\theta_{M(0)=m}/\theta_{M(1)=m}$ for the original LFA units for estimating $E[Y(1, M(0))]$ and is $\theta_{M(1)=m}/\theta_{M(0)=m}$ for the duplicate control units for estimating $E[Y(0, M(1))]$.

Step 7. Estimate the causal effects. Regardless of the distribution of the multicategory mediator, the outcome model is specified the same as that in model (11.1) or (11.2).

From analyzing model (11.1), the estimated direct effect is 1.24 (SE = 0.81, $t = 1.54$, $p = 0.13$); the estimated indirect effect is -0.86 (SE = 0.37, $t = -2.31$, $p = 0.02$). The magnitude of these results is similar to the decomposition of the total effect earlier when employment was measured on a binary scale. However, by considering employment on a three-category scale, the treatment effect decomposition becomes more precise as shown in a smaller standard error, making it possible to detect a negative indirect effect that reaches statistical significance.

12.1.1.2   Nonparametric RMPW procedure

When studying the three-category employment as a mediator, the nonparametric procedure is similar to its parametric counterpart in Steps 1, 2, 3, and 7. These two procedures are different in Steps 4, 5, and 6.

Step 4. Marginal mean weighting through stratification (MMWS) rather than IPTW is employed to remove the confounding of mediator–outcome relationships when one checks balance in covariate distribution across the six treatment-by-mediator combinations. As noted in Step 2, adjusting for the balancing scores $\varphi^{(M(1))}$ and $\varphi^{(M(0))}$ removes bias associated with the observed pretreatment covariates. One may rank and stratify the sampled units by $\varphi^{(M(1))}$ and then rank and subdivide each stratum by $\varphi^{(M(0))}$. Within stratum s, the MMWS for a unit in treatment group z and employment category m is

$$MMWS = \frac{pr(M = m|Z = z)}{pr(M = m|Z = z, S = s)}$$

for $z = 0$, 1 and for $m = 0$, 1, 2.

Step 5. MMWS is applied to regression model (12.1) for estimating the mediator effect on the outcome under each treatment condition.

Step 6. After creating a duplicate set, compute

$$RMPW = \frac{pr(M = m|Z = 0, S = s)}{pr(M = m|Z = 1, S = s)}$$

for the original LFA units for estimating $E[Y(1, M(0))]$, and compute

$$RMPW = \frac{pr(M = m|Z = 1, S = s)}{pr(M = m|Z = 0, S = s)}$$

for the duplicate control units for estimating $E[Y(0, M(1))]$. The weight is again 1.0 for the original control units and the duplicate LFA units.

12.1.2   Extensions to a continuous mediator

A continuous mediator, if measured reliably, usually contains more fine-grained information than its categorical counterpart. For example, NEWWS administrative records provide information about the number of quarters a welfare recipient was employed during the 2 years after randomization. The existing methods for causal mediation analysis mostly require specifying the functional form of the relationship between the outcome and the continuous mediator and additionally require that the relationship be linear. The parametric RMPW method is proposed here as an alternative that builds on a distributional assumption of the mediator under each treatment condition. The following is the example of a normally distributed mediator.

Suppose that within levels of pretreatment covariates $\mathbf{X} = \mathbf{x}$, the mediator values are normally distributed, that is, $M(z) \sim N\left(\mu_{z|\mathbf{x}}, \sigma^2_{z|\mathbf{x}}\right)$ for $z = 0$, 1. For each individual unit in the treated group, we can predict the conditional mean mediator under the experimental condition

$\mu_{1|x} = E[M(1)|X=x]$ through regressing M on X in the treated group. Applying the model parameters to each individual in the control group, we can predict the conditional mean mediator under the counterfactual experimental condition. In parallel to this, we then specify a model regressing M on X in the control group and predict, for each individual in the control group, the conditional mean mediator under the control condition $\mu_{0|x} = E[M(0)|X=x]$. Applying the model parameters to each individual in the treated group, we can predict the conditional mean mediator under the counterfactual control condition. Additionally, through regressing M^2 on X in each treatment group and applying the same model to the alternative group, we obtain $\phi_{1|x} = E[M^2(1)|X=x]$ and $\phi_{0|x} = E[M^2(0)|X=x]$ for each individual. Hence, the conditional variance can be computed as $\sigma^2_{z|x} = \varphi_{z|x} - \mu^2_{z|x}$ for $z=0$, 1. We then estimate, within levels of pretreatment covariates $X=x$, the density for a given mediator value m under treatment z as follows:

$$\theta_{M(z)=m} = p^{(z)}(M=m|Z=z,X=x) = \left(2\pi\sigma^2_{z|x}\right)^{-\frac{1}{2}} \exp\left[-\frac{\left(m-\mu_{z|x}\right)^2}{2\sigma^2_{z|x}}\right], \text{ for } z=0, 1.$$

Here, $p^{(1)}(\cdot)$ and $p^{(0)}(\cdot)$ represent the density functions under the experimental condition and the control condition, respectively.

To estimate $E[Y(1, M(0))]$, the weight for the original LFA units displaying mediator value m is $\theta_{M(0)=m}/\theta_{M(1)=m}$, which can be simplified as

$$\text{RMPW} = \frac{\sigma_{1|x}}{\sigma_{0|x}} \exp\left[\frac{\left(m-\mu_{1|x}\right)^2}{2\sigma^2_{1|x}} - \frac{\left(m-\mu_{0|x}\right)^2}{2\sigma^2_{0|x}}\right].$$

To estimate $E[Y(0, M(1))]$, the weight for the duplicate control units displaying mediator value m is $\theta_{M(1)=m}/\theta_{M(0)=m}$, which can be simplified as

$$\text{RMPW} = \frac{\sigma_{0|x}}{\sigma_{1|x}} \exp\left[\frac{\left(m-\mu_{0|x}\right)^2}{2\sigma^2_{0|x}} - \frac{\left(m-\mu_{1|x}\right)^2}{2\sigma^2_{1|x}}\right].$$

The weight is again 1.0 for the duplicate LFA units and the original control units. The subsequent procedure is the same as that for analyzing a multicategory mediator. To check balance, one may regress each pretreatment covariate on the treatment, the mediator, and their interaction in the weighted data and expect all three regression coefficients to be approximately zero.

Alternatively, one may use the inverse probability weighting (IPW) method (Huber, 2014) discussed previously in section 10.5.1 in Chapter 10. Although IPW is mathematically equivalent to RMPW, by modeling the treatment as a function of the observed mediator and the pretreatment covariates, the IPW method avoids making distributional assumptions about the mediators.

12.1.3   Extensions to a binary outcome

In the NEWWS Riverside application, a binary outcome can be used to indicate whether an individual's depression score has exceeded a threshold of clinical significance (i.e., ≥ 10 on a 36-point scale). Let $Y = 1$ if an individual's depressive symptoms at the 2-year follow-up were clinically significant and 0 otherwise. The population mean of this binary outcome $E(Y)$ represents the proportion of individuals in the population who would display clinically significant depressive symptoms, that is, $\text{pr}(Y = 1)$. Because odds ratios commonly used in epidemiologic statistics are undefined when a risk (e.g., the probability of being depressed) is 0 or 1 and hence are not always biologically interpretable (Greenland, 1987; Greenland, Robins, and Pearl, 1999; Miettinen and Cook, 1981), following a tradition in public health research, we define a causal effect as risk difference also known as difference in incidence proportions. We will show that the RMPW strategy is considerably simpler in computation while broader in its applicability than most alternative methods for binary outcomes reviewed in Chapter 10 (e.g., Huang *et al.*, 2004; Tchetgen Tchetgen, 2013; VanderWeele and Vansteelandt, 2010).

The total effect of the treatment assignment is simply the change in the proportion of individuals with clinically significant depressive symptoms when the entire population is hypothetically assigned to LFA rather than to the control condition, that is, $\text{pr}(Y(1) = 1) - \text{pr}(Y(0) = 1)$. In the current randomized experiment, the total effect can be estimated as the sample difference in the proportion of such individuals between the LFA group and the control group, that is, $\text{pr}(Y = 1 | Z = 1) - \text{pr}(Y = 1 | Z = 0)$. In the NEWWS sample, 30% of the control units and a roughly same percentage of the LFA units reported depressive symptoms above the threshold at the 2-year follow-up. Under the sequential ignorability, parametric or nonparametric RMPW as summarized in Tables 11.3 and 11.4 can be employed to estimate the population average counterfactual outcomes $E[Y(1, M(0))]$ and $E[Y(0, M(1))]$. In this application, the parametric and the nonparametric RMPW procedures generate similar results. The natural direct effect, estimated as the difference in the proportion of individuals with clinically significant depressive symptoms between the weighted LFA group and the unweighted control group, is 0.05 (i.e., 5% higher) and is not statistically significant (SE = 0.04, $t = 1.17$, $p = 0.24$); the natural indirect effect, estimated as the difference in the proportion of such individuals between the unweighted LFA group and the weighted LFA group, is −0.02 (i.e., 2% lower) and is also not significant (SE = 0.02, $t = -1.05$, $p = 0.29$); and the pure indirect effect, estimated as the proportion difference between the weighted control group and the unweighted control group, is close to zero. The aforementioned results seem to suggest that the treatment assignment and the subsequent change in employment experience did not influence participants who would be clinically and perhaps chronically depressed.

12.2   RMPW extensions to alternative research designs

The discussion of mediation analysis in this book so far has been restricted to data from experimental research in which individuals are assigned at random to either an experimental condition or a control condition. This section will show that in combination with weighting adjustment for treatment selection, the RMPW strategy can be readily extended to quasiexperimental data. We then examine the applicability of RMPW to data from two types of multilevel experimental designs often employed in social science research. Given that individuals are typically nested in clusters such as families, schools, neighborhoods, and firms, in a cluster randomized design, clusters rather than individuals are assigned at random to different treatment conditions. Alternatively, taking each cluster as an experimental site, the experimenter may assign individuals within a cluster at random to different treatment conditions. As discussed in Chapter 6, the latter is analogous to a randomized block design in which the sites can be viewed as blocks. However, causal mediation analysis becomes considerably more challenging in these multilevel settings than in single-level settings.

12.2.1   Extensions to quasiexperimental data

A randomized experiment, especially those with a longitudinal design, often suffers from nonrandom attrition such that, among those in the remaining sample, the experimental group and the control group may become systematically different. When randomization is simply unfeasible, researchers typically analyze quasiexperimental data for evaluating treatment effects and for investigating mediation mechanisms. In all these cases, a wide array of statistical techniques is now available for reducing selection bias in treatment effect estimation. For example, IPTW and MMWS can be employed to equate the observed pretreatment composition between the experimental group and the control group. These techniques have been discussed in detail in Chapter 4.

To proceed with mediation analysis, one may apply the parametric RMPW procedure to the data after adjusting for treatment selection through IPTW. For an individual under treatment condition z, the IPTW is

$$\frac{\text{pr}(Z=z)}{\theta_z},$$

where $\theta_z = \theta_z(\mathbf{x}) = \text{pr}(Z=z|\mathbf{X}=\mathbf{x})$. When implementing the parametric RMPW procedure, one may carry out Step 2 "specifying the propensity score models" and Step 3 "identifying the common support" with the IPTW-adjusted data. In Step 4 "checking balance" and Step 5 "estimating the mediator effect on the outcome," the IPTW for treatment selection adjustment can be multiplied by the IPTW for mediator value selection adjustment. For an individual displaying mediator value m under treatment condition z, the product of these two weights is

$$\frac{\text{pr}(Z=z)}{\theta_z} \times \frac{\text{pr}(M=m|Z=z)}{\theta_{M(z)=m}}.$$

After computing RMPW in Step 6, the IPTW for treatment selection adjustment is multiplied by the parametric RMPW. For example, for an individual in the original experimental group displaying mediator value m, the weights is

$$\frac{\text{pr}(Z=1)}{\theta_1} \times \frac{\theta_{M(0)=m}}{\theta_{M(1)=m}}.$$

The weight is simply $\frac{\text{pr}(Z=0)}{\theta_0}$ for those in the control group and is $\frac{\text{pr}(Z=1)}{\theta_1}$ for those in the duplicate experimental group. The weight is then applied to the outcome model (11.1) in Step 7. In practice, all the propensity scores are to be estimated from the sample data.

In parallel, when the nonparametric procedure is employed, one may use MMWS to adjust for selection in treatment assignment in Steps 2 and 3. As discussed in Chapter 4, after stratifying the sample on the estimated propensity score for treatment assignment, the MMWS for an individual under treatment condition z in stratum s_z is

$$\frac{\text{pr}(Z=z)}{\text{pr}(Z=z|s_z)}.$$

Here, $\text{pr}(Z=z|s_z)$ is the proportion of individuals assigned to treatment condition z in stratum s_z. MMWS may also be used to adjust for selection in mediator value assignment in Steps 4 and 5. Chapter 11 has described a procedure that cross-stratifies a sample by the propensity score for mediator value assignment under the experimental condition and that under the control condition. Consider an individual who has been assigned to treatment condition z and has displayed mediator value m. If this individual is in stratum s_z under the stratification on the propensity score for treatment assignment and is in stratum $s_{m|z}$ under the cross-stratification on the two propensity scores for mediator value assignment, the MMWS for the individual is

$$\frac{\text{pr}(Z=z)}{\text{pr}(Z=z|s_z)} \times \frac{\text{pr}(M=m|Z=z)}{\text{pr}(M=m|Z=z,s_{m|z})}.$$

Here, $\text{pr}(M=m|Z=z,s_{m|z})$ is the proportion of individuals displaying mediator value m in stratum $s_{m|z}$ under treatment condition z. Finally, to estimate the natural direct and indirect effects, the nonparametric RMPW computed in Step 6 is multiplied by the MMWS for treatment selection adjustment. For an individual displaying mediator value m in the original experimental group, this product is

$$\frac{\text{pr}(Z=1)}{\text{pr}(Z=1|s_z)} \times \frac{\text{pr}(M=m|Z=0,s_{m|z})}{\text{pr}(M=m|Z=1,s_{m|z})}.$$

The weight is simply $\dfrac{\text{pr}(Z=0)}{\text{pr}(Z=0|s_z)}$ for those in the control group and is $\dfrac{\text{pr}(Z=1)}{\text{pr}(Z=1|s_z)}$ for those in the duplicate experimental group. This product is applied to the outcome model in Step 7. Hong and Nomi (2012) provided an example of combining the nonparametric RMPW with MMWS in analyzing quasiexperimental data. This application will be discussed in greater detail in Chapter 15.

12.2.2   Extensions to data from cluster randomized trials

Cluster randomized trials are particularly suitable for evaluating interventions targeted at social institutions such as schools, firms, or communities in which individuals cluster. Through changing the organizational structure, process, or group dynamics, the ultimate goals of such interventions are often to improve individual performance or well-being. In this type of experimental design, clusters are assigned at random to the intervention or the control condition. While individual-level outcomes are typically of interest, the theorized mediator could be either at the cluster or the individual level. Here, we focus on a cluster-level mediator only. The RMPW procedure is similar for an individual-level mediator when there is no spill-over among individuals in the same cluster. Chapter 15 will discuss possible within-cluster spill-over as a part of the causal mediation mechanism when the treatment and/or the mediator is at the individual level.

12.2.2.1 Application example and research questions

Induction programs for beginning teachers are typically administered by schools or school districts. An evaluation of innovative comprehensive teacher induction (CTI) programs in the United States adopted a design that assigned schools at random to either the experimental group or the control group. The researchers expected that by greatly intensifying induction activities including regular consultations with mentors under the CTI programs, the beginning teachers in the experimental schools would improve their teaching performance more than their counterparts in the control schools at the end of the treatment year (Glazerman $et\ al.$, 2010). Even though it was reported that the beginning teachers' experience with intensive induction was generally more prevalent in the experimental schools than in the control schools, some control schools nonetheless offered intensive induction, which reflects natural variation in induction practices across the control schools. The key research question is whether the experience with intensive induction mediated the CTI program effect on the beginning teachers' teaching performance as initially hypothesized.

To be specific, let Y_{ij} denote the outcome of individual teacher i in school j measured on a continuous scale; let $Z_j = 1$ if school j was assigned to the experimental group and 0 otherwise; and let $M_j = 1$ if the beginning teachers in school j received intensive induction and 0 otherwise. For simplicity, here, we focus on the case of a binary mediator at the cluster level and a continuous outcome at the individual level. Nonetheless, the RMPW procedure presented in the following applies to categorical and continuous mediators and binary outcomes both measured at the individual level.

To define the causal effects of interest, we assume that teachers did not switch schools as a result of their treatment assignment and that a teacher's teaching performance at the end of the treatment year could not be influenced by the treatment assignment of other schools. The first is an assumption of "intact clusters" and the second a "cluster-level stable unit treatment value assumption (SUTVA)" discussed in the literature on causal inference with multilevel data (Hong and Raudenbush, 2006; VanderWeele, 2010). Possible violations of these two assumptions due to spill-over between, for example, adjacent clusters will be discussed in Chapter 14. Under these two assumptions, the individual-specific total effect of the treatment on the outcome is $Y_{ij}(1) - Y_{ij}(0)$ since Z_{ij} and Z_j are equivalent in a cluster randomized design. Similarly, the treatment effect on the mediator in school j is $M_j(1) - M_j(0)$.

Because the CTI programs contain multiple elements, the causal mediation mechanism is multifaceted. The individual-specific natural direct effect $Y_{ij}(1, M_j(0)) - Y_{ij}(0, M_j(0))$ indicates whether being assigned at random to the CTI programs could have an impact on the teacher's performance without changing the induction experience. The individual-specific natural indirect effect $Y_{ij}(1, M_j(1)) - Y_{ij}(1, M_j(0))$ indicates the amount of program impact on a teacher's performance under the experimental condition solely attributable to the program-induced change in the induction experience. The natural indirect effect can be further decomposed into two pieces. The first is the pure indirect effect $Y_{ij}(0, M_j(1)) - Y_{ij}(0, M_j(0))$ indicating the amount of program impact on a teacher's performance under the control condition solely attributable to the program-induced change in the induction experience;

the second is the natural treatment-by-mediator interaction effect indicating whether the intensive induction one would receive under the CTI programs as opposed to the induction experience one would have under the control condition is more effective if the teacher was actually employed in an experimental school rather than a control school. Each of the four population average potential outcomes $E[Y(1, M(1))]$, $E[Y(0, M(0))]$, $E[Y(1, M(0))]$, and $E[Y(0, M(1))]$ is an average over all the beginning teachers in a school and over all the schools in the population.

12.2.2.2   Estimation of the total effect

Because of random treatment assignment, the mean observed outcome of the experimental group $E[Y|Z = 1]$ is unbiased for $E[Y(1, M(1))]$. Given that the observations of beginning teachers were nested in schools, an optimal estimate of $E[Y|Z = 1]$ takes into account the interdependence between individuals in the same school and the precision of each school mean outcome estimate. One may analyze a mixed-effects model specified as follows for the sample of teachers in the experimental schools:

$$Y_{ij} = \gamma + u_j + e_{ij},$$
$$u_j \sim N(0, \tau), \quad e_{ij} \sim N(0, \sigma^2). \tag{12.2}$$

Here, γ is the mean observed outcome of the experimental group and is unbiased for $E[Y(1, M(1))]$ under randomization; u_j is the random increment of school j; and e_{ij} is the random increment of teacher i in school j in the outcome Y_{ij}. Typically, for a continuous outcome, u_j is assumed to be normally distributed with mean 0 and variance τ; e_{ij} is assumed to be normally distributed with mean 0 and variance σ^2. This is equivalent to a one-way ANOVA model that partitions the total variation into the between-school variation and the within-school variation. Statistical software programs for multilevel data generate a maximum likelihood estimate of γ along with its standard error as well as maximum likelihood estimates of τ and σ^2. A similar analysis can be done with the outcome data from the control group, which generates an estimate of $E[Y(0, M(0))]$.

A contrast between the mean observed outcome of the experimental group and that of the control group is unbiased for the population average total effect of the treatment on the outcome. The following mixed-effects model estimates the mean outcome of the control group γ_0 and the mean contrast γ_1 between the two treatment groups as well as the between-school variation $\text{Var}(u_j) = \tau$ and the within-school variation of the outcome $\text{Var}(e_{ij}) = \sigma^2$. The latter is typically assumed to be constant across the two treatment groups, though one may easily estimate $\sigma^2(1)$ and $\sigma^2(0)$ for the experimental group and the control group, respectively:

$$Y_{ij} = \gamma_0 + \gamma_1 Z_j + u_j + e_{ij},$$
$$u_j \sim N(0, \tau), \quad e_{ij} \sim N(0, \sigma^2). \tag{12.3}$$

12.2.2.3   RMPW analysis of causal mediation mechanisms

The variation across experimental schools and that across control schools in whether the beginning teachers received intensive induction are at least partly associated with some systematic pretreatment differences between the schools such as principal leadership and school resources denoted by $\mathbf{G}_j$ for school j. Hence, one may estimate a school's propensity of having its beginning teachers experience intensive induction under the experimental condition $\theta_{M(1)} = \theta_{M(1)}(\mathbf{g}) = \mathrm{pr}(M(1) = 1|Z = 1, \mathbf{G} = \mathbf{g})$ and that under the control condition $\theta_{M(0)} = \theta_{M(0)}(\mathbf{g}) = \mathrm{pr}(M(0) = 1|Z = 0, \mathbf{G} = \mathbf{g})$ each as a function of $\mathbf{g}$ through analyzing a logistic regression in each treatment group. One may then examine, with statistical adjustment for $\theta_{M(1)}$ and $\theta_{M(0)}$ such as through IPTW, whether the observed pretreatment school characteristics $\mathbf{G}$ are balanced across the four treatment-by-mediator combinations.

What would be the population average potential outcome if all schools were assigned to the experimental condition yet all beginning teachers would experience the same induction as they would have had under the control condition? To estimate $E[Y(1, M(0))]$, we assign an RMPW equal to $\theta_{M(0)}/\theta_{M(1)}$ to experimental schools in which the beginning teachers actually experienced intensive induction and assign an RMPW equal to $(1-\theta_{M(0)})/(1-\theta_{M(1)})$ to the experimental schools in which the beginning teachers did not experience intensive induction. One may fit a propensity score model to each treatment group as a function of pretreatment school characteristics $\mathbf{g}$. The coefficients are then used for estimating or predicting a pair of propensity scores for each experimental school. After creating a duplicate of each experimental school indicated by $D1 = 1$, we let the original experimental and control schools be indicated by $D1 = 0$. One then applies the aforementioned RMPW to each original experimental school.

What would be the population average potential outcome if all schools were assigned to the control condition yet all beginning teachers would experience the same induction as they would have had under the experimental condition? The estimation of $E[Y(0, M(1))]$ involves duplicating the control schools and creating a dummy indicator $D0$ that takes value 1 for the duplicate control schools and 0 for the original control schools as well as the original and duplicate experimental schools. The RMPW is $\theta_{M(1)}/\theta_{M(0)}$ for the duplicate control schools in which the beginning teachers actually experienced intensive induction and is $(1-\theta_{M(1)})/(1-\theta_{M(0)})$ for the duplicate control schools in which the beginning teachers actually did not experience intensive induction. The weight is 1.0 for the duplicate experimental schools and the original control schools.

The following two-level model decomposes the total treatment effect into a natural direct effect represented by $\gamma^{(DE.0)}$ and a natural indirect effect represented by $\gamma^{(IE.1)}$. The model additionally estimates a pure indirect effect represented by $\gamma^{(IE.0)}$. One may then test the treatment-by-mediator interaction effect represented by $\gamma^{(IE.1)} - \gamma^{(IE.0)}$:

$$\text{Level 1} \quad Y_{ij} = \beta_j + e_{ij},$$
$$e_{ij} \sim N(0, \sigma^2);$$

$$\text{Level 2} \quad \beta_j = \gamma_0 + \gamma^{(DE.0)}Z_j + \gamma^{(IE.1)}D1_j + \gamma^{(IE.0)}D0_j + u_j,$$
$$u_j \sim N(0, \tau). \tag{12.4}$$

Here, σ^2 is the within-school variance of the outcome, while τ is the between-school variance of the outcome conditioning on treatment group membership and duplication status. However, due to the duplication of schools, standard multilevel software programs do not decompose the variance appropriately, which would lead to incorrect estimation of the model-based standard errors. One may obtain robust standard errors for the causal effect estimates and may use bootstrapping to construct a confidence interval for each causal effect.

Chapter 11 has shown in Appendix 11.A the computation of cluster-robust standard errors, which can be extended to the multilevel setting. Finally, estimates of σ^2 and τ can be obtained from a standard two-level analysis of the intent-to-treat (ITT) effect of the treatment on the outcome. An alternative to maximum likelihood estimation is method-of-moments estimation which does not require the errors to be normally distributed.

12.2.2.4   Identification assumptions

The multilevel RMPW strategy presented here assumes intact clusters and cluster-level SUTVA and requires the sequential ignorability. To be specific, it assumes that, conditional on the observed pretreatment covariates $\mathbf{G}$, the treatment assignment Z is independent of the potential mediators $M(1)$ and $M(0)$ and the potential outcomes $Y(1, M(1))$, $Y(0, M(0))$, $Y(1, M(0))$, and $Y(0, M(1))$ and that the mediator value assignment $M(z)$ under treatment z for $z = 0, 1$ is independent of all the potential outcomes. The RMPW method, however, makes no other identification assumptions required by multilevel path analysis or multilevel SEM as discussed in the following. Furthermore, RMPW minimizes the need to specify the outcome model.

12.2.2.5   Contrast with multilevel path analysis and SEM

Cluster randomized trials with a cluster-level treatment, a cluster-level mediator, and an individual-level outcome have been categorized as a 2–2–1 design in the literature (Krull and MacKinnon, 2001). Extending the path analysis approach to multilevel data, past researchers (Krull and MacKinnon, 2001; Pituch, Stapleton, and Kang, 2006) have recast the outcome model as a two-level model by including a cluster random intercept for individual i in cluster j, while the mediator model remains a single-level model:

$$M_j = \alpha_0 + aZ_j + e_{Mj}, \quad e_{Mj} \sim N\left(0, \sigma_M^2\right);$$

$$Y_{ij} = \beta_0 + bM_j + cZ_j + u_j + e_{Yij} \quad u_j \sim N(0, \tau), \quad e_{Yij} \sim N\left(0, \sigma_Y^2\right).$$

These models may include pretreatment covariates $\mathbf{G}$ that confound the mediator–outcome relationships.

As the indirect effect estimate $\hat{a}\hat{b}$ is in general not normally distributed, for statistical inference, Pituch, Stapleton, and Kang (2006) recommended using bootstrapping to generate the confidence interval or using empirically simulated critical values to construct asymmetric confidence limits for the estimate (MacKinnon, Lockwood, and Williams, 2004; Meeker, Cornwell, and Aroian, 1981). Preacher and colleagues (2010, 2011) proposed using a multilevel SEM framework to encompass the aforementioned models by specifying the joint distribution of M and Y. In the case of a 2–2–1 design, a multilevel SEM analysis is expected to generate the same results as the aforementioned two-step analysis except that the former can incorporate measurement models when the treatment, the mediator, or the outcome are latent constructs.

The aforementioned analyses have assumed, first of all, intact clusters and cluster-level SUTVA. The identification assumptions include all those required by path analysis as listed in Table 10.2. These include the sequential ignorability, no treatment-by-mediator interaction, and the functional form of each model. The assumption of no treatment-by-mediator interaction is especially worrisome in the current application because it is contradictory with the theoretical hypothesis that intensive induction offered by the CTI programs is expected to be more effective than that offered under the control condition.

12.2.2.6   Contrast with multilevel prediction models

VanderWeele (2010) identified the natural direct effect and the natural indirect effect by the observable quantities as follows:

$$E[Y(1,M(0)) - Y(0,M(0))] = \Sigma_g \Sigma_m [E(Y|Z=1, M=m, \mathbf{G}=\mathbf{g}) - E(Y|Z=0, M=m, \mathbf{G}=\mathbf{g})] \times$$
$$\mathrm{pr}(M=m|Z=0, \mathbf{G}=\mathbf{g}) \times \mathrm{pr}(\mathbf{G}=\mathbf{g});$$

$$E[Y(1,M(1)) - Y(1,M(0))] = \Sigma_g \Sigma_m E(Y|Z=1, M=m, \mathbf{G}=\mathbf{g}) \times [\mathrm{pr}(M=m|Z=1, \mathbf{G}=\mathbf{g}) -$$
$$\mathrm{pr}(M=m|Z=0, \mathbf{G}=\mathbf{g})] \times \mathrm{pr}(\mathbf{G}=\mathbf{g})$$

The estimation of $E(Y|Z=1, M=m, \mathbf{G}=\mathbf{g})$ and $E(Y|Z=0, M=m, \mathbf{G}=\mathbf{g})$ involves specifying a multilevel outcome model regressing Y on Z, M, and $\mathbf{G}$, possibly accommodating a Z-by-M interaction. The estimation of $\mathrm{pr}(M=m|Z=1, \mathbf{G}=\mathbf{g})$ and $\mathrm{pr}(M=m|Z=0, \mathbf{G}=\mathbf{g})$ involves specifying a separate regression model regressing M on Z and $\mathbf{G}$. Note that the $\mathbf{G}$–Y relationships are not necessarily linear and may differ across the treatment and mediator levels; similarly, the $\mathbf{G}$–M relationships may differ across the treatment conditions. This strategy requires predicting, for each individual in the sample:

1. The potential outcome under the treatment that one actually was assigned to and under the mediator value m that one actually displayed

2. The counterfactual outcome under the alternative treatment that one was not assigned to and under the mediator value m the same as that one displayed under the actual treatment assignment

3. The probability that one would display mediator value m under the treatment that one was actually assigned to

4. The counterfactual probability that one would display mediator value m under the alternative treatment that one was not assigned to

After summing these predictive quantities over all the individuals in the sample, VanderWeele suggested calculating confidence intervals through bootstrapping.

In addition to the assumptions of intact clusters and cluster-level SUTVA, the key identification assumption, clarified by VanderWeele (2010), is the sequential ignorability. Hence, this strategy avoids the no treatment-by-mediator interaction assumption. Yet possible misspecifications of the functional form of either the outcome model or the mediator model would nonetheless bias the estimation results. In contrast, the RMPW method avoids the consequences of misspecifying the outcome model, while the nonparametric RMPW procedure also alleviates the consequences of misspecifying the mediator model.

12.2.3   Extensions to data from multisite randomized trials

Multisite randomized trials are another widely used multilevel experimental design. Typically, a sample of experimental sites is drawn and a randomized experiment is replicated across all the sites. In many cases, the treatment, the mediator, and the outcome are all measured at the individual level, which have been categorized as 1–1–1 designs (Krull and MacKinnon, 2001). The evaluation of the Head Start programs for increasing school readiness of children from low-income families in the United States provided such an example. The Head Start Impact Study (HSIS) from 2002 to 2006 was conducted with a nationally representative sample of about 380 sites and included nearly 5000 children. Newly entering 3- and 4-year-old children eligible for Head Start were assigned at random at each site either to an experimental group that had access to Head Start program services or to a control group that did not have such access but could enroll in other programs or services selected by their parents.

12.2.3.1   Research questions and causal parameters

Research has consistently shown that home literacy activities promote child language and literacy development and reduce behavior problems among low-income families (Bradley *et al.*, 2001; Snow, Burns, and Griffin, 1998). Head Start pioneered a two-generation approach to early childhood education and offered parenting education emphasizing cognitive stimulation among other parenting skills. HSIS reported program impacts on both parenting practices and child outcomes. For example, at the end of the Head Start year, parents of 3-year-olds in the Head Start group were more likely to read to their children when compared with parents in the control group. In the meantime, Head Start programs showed positive impacts on 3-year-olds' language and literacy development including their vocabulary score. An immediate research question is whether the Head Start-induced increase in parent reading to child mediated at least partially the Head Start program effect on child vocabulary development.

Again for simplicity, we focus on the case of a binary mediator and a continuous outcome. Let $Z_{ij} = 1$ if child i at site j was assigned at random to a Head Start program and 0 otherwise; let $M_{ij} = 1$ if the parent read frequently to the child during the Head Start year and 0 otherwise. We tentatively assume that possible interactions between parents and between children at the same site had no impact on either how frequently a parent read to a child or how well the child performed on a vocabulary test. Under this assumption, whether or not a parent read frequently to a child would not depend on whether the children of other parents at the same site were enrolled in Head Start; similarly, a child's vocabulary development would depend neither on whether other children at the same site were enrolled in Head Start nor on whether other parents at the same site read frequently to their children. In essence, this is the assumption of individual-level SUTVA within each site. Chapter 14 will discuss a relaxation of these assumptions when the spill-over between individuals at the same site is likely and is of scientific interest. Because each site can be viewed as a cluster, we additionally assume "intact clusters" and "cluster-level SUTVA" such that the mediators and outcomes of individuals at one site cannot be affected by the treatment assignments and the mediator value assignments of individuals at other sites.

Under this framework, child i at site j has two potential mediators and four potential outcomes: $M_{ij}(1)$ is the potential frequency of parent reading to child if the child was assigned to Head Start; $M_{ij}(0)$ is the potential frequency of parent reading to child if the child was assigned to the control condition; $Y_{ij}(1, M_{ij}(1))$ is the child's vocabulary score at the end of the Head Start year if the child was assigned to Head Start; $Y_{ij}(0, M_{ij}(0))$ is the child's vocabulary score if the child was assigned to the control condition; $Y_{ij}(1, M_{ij}(0))$ is the child's vocabulary score if assigned to Head Start yet the parent would read to the child, perhaps counterfactually, with the same frequency as that under the control condition; and $Y_{ij}(0,$

$M_{ij}(1))$ is the child's vocabulary score if assigned to the control condition yet the parent would read to the child, perhaps counterfactually, with the same frequency as that under Head Start.

The natural direct effect, defined as $\delta_{ij}^{(DE.0)} = Y_{ij}(1, M_{ij}(0)) - Y_{ij}(0, M_{ij}(0))$, therefore represents the Head Start impact on child vocabulary without changing the frequency of parent reading to child. The natural indirect effect, defined as $\delta_{ij}^{(IE.1)} = Y_{ij}(1, M_{ij}(1)) - Y_{ij}(1, M_{ij}(0))$, represents the change in child vocabulary under Head Start attributable to the program-induced change in parent reading to child. The pure indirect effect, defined as $\delta_{ij}^{(IE.0)} = Y_{ij}(0, M_{ij}(1)) - Y_{ij}(0, M_{ij}(0))$, represents the potential amount of change in child vocabulary under the control condition that could have been caused by the program-induced change in parent reading to child. One may hypothesize that with the parenting education offered by the Head Start program, the parent might learn to read to the child more effectively than he or she otherwise might under the control condition, in which case there would be a positive treatment-by-mediator interaction effect defined as $\delta_{ij}^{(IE.1)} - \delta_{ij}^{(IE.0)}$. All these causal effects can be defined for the entire population of 3-year-old children eligible for Head Start.

12.2.3.2 Estimation of the total effect and its between-site variation

With data from a multisite randomized trial, the mean contrast in the observed outcome between the Head Start group and the control group at site j is clearly unbiased for the ITT effect of the treatment on the outcome at that particular site denoted by $\beta_{1j}^{(ITT)}$. To estimate the average and the variation of this total treatment effect over the population of sites, one may specify a two-level model:

$$
\begin{aligned}
\text{Level 1:} \quad & Y_{ij} = \beta_{0j} + \beta_{1j}^{(ITT)} Z_{ij} + e_{ij}, \\
& e_{ij} \sim N(0, \sigma^2); \\
\text{Level 2:} \quad & \beta_{0j} = \gamma_0 + u_{0j}, \\
& \beta_{1j}^{(ITT)} = \gamma_1^{(ITT)} + u_{1j}^{(ITT)}, \\
& \begin{pmatrix} u_{0j} \\ u_{1j}^{(ITT)} \end{pmatrix} \sim N \left(\begin{pmatrix} 0 \\ 0 \end{pmatrix}, \begin{pmatrix} \tau_{00} & \tau_{01} \\ \tau_{10} & \tau_{11}^{(ITT)} \end{pmatrix} \right).
\end{aligned}
\tag{12.5}
$$

Here, β_{0j} is the mean outcome of the control group at site j; $\beta_{1j}^{(ITT)}$ is unbiased for the total treatment effect at site j under the randomized design. The site-specific increment $u_{1j}^{(ITT)}$ to the average total effect $\gamma_1^{(ITT)}$ has a mean zero and a variance $\tau_{11}^{(ITT)}$ representing the between-site variation in the total effect. The within-site variance σ^2 conditioning on treatment group membership does not have to be constant across the treatment groups and across sites.

However, as discussed in Chapter 6, a challenge arises in estimation when the site-specific probability of treatment assignment $P_j = \mathrm{pr}(Z_{ij} = 1 | j)$ varies systematically across the sites. For example, one may anticipate that at those sites with a high concentration of poverty and low provision of early childcare resources, the demand for Head Start will be relatively higher, leading to a lower P_j. In other words, eligible children at those sites will have a lower probability of being assigned to Head Start. In the meantime, the site mean outcome of the control group β_{0j} tends to be relatively lower. When P_j is omitted from the level 2 model and is therefore hidden in the random effect u_{0j}, it will contribute to a positive association between Z_{ij} and u_{0j}. It is well-known in the regression context that the treatment effect cannot be identified if the treatment indicator is not independent of the error terms. We may view P_j as a confounder that will bias the estimation of $\gamma_1^{(\mathrm{ITT})}$ and $\tau_{11}^{(\mathrm{ITT})}$. A number of strategies have been proposed by past researchers for bias removal. These include fixed-effects models and mixed-effects models with centering at the site mean or with covariance adjustment for the site-specific probability of treatment assignment. Yet as shown in Chapter 6, when the treatment effect is heterogeneous across the sites, none of these strategies can successfully remove all the bias. A novel yet simple solution is to equate the probability of treatment assignment across all the sites through weighting. The weight operates as a sample weight and hence depends on the sampling design and the population to which the sample results are to be generalized (Raudenbush, 2014).

When the sample results are to be generalized to the population of individual children of interest, an ideal design would select a random sample of sites and a random sample of individual children within each sampled site. Assuming equal sampling probability at the individual level, the sample size at each site n_j is proportional to the actual site size and varies naturally across the sites. One then uses weighting to approximate a multisite design with a constant probability of treatment assignment across all the sites while allowing the sample size to vary naturally from site to site. As shown in Chapter 6, the weight to be applied is

$$W_{ij} = Z_{ij} \times \frac{P}{P_j} + \left(1 - Z_{ij}\right) \times \frac{1-P}{1-P_j}. \tag{12.6}$$

When applied to the two-level model specified in (12.5), the weight will be partitioned into a level 2 weight and a conditional level 1 weight. One obtains an unbiased estimate of the population average total effect of Head Start relative to the control condition. Additionally, $\hat{\tau}_{11}$ estimates the between-site variation in this total effect.

12.2.3.3 RMPW analysis of causal mediation mechanisms

In order to decompose the total effect into the natural direct effect and the natural indirect effect and, when it is of interest, to further decompose the latter into the pure indirect effect and the natural treatment-by-mediator indirect effect, as before, we aim to estimate four population average potential outcomes $E[Y(1, M(1))]$, $E[Y(0, M(0))]$, $E[Y(1, M(0))]$, and $E[Y(0, M(1))]$. The first two can be estimated through reparameterizing model (12.5):

$$\begin{aligned}
\text{Level 1:} \quad & Y_{ij} = \left(1 - Z_{ij}\right)\beta_{0j} + \beta_{1j}Z_{ij} + e_{ij}, \\
& e_{ij} \sim N(0, \sigma^2). \\
\text{Level 2:} \quad & \beta_{0j} = \gamma_0 + u_{0j}, \\
& \beta_{1j} = \gamma_1 + u_{1j}, \\
& \begin{pmatrix} u_{0j} \\ u_{1j} \end{pmatrix} \sim N\left(\begin{pmatrix} 0 \\ 0 \end{pmatrix}, \begin{pmatrix} \tau_{00} & \tau_{01} \\ \tau_{10} & \tau_{11} \end{pmatrix} \right).
\end{aligned} \tag{12.7}$$

Applying the weight specified in (12.6), one obtains $\hat{\gamma}_0$ as an unbiased estimate of $E[Y(0, M(0))]$ and $\hat{\gamma}_1$ as an unbiased estimate of $E[Y(1, M(1))]$.

As is true with other experimental designs in which only the treatment is randomized, potential confounding of the mediator–outcome relationship is of primary concern in the causal mediation analysis. For example, because parents of better education tend to read to their children more often, it is essential to tease apart the benefit of having better-educated parents in general and the benefit of increasing parent reading to child as a result of enrollment in Head Start. Other individual-level pretreatment covariates such as demographics, family structure, and parents' work schedule and site-level pretreatment covariates such as community culture and the availability of high-quality alternative childcare may also predict both the frequency of the parent reading to the child and child vocabulary development. Let $\theta_{M(1)} = \mathrm{pr}(M = m|Z = 1, \mathbf{X} = \mathbf{x}, j)$ denote the propensity that a child at site j would be read to by the parent with frequency level m if assigned to Head Start; and let $\theta_{M(0)} = \mathrm{pr}(M = m|Z = 0, \mathbf{X} = \mathbf{x}, j)$ denote the child's propensity to be read to by the parent with frequency level m if assigned to the control condition. Here, $\mathbf{X}$ denotes a vector of individual-level pretreatment covariates. One may specify a two-level propensity score model for child i at site j in the form of a mixed-effects model as follows for each of the two treatment groups $z = 0, 1$:

$$\ln\left(\frac{\theta_{M(z)}}{1 - \theta_{M(z)}}\right) = \alpha_0^{(z)} + \boldsymbol{\alpha}_1^{(z)}\mathbf{X}_{ij} + u_j^{(z)},$$

$$u_j^{(z)} \sim N\left(0, \tau^{(z)}\right).$$

The site-specific random intercepts $u_j^{(1)}$ under Head Start and $u_j^{(0)}$ under the control condition capture the impacts of site-level characteristics such as community culture and the quality of alternative childcare on the parent reading to the child under the respective treatment conditions. If the predictive relationship between a pretreatment covariate X_{ij} and the mediator M_{ij} varies across the sites under a given treatment, one may include a corresponding random slope in the propensity score model.

If, in analyzing model (12.7), we instead assign the weight $(P/P_j) \times (\theta_{M(0)}/\theta_{M(1)})$ to a Head Start child at site j who was read to by the parent with frequency level m, RMPW will transform the mediator value distribution in the Head Start group to resemble that in the control group. One thereby obtains an estimate of the population average counterfactual outcome $E[Y(1, M(0))]$.

Similarly, if we assign the weight $[(1-P)/(1-P_j)] \times (\theta_{M(1)}/\theta_{M(0)})$ to a control child at site j who was read to by the parent with frequency level m, RMPW will transform the mediator value distribution in the control group to resemble that in the Head Start group. One thereby obtains an estimate of the population average counterfactual outcome $E[Y(0, M(1))]$.

The general weighting scheme at level 1, applicable to multivalued mediators, is summarized in Table 12.1. To make statistical inference for the contrasts between these estimated population average potential outcomes, one may combine the original sample with a duplicate Head Start group indicated by $D1 = 1$ and a duplicate control group indicated by $D0 = 1$.

Table 12.1 Parametric RMPW for analyzing data from multisite randomized trials

	$E[Y(0, M(0))]$	$E[Y(1, M(1))]$	$E[Y(1, M(0))]$	$E[Y(0, M(1))]$
Z	0	1	1	0
$D1$	0	0	1	0
$D0$	0	1	0	1
W	$\dfrac{1-P}{1-P_j}$	$\dfrac{P}{P_j}$	$\dfrac{P}{P_j} \times \dfrac{\theta_{M(0)}}{\theta_{M(1)}}$	$\dfrac{1-P}{1-P_j} \times \dfrac{\theta_{M(1)}}{\theta_{M(0)}}$

As before, a structural model decomposes the total treatment effect into a natural direct effect and a natural indirect effect:

$$Y_{ij} = \gamma_0 + \gamma^{(DE.0)}Z_{ij} + \gamma^{(IE.1)}D1_{ij} + \gamma^{(IE.0)}D0_{ij} + \varepsilon_{ij}, \tag{12.8}$$

where $\varepsilon_{ij} = u_j + e_{ij}$. When the primary interest is in the point estimates of the causal effects only, one may estimate the coefficients γ_0, $\gamma^{(DE.0)}$, $\gamma^{(IE.1)}$, and $\gamma^{(IE.0)}$ via weighted least squares and compute the cluster-robust standard errors that accommodate the duplicates within individuals and the clustering of individuals within sites in the variance–covariance matrix. Bootstrapping can be employed for obtaining the confidence interval for each causal effect.

A major advantage of a multisite randomized experiment is that it enables one to investigate possible heterogeneity in the causal effects across sites. For example, one may hypothesize that the natural direct effect, primarily attributable to the instruction that children received at childcare settings, might differ by site because the quality of instruction in Head Start programs might not be consistent and because alternative childcare and their quality available to the control children might be vastly different across the sites. The natural indirect effect of the program might also differ by site possibly due to differential amount of program emphasis on parenting education in Head Start or in alternative childcare programs available to the control children. If a Head Start program that displayed a higher instructional quality was also generally more successful in enhancing literacy activities at home, then the natural direct effect and the natural indirect effect may be positively correlated.

Qin and Hong (2014) presented a strategy for estimating the between-site variation of the causal effects. Let $\delta_j^{(DE.0)} = E\left[\delta_{ij}^{(DE.0)}|j\right]$ and $\delta_j^{(IE.1)} = E\left[\delta_{ij}^{(IE.1)}|j\right]$ denote the average natural direct effect and natural indirect effect at site j. Estimates of these site-specific causal effects along with their sampling variability $Var\left(\widehat{\delta}_j^{(DE.0)}\right)$ and $Var\left(\widehat{\delta}_j^{(IE.1)}\right)$ can be obtained through a weighted least squares analysis or a weighted method-of-moments analysis of the data at site j. Most importantly, this strategy makes use of the fact that the total amount of between-site variability in the sample estimate of the site-specific direct effect $G^{(DE.0)} = Var\left(\widehat{\delta}_j^{(DE.0)} - \delta^{(DE.0)}\right)$ is a sum of two components. The first is the between-site variability of the site-specific direct effect $\tau^{(DE.0)} = Var\left(\delta_j^{(DE.0)} - \delta^{(DE.0)}\right)$, a quantity of scientific interest; the second is the pure sampling variability $V^{(DE.0)} = Var\left(\widehat{\delta}_j^{(DE.0)} - \delta_j^{(DE.0)}\right)$. Hence one may obtain $\widehat{\tau}^{(DE.0)}$ through subtracting $\widehat{V}^{(DE.0)}$ from $\widehat{G}^{(DE.0)}$. Similarly, $G^{(IE.1)} = Var$

$\left(\widehat{\delta}_j^{(IE.1)} - \delta^{(IE.1)}\right)$ is a sum of $\tau^{(IE.1)}$ and $V^{(IE.1)}$; and hence $\widehat{\tau}^{(IE.1)} = \widehat{G}^{(IE.1)} - \widehat{V}^{(IE.1)}$. In each case, a Chi square test is useful for determining whether the between-site variance of the causal effect can be distinguished from zero. One may also decompose $Cov\left(\widehat{\delta}_j^{(DE.0)}, \widehat{\delta}_j^{(IE.1)}\right)$ into the between-site covariance of the site-specific direct and indirect effects $Cov\left(\delta_j^{(DE.0)}, \delta_j^{(IE.0)}\right)$ and the covariance purely due to sampling and then estimate the former through subtraction. Results from Monte Carlo simulations have shown that not only are the estimated population average direct effect and indirect effect unbiased but also the variance of the site-specific direct effect, that of the site-specific indirect effect, and their covariance are estimated with minimal bias except when the sample size at each site is extremely small.

12.2.3.4   Identification assumptions

All the analytic strategies for multisite randomized trials discussed in this section assume intact clusters (i.e., sites), cluster-level SUTVA, and within-cluster SUTVA. However, the multilevel RMPW strategy requires considerably fewer assumptions than the existing alternatives. In analyses of data from multisite randomized trials, the key assumption required by the multilevel RMPW approach is the sequential ignorability. A multisite randomized design ensures that treatment assignment Z is independent of the two potential mediators $M(1)$ and $M(0)$ and all four potential outcomes $Y(1, M(1))$, $Y(1, M(0))$, $Y(0, M(1))$, and $Y(0, M(0))$ when the site membership is given. The sequential ignorability additionally states that within levels of the observed pretreatment covariates, the mediator is as if randomized under each treatment, a strong assumption that may not hold if the observed pretreatment covariates do not account for all the selection of mediator values. However, as shown in the following, the multilevel RMPW approach is considerably more flexible than the multilevel path analysis methods as the former allows for treatment-by-mediator interaction and in the meantime makes no assumptions about the functional form of the outcome model.

12.2.3.5   Contrast with multilevel path analysis and SEM

Some attempts have been made in the past decade to investigate the causal mediation mechanisms in multisite trials with 1–1–1 designs. Kenny, Korchmaros, and Bolger (2003) extended Baron and Kenny's (1986) path analysis models (8.1) and (8.2) directly to such multilevel settings by letting the treatment effect on the mediator at site j, denoted by a_j in the mediator model, have a random variation across the sites rather than assuming it to be a constant. Similarly, in the outcome model, the mediator effect on the outcome conditioning on the treatment at site j, denoted by b_j, and the treatment effect on the outcome conditioning on the mediator at site j, denoted by c_j, are each allowed to vary at random across the sites:

$$M_{ij} = \alpha_0 + a_j Z_{ij} + u_{M0j} + e_{Mij}, \quad \mathbf{u}_{Mj} \sim N(0, \tau_M), \quad e_{Mij} \sim N(0, \sigma_M^2);$$

$$Y_{ij} = \beta_0 + b_j M_{ij} + c_j Z_{ij} + u_{Y0j} + e_{Yij}, \quad \mathbf{u}_{Yj} \sim N(0, \tau_Y), \quad e_{Yij} \sim N(0, \sigma_Y^2).$$

Here, $a_j = a + u_{aj}$, $b_j = b + u_{bj}$, and $c_j = c + u_{cj}$; $\mathbf{u}_{Mj} = \left(u_{M0j} \ u_{aj}\right)^{\mathrm{T}}$ and, correspondingly, τ_M is a 2×2 variance–covariance matrix; and $\mathbf{u}_{Yj} = \left(u_{Y0j} \ u_{bj} \ u_{cj}\right)^{\mathrm{T}}$ and, correspondingly, τ_Y is a 3×3 variance–covariance matrix. Under a series of identification assumptions discussed in the following, $\hat{a}_j\hat{b}_j$ estimates the site-specific natural indirect effect, while $\hat{c}_j$ estimates the site-specific natural direct effect. However, a major difficulty arises in estimating the average natural indirect effect and its standard error. This is because, as Kenny and his colleagues pointed out,

$$E\left(a_jb_j\right)=E\left(a_j\right)\times E\left(b_j\right)+\mathrm{Cov}\left(a_j,b_j\right).$$

Because a_j is estimated in the mediator model, while b_j is estimated in the outcome model, these two separate analyses do not generate an estimate of the covariance between a_j and b_j. Nor do these analyses generate an estimate of the between-site variation of the natural indirect effect denoted by $\mathrm{Var}(a_jb_j)$.

A solution proposed by Bauer, Preacher, and Gil (2006) is to combine the mediator model and the outcome model into a single-step multivariate two-level model as follows:

$$R_{ij}=I_{Mij}\left(\alpha_0+a_jZ_{ij}+u_{M0j}\right)+I_{Yij}\left(\beta_0+b_jM_{ij}+c_jZ_{ij}+u_{Y0j}\right)+e_{ij},$$

$$\mathbf{u}_j\sim N(0,\tau),\ \ e_{ij}\sim N\left(0,\sigma^2\right).$$

In the level 1 data file, M_{ij} and Y_{ij} are stacked for individual i at site j. Here, I_{Mij} is an indicator for the mediator model that takes value 1 when $R_{ij}=M_{ij}$ and takes value 0 otherwise; I_{Yij} is an indicator for the outcome model that takes value 1 when $R_{ij}=Y_{ij}$ and takes value 0 otherwise; and $\mathbf{u}_j=\left(u_{M0j}\ \ u_{aj}\ \ u_{Y0j}\ \ u_{bj}\ \ u_{cj}\right)^{\mathsf{T}}$ and, correspondingly, τ_Y is a 5×5 variance–covariance matrix that includes, among other elements, $\mathrm{Var}\left(u_{a_j}\right)$, $\mathrm{Var}\left(u_{b_j}\right)$, $\mathrm{Var}\left(u_{c_j}\right)$, and $\mathrm{Cov}\left(u_{a_j},u_{b_j}\right)$. Note that $\mathrm{Var}\left(u_{a_j}\right)=\mathrm{Var}\left(a_j\right)$, $\mathrm{Var}\left(u_{b_j}\right)=\mathrm{Var}\left(b_j\right)$, and $\mathrm{Cov}\left(u_{a_j},u_{b_j}\right)=\mathrm{Cov}\left(a_j,b_j\right)$. And finally, $e_{ij}=I_{Mij}e_{Mij}+I_{Yij}e_{Yij}$ and therefore $\sigma^2=I_{Mij}\sigma_M^2+I_{Yij}\sigma_Y^2$. The sample estimates of a, b, $\mathrm{Var}(a_j)$, $\mathrm{Var}(b_j)$, and $\mathrm{Cov}(a_j, b_j)$ provide all the elements one needs for estimating the population average natural indirect effect $E(a_jb_j)$ and the between-site variance of the natural indirect effect $\mathrm{Var}(a_jb_j)$. According to Kenny, Korchmaros, and Bolger (2003), when a_j and b_j are normally distributed, one would have that

$$\mathrm{Var}\left(a_jb_j\right)=b^2\mathrm{Var}\left(a_j\right)+a^2\mathrm{Var}\left(b_j\right)+\mathrm{Var}\left(a_j\right)\mathrm{Var}\left(b_j\right)+2ab\mathrm{Cov}\left(a_j,b_j\right)+\left[\mathrm{Cov}\left(a_j,b_j\right)\right]^2.$$

Bauer and colleagues (2006) also derived the standard error for the estimated population average natural indirect effect.

However, the aforementioned multilevel path analysis models for 1–1–1 designs share a common problem. In the mediator model, the relationship between Z_{ij} and M_{ij} within a site is assumed to be the same as the relationship between site mean treatment $\bar{Z}_j$ and site mean mediator $\bar{M}_j$. In the outcome model, the relationship between M_{ij} and Y_{ij} conditioning on Z_{ij} within a site is assumed to be the same as the conditional relationship between site mean mediator $\bar{M}_j$ and site mean outcome $\bar{Y}_j$. In parallel, the relationship between Z_{ij} and Y_{ij} conditioning on M_{ij} within a site is assumed to be the same as the conditional relationship between site mean treatment $\bar{Z}_j$ and site mean outcome $\bar{Y}_j$.

Zhang, Zyphur, and Preacher (2009) revealed this conflation of between-group and within-group effects as a problem of model misspecification that could bias the causal effect estimates. Their solution was to decompose each causal effect into a within-site element and a between-site element by centering each level 1 predictor at its site mean and then reintroducing the site mean as a level 2 predictor:

$$M_{ij} = \alpha_0 + a_j(Z_{ij} - \bar{Z}_j) + a^*\bar{Z}_j + u_{M0j} + e_{Mij}, \quad \mathbf{u}_{Mj} \sim N(0, \tau_M), \quad e_{Mij} \sim N(0, \sigma_M^2);$$

$$Y_{ij} = \beta_0 + b_j(M_{ij} - \bar{M}_j) + b^*\bar{M}_j + c_j(Z_{ij} - \bar{Z}_j) + c^*\bar{Z}_j + u_{Y0j} + e_{Yij}, \quad \mathbf{u}_{Yj} \sim N(0, \tau_Y), \quad e_{Yij} \sim N(0, \sigma_Y^2).$$

According to these authors, $\hat{a}_j\hat{b}_j$ estimates the within-site natural indirect effect, while $\hat{a}^*\hat{b}^*$ estimates the between-site natural indirect effect. Similarly, $\hat{c}_j$ estimates the within-site natural direct effect, while $\hat{c}^*$ estimates the between-site natural direct effect. However, as pointed out earlier, conditioning on the site means $\bar{Z}_j$ and $\bar{M}_j$, the analysis generates the conditional variances of the site-specific natural indirect effect and of the site-specific natural direct effect rather than the unconditional variances. Moreover, Chapter 6 has shown that conditioning on $\bar{Z}_j$ leads to a biased estimate of the ITT effect.

The identification assumptions required by the multilevel path analysis methods are mostly implied by the structural form of the mediator model and that of the outcome model. These methods all require the sequential ignorability. Selection bias associated with the observed pretreatment covariates is reduced through covariance adjustment for such covariates in the two models, which requires correct specifications of the functional form of both models. The relationships among the treatment, the mediator, and the outcome are required to be linear additive, and hence, there must be no treatment-by-mediator interaction.

12.3   Alternative decomposition of the treatment effect

In this last section of the chapter, we discuss an alternative way of decomposing the total treatment effect in which the treatment-by-mediator interaction effect is conceptualized as a component of the direct effect rather than a component of the indirect effect. The decomposition of the total treatment effect into the natural direct effect $E[Y(1, M(0)) - Y(0, M(0))]$ and the natural indirect effect $E[Y(1, M(1)) - Y(1, M(0))]$ as defined by Pearl (2001) is not unique. In some applications, researchers might be primarily interested in investigating whether another potential mediator, even if unmeasured, might explain the treatment-by-mediator interaction. Chapter 9 presented a relatively simple example in which encouragement to study was expected to increase study hours, which in turn was expected to improve test performance. Suppose that the encouragement would increase the average amount of study from 4 to 8 hours/day. Moreover, suppose that students study more attentively if encouraged than if not encouraged. Some may argue that the increased attentiveness is a second potential mediator parallel to study hours and that this second mediator may explain why the effect of encouragement on test performance may depend on the amount of study. To be specific, the benefit of attentively studying 8 hours/day under encouragement as opposed to studying 8 hours/day with low attention due to the lack of encouragement is expected to be greater than the benefit of attentively studying 4 hours/day as opposed to studying 4 hours/day with low attention. The researcher would then have interest in "the total direct effect" defined as $E[Y(1, M(1)) - Y(0, M(1))]$ and would decompose it into "the pure direct effect" $E[Y(1, M(0)) - Y(0, M(0))]$ (Robins and Greenland, 1992) and the natural treatment-by-mediator interaction effect $E[Y(1, M(1)) - Y(0, M(1))] - E[Y(1, M(0)) - Y(0, M(0))]$. In this

Table 12.2   Parametric RMPW for the second set of treatment effect decomposition

	$E[Y(0, M(0))]$	$E[Y(1, M(1))]$	$E[Y(1, M(0))]$	$E[Y(0, M(1))]$
$D0$	0	1	0	1
$Z \times D0$	0	1	0	0
$Z \times D1$	0	0	1	0
W	1.0	1.0	$\dfrac{\theta_{M(0)}}{\theta_{M(1)}}$	$\dfrac{\theta_{M(1)}}{\theta_{M(0)}}$

example, the total direct effect would indicate whether encouragement would raise test performance if students would study 8 hours/day on average regardless of treatment assignment. The pure direct effect would indicate whether encouragement would raise test performance if students would study 4 hours/day on average. Due to the increase in attentiveness under encouragement, the direct effect of encouragement with 8 hours of study per day on average is expected to be greater than the direct effect with 4 hours of study per day on average. This difference is the "natural treatment-by-mediator interaction effect."

To estimate these causal effects, only a slight modification of the RMPW procedure in Steps 6 and 7 is required. In Step 6, the data are again reconstructed to include a duplicate set. As before, $D1$ is a dummy indicator that takes value 1 for the duplicate of an experimental unit and 0 for all other units, original or duplicate. Yet unlike the first set of decomposition, $D0$ is now a dummy indicator that takes value 1 not only for the duplicate of a control unit but also for an original experimental unit; $D0$ is 0 for the original control units as well as the duplicate experimental units. The parametric RMPW is $\theta_{M(0)=m}/\theta_{M(1)=m}$ for the duplicate experimental units whose $D1 = 1$, $D0 = 0$, and $M(1) = m$ and is $\theta_{M(1)=m}/\theta_{M(0)=m}$ for the duplicate control units whose $D1 = 0$, $D0 = 1$, and $M(0) = m$. The weight is 1.0 for the original control units and the original LFA units. The weighting scheme is summarized in Table 12.2. One then analyzes an outcome model specified as follows:

$$Y = \gamma^{(0)} + \gamma^{(IE.0)} D0 + \gamma^{(DE.1)} Z \times D0 + \gamma^{(DE.0)} Z \times D1 + e.$$

When the strong ignorability assumption holds, $\gamma^{(IE.0)}$ is unbiased for the pure indirect effect, $\gamma^{(DE.1)}$ is unbiased for the total direct effect, and $\gamma^{(DE.0)}$ is unbiased for the pure direct effect. Hence, $\gamma^{(DE.1)} - \gamma^{(DE.0)}$ is unbiased for the natural treatment-by-mediator interaction effect with the standard error of its estimate equal to $\mathrm{Var}\left(\hat{\gamma}^{(DE.1)}\right) + \mathrm{Var}\left(\hat{\gamma}^{(DE.0)}\right) - 2\mathrm{Cov}\left(\hat{\gamma}^{(DE.1)}, \hat{\gamma}^{(DE.0)}\right)$.

It is easy to show that the natural treatment-by-mediator interaction effect under this second decomposition $E[Y(1, M(1)) - Y(0, M(1))] - E[Y(1, M(0)) - Y(0, M(0))]$ is equivalent to that under the first decomposition $E[Y(1, M(1)) - Y(1, M(0))] - E[Y(0, M(1)) - Y(0, M(0))]$. In general, the treatment-induced change in the focal mediator and the existence of other potential mediators parallel to the focal mediator may jointly contribute to this interaction effect. Developing theoretical reasoning about the natural treatment-by-mediator interaction effect when it is relevant and generating empirical evidence to test the theoretical hypothesis are therefore crucial for advancing knowledge.

References

Baron, R.M. and Kenny, D.A. (1986). The moderator-mediator variable distinction in social psychological research: conceptual, strategic, and statistical considerations. *Journal of Personality and Social Psychology*, *51*, 1173–1182.

Bauer, D.J., Preacher, K.J. and Gil, K.M. (2006). Conceptualizing and testing random indirect effects and moderated mediation in multilevel models: new procedures and recommendations. *Psychological Methods*, *11*, 142–163.

Bradley, R.H., Corwyn, R.F., Burchinal, M. *et al.* (2001). The home environments of children in the United States part II: relations with behavioral development through age thirteen. *Child Development*, *72*(6), 1868–1886.

Glazerman, S., Isenberg, E., Dolfin, S. *et al.* (2010). Impacts of Comprehensive Teacher Induction: Final Results from a Randomized Controlled Study *(NCEE 2010-4027)*, U.S. Department of Education, Institute of Education Sciences, National Center for Education Evaluation and Regional Assistance, Washington, DC.

Greenland, S. (1987). Interpretation and choice of effect measures in epidemiologic analyses. *American Journal of Epidemiology*, *125*(5), 761–768.

Greenland, S., Robins, J.M. and Pearl, J. (1999). Confounding and collapsibility in causal inference. *Statistical Science*, *14*(1), 29–46.

Hong, G. and Nomi, T. (2012). Weighting methods for assessing policy effects mediated by peer change. *Journal of Research on Educational Effectiveness, special issue on the statistical approaches to studying mediator effects in education research*, *5*(3), 261–289.

Hong, G. and Raudenbush, S.W. (2006). Evaluating kindergarten retention policy: a case study of causal inference for multi-level observational data. *Journal of the American Statistical Association*, *101* (475), 901–910.

Huang, B., Sivaganesan, S., Succop, P. and Goodman, E. (2004). Statistical assessment of mediational effects for logistic mediational models. *Statistics in Medicine*, *23*, 2713–2728.

Huber, M. (2014). Identifying causal mechanisms (primarily) based on inverse probability weighting. *Journal of Applied Econometrics*, *29*(6), 920–943.

Imai, K., Keele, L. and Tingly, D. (2010). A general approach to causal mediation analysis. *Psychological Methods*, *15*(4), 309–334.

Imai, K., Keele, L. and Yamamoto, T. (2010). Identification, inference and sensitivity analysis for causal mediation effects. *Statistical Science*, *25*(1), 51–71.

Kenny, D.A., Korchmaros, J.D. and Bolger, N. (2003). Lower level mediation in multilevel models. *Psychological Methods*, *8*(2), 115–128.

Krull, J.L. and MacKinnon, D.P. (2001). Multilevel modeling of individual and group level mediated effects. *Multivariate Behavioral Research*, *36*(2), 249–277.

MacKinnon, D.P., Lockwood, C.M. and Williams, J. (2004). Confidence limits for the indirect effect: distribution of the product and resampling methods. *Multivariate Behavioral Research*, *39*, 99–128.

Meeker, W.Q., Cornwell, L.W. and Aroian, L.A. (1981). *Selected Tables in Mathematical Statistics, Vol. VII: The Product of Two Normally Distributed Random Variables*, American Mathematical Society, Providence, RI.

Miettinen, O.S. and Cook, E.F. (1981). Confounding: essence and detection. *American Journal of Epidemiology*, *114*(4), 595–603.

Pearl, J. (2001). *Direct and Indirect Effects*. Proceedings of the Seventeenth Conference on Uncertainty in Artificial Intelligence. Morgan Kaufmann, San Francisco, CA, pp. 411–420.

Pituch, K.A., Stapleton, L.M. and Kang, J.Y. (2006). A comparison of single sample and bootstrap methods to assess mediation in cluster randomized trials. *Multivariate Behavioral Research*, *41*(3), 367–400.

Preacher, K.J., Zyphur, M.J. and Zhang, Z. (2010). A general multilevel SEM framework for assessing multilevel mediation. *Psychological Methods*, *15*(3), 209–233.

Preacher, K.J., Zhang, Z. and Zyphur, M.J. (2011). Alternative methods for assessing mediation in multilevel data: the advantages of multilevel SEM, structural equation modeling. *A Multidisciplinary Journal*, *18*(2), 161–182.

Qin, X. and Hong, G. (2014). Causal mediation analysis in multi-site trials: An application of ratio-of-mediator-probability weighting to the Head Start Impact Study. In *JSM Proceedings*. Social Statistics Section. Alexandria, VA: American Statistical Association, pp. 912–926.

Raudenbush, S.W. (2014). *Random coefficient models for multi-site randomized trials with inverse probability of treatment weighting* (Working paper). Chicago, IL: University of Chicago.

Robins, J.M. and Greenland, S. (1992). Identifiability and exchangeability for direct and indirect effects. *Epidemiology*, *3*(2), 143–155.

Snow, C.E., Burns, M.S. and Griffin, P. (eds) (1998). *Preventing Reading Difficulties in Young Children*, National Academy Press, Washington, DC.

Tchetgen Tchetgen, E.J. (2013). Inverse odds ratio-weighted estimation for causal mediation analysis. *Statistics in Medicine*, *32*(26), 4567–4580.

VanderWeele, T.J. (2010). Direct and indirect effects for neighborhood-based clustered and longitudinal data. *Sociological Methods & Research*, *38*(4), 515–544.

VanderWeele, T. and Vansteelandt, S. (2010). Odds ratios for mediation analysis for a dichotomous outcome. *American Journal of Epidemiology*, *171*(12), 1339–1348.

Zhang, Z., Zyphur, M.J. and Preacher, K.J. (2009). Testing multilevel mediation using hierarchical linear models, problems and solutions. *Organizational Research Methods*, *12*(4), 695–719.

13

RMPW extensions to studies of complex mediation mechanisms

This final chapter in the unit on mediation analyses discusses further extensions of ratio-of-mediator-probability weighting (RMPW) to studies of mediation mechanisms moderated by individual or contextual pretreatment characteristics. We also discuss RMPW extensions to applications that involve more than one mediator. Researchers in the past have mostly employed path analysis and structural equation modeling (SEM) in investigating complex mediation mechanisms. Under a series of model-based assumptions, each causal effect of interest has been defined as a function of path coefficients. Yet the causal meaning of these coefficients often remains ambiguous or dubious. This chapter emphasizes decomposing the total treatment effect into causal effects defined in terms of potential outcomes. Under the identification assumptions to be explicated in the following, RMPW is utilized to estimate each of the population average counterfactual outcomes. The outcome models remain nonparametric despite the increasing complexity of the mediation mechanisms. Therefore, the RMPW analytic procedures minimize the reliance on arbitrary assumptions about the structural form of the outcome model.

13.1 RMPW extensions to moderated mediation

As discussed in Chapter 12, by examining heterogeneity in not only the total treatment effect but also the natural indirect effect and the natural direct effect across multiple experimental sites, multisite randomized trials provide opportunities for revealing possible site-level moderation of the mediation mechanisms. In a multisite randomized trial, a statistically significant between-site difference in the indirect effect or the direct effect would lend support to theoretical hypotheses with regard to *whether* the hypothesized mechanism represented by the indirect effect or any unspecified mechanism captured by the direct effect operates differently across the sites. Although the potential moderators remain implicit in such analyses, one may further investigate whether some particular site characteristics such as differences in local contexts, in participant composition, or in treatment implementation provide explanations for the between-site differences in the mediation mechanisms (Weiss, Bloom, & Brock, 2014).

Causality in a Social World: Moderation, Mediation and Spill-over, First Edition. Guanglei Hong.
© 2015 John Wiley & Sons, Ltd. Published 2015 by John Wiley & Sons, Ltd.

Moderated mediation has been an intriguing topic in the past literature. In both single-level and multilevel studies, an explicit moderator defines subpopulations between which the treatment effect and the mediation mechanisms may differ. Muller, Judd, and Yzerbyt (2005) emphasized the pretreatment nature of the moderator and described it as a measure of some stable individual or contextual differences that are assumed not to be affected by the treatment yet are related to different mediation mechanisms. For example, in the National Evaluation of Welfare-to-Work Strategies (NEWWS) Riverside study discussed in Chapter 11, more than a third of the welfare recipients had been teen parents in the past and appeared to respond differently to the employment mandate than those who had never become teen parents. Hong, Deutsch, and Hill (2011) found that the teen mothers, if assigned at random to the labor force attachment (LFA) program that emphasized active participation in job search rather than to the control condition that guaranteed cash assistance, displayed a greater improvement in employment status than did the nonteen mothers during the two years after randomization. Interestingly, the assignment to LFA appeared to have increased nonteen mothers' depressive symptoms while showing no such impact on the teen mothers. The researchers tested whether the treatment effect on maternal depression mediated through employment depended on past teen parenthood status. We explain the application of the RMPW analytic procedure in the following and contrast it with the path analysis/SEM approach. We then review the principal stratification method that has been proposed in the past literature for investigating moderated mediation.

13.1.1 RMPW analytic procedure for estimating and testing moderated mediation

Let $V = 1$ represent a teen parent and $V = 0$ represent a nonteen parent. Because the treatment was randomized, within each of these two subpopulations, the treatment assignment is independent of the potential mediators and the potential outcomes. It is essential that, within each subpopulation, the mediator value assignment is independent of the potential outcomes under each treatment condition for those sharing the same observed pretreatment characteristic. In other words, the assumption of sequential ignorability within each subpopulation must be invoked when we apply the RMPW method for causal mediation analysis.

To investigate teen parenthood as a potential moderator, conventionally, one would conduct two-group comparisons in SEM, a strategy that has limitations when the treatment and the mediator have an interaction effect in either or both subpopulations. The RMPW strategy avoids this constraint. The first six steps of the parametric RMPW procedure are to be applied within each subpopulation. Details of these steps have been explained in Chapter 11. In Step 7, the outcome model now includes two submodels, one for teen mothers and the other for nonteen mothers:

$$Y = V \times \left(\gamma_{V1}^{(0)} + \gamma_{V1}^{(DE.0)} Z + \gamma_{V1}^{(IE.1)} D1 \right) + (1 - V) \times \left[\gamma_{V0}^{(0)} + \gamma_{V0}^{(DE.0)} Z + \gamma_{V0}^{(IE.1)} D1 \right] + e.$$

When the sequential ignorability assumption holds, $\gamma_{V1}^{(DE.0)}$ and $\gamma_{V1}^{(IE.1)}$ are unbiased for the natural direct effect and the natural indirect effect, respectively, for teen mothers; $\gamma_{V0}^{(DE.0)}$ and $\gamma_{V0}^{(IE.1)}$ are unbiased for the natural direct effect and the natural indirect effect, respectively, for nonteen mothers. Additionally, one may test whether $\gamma_{V1}^{(DE.0)} = \gamma_{V0}^{(DE.0)}$ and $\gamma_{V1}^{(IE.1)} = \gamma_{V0}^{(IE.1)}$ by simply reparameterizing the aforementioned model as follows:

$$Y = \gamma_{V0}^{(0)} + \gamma_{V0}^{(DE.0)}Z + \gamma_{V0}^{(IE.1)}D1 + \gamma_{V1-0}^{(0)}V + \gamma_{V1-0}^{(DE.0)}V \times Z + \gamma_{V1-0}^{(IE.1)}V \times D1 + e,$$

where $\gamma_{V1-0}^{(DE.0)} = \gamma_{V1}^{(DE.0)} - \gamma_{V0}^{(DE.0)}$ is the average difference between teen mothers and nonteen mothers in the natural direct effect and $\gamma_{V1-0}^{(IE.1)} = \gamma_{V1}^{(IE.1)} - \gamma_{V0}^{(IE.1)}$ is the average difference between these two subpopulations in the natural indirect effect. We provide on the publisher's website the syntax files as well as a user-friendly RMPW software for conducting moderated mediation analysis. The web address is http://www.wiley.com/go/social_world.

Even though the estimated difference in the average natural direct effect between teen mothers and nonteen mothers amounts to nearly one-third of a standard deviation of the outcome, the relatively small sample size constrains the statistical power for distinguishing this interaction effect from zero ($\hat{\gamma}_{V1-0}^{(DE.0)} = -2.61$, SE = 1.75, $t = -1.49$, $p = 0.14$). Hence, one cannot conclude that without the policy-induced increase in employment, the assignment to the LFA program rather than to the control condition would have worsened depressive symptoms among nonteen mothers ($\hat{\gamma}_{V0}^{(DE.0)} = 2.31$, SE = 1.31, $t = 1.76$, $p = 0.08$) but not among teen mothers ($\hat{\gamma}_{V1}^{(DE.0)} = -0.29$, SE = 1.16, $t = -0.25$, $p = 0.80$). The natural indirect effect appears to be similar for these two subpopulations ($\hat{\gamma}_{V1-0}^{(IE.1)} = 0.31$, SE = 0.92, $t = 0.33$, $p = 0.74$).

13.1.2   Path analysis/SEM approach to analyzing moderated mediation

Researchers have extended the path analysis/SEM framework to test moderated direct and indirect effects (Morgan-Lopez et al., 2003; Morgan-Lopez and MacKinnon, 2006). For instance, James and Brett (1984) described an example of mediation mechanisms hypothesized to differ by subpopulations of individuals. To be specific, the mediator M is regressed not only on treatment assignment Z but also on subpopulation indicator V and the interaction ZV; the outcome Y is regressed on M and Z and, additionally, on V, MV, and ZV:

$$M = \alpha_0 + a_1 Z + a_2 V + a_3 ZV + e_M;$$

$$Y = \beta_0 + b_1 M + b_2 MV + c_1 Z + c_2 V + c_3 ZV + e_Y.$$

For individuals in subpopulation $V = v$, the treatment effect on the mediator is represented by $a_1 + a_3 v$; the mediator effect on the outcome under a given treatment is represented by $b_1 + b_2 v$ and is assumed to be constant across different treatment conditions; the direct effect of the treatment on the outcome at a given mediator value is represented by $c_1 + c_3 v$ and is assumed to be constant across different mediator values.

Analyzing the same set of path analysis models, Preacher, Rucker, and Hayes (2007) presented procedures with a focus on estimating and testing moderated indirect effects, which they called "conditional indirect effects." When the treatment effect on the mediator and the mediator effect on the outcome both differ across subpopulations indicated by V, following Muller, Judd, and Yzerbyt (2005), Preacher and colleagues estimated the indirect effect as $(\hat{a}_1 + \hat{a}_3 v)(\hat{b}_1 + \hat{b}_2 v)$ for those in subpopulation v. This representation of the indirect effect requires the assumption that the treatment effect on the mediator is independent of the mediator effect on the outcome. To test whether the indirect effect differs between two distinct subpopulations $V = 1$ versus $V = 0$, one needs to estimate the between-subpopulation difference as $\hat{a}_1\hat{b}_2 + \hat{b}_1\hat{a}_3 + \hat{a}_3\hat{b}_2$ and obtain its standard error through bootstrapping or the delta method. However, as their simulations showed, neither method arrived at an acceptable type I error rate. The mediator model and the outcome model specified earlier will become especially cumbersome in the presence of a relatively large number of confounding covariates $\mathbf{X}$ and especially if the $\mathbf{X}$–M relationships and the $\mathbf{X}$–Y relationships differ across the subpopulations. And as pointed out earlier, the results from analyzing the aforementioned path analysis models will be biased if there is a treatment-by-mediator interaction in one or more subpopulations. These limitations can effectively be overcome by the RMPW method.

13.1.3 Principal stratification and moderated mediation

When subpopulations are not identified a priori, some researchers argued that distinctions between subpopulations may be reflected in distinct mediator values (James and Brett, 1984). However, using the observed mediator values as subpopulation indicators is problematic and can produce misleading results. This is because mediators are intermediate outcomes of the treatment. The treated units and the control units who display the same mediator values may not belong to the same subpopulation. In the NEWWS Riverside example, some of the participants who were assigned to the LFA program and became employed might have remained unemployed if assigned to the control condition instead and therefore do not belong to the same subpopulation with the employed control units.

The principal stratification framework (Frangakis and Rubin, 2002) seems to be particularly useful for clarifying the aforementioned confusion. This framework represents subpopulations of individuals by a set of their potential mediators $M(1)$ and $M(0)$. When the treatment is given, $M(1)$ and $M(0)$ are each strictly a function of individual pretreatment characteristics. Hence, principal stratification characterizes pretreatment differences among individuals. In the simplest case, those who would be employed if assigned to LFA and would be unemployed if assigned to the control condition constitute one subpopulation and are often labeled as "compliers," those who would be employed regardless of treatment assignment constitute a second subpopulation and are often labeled as "always-takers," those who would be unemployed regardless of treatment assignment constitute a third subpopulation and are often labeled as "never-takers," and so on. One may employ a Bayesian approach to empirically assess each individual's probability of being a member of each subpopulation given the observed pretreatment information (Gallop et al., 2009; Jin and Rubin, 2008, 2009; Page, 2012), conduct a mediation analysis within each subpopulation, and compare across the subpopulations. This is a strategy yet to be explored by statisticians and applied researchers.

13.2   RMPW extensions to concurrent mediators

A comprehensive intervention program typically contains multiple elements that are intended to produce a desired outcome through multiple pathways. Hence, more often than not, theoretical considerations of causal mediation mechanisms involve more than one mediator. Basic patterns of complex mechanisms include concurrent mediators and consecutive mediators. Two mediators are *concurrent* if they are independent of each other under a given treatment condition; that is, for a given individual assigned to a certain treatment, a change in one mediator is not necessarily accompanied by a change in the other. Two mediators are consecutive if one precedes the other logically and temporally and if a change in the first mediator would lead to a change in the second. Each pattern may have some variants; and the two patterns may be combined in multiple ways. Using the Head Start Impact Study to illustrate, in the following, we examine several theoretical examples, define the causal effects in each case, and discuss possible applications of the RMPW strategy in contrast with the existing path analysis/SEM strategies.

A two-generation approach by design, Head Start programs attempt to improve the school readiness skills of preschool-aged children from low-income households by simultaneously providing high-quality nonparental care at Head Start centers and improving the quality of parental care. Most children assigned to the control condition attended a non-Head Start day-care center or were cared for by relatives or babysitters at a certain point. The control children may experience nonparental care and parental care with a relatively lower quality and therefore may develop school readiness skills at a lower level in comparison with the Head Start children. In theory, the quality of nonparental care and that of parental care can be viewed as *concurrent mediators* of the treatment effect on the outcome. For simplicity, here, we consider the population at a single site in which the Head Start children and the control children received both nonparental care and parental care. Chapter 12 has discussed estimation strategies for causal mediation analysis with data from multisite randomized trials.

13.2.1 Treatment effect decomposition

As before, let $z = 1$ if a child was assigned at random to a Head Start program and 0 otherwise. Let $M1(z)$ be the first mediator representing the quality of nonparental care for the child under treatment z; and let $M2(z)$ be the second mediator representing the quality of parental care under treatment z for $z = 0,1$. Finally, let $Y(z, M1(z), M2(z))$ denote the potential outcome of the child (e.g., vocabulary at kindergarten entry) under treatment z. Hence, the total effect of the treatment on the outcome is

$$Y(1, M1(1), M2(1)) - Y(0, M1(0), M2(0)).$$

13.2.1.1 Treatment effect decomposition without between-mediator interaction

Figure 13.1 shows two concurrent mediators on parallel pathways connecting the treatment and the outcome. For the simplicity of presentation, the figure has tentatively omitted potential confounders of the $M1$–Y relationship and of the $M2$–Y relationship. The total effect of the treatment can be decomposed into three parts: a natural indirect effect transmitted through the quality of parental care $M2$ if a child was assigned to Head Start,

$$Y(1, M1(1), M2(1)) - Y(1, M1(1), M2(0)); \tag{13.1}$$

a second natural indirect effect transmitted through the quality of nonparental care $M1$ should Head Start fail to improve parental care,

$$Y(1, M1(1), M2(0)) - Y(1, M1(0), M2(0)); \tag{13.2}$$

and a natural direct effect of the treatment transmitted through neither $M1$ nor $M2$,

$$Y(1, M1(0), M2(0)) - Y(0, M1(0), M2(0)). \tag{13.3}$$

The decomposition as described above involves four potential outcomes: $Y(1, M1(1), M2(1))$, $Y(1, M1(1), M2(0))$, $Y(1, M1(0), M2(0))$, and $Y(0, M1(0), M2(0))$. Here, $Y(1, M1(1),$

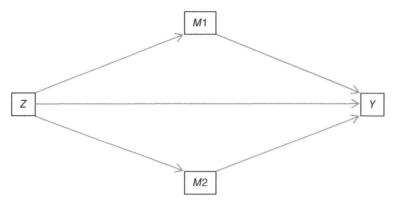

Figure 13.1   Two concurrent mediators

Table 13.1   Glossary of potential outcomes associated with two concurrent mediators

Potential outcome	Definition
$Y(1, M1(1), M2(1))$	Potential outcome if an individual is treated
$Y(1, M1(1), M2(0))$	Potential outcome if an individual is treated yet the treatment fails to change the value of the second mediator
$Y(1, M1(0), M2(1))$	Potential outcome if an individual is treated yet the treatment fails to change the value of the first mediator
$Y(1, M1(0), M2(0))$	Potential outcome if an individual is treated yet the treatment changes the value of neither the first nor the second mediator
$Y(0, M1(0), M2(0))$	Potential outcome if an individual is untreated

$M2(0))$ is the potential outcome if a child is assigned to Head Start yet the program counterfactually fails to improve the quality of parental care; $Y(1, M1(0), M2(0))$ is the potential outcome if a child is assigned to Head Start yet the program counterfactually fails to improve the quality of nonparental care and parental care. Table 13.1 provides a glossary for the potential outcomes associated with two concurrent mediators.

The aforementioned causal effects are defined for each individual in the population. The corresponding population average causal effects can be obtained in each case by taking an average over all the individuals in the population. Table 13.2 shows the decomposition of the population average total effect into an indirect effect mediated by $M2$, an indirect effect mediated by $M1$, and a direct effect. The notation for each causal effect is listed in the last column of this table. For example, a superscript "$(IE(1).1.0)$" indicates that this is the natural indirect effect mediated by $M1$ under the experimental condition $z = 1$ when the other mediator is counterfactually unchanged by the treatment; a superscript "$(IE(2).1.1)$" indicates that this is the natural indirect effect mediated by $M2$ under the experimental condition $z = 1$ when the other mediator is changed by the treatment.

The decomposition is not unique and can be equated with an alternative set of decomposition when there is no interaction between the two mediators, that is, if the transmission of the treatment effect through one mediator does not depend on the value of the concurrent mediator. When this is the case, the natural indirect effect transmitted through the quality

Table 13.2   Decomposition of the population average total effect with two noninteracting concurrent mediators

$E[Y(1,M1(1),M2(1))-Y(0,M1(0),M2(0))]$ $= E[Y(1,M1(0),M2(0))-Y(0,M1(0),M2(0))]$	Total treatment effect Natural direct effect of the treatment	$\gamma^{(DE,0)}$
$+E[Y(1,M1(1),M2(0))-Y(1,M1(0),M2(0))]$	Natural indirect effect mediated by $M1$	$\gamma^{(IE(1),1,0)}$
$+E[Y(1,M1(1),M2(1))-Y(1,M1(1),M2(0))]$	Natural indirect effect mediated by $M2$	$\gamma^{(IE(2),1,1)}$

of parental care $M2$ for a child assigned to Head Start will remain the same even if the treatment perhaps counterfactually fails to improve the quality of nonparental care. That is, (13.1) will be equal to

$$Y(1,M1(0),M2(1))-Y(1,M1(0),M2(0)).$$

Here, $Y(1,M1(0),M2(1))$ is the potential outcome if a child is assigned to Head Start yet the program counterfactually fails to improve the quality of nonparental care $M1$. Similarly, without an interaction between the two mediators, the natural indirect effect transmitted through the quality of nonparental care $M1$ will remain the same regardless of whether Head Start improves parental care $M2$. That is, (13.2) will be equal to

$$Y(1,M1(1),M2(1))-Y(1,M1(0),M2(1)).$$

The direct effect stays the same as that defined in (13.3). This alternative set of decomposition involves the potential outcome $Y(1,M1(0),M2(1))$ rather than $Y(1,M1(1),M2(0))$. Importantly, both sets of decomposition allow the relationship between each mediator and the outcome to depend on the treatment condition. Imai and Yamamoto (2013) similarly defined the indirect effect with respect to $M1$, the indirect effect with respect to $M2$, and the direct effect.

13.2.1.2   Treatment effect decomposition with between-mediator interaction

One may suspect that the treatment effect mediated by nonparental care would be enhanced if the treatment could simultaneously improve the quality of parental care. For example, literacy skills that a child acquires in a Head Start program will likely be reinforced if the parent reads to the child at home on a daily basis. Under this hypothesis, the treatment effect mediated by nonparental care $M1$ when the treatment successfully improves the quality of parental care $M2$ is expected to be greater than the treatment effect mediated by $M1$ when the treatment fails to improve $M2$, that is,

$$Y(1,M1(1),M2(1))-Y(1,M1(0),M2(1))>Y(1,M1(1),M2(0))-Y(1,M1(0),M2(0)).$$

Table 13.3   Decomposition of the population average total effect with two interacting concurrent mediators

$E[Y(1,M1(1),M2(1))-Y(0,M1(0),M2(0))]$	Total treatment effect	
$=E[Y(1,M1(0),M2(0))-Y(0,M1(0),M2(0))]$	Natural direct effect of the treatment	$\gamma^{(DE.0)}$
$+E[Y(1,M1(1),M2(0))-Y(1,M1(0),M2(0))]$	Natural indirect effect mediated by $M1$	$\gamma^{(IE(1).1.0)}$
$+E[Y(1,M1(0),M2(1))-Y(1,M1(0),M2(0))]$	Natural indirect effect mediated by $M2$	$\gamma^{(IE(2).1.0)}$
$+\{E[Y(1,M1(1),M2(1))-Y(1,M1(1),M2(0))]$ $-E[Y(1,M1(0),M2(1))-Y(1,M1(0),M2(0))]\}$	Natural indirect effect mediated by the $M1$-by-$M2$ interaction	$\gamma^{(IE(2).1.1)}-$ $\gamma^{(IE(2).1.0)}$

One may view the difference between these two quantities as the treatment effect mediated by the interaction effect of $M1$ and $M2$ on Y.

In the presence of $M1$ and $M2$ interaction, the natural indirect effect mediated by $M1$ when Head Start has an impact on $M2$ may be further decomposed into two components. The first is the natural indirect effect mediated by $M2$ when Head Start hypothetically shows no impact on $M1$ and is defined as

$$Y(1,M1(0),M2(1))-Y(1,M1(0),M2(0));   (13.4)$$

and the second is the natural indirect effect mediated by the interaction effect of $M1$ and $M2$ and is defined as the difference between (13.1) and (13.4):

$$[Y(1,M1(1),M2(1))-Y(1,M1(1),M2(0))]-[Y(1,M1(0),M2(1))-Y(1,M1(0),M2(0))].   (13.5)$$

Accordingly, one may define the population average causal effects mediated by $M1$, by $M2$, and by their interaction in addition to the population average direct effect. Table 13.3 shows the decomposition of the population average total effect of the treatment on the outcome in the presence of $M1$-by-$M2$ interaction. The decomposition involves the estimation of five population average potential outcomes:

$$E[Y(1,M1(1),M2(1))];$$

$$E[Y(1,M1(1),M2(0))];$$

$$E[Y(1,M1(0),M2(1))];$$

$$E[Y(1,M1(0),M2(0))];$$

$$E[Y(0,M1(0),M2(0))].$$

13.2.2   Identification assumptions

The key identification assumptions again can be summarized in terms of the sequential ignorability. That is, we assume (i) that the treatment assignment Z is independent of all the potential outcomes and of all the potential mediators given the observed pretreatment covariates $\mathbf{X}$, (ii) that the mediator value assignment $M1$ is independent of all the potential outcomes given Z and $\mathbf{X}$, and (iii) that the mediator value assignment $M2$ is independent of all the potential outcomes given Z and $\mathbf{X}$. Assumptions (ii) and (iii) imply that, given the observed pretreatment covariates $\mathbf{X}$, the relationship between each mediator and the outcome under each treatment condition should not be confounded by unobserved pretreatment covariates; nor should it be confounded by unobserved or observed posttreatment covariates. These are, of course, very strong assumptions and may not hold in many cases especially if the observed pretreatment covariates fail to explain most of the variation in the outcome. In the current example, the observed pretreatment covariates may include baseline child and family characteristics as well as site characteristics such as the effectiveness of the Head Start programs and of the non-Head Start programs in the past year. Additionally, we invoke assumption (iv) that, given Z and $\mathbf{X}$, $M1$ and $M2$ are independent of each other. For the decomposition described earlier, these assumptions are summarized in Table 13.4. In an experiment in which only the treatment has been randomized, assumptions (ii) and (iii) are not directly testable and therefore must be scrutinized on scientific grounds; only assumption (iv) can be partially examined under each treatment by assessing the covariance between $M1$ and $M2$ within levels of the observed pretreatment covariates. Nonetheless, the RMPW approach makes considerably fewer assumptions than the linear SEM approach and the multivariate IV approach to be discussed in Sections 13.2.4 and 13.2.5.

13.2.3   RMPW procedure

Due to the randomization of treatment assignment, the population average potential outcome under Head Start $E[Y(1, M1(1), M2(1))]$ can be estimated without bias from the observed mean outcome of the Head Start group; similarly, the population average potential outcome under the control condition $E[Y(0, M1(0), M2(0))]$ can be estimated without bias from the observed mean outcome of the control group. The initial randomization would also enable the researcher to obtain unbiased estimates of the treatment effect on nonparental care $E[M1(1) - M1(0)]$ and of the treatment effect on parental care $E[M2(1) - M2(0)]$. The question is how to employ the RMPW strategy to estimate the population average counterfactual outcomes.

Suppose that, following the initial randomization of treatment assignment, individual children in each treatment group were hypothetically assigned at random to either high-quality nonparental care or low-quality nonparental care and to either high-quality parental care or low-quality parental care. This second randomization in the sequence is a factorial design that may reflect the Head Start impacts on the mediators through designating a higher probability of receiving high-quality nonparental care and a higher probability of receiving high-quality parental care to the Head Start children rather than to the control children. The following sections describe the RMPW strategy for estimating the population average counterfactual outcomes under this simplified sequential randomization design and then derive the weight needed for estimation when the mediators are not randomized.

Table 13.4 Sequential ignorability required by the RMPW strategy for two concurrent mediators

Assumption	Assignment of	Independent of	Conditional on
(i)	Z	$Y(1, M1(1), M2(1)), Y(1, M1(1), M2(0)),$ $Y(1, M1(0), M2(0)), Y(0, M1(0), M2(0)), Y(1, M1(0), M2(1)),$ $M1(1), M1(0),$ $M2(1), M2(0)$	$\mathbf{X} = \mathbf{x}$
(ii)	$M1(1), M1(0)$	$Y(1, M1(1), M2(1)), Y(1, M1(1), M2(0)),$ $Y(1, M1(0), M2(0)), Y(0, M1(0), M2(0)), Y(1, M1(0), M2(1))$	$Z = z, \mathbf{X} = \mathbf{x}$
(iii)	$M2(1), M2(0)$	$Y(1, M1(1), M2(1)), Y(1, M1(1), M2(0)),$ $Y(1, M1(0), M2(0)), Y(0, M1(0), M2(0)), Y(1, M1(0), M2(1))$	$Z = z, \mathbf{X} = \mathbf{x}$
(iv)	$M1(z)$	$M2(z)$	$Z = z, \mathbf{X} = \mathbf{x}$

13.2.3.1   Estimating $E[Y(1, M1(1), M2(0))]$

To estimate the average potential outcome when the entire population is assigned to Head Start yet parental care remains at the same quality as that under the control condition, under the hypothetical sequential randomization, every Head Start child who has actually received parental care at quality level $m2$ is assigned the weight:

$$\frac{\text{pr}(M2(0) = m2|Z=0)}{\text{pr}(M2(1) = m2|Z=1)}.$$

Here, the denominator is the child's probability of experiencing parental care at quality level $m2$ under the Head Start program that the child was actually assigned to; the numerator is the child's counterfactual probability of experiencing parental care at quality level $m2$ under the control condition. After weighting, the distribution of the quality of parental care in the Head Start group will resemble that in the control group. Because the assignment to high-quality parental care is independent of the assignment to high-quality nonparental care, the aforementioned weighting will not change the distribution of $M1$ (i.e., the quality of nonparental care) in the Head Start group.

To remove selection bias when only the treatment is randomized, we predict a Head Start child's probability of receiving parental care at quality level $m2$ under each treatment condition, denoted by $\theta_{M2(z)=m2} = \theta_{M2(z)=m2}(\mathbf{x}) = \text{pr}(M2(z) = m2|Z=z, \mathbf{X}=\mathbf{x})$ for $z = 0,1$, as a function of the observed pretreatment covariates $\mathbf{x}$ that may confound the $M2$–Y relationship. Appendix 13.A(a) derives the following weight for a Head Start child who has received parental care at quality level $m2$:

$$\frac{\theta_{M2(0)=m2}}{\theta_{M2(1)=m2}}.$$

The distribution of $M2$ (i.e., the quality of parental care) in the Head Start group will resemble that in the control group after weighting under the assumption that, within levels of the pretreatment covariates, $M2$ is independent of $M1$ under each treatment condition.

13.2.3.2   Estimating $E[Y(1, M1(0), M2(0))]$

This is the average potential outcome when the entire population is assigned to Head Start yet both nonparental care and parent care counterfactually remain at the same quality as that under the control condition. To estimate this population average potential outcome under the hypothetical sequential randomization, every Head Start child who has actually received nonparental care at quality level $m1$ and parental care at quality level $m2$ is assigned the following weight:

$$\frac{\text{pr}(M1(0) = m1|Z=0)}{\text{pr}(M1(1) = m1|Z=1)} \times \frac{\text{pr}(M2(0) = m2|Z=0)}{\text{pr}(M2(1) = m2|Z=1)}.$$

Here, the denominator is the child's joint probability of experiencing a pair of mediator values $m1$ and $m2$ when assigned to the Head Start program; the numerator is the child's counterfactual joint probability of displaying $m1$ and $m2$ if assigned to the control condition instead. After weighting, the joint distribution of $M1$ and $M2$ in the Head Start group will resemble that in the control group.

To remove selection bias when only the treatment is randomized, we predict each mediator probability under each treatment condition as a function of the observed pretreatment covariates $\mathbf{x}$ that may confound the $M1$–Y relationship, the $M2$–Y relationship, or both. The weight is

$$\frac{\theta_{M1(0)=m1}}{\theta_{M1(1)=m1}} \times \frac{\theta_{M2(0)=m2}}{\theta_{M2(1)=m2}},$$

where $\theta_{M1(z)=m1} = \theta_{M1(z)=m1}(\mathbf{x}) = \mathrm{pr}(M1(z)=m1|Z=z, \mathbf{X}=\mathbf{x})$ denotes a child's conditional probability of receiving nonparental care at quality level $m1$ under treatment condition z for $z=0,1$ and $\theta_{M2(z)=m2}$ is the same as that defined previously. Appendix 13.A(b) shows the derivation of the weight.

13.2.3.3   Causal effect estimation with noninteracting concurrent mediators

To implement the weighting strategy as described earlier, one would fit to each treatment group a propensity score model for $M1$ and a second propensity score model for $M2$ and then apply the control group models to the Head Start group. One thereby estimates every Head Start child's propensity scores $\theta_{M1(1)=m1}$ and $\theta_{M2(1)=m2}$ and predicts the child's counterfactual propensity scores $\theta_{M1(0)=m1}$ and $\theta_{M2(0)=m2}$. To avoid the potential problem of overfitting the prediction model, again one may resort to cross-validation or leave-one-out bootstrap. One then generates two duplicate sets of the experimental group indicated by dummies $D1\_1$ and $D1\_2$ to be merged with the original experimental group and the control group. These four groups are weighted according to the scheme shown in Table 13.5. The weighted outcome model takes the following basic form:

$$Y = \gamma_0 + \gamma^{(DE.0)}Z + \gamma^{(IE(1).1.0)}D1\_1 + \gamma^{(IE(2).1.1)}D1\_2 + e. \tag{13.6}$$

Under the identification assumptions explicated in Section 13.2.2, γ_0 is unbiased for the population average potential outcome under the control condition; $\gamma^{(DE.0)}$ is unbiased for the natural direct effect of being assigned to Head Start rather than to the control condition as

Table 13.5   Parametric RMPW for treatment effect decomposition with two noninteracting concurrent mediators

	Z	$D1\_1$	$D1\_2$	W
$E[Y(0, M1(0), M2(0))]$	0	0	0	1.0
$E[Y(1, M1(0), M2(0))]$	1	0	0	$\dfrac{\theta_{M1(0)=m1}}{\theta_{M1(1)=m1}} \times \dfrac{\theta_{M2(0)=m2}}{\theta_{M2(1)=m2}}$
$E[Y(1, M1(1), M2(0))]$	1	1	0	$\dfrac{\theta_{M2(0)=m2}}{\theta_{M2(1)=m2}}$
$E[Y(1, M1(1), M2(1))]$	1	1	1	1.0

defined in (13.3) when the mediators are taking values as they would have under the control condition; $\gamma^{(IE(1).1.0)}$ is unbiased for the natural indirect effect of $M1$ (i.e., the quality of non-parental care) under Head Start as defined in (13.2) when $M2$ (i.e., the quality of parental care) is counterfactually unchanged by Head Start; and $\gamma^{(IE(2).1.1)}$ estimates the natural indirect effect of $M2$ under Head Start as defined in (13.1) when M1 under the control condition counter-factually shows the same value as that under Head Start. Finally, $\gamma^{(IE(1).1.0)} + \gamma^{(IE(2).1.1)}$ is unbi-ased for the sum of the natural indirect effect mediated by $M1$ and that by $M2$ with a standard error equal to the square root of

$$\operatorname{Var}\left(\hat{\gamma}^{(IE(1).1.0)}\right) + \operatorname{Var}\left(\hat{\gamma}^{(IE(2).1.1)}\right) + 2\operatorname{Cov}\left(\hat{\gamma}^{(IE(1).1.0)}, \hat{\gamma}^{(IE(2).1.1)}\right).$$

13.2.3.4   Estimating $E[Y(1, M1(0), M2(1))]$

In order to test whether the treatment effect is partially mediated by an interaction between two mediators, one needs to estimate an additional population average potential outcome when the entire population is assigned to Head Start yet nonparental care remains at the same quality as that under the control condition. The estimation will involve assigning the following weight to every Head Start child:

$$\frac{\theta_{M1(0) = m1}}{\theta_{M1(1) = m1}},$$

where $\theta_{M1(z) = m1}$ for $z = 0,1$ is the same as that defined earlier. See Appendix 13.A(c) for the derivation of the weight. The distribution of $M1$ (i.e., the quality of nonparental care) in the weighted Head Start group will resemble that in the control group. Within levels of the pre-treatment covariates, when $M1$ is independent of $M2$ under each treatment condition, the weighting will not change the distribution of $M2$ (i.e., the quality of parental care) in the Head Start group.

13.2.3.5   Causal effect estimation with interacting concurrent mediators

When the two concurrent mediators interact in transmitting the treatment effect on the out-come, one would create a third duplicate set of the experimental group and merge it with the earlier data. The weighted outcome model is then modified as follows to further decom-pose the total effect of the treatment:

$$Y = \gamma_0 + \gamma^{(DE.0)}Z + \gamma^{(IE(1).1.0)}D1\_1 + \gamma^{(IE(2).1.0)}D1\_2 + \gamma^{(IE(2).1.1)}D1\_3 + e.$$

The weighting scheme is summarized in Table 13.6. Here, γ_0, $\gamma^{(DE.0)}$, $\gamma^{(IE(1).1.0)}$, and $\gamma^{(IE(2).1.1)}$ are the same as those defined in model (13.6). Specifically, under the identification assumptions explicated earlier, $\gamma^{(IE(1).1.0)}$ is unbiased for the natural indirect effect of $M1$ (i.e., the quality of nonparental care) under Head Start as defined in (13.2) when $M2$ (i.e., the quality of parental care) is counterfactually unchanged by Head Start; $\gamma^{(IE(2).1.1)}$ is unbiased for the

Table 13.6   Parametric RMPW for treatment effect decomposition with two interacting concurrent mediators

	Z	$D1\_1$	$D1\_2$	$D1\_3$	W
$E[Y(0, M1(0), M2(0))]$	0	0	0	0	1.0
$E[Y(1, M1(0), M2(0))]$	1	0	0	0	$\dfrac{\theta_{M1(0)=m1}}{\theta_{M1(1)=m1}} \times \dfrac{\theta_{M2(0)=m2}}{\theta_{M2(1)=m2}}$
$E[Y(1, M1(1), M2(0))]$	1	1	0	0	$\dfrac{\theta_{M2(0)=m2}}{\theta_{M2(1)=m2}}$
$E[Y(1, M1(0), M2(1))]$	1	0	1	0	$\dfrac{\theta_{M1(0)=m1}}{\theta_{M1(1)=m1}}$
$E[Y(1, M1(1), M2(1))]$	1	1	1	0	1.0

natural indirect effect of $M2$ under Head Start as defined in (13.1) when $M1$ presumably has been improved by Head Start. Additionally, $\gamma^{(IE(2).1.0)}$ is unbiased for the natural indirect effect of $M2$ under Head Start as defined in (13.4) when $M1$ is counterfactually unchanged by Head Start. The estimation of these regression coefficients under weighting and the computation of the standard errors are similarly to those shown in Appendix 11.A. Additionally, $\gamma^{(IE(2).1.1)} - \gamma^{(IE(2).1.0)}$ is unbiased for the natural indirect effect mediated by the interaction between $M1$ and $M2$ as defined in (13.5). The standard error of the estimate of this last causal effect is the square root of

$$\mathrm{Var}\left(\widehat{\gamma}^{(IE(2).1.1)}\right) + \mathrm{Var}\left(\widehat{\gamma}^{(IE(2).1.0)}\right) - 2\mathrm{Cov}\left(\widehat{\gamma}^{(IE(2).1.1)}, \widehat{\gamma}^{(IE(2).1.0)}\right).$$

In the presence of a nonzero interaction between $M1$ and $M2$ that mediates the treatment effect on the outcome, if one analyzes model (13.6) instead, an estimate of $\gamma^{(IE(2).1.1)}$ will be the estimated sum of $E[Y(1, M1(0), M2(1)) - Y(1, M1(0), M2(0))]$ and $E[Y(1, M1(1), M2(1)) - Y(1, M1(0), M2(1))] - E[Y(1, M1(1), M2(0)) - Y(1, M1(0), M2(0))]$. The first part of this sum is the natural indirect effect of $M2$ as defined in (13.4) and the second part is the natural indirect effect of $M1$-by-$M2$ interaction equivalent to that defined in (13.5).

The RMPW strategy can be extended to analyses with three or more concurrent mediators once the data analyst defines the causal effects of interest. The outcome model will generate estimates of the natural direct effect and the natural indirect effect transmitted through each mediator along with the standard errors. Extensions can also be made to multivalued mediators and binary outcomes.

13.2.4   Contrast with the linear SEM approach

Adopting the linear additive SEM framework instead, researchers have proposed analyzing three structural models for two mediators:

$$M1 = \alpha_{M1} + a_1 Z + e_{M1};$$

$$M2 = \alpha_{M2} + a_2 Z + e_{M2};$$

$$Y = \beta_0 + b_1 M1 + b_2 M2 + cZ + e_Y.$$

Under the assumption that all three models are correctly specified, c is viewed as the natural direct effect of the treatment on the outcome, $a_1 b_1$ as the natural indirect effect transmitted through $M1$, and $a_2 b_2$ as the natural indirect effect transmitted through $M2$. The standard error of the sample estimate $\hat{a}_1 \hat{b}_1$ is the square root of $\hat{a}_1^2 \mathrm{Var}(\hat{b}_1) + \hat{b}_1^2 \mathrm{Var}(\hat{a}_1)$, while the standard error of the sample estimate $\hat{a}_2 \hat{b}_2$ is the square root of $\hat{a}_2^2 \mathrm{Var}(\hat{b}_2) + \hat{b}_2^2 \mathrm{Var}(\hat{a}_2)$. To determine whether the total indirect effect, represented as $a_1 b_1 + a_2 b_2$, is zero, the multivariate delta method has been employed in the past to compute the asymptotic standard error of its sample estimate as the square root of the following quantity (Bollen, 1989):

$$\mathrm{Var}(\hat{a}_1 \hat{b}_1 + \hat{a}_2 \hat{b}_2) = \mathrm{Var}(\hat{a}_1 \hat{b}_1) + \mathrm{Var}(\hat{a}_2 \hat{b}_2) + 2[\hat{a}_1 \hat{a}_2 \mathrm{Cov}(\hat{b}_1, \hat{b}_2) + \hat{b}_1 \hat{b}_2 \mathrm{Cov}(\hat{a}_1, \hat{a}_2)].$$

The inference is applicable only when the estimates $\hat{a}_1 \hat{b}_1$ and $\hat{a}_2 \hat{b}_2$ are multivariate normal. Others (Briggs, 2006; Preacher and Hayes, 2008; Williams and MacKinnon, 2008) have recommended using bootstrapping instead which does not require the multivariate normality assumption.

The SEM approach, similar to the RMPW strategy, assumes the sequential ignorability. However, the SEM approach requires a series of additional assumptions. It generally assumes that the mediator–outcome relationships are all linear and that there is no interaction between the treatment and either mediator, nor is there an interaction between the mediators in the outcome model. The estimate of each natural indirect effect takes much more complex forms when these assumptions are relaxed. Finally, covariance adjustment for pretreatment covariates $\mathbf{X}$ in each model imposes assumptions about the functional form of the covariate–mediator and covariate–outcome relationships, including the assumptions that there are no interactions between the treatment and $\mathbf{X}$ and between each mediator and $\mathbf{X}$. A violation of any of these assumptions will lead to bias in the causal effect estimation.

13.2.5   Contrast with the multivariate IV approach

A number of studies published in the recent years (Duncan, Morris, and Rodrigues, 2011; Kling, Liebman, and Katz, 2007; Raudenbush, Reardon, and Nomi, 2012) have represented attempts to identify the impacts of multiple concurrent mediators on an outcome of interest in multisite randomized trials by using site-by-treatment interactions as instrumental variables. Suppose that a study includes J experimental sites denoted by $j = 1, \ldots, J$. The multivariate IV approach takes advantage of the fact that one may estimate the site-specific total effect of the treatment on the outcome denoted by c'_j and the site-specific treatment effects on the concurrent mediators $M1$ and $M2$ denoted by a_{1j} and a_{2j}, respectively. The goal is to obtain unbiased estimates of the population average effects of $M1$ and $M2$ on Y denoted by b_1 and b_2, respectively. Let b_{1j} and b_{2j} denote the respective site-specific effects of these concurrent mediators on the outcome. Here, c'_j, a_{1j}, a_{2j}, b_{1j}, and b_{2j} are random variables that can take different values across the sites in the population. This approach requires that the treatment effect on the outcome is completely transmitted through $M1$ and $M2$ (i.e., the exclusion restriction). Under a number of additional identification assumptions delineated below, we have that

$$c'_j = a_{1j}b_{1j} + a_{2j}b_{2j}$$
$$= b_1 a_{1j} + (b_{1j} - b_1)a_{1j} + b_2 a_{2j} + (b_{2j} - b_2)a_{2j}$$
$$= b_1 a_{1j} + b_2 a_{2j} + e_j,$$

where $e_j = (b_{1j} - b_1)a_{1j} + (b_{2j} - b_2)a_{2j}$. By regressing c'_j on a_{1j} and a_{2j}, one obtains sample estimates of b_1 and b_2. Let a_1 and a_2 denote the population average effects of the treatment on $M1$ and $M2$, respectively. The treatment randomization ensures that a_1 and a_2 can be estimated without bias. One may then compute $\hat{a}_1 \hat{b}_1$ as an estimate of the natural indirect effect mediated by $M1$ and $\hat{a}_2 \hat{b}_2$ as an estimate of the natural indirect effect mediated by $M2$.

Instead of assuming the sequential ignorability, the multivariate IV approach assumes that Z is ignorable within each experimental site, an assumption easy to satisfy in a multisite randomized trial. Hence, it may appear that this approach does not require $M1$ and $M2$ to be ignorable under each treatment condition. However, as Reardon and Raudenbush (2013) have shown, this approach works under the condition that there is no within-site covariance between the individual-specific treatment effect on each mediator and the individual-specific effect of each mediator on the outcome. This condition requires that $\text{Cov}(a_{1i}, b_{1i}) = \text{Cov}(a_{2i}, b_{2i}) = 0$ at the individual level within each experimental site. An additional condition is that the site-specific treatment effect on each mediator is independent of the site-specific effect of each mediator on the outcome, that is, $\text{Cov}(a_{1j}, b_{1j}) = \text{Cov}(a_{2j}, b_{2j}) = 0$, which ensures that $E(e_j) = 0$. Reardon and Raudenbush (2013) considered this a nontrivial assumption. In the Head Start example, at the experimental sites where the quality of Head Start programs was considerably higher than that of the alternative childcare available to the control children, it seems likely that nonparental care would play a significant role in promoting children's school readiness skills due to the deprivation of other resources at such sites. The aforementioned reasoning suggests a positive covariance between a_{1j} and b_{1j}. The zero covariance assumptions are met when the mediator value assignment under each treatment condition is independent of the potential outcomes within and across the treatment conditions. Hence, one may argue that the multivariate IV approach again relies on the sequential ignorability. The zero covariance assumptions could perhaps be made plausible conditioning on individual and site characteristics if such covariates are entered into each mediator model and the outcome model, which will then raise issues, however, with regard to the functional relationships involving the covariates. Finally, in addition to the exclusion restriction, this approach assumes a linear treatment–mediator relationship and a linear mediator–outcome relationship for each mediator. According to Reardon and Raudenbush (2013), when regressing c'_j on a_{1j} and a_{2j}, the design matrix will have sufficient rank only if there are at least as many sites as mediators, if the treatment has a constant effect on at most one mediator across sites, and if the mediators are linearly independent.

13.3   RMPW extensions to consecutive mediators

In most intervention studies, there is a time span of months or years between the initial treatment assignment and the collection of outcome data. Hence, the intermediate process may involve multiple mediators that affect one another in a sequential manner. For example, the Head Start Impact Study evaluated the long-term effect of being assigned to a Head Start program versus the control condition on children's academic achievement at the end of kindergarten. In theory, the treatment effect on a child's kindergarten outcome would be mediated initially by the quality of nonparental care during the preschool years and subsequently by the quality of the kindergarten program that the child attended. High-quality nonparental care received by a Head Start child would likely show long-term benefit if it was to be followed by high-quality instruction in kindergarten. However, if kindergarten instruction failed to build on a child's strength previously gained from a high-quality Head Start program, the benefit of high-quality nonparental care in preschool would possibly dissipate over the kindergarten year.

13.3.1   Treatment effect decomposition

We use Z to denote treatment assignment and Y for the kindergarten outcome. Y could be a child's achievement score measured at the end of the kindergarten year, a gain score over the preschool and kindergarten years, or a growth parameter estimated through longitudinal modeling of repeated assessments of achievement over these same years. The first mediator $M1$ denotes the quality of nonparental care prior to kindergarten, while the subsequent mediator $M2$ denotes the quality of kindergarten instruction. Figure 13.2 illustrates the mechanisms of interest here. Although path diagrams appear to be intuitive, the decomposition of the total treatment effect eventually becomes transparent when the causal effects are defined in terms of potential outcomes.

In the current example, $M1$ is an intermediate outcome of Z; $M2$ an intermediate outcome of Z and $M1$; and Y the final outcome of Z, $M1$, and $M2$. Hence, a child assigned to the Head Start program indicated by $z = 1$ would have prekindergarten nonparental care at quality level $M1(1)$, kindergarten instruction at quality level $M2(1, M1(1))$, and kindergarten outcome $Y(1, M1(1),$ $M2(1, M1(1)))$. The same child, if assigned to the control condition ($z = 0$) instead, would have prekindergarten nonparental care at quality level $M1(0)$, kindergarten instruction at quality level $M2(0, M1(0))$, and kindergarten outcome $Y(0, M1(0), M2(0, M1(0)))$. The difference between these two potential outcomes defines the *total treatment effect* on the final outcome:

$$Y(1, M1(1), M2(1, M1(1))) - Y(0, M1(0), M2(0, M1(0))).$$

Before one proceeds, it is important to empirically examine whether $M1$ mediates the effect of Z on $M2$, whether $M2$ mediates the effect of Z on Y, and whether $M2$ mediates the effect of $M1$ on Y. Moreover, in each of these mediation analyses, empirical evidence

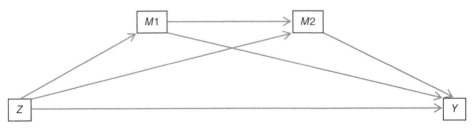

Figure 13.2   Two consecutive mediators

can be generated with regard to whether Z has a direct effect on $M2$ and a direct effect on Y and whether $M1$ has a direct effect on Y. The RMPW strategy described in Chapters 11 and 12 can be readily applied in these analyses. Suppose that the identification assumptions hold and that the data have sufficient statistical power. If each of these analyses detects a nonzero natural indirect effect and a nonzero natural direct effect, then we will proceed with decomposing the total treatment effect of Z on Y into the pathways shown in Figure 13.2. If, on the contrary, some of these effects are zero, one may simplify the path diagram accordingly. In the following, we discuss, in the context of the Head Start evaluation, the decomposition of the total treatment effect useful for addressing a series of theoretical questions.

13.3.1.1   Natural direct effect of the treatment on the outcome

Is improving the quality of educational institutions including preschool and kindergarten programs the only mechanisms through which Head Start contributed to low-income children's academic success? One may reason that if Head Start parents organized themselves into a social network, shared resources, and provided mutual support in child-rearing throughout their children's preschool and kindergarten years, this might constitute an unspecified pathway captured by the *natural direct effect* of the treatment on the outcome, a causal effect *mediated neither by M1 nor by M2*:

$$Y(1,M1(0),M2(0,M1(0)))-Y(0,M1(0),M2(0,M1(0))). \tag{13.7}$$

13.3.1.2   Natural indirect effect mediated by $M1$ only

If a Head Start program offered nonparental care of a higher quality than what was typically available under the control condition, one interesting research question is whether the benefit due to the program-induced improvement in nonparental care, represented by the difference between $M1(1)$ and $M1(0)$, would sustain and manifest in the kindergarten outcome even if the quality of kindergarten instruction was counterfactually unaffected by the treatment assignment. This causal effect is the natural indirect effect mediated by $M1$ but not by $M2$ and can be interpreted as the long-term impact of Head Start solely attributable to the quality change in preschool nonparental care:

$$Y(1,M1(1),M2(0,M1(0)))-Y(1,M1(0),M2(0,M1(0))). \tag{13.8}$$

13.3.1.3   Natural indirect effect mediated by $M2$ only

The treatment effect on the outcome was possibly mediated by $M2$ bypassing $M1$. Head Start children might find access to high-quality kindergarten instruction just by virtue of having been assigned to Head Start in the past even though the quality of their preschool nonparental care was no better than what would have been available under the control condition. One would then ask what the value of the enhanced quality of kindergarten instruction is for Head Start children. This is the *natural indirect effect mediated by M2 but not by M1* when the quality of kindergarten instruction for a Head Start child changes from $M2(0, M1(0))$ to $M2(1, M1(1))$. The causal effect is defined as follows:

$$Y(1,M1(0),M2(1,M1(1)))-Y(1,M1(0),M2(0,M1(0))). \tag{13.9}$$

13.3.1.4   Natural indirect effect mediated by an *M1*-by-*M2* interaction

One may further hypothesize that if a child was initially assigned to Head Start, the impact of kindergarten instruction on the outcome would be greater should the treatment assignment already change the quality of preschool nonparental care favorably and that the impact of kindergarten instruction would be lesser should the treatment assignment fail to improve preschool nonparental care. This hypothesis points to an interaction effect between $M1$ and $M2$ on the outcome under Head Start represented as the difference between two causal effects: the first is the natural indirect effect of kindergarten instruction $M2$ when preschool nonparental care is a potential outcome of Head Start denoted by $M1(1)$; the second is the natural indirect effect of $M2$ when $M1$ is a potential outcome of the control condition denoted by $M1(0)$:

$$E[Y(1,M1(1),M2(1,M1(1)))-Y(1,M1(1),M2(0,M1(0)))]$$
$$-E[Y(1,M1(0),M2(1,M1(1)))-Y(1,M1(0),M2(0,M1(0)))]. \tag{13.10}$$

The total effect of the treatment on the outcome is now decomposed into a natural direct effect as defined in (13.7), a natural indirect effect of $M1$ as defined in (13.8) when $M2$ is counterfactually unaffected by the treatment, a natural indirect effect of $M2$ as defined in (13.9) when $M1$ is counterfactually unaffected by the treatment, and the natural indirect effect of $M1$-by-$M2$ interaction as defined in (13.10). The decomposition is summarized in Table 13.7. This decomposition involves five population average potential outcomes:

$$E[Y(1,M1(1),M2(1,M1(1)))],$$
$$E[Y(1,M1(0),M2(1,M1(1)))],$$
$$E[Y(1,M1(1),M2(0,M1(0)))],$$
$$E[Y(1,M1(0),M2(0,M1(0)))],$$
$$E[Y(0,M1(0),M2(0,M1(0)))].$$

13.3.1.5   Treatment-by-mediator interactions

In addition to an $M1$-by-$M2$ interaction effect on the outcome as defined in (13.10), one may further hypothesize a Z-by-$M1$ interaction effect and a Z-by-$M2$ interaction effect if preschool nonparental care and kindergarten instruction each influences the end-of-kindergarten outcome differently across different treatment conditions. The Z-by-$M1$ interaction effect is defined by

$$E[Y(1,M1(1),M2(0,M1(0)))-Y(1,M1(0),M2(0,M1(0)))]$$
$$-E[Y(0,M1(1),M2(0,M1(0)))-Y(0,M1(0),M2(0,M1(0)))]; \tag{13.11}$$

and the Z-by-$M2$ interaction effect is defined by

$$E[Y(1,M1(0),M2(1,M1(1)))-Y(1,M1(0),M2(0,M1(0)))]$$
$$-E[Y(0,M1(0),M2(1,M1(1)))-Y(0,M1(0),M2(0,M1(0)))]. \tag{13.12}$$

Table 13.7 Decomposition of the population average total effect with two interacting consecutive mediators

$$E[Y(1,M1(1),M2(1,M1(1))) - Y(0,M1(0),M2(0,M1(0)))]$$

$= E[Y(1,M1(0),M2(0,M1(0))) - Y(0,M1(0),M2(0,M1(0)))]$	Total treatment effect	
$+\, E[Y(1,M1(1),M2(0,M1(0))) - Y(1,M1(0),M2(0,M1(0)))]$	Natural direct effect of the treatment	$\gamma^{(DE.0)}$
$+\, E[Y(1,M1(0),M2(1,M1(1))) - Y(1,M1(0),M2(0,M1(0)))]$	Natural indirect effect mediated by $M1$ only	$\gamma^{(IE(1).1.0)}$
$+\, \{E[Y(1,M1(1),M2(1,M1(1))) - Y(1,M1(1),M2(0,M1(0)))]$	Natural indirect effect mediated by $M2$	$\gamma^{(IE(2).1.0)}$
$-\, E[Y(1,M1(0),M2(1,M1(1))) - Y(1,M1(0),M2(0,M1(0)))]\}$	Natural indirect effect mediated by an $M1$-by-$M2$ interaction	$\gamma^{(IE(2).1.1)} - \gamma^{(IE(2).1.0)}$

Table 13.8 Decomposition of the population average total effect with two interacting consecutive mediators and treatment-by-mediator interactions

$$E[Y(1,M1(1),M2(1,M1(1))) - Y(0,M1(0),M2(0,M1(0)))]$$

$= E[Y(1,M1(0),M2(0,M1(0))) - Y(0,M1(0),M2(0,M1(0)))]$	Total treatment effect	
$+\, E[Y(0,M1(1),M2(0,M1(0))) - Y(0,M1(0),M2(0,M1(0)))]$	Natural direct effect of the treatment	$\gamma^{(DE.0)}$
$+\, E[Y(0,M1(0),M2(1,M1(1))) - Y(0,M1(0),M2(0,M1(0)))]$	Natural indirect effect of $M1$	$\gamma^{(IE(1).0.0)}$
$+\, \{E[Y(1,M1(1),M2(0,M1(0))) - Y(1,M1(0),M2(0,M1(0)))]$	Natural indirect effect of $M2$	$\gamma^{(IE(2).0.0)}$
$-\, E[Y(0,M1(1),M2(0,M1(0))) - Y(0,M1(0),M2(0,M1(0)))]\}$	Natural indirect effect mediated by a Z-by-$M1$ interaction	$\gamma^{(IE(1).1.0)} - \gamma^{(IE(1).0.0)}$
$+\, \{E[Y(1,M1(0),M2(1,M1(1))) - Y(1,M1(0),M2(0,M1(0)))]$	Natural indirect effect mediated by a Z-by-$M2$ interaction	$\gamma^{(IE(2).1.0)} - \gamma^{(IE(2).0.0)}$
$-\, E[Y(0,M1(0),M2(1,M1(1))) - Y(0,M1(0),M2(0,M1(0)))]\}$		
$+\, \{E[Y(1,M1(1),M2(1,M1(1))) - Y(1,M1(1),M2(0,M1(0)))]$	Natural indirect effect mediated by an $M1$-by-$M2$ interaction	$\gamma^{(IE(2).1.1)} - \gamma^{(IE(2).1.0)}$
$-\, E[Y(1,M1(0),M2(1,M1(1))) - Y(1,M1(0),M2(0,M1(0)))]\}$		

The total effect of the treatment on the outcome is now decomposed into six components as summarized in Table 13.8. These include a natural direct effect as defined in (13.7), a natural indirect effect of $M1$ under the control condition defined as

$$E[Y(0,M1(1),M2(0,M1(0)))-Y(0,M1(0),M2(0,M1(0)))], \qquad (13.13)$$

a natural indirect effect of $M2$ under the control condition defined as

$$E[Y(0,M1(0),M2(1,M1(1)))-Y(0,M1(0),M2(0,M1(0)))], \qquad (13.14)$$

in addition to a Z-by-M1 interaction effect, a Z-by-M2 interaction effect, and an M1-by-M2 interaction effect. The decomposition now involves the following seven potential outcomes:

$$E[Y(1,M1(1),M2(1,M1(1)))],$$
$$E[Y(1,M1(1),M2(0,M1(0)))],$$
$$E[Y(1,M1(0),M2(1,M1(1)))],$$
$$E[Y(1,M1(0),M2(0,M1(0)))],$$
$$E[Y(0,M1(0),M2(0,M1(0)))],$$
$$E[Y(0,M1(1),M2(0,M1(0)))],$$
$$E[Y(0,M1(0),M2(1,M1(1)))].$$

Under the identification assumptions to be clarified in the next section, we can estimate all these potential outcomes by employing the RMPW strategy.

13.3.2 Identification assumptions

Applying RMPW to studies involving consecutive mediators requires stronger assumptions about the sequential ignorability than other mediation problems discussed earlier. Specifically, to obtain unbiased estimates of the causal effects requires the following:

(i) Treatment assignment Z is independent of all the potential outcomes and the potential mediators given the observed pretreatment covariates $\mathbf{X}$.

(ii) Given treatment assignment Z and the observed pretreatment covariates $\mathbf{X}$, the assignment of the first mediator $M1$ is independent of all the potential outcomes and the potential mediators.

(iii*) Given Z, $M1$, and $\mathbf{X}$, the assignment of the second mediator $M2$ is independent of all the potential outcomes.

For the decompositions described earlier, these assumptions are summarized in Table 13.9. When all these three assumptions are met, one may view the data as generated from a three-stage sequential randomized experiment within levels of the observed pretreatment covariates $\mathbf{X} = \mathbf{x}$. Specifically, individuals are assigned at random to the treatment $Z = 1$ or the control condition $Z = 0$ at the first stage of the experiment; those in the same treatment

Table 13.9 Sequential ignorability required by the RMPW strategy for two consecutive mediators

Assumption	Assignment of	Independent of	Conditional on
(i)	Z	$Y(1, M1(1), M2(1, M1(1)), Y(0, M1(1), M2(0, M1(0)))$ $Y(0, M1(0), M2(1, M1(1)))$ $Y(1, M1(1), M2(0, M1(0)), Y(1, M1(0), M2(1, M1(1)))$ $Y(1, M1(0), M2(0, M1(0)), Y(0, M1(0), M2(0, M1(0)))$ $M1(1), M1(0)$ $M2(1, M1(1)), M2(0, M1(0))$	$\mathbf{X = x}$
(ii)	$M1(1), M1(0)$	$Y(1, M1(1), M2(1, M1(1)), Y(0, M1(1), M2(0, M1(0)))$ $Y(0, M1(0), M2(1, M1(1)))$ $Y(1, M1(1), M2(0, M1(0)), Y(1, M1(0), M2(1, M1(1)))$ $Y(1, M1(0), M2(0, M1(0)), Y(0, M1(0), M2(0, M1(0)))$ $M2(1, M1(1)), M2(0, M1(0))$	$Z = z, \mathbf{X = x}$
(iii*)	$M2(1, m1), M2(0, m1)$	$Y(1, M1(1), M2(1, M1(1)), Y(0, M1(1), M2(0, M1(0)))$ $Y(0, M1(0), M2(1, M1(1)))$ $Y(1, M1(1), M2(0, M1(0)), Y(1, M1(0), M2(1, M1(1)))$ $Y(1, M1(0), M2(0, M1(0)), Y(0, M1(0), M2(0, M1(0)))$	$Z = z, M1(z) = m1, \mathbf{X = x}$

group z are assigned at random to different values of the first mediator $M1(z)$ at the second stage for $z = 0,1$; those in the same treatment group z and with the same mediator value $m1$ are assigned at random to different values of the second mediator $M2(z, m1)$ at the third stage for $z = 0,1$ and for all possible values of $m1$.

In the Head Start evaluation presented earlier, assumption (i) is warranted by the random assignment of children to Head Start or control within each experimental site. Assumption (ii) requires that, within each treatment group and among those who shared the same pretreatment characteristics, the assignment of children to preschool high-quality versus low-quality nonparental care was "as if random" and was not confounded by other unobserved or observed factors including those that could have been caused by the treatment assignment. Assumption (iii*) further requires that, within each treatment group and among those who received preschool nonparental care at the same quality level and shared the same pretreatment characteristics, the assignment of children to high-quality versus low-quality kindergarten instruction was "as if random" and was not confounded by other unobserved or observed factors including those that could have been caused by the treatment assignment and those that could have been caused by the quality of preschool nonparental care. Assumption (iii*) is different from assumption (iii) required in the analyses of two concurrent mediators because, when $M1$ and $M2$ are concurrent, the ignorability of $M2$ is not conditional on the value of $M1$.

Assumption (iii*) may seem difficult to satisfy because a Head Start program, especially one that provides high-quality care, might change other aspects of a child's life such as increasing parental involvement and improving parental care during the preschool years. Parents empowered by the Head Start programs might subsequently seek involvement in elementary school and manage to secure high-quality kindergarten instruction with a higher likelihood than control parents might. Nonetheless, the plausibility of assumption (iii*) can be increased by two alternative strategies. The first strategy, sometimes feasible in an experimental study in collaboration with school districts and schools, is to assign children at random to kindergarten programs and assign teachers at random to kindergarten classes; the second strategy involves measuring, in addition to $M1$, all the other possible mediators in the pathways from Z to $M2$ and explicitly defining and estimating the natural indirect effects that are transmitted by $M2$ through these additional pathways.

13.3.3 RMPW procedure

Under the randomization of treatment assignment, the observed mean outcome of the Head Start group and that of the control group are unbiased estimates of the population average potential outcome associated with Head Start $E[Y(1, M1(1), M2(1, M1(1)))]$ and that associated with the control condition $E[Y(0, M1(0), M2(0, M1(0)))]$, respectively. The RMPW strategy can be employed for estimating the other five population average potential outcomes.

13.3.3.1 Estimating $E[Y(1, M1(0), M2(0, M1(0)))]$

The population average natural direct effect involves the population average potential outcome when the entire population is assigned to Head Start and yet the treatment counterfactually generates no impact on the quality of either preschool nonparental care or kindergarten instruction. According to the derivation shown in Appendix 13.B(a), for a Head Start child experiencing nonparental care at quality level $m1$ and subsequently experiencing kindergarten instruction at quality level $m2$, the weight is

$$\frac{\theta_{M1(0)=m1}}{\theta_{M1(1)=m1}} \times \frac{\theta_{M2(0,m1)=m2}}{\theta_{M2(1,m1)=m2}}, \tag{13.15}$$

where

$$\theta_{M1(z)=m1} = \theta_{M1(z)=m1}(\mathbf{x}) = \mathrm{pr}(M1(z)=m1 \mid Z=z, \mathbf{X}=\mathbf{x}),$$

$$\theta_{M2(z,m1)=m2} = \theta_{M2(z,m1)=m2}(\mathbf{x}) = \mathrm{pr}(M2(z,m1)=m2 \mid Z=z, M1(z)=m1, \mathbf{X}=\mathbf{x})$$

for $z=0,1$. Here, $\theta_{M1(z)=m1}$ is the child's conditional probability of receiving preschool nonparental care at quality level $m1$ under treatment condition z; $\theta_{M2(z,m1)=m2}$ is the child's conditional probability of receiving kindergarten instruction at quality level $m2$ if the child has been assigned to treatment condition z and has received preschool nonparental care at quality level $m1$ for $z=0,1$ and for all possible values of $m1$.

Suppose that $M1$ and $M2$ are both binary taking value 1 for high quality and 0 for low quality. One may fit a logistic regression model for $M1$ as a function of the pretreatment characteristics $\mathbf{x}$ in each treatment group. The model fitted to the Head Start group estimates every Head Start child's propensity of receiving high-quality nonparental care while attending Head Start, which is denoted by $\theta_{M1(1)=1}$. The model fitted to the control group, when applied to the Head Start group, predicts every Head Start child's counterfactual propensity of receiving high-quality preschool nonparental care under the control condition, which is denoted by $\theta_{M1(0)=1}$.

A Head Start child's propensity of receiving high-quality kindergarten instruction after having received preschool nonparental care at quality level $m1$ is denoted by $\theta_{M2(1,m1)=1}$. One may fit a logistic regression model for $M2$, again as a function of the pretreatment characteristics $\mathbf{x}$, to the subsample of Head Start children who actually received preschool nonparental care at the same quality level $m1$ as the focal child. The same child would have a counterfactual propensity of receiving high-quality kindergarten instruction after having supposedly received preschool nonparental care at quality level $m1$ under the control condition, which is denoted by $\theta_{M2(0,m1)=1}$. To predict this counterfactual propensity score, one may fit a logistic regression model for $M2$, as a function of the pretreatment characteristics $\mathbf{x}$, to the subsample of control children who actually received preschool nonparental care at quality level $m1$. One may then apply the fitted model to the focal child in the Head Start group.

The decomposition then requires merging one duplicate set of the experimental group indicated by $D1$ with the original experimental group and the control group. The original experimental group is to be weighted by (13.15). The weighted outcome model is simply

$$Y = \gamma_0 + \gamma^{(DE.0)}Z + \gamma^{(IE.1)}D1 + e.$$

Under the identification assumptions clarified in Section 13.3.2, γ_0 is unbiased for the population average potential outcome under the control condition; $\gamma^{(DE.0)}$ is unbiased for the population average natural direct effect mediated by neither $M1$ nor $M2$ as defined in (13.7) (note: as before, the superscript $DE.0$ indicates the natural direct effect of the treatment when both mediators take values as they would have under the control condition); and $\gamma^{(IE.1)}$ is unbiased for the population average total indirect effect mediated by $M1$ and $M2$ under Head Start, which is a sum of (13.8), (13.9), and (13.10) for the entire population. The estimation of these causal effects and their standard errors are the same as those shown in Appendix 11.A.

13.3.3.2   Estimating $E[Y(1, M1(1), M2(0, M1(0)))]$

Alternatively, one may argue that the Head Start-induced improvement in the quality of preschool nonparental care could independently contribute to a child's kindergarten outcome even if kindergarten instruction was not responsive to the child's readiness skills. This would require estimating the population average potential outcome if all eligible children were assigned to Head Start yet receiving kindergarten instruction at a quality level they would have under the control condition. A Head Start child actually experiencing nonparental care at quality level $m1$ and kindergarten instruction at quality level $m2$ is to be weighted simply by

$$\frac{\theta_{M2(0,m1)=m2}}{\theta_{M2(1,m1)=m2}}, \tag{13.16}$$

where, as before, $\theta_{M2(z,m1)=m2}=\theta_{M2(z,m1)=m2}(\mathbf{x})=\mathrm{pr}(M2(z,m1)=m2|Z=z,M1(z)=m1,\mathbf{X}=\mathbf{x})$ for $z=0,1$. See Appendix 13.B(b) for the derivation of the weight.

13.3.3.3   Estimating $E[Y(1, M1(0), M2(1, M1(1)))]$

To estimate the treatment effect mediated by $M2$ only, the analysis involves estimating the population average counterfactual outcome when all children are assigned to Head Start yet counterfactually experiencing preschool nonparental care at the same quality level as that under the control condition. As shown in Appendix 13.B(c), under the identification assumptions clarified earlier, an unbiased estimate of $E[Y(1, M1(0), M2(1, M1(1)))]$ can be obtain by assigning the following weight to the experimental group:

$$\frac{\theta_{M1(0)=m1}}{\theta_{M1(1)=m1}}, \tag{13.17}$$

where, as defined earlier, $\theta_{M1(z)=m1}=\theta_{M1(z)=m1}(\mathbf{x})=\mathrm{pr}(M1(z)=m1|Z=z,\mathbf{X}=\mathbf{x})$ for $z=0,1$.

The contrast between $E[Y(1, M1(1), M2(0, M1(0)))]$ and $E[Y(1, M1(0), M2(0, M1(0)))]$ is the natural indirect effect mediated by $M1$ only as defined in (13.8); and the contrast between $E[Y(1, M1(0), M2(1, M1(1)))]$ and $E[Y(1, M1(0), M2(0, M1(0)))]$ is the natural indirect effect mediated by $M2$ only as defined in (13.9). To estimate these two natural indirect effects, we merge two duplicate sets of the experimental group with the original experimental group and the control group. The weight is assigned as follows: the RMPW specified in (13.15) is to be assigned to the original experimental group indicated by $D1\_1=0$ and

Table 13.10 Parametric RMPW for treatment effect decomposition with two consecutive mediators

	Z	$D1\_1$	$D1\_2$	W
$E[Y(0, M1(0), M2(0, M1(0)))]$	0	0	0	1.0
$E[Y(1, M1(0), M2(0, M1(0)))]$	1	0	0	$\dfrac{\theta_{M1(0)=m1}}{\theta_{M1(1)=m1}} \times \dfrac{\theta_{M2(0,m1)=m2}}{\theta_{M2(1,m1)=m2}}$
$E[Y(1, M1(1), M2(0, M1(0)))]$	1	1	0	$\dfrac{\theta_{M2(0,m1)=m2}}{\theta_{M2(1,m1)=m2}}$
$E[Y(1, M1(0), M2(1, M1(1)))]$	1	0	1	$\dfrac{\theta_{M1(0)=m1}}{\theta_{M1(1)=m1}}$

$D1\_2 = 0$ for estimating $E[Y(1, M1(0), M2(0, M1(0)))]$; the RMPW specified in (13.16) is to be assigned to the first duplicate experimental group indicated by $D1\_1 = 1$ and $D1\_2 = 0$ for estimating $E[Y(1, M1(1), M2(0, M1(0)))]$; the RMPW specified in (13.17) is to be assigned to the second duplicate experimental group indicated by $D1\_1 = 0$ and $D1\_2 = 1$ for estimating $E[Y(1, M1(0), M2(1, M1(1)))]$. The control group is weighted simply by 1.0 for estimating $E[Y(0, M1(0), M2(0, M1(0)))]$. This weighting scheme is summarized in Table 13.10. The weighted outcome model is

$$Y = \gamma_0 + \gamma^{(DE.0)}Z + \gamma^{(IE(1).1.0)}D1\_1 + \gamma^{(IE(2).1.0)}D1\_2 + e. \tag{13.18}$$

As before, $\widehat{\gamma}_0$ estimates the population average potential outcome under the control condition; $\widehat{\gamma}^{(DE.0)}$ estimates the population average natural direct effect mediated by neither $M1$ nor $M2$ as defined in (13.7); $\widehat{\gamma}^{(IE(1).1.0)}$ estimates the population average natural indirect effect under Head Start mediated by $M1$ but not by $M2$ as defined in (13.8); and $\widehat{\gamma}^{(IE(2).1.0)}$ estimates the population average natural indirect effect under Head Start mediated by $M2$ but not by $M1$ as defined in (13.9).

To estimate the natural indirect effect mediated by an $M1$-by-$M2$ interaction, the analysis will then involve a third duplicate set of the experimental group indicated by $D1\_3$ and weighted simply by 1.0. The weighted outcome model is specified as follows with the weighting scheme summarized in Table 13.11:

$$Y = \gamma_0 + \gamma^{(DE.0)}Z + \gamma^{(IE(1).1.0)}D1\_1 + \gamma^{(IE(2).1.0)}D1\_2 + \gamma^{(IE(2).1.1)}D1\_3 + e. \tag{13.19}$$

One obtains $\widehat{\gamma}^{(IE(2).1.1)}$ estimating the natural indirect effect mediated by $M2$ under Head Start when $M1$ takes values associated with the experimental condition. Therefore, $\widehat{\gamma}^{(IE(2).1.1)} - \widehat{\gamma}^{(IE(2).1.0)}$ estimates the natural indirect effect mediated by the interaction between $M1$ and $M2$ with its standard error equal to the square root of $\text{Var}\left(\widehat{\gamma}^{(IE(2).1.1)}\right) + \text{Var}\left(\widehat{\gamma}^{(IE(2).1.0)}\right) - 2\text{Cov}\left(\widehat{\gamma}^{(IE(2).1.1)}, \widehat{\gamma}^{(IE(2).1.0)}\right)$.

Table 13.11   Parametric RMPW for treatment effect decomposition with two interacting consecutive mediators

	Z	$D1\_1$	$D1\_2$	$D1\_3$	W
$E[Y(0, M1(0), M2(0, M1(0)))]$	0	0	0	0	1.0
$E[Y(1, M1(0), M2(0, M1(0)))]$	1	0	0	0	$\dfrac{\theta_{M1(0)=m1}}{\theta_{M1(1)=m1}} \times \dfrac{\theta_{M2(0,m1)=m2}}{\theta_{M2(1,m1)=m2}}$
$E[Y(1, M1(1), M2(0, M1(0)))]$	1	1	0	0	$\dfrac{\theta_{M2(0,m1)=m2}}{\theta_{M2(1,m1)=m2}}$
$E[Y(1, M1(0), M2(1, M1(1)))]$	1	0	1	0	$\dfrac{\theta_{M1(0)=m1}}{\theta_{M1(1)=m1}}$
$E[Y(1, M1(1), M2(1, M1(1)))]$	1	1	0	1	1.0

13.3.3.4 Estimating $E[Y(0, M1(1), M2(0, M1(0)))]$ and $E[Y(0, M1(0), M2(1, M1(1)))]$

Two additional potential outcomes are involved in estimating the treatment interaction with the first mediator and its interaction with the second mediator. To estimate $E[Y(0, M1(1), M2(0, M1(0)))]$, we simply need to use weighting to transform the $M1$ distribution in the control group such that it resembles the $M1$ distribution in the experimental group. The weight to be applied to the control group for this purpose is

$$\frac{\theta_{M1(1)=m1}}{\theta_{M1(0)=m1}}. \tag{13.20}$$

To estimate $E[Y(0, M1(0), M2(1, M1(1)))]$, we may transform the $M2$ distribution in the control group such that it resembles the $M2$ distribution in the experimental group by applying to the control group the following weight:

$$\frac{\theta_{M2(1,m1)=m2}}{\theta_{M2(0,m1)=m2}}. \tag{13.21}$$

The derivations are similar to those shown in Appendix 13.B(c) and Appendix 13.B(b) and therefore are omitted here.

To decompose the total treatment effect into a natural direct effect, a natural indirect effect of $M1$, a natural indirect effect of $M2$, a Z-by-$M1$ interaction effect, a Z-by-$M2$ interaction effect, and an $M1$-by-$M2$ interaction effect as shown in Table 13.8, we need to estimate seven potential outcomes, four under the experimental condition and three under the control condition. Hence, we additionally generate two duplicate sets of the control group denoted by $D0\_1$ and $D0\_2$ weighted by (13.20) and (13.21), respectively. We then analyze the following model:

$$Y = \gamma_0 + \gamma^{(DE.0)}Z + \gamma^{(IE(1).0.0)}D0\_1 + \gamma^{(IE(2).0.0)}D0\_2$$

$$+ \gamma^{(IE(1).1.0)}D1\_1 + \gamma^{(IE(2).1.0)}D1\_2 + \gamma^{(IE(2).1.1)}D1\_3 + e. \tag{13.22}$$

The weighting scheme is summarized in Table 13.12. Here, $\widehat{\gamma}^{(IE(1).0.0)}$ estimates the natural indirect effect of $M1$ under the control condition; and $\widehat{\gamma}^{(IE(2).0.0)}$ estimates the natural indirect

Table 13.12 Parametric RMPW for treatment effect decomposition with two interacting consecutive mediators and treatment-by-mediator interactions

	Z	$D0\_1$	$D0\_2$	$D1\_1$	$D1\_2$	$D1\_3$	W
$E[Y(0, M1(0), M2(0, M1(0)))]$	0	0	0	0	0	0	1.0
$E[Y(0, M1(1), M2(0, M1(0)))]$	0	1	0	0	0	0	$\dfrac{\theta_{M1(1)=m1}}{\theta_{M1(0)=m1}}$
$E[Y(0, M1(0), M2(1, M1(1)))]$	0	0	1	0	0	0	$\dfrac{\theta_{M2(1,m1)=m2}}{\theta_{M2(0,m1)=m2}}$
$E[Y(1, M1(0), M2(0, M1(0)))]$	1	0	0	0	0	0	$\dfrac{\theta_{M1(0)=m1}}{\theta_{M1(1)=m1}} \times \dfrac{\theta_{M2(0,m1)=m2}}{\theta_{M2(1,m1)=m2}}$
$E[Y(1, M1(1), M2(0, M1(0)))]$	1	0	0	1	0	0	$\dfrac{\theta_{M2(0,m1)=m2}}{\theta_{M2(1,m1)=m2}}$
$E[Y(1, M1(0), M2(1, M1(1)))]$	1	0	0	0	1	0	$\dfrac{\theta_{M1(0)=m1}}{\theta_{M1(1)=m1}}$
$E[Y(1, M1(1), M2(1, M1(1)))]$	1	0	0	1	0	1	1.0

effect of $M2$ under the control condition. The other coefficients are the same as those defined in model (13.19).

13.3.4   Contrast with the linear SEM approach

Applying the linear additive SEM framework to studies of two consecutive mediators with a single treatment at the baseline and a single outcome at the end of the study, researchers (Hayes, 2013) have recommended analyzing the following structural models:

$$M1 = \alpha_{M1} + a_1 Z + e_{M1};$$

$$M2 = \alpha_{M2} + a_2 Z + d_1 M1 + e_{M2};$$

$$Y = \beta_0 + b_1 M1 + b_2 M2 + cZ + e_Y.$$

Assuming that all the relationships are linear additive and assuming the sequential ignorability of Z, $M1$, and $M2$, one may interpret c as the natural direct effect of Z on Y mediated by neither $M1$ nor $M2$, $a_1 b_1$ as the natural indirect effect of Z on Y transmitted through $M1$ but not through $M2$, $a_2 b_2$ as the natural indirect effect transmitted through $M2$ but not through $M1$, and $a_1 d_1 b_2$ as the natural indirect effect transmitted through $M1$ and $M2$ consecutively. Hence, the total indirect effect is $a_1 b_1 + a_2 b_2 + a_1 d_1 b_2$. Assuming that all the estimated coefficients and their products are normally distributed, one may obtain the standard errors for $\hat{a}_1 \hat{b}_1$ and $\hat{a}_2 \hat{b}_2$ in the same way as that described in Section 13.2.4. The standard error for $\hat{a}_1 \hat{d}_1 \hat{b}_2$ is the square root of

$$\hat{a}_1^2 \hat{d}_1^2 Var(\hat{b}_2) + \hat{a}_1^2 \hat{b}_2^2 Var(\hat{d}_1) + \hat{d}_1^2 \hat{b}_2^2 Var(\hat{a}_1).$$

Because simulations have shown that the normality assumption for each product of coefficients is very implausible, the bootstrapping method has been advocated for constructing confidence intervals (Taylor, MacKinnon, and Tein, 2008).

Others (Cole and Maxwell, 2003; MacKinnon, 2008) have proposed autoregressive models for time-varying treatments, time-varying mediators, and time-varying outcomes, typically assuming that the treatment effect on the mediator and the mediator effect on the outcome within each time interval do not change over multiple time intervals. Further model-based assumptions are required for making covariance adjustment for confounders as neither $M1$ nor $M2$ is typically randomized. Cole and Maxwell (2003) discussed some potential pitfalls in research design, measurement, and analysis often seen in past mediation studies using longitudinal psychology data. However, their discussion failed to highlight the following SEM assumptions.

The SEM strategy of estimating each indirect effect as a product of path coefficients assumes that the causal effects are constant for units in a homogeneous population. In a heterogeneous population, let A_1, A_2, B_1, B_2, and D_1 denote the random effects with population averages a_1, a_2, b_1, b_2, and d_1, respectively. It is assumed that $Cov(A_1, B_1)$, $Cov(A_2, B_2)$, $Cov(A_1, D_1)$, and $Cov(A_1 D_1, B_2)$ are all zero. In the context of the Head Start Impact Study, some children were from households with relatively limited resources for education and therefore had few options for high-quality nonparental care. Such children were expected to receive a greater improvement in the quality of nonparental care as a direct result of being assigned to Head Start rather than to the control condition in comparison with their peers who could find access to high-quality non-Head Start childcare alternatives. The aforementioned assumptions imply that, for children from the former group in comparison with those from the latter group, a fixed amount of improvement in the quality of preschool nonparental care was not expected to lead to a greater (or lesser) improvement in the quality of kindergarten instruction; nor was it expected to lead to a great (or lesser) improvement in the kindergarten outcome. Additionally, even if children in the former group received a greater improvement than those in the latter group in the quality of kindergarten instruction as either a direct or an indirect result of attending Head Start, it is assumed that a fixed amount of improvement in kindergarten instruction was not expected to generate a larger impact on the kindergarten outcome of the former group than of the latter group. These assumptions do not seem to be tenable given that educational opportunities and academic growth of children from more disadvantaged families are more reliant on the quality of the educational institutions in preschool and kindergarten years. One may argue that these zero covariance assumptions become somewhat plausible only after controlling for all the relevant pretreatment covariates.

Moreover, as before, the SEM approach generally has difficulty accommodating the Z-by-$M1$ interaction, the Z-by-$M2$ interaction, and the $M1$-by-$M2$ interaction. This is because, with these interactions in the models, the coefficients do not correspond to the causal parameters of interest. Without these interactions, however, the models assume that experiencing high-quality rather than low-quality preschool nonparental care under Head Start has the same impact as that under the control condition and that experiencing high-quality rather than low-quality kindergarten instruction following the Head Start treatment has the same impact as that following the control treatment for a child's kindergarten outcome. And finally, it is assumed that the impact of kindergarten instruction on the kindergarten outcome does not depend on a child's past treatment history and mediator history. These assumptions simply rule out the possibility that the potential benefit of attending Head Start and receiving high-quality preschool nonparental care could be amplified by subsequently receiving high-quality kindergarten instruction.

13.3.5   Contrast with the sensitivity-based estimation of bounds for causal effects

In investigating complex causal mechanisms involving at least two consecutive mediators, Imai and Yamamoto (2013) discussed an alternative set of treatment effects. With their primary interest in $M2$ as a focal mediator, these researchers proposed a procedure that decomposes the total treatment effect on the outcome into only two components under the identification assumption of sequential ignorability. The first is the sum of the natural direct effect of the treatment and the natural indirect effect mediated by $M1$ only—that is, the effect of the treatment not mediated by $M2$; the second is the natural indirect effect mediated by $M2$.

These researchers further assumed that two linear structural equation models apply to the data, one for $M2$ as a function of Z, $M1$, and the Z-by-$M1$ interaction, and the other for Y as a function of Z, $M1$, $M2$, the Z-by-$M1$ interaction, and the Z-by-$M2$ interaction. In these two theoretical models, the coefficients are allowed to vary across individuals. They showed that when the Z-by-$M2$ interaction is zero, the causal effects of interest are identified under the conventional linear SEM framework. When the Z-by-$M2$ interaction is nonzero but is homogeneous in the population, the causal effects are nonetheless identifiable. When the Z-by-$M2$ interaction is heterogeneous, however, the estimated causal effects change as a function of two sensitivity parameters for which empirical knowledge is generally limited. As a result, the analysis tends to generate relatively wide bounds for the causal effects of interest, making it difficult to reject the null hypotheses even when they are false.

Imai and Yamamoto (2013) further stated that the natural indirect effect mediated by $M1$ but not $M2$ cannot be identified in the presence of an $M1$-by-$M2$ interaction effect and can only be bounded within the plausible range of some sensitivity parameters similar to those discussed earlier. Other researchers including Albert and Nelson (2011) concluded that the natural indirect effect mediated by $M1$ but not $M2$ is not identifiable. This is because the estimation strategy proposed by these authors involves the relationship between two counterfactuals $M1(1)$ and $M1(0)$. Albert and Nelson suggested that a certain value for the unknown relationship should be assumed and that the impact of the assumption should be assessed through a sensitivity analysis. Viewing $M1$ as a potential confounder of the causal relationship between $M2$ and the outcome, Huber (2014) proposed an alternative weighting approach, reviewed in Chapter 10, that identifies the natural direct effect of the treatment under the sequential ignorability and the additional assumption that treatment assignment is independent of potential values of $M2$ and of potential outcomes conditioning on the observed pretreatment covariates and also conditioning on $M1$. The natural indirect effect mediated by $M2$, however, remains unidentified without further specifying a linear additive outcome model with regard to the relationship between $M2$ and Y. We have shown in the previous sections that many of these constraints can possibly be removed by the RMPW approach.

13.4   Discussion

Advances of theoretical knowledge in social sciences eventually depend on empirical understanding of complex causal mechanisms. Concurrent mediators are prevalent in interventions with multiple treatment components. When there is a lag time between the treatment assignment and the measurement of the outcome, the causal mechanisms often involve consecutive mediators. Moreover, an earlier mediator such as $M1$, if not taken into account, might play the role of a posttreatment covariate confounding the relationship between $M2$ and Y. Hence, a theoretical elaboration of the causal relationships among all the possible mediators is an essential starting point in such mediation analyses. As the number of causal effects to be estimated increases in a given analysis, however, the statistical power may correspondingly decrease. In this regard, theoretical elaboration prior to data collection also provides the basis for determining the minimal sample size for a planned analysis. Future research will need to develop tools that will allow users to assess statistical power in relation to sample size, taking into account the anticipated effect size of each causal effect and the error variance.

Appendix 13.A: Derivation of RMPW for estimating population average counterfactual outcomes of two concurrent mediators

(a) Estimating $E[Y(1, M1(1), M2(0))]$,

$$E[Y(1,M1(1),M2(0))]\equiv E\{E[Y(1,M1(1),M2(0))|\mathbf{X}]\},$$

which is equal to $E\{E[Y(1,M1(1),M2(0))|Z=1,\mathbf{X}]\}$ because Assumption (i) is met given that the treatment was randomized. When assumption (iv) holds, $M1(1)$ and $M2(0)$ are independent of each other given z and $\mathbf{x}$. Additionally, under assumptions (ii) and (iii) that, given z and $\mathbf{x}$, the assignments of $M1$ and $M2$ are independent of the potential outcomes within and across the treatment conditions, the above is equal to

$$\int_{\mathbf{x}}\int_{m1}\int_{m2}\int_{y} y \times f(Y(z,m1,m2)=y|Z=1,M1(1)=m1,M2(0)=m2,\mathbf{X}=\mathbf{x})$$

$$\times \mathrm{pr}(M1(1)=m1|Z=1,\mathbf{X}=\mathbf{x}) \times \mathrm{pr}(M2(0)=m2|Z=1,\mathbf{X}=\mathbf{x}) \times h(\mathbf{X}=\mathbf{x})dydm2dm1d\mathbf{x}$$

$$=\int_{\mathbf{x}}\int_{m1}\int_{m2}\int_{y} y \times [(f(Y(z,m1,m2)=y,M1(1)=m1|Z=1,M2(0)=m2,\mathbf{X}=\mathbf{x}))/(\mathrm{pr}(M1(1)=m1|Z=1,\mathbf{X}=\mathbf{x}))]$$

$$\times \mathrm{pr}(M1(1)=m1|Z=1,\mathbf{X}=\mathbf{x}) \times \mathrm{pr}(M2(0)=m2|Z=1,\mathbf{X}=\mathbf{x}) \times h(\mathbf{X}=\mathbf{x})dydm2dm1d\mathbf{x}$$

$$=\int_{\mathbf{x}}\int_{m1}\int_{m2}\int_{y} y \times [(f(Y(z,m1,m2)=y,M1(1)=m1|Z=1,M2(1)=m2,\mathbf{X}=\mathbf{x})/(\mathrm{pr}(M1(1)=m1|Z=1,\mathbf{X}=\mathbf{x})))]$$

$$\times \mathrm{pr}(M1(1)=m1|Z=1,\mathbf{X}=\mathbf{x}) \times \mathrm{pr}(M2(0)=m2|Z=0,\mathbf{X}=\mathbf{x}) \times h(\mathbf{X}=\mathbf{x})dydm2dm1d\mathbf{x}$$

$$=\int_{\mathbf{x}}\int_{m1}\int_{m2}\int_{y} y \times f(Y(z,m1,m2)=y|Z=1,M1(1)=m1,M2(1)=m2,\mathbf{X}=\mathbf{x})$$

$$\times \mathrm{pr}(M1(1)=m1|Z=1,\mathbf{X}=\mathbf{x}) \times \mathrm{pr}(M2(0)=m2|Z=0,\mathbf{X}=\mathbf{x}) \times h(\mathbf{X}=\mathbf{x})dydm2dm1d\mathbf{x}$$

$$=\int_{\mathbf{x}}\int_{m1}\int_{m2}\int_{y} y \times f(Y(z,m1,m2)=y|Z=1,M1(1)=m1,M2(1)=m2,\mathbf{X}=\mathbf{x}) \times \mathrm{pr}(M1(1)=m1|Z=1,\mathbf{X}=\mathbf{x})$$

$$\times \mathrm{pr}(M2(1)=m2|Z=1,\mathbf{X}=\mathbf{x}) \times \frac{\mathrm{pr}(M2(0)=m2|Z=0,\mathbf{X}=\mathbf{x})}{\mathrm{pr}(M2(1)=m2|Z=1,\mathbf{X}=\mathbf{x})} \times h(\mathbf{X}=\mathbf{x})dydm2dm1d\mathbf{x}$$

$$=E(Y^*|Z=1)$$

where $Y^* = WY$ and $W = \dfrac{\mathrm{pr}(M2(0)=m2|Z=0,\mathbf{X}=\mathbf{x})}{\mathrm{pr}(M2(1)=m2|Z=1,\mathbf{X}=\mathbf{x})} = \dfrac{\theta_{M2(0)=m2}}{\theta_{M2(1)=m2}}.$

(b) Estimating $E[Y(1, M1(0), M2(0))]$,

$$E[Y(1,M1(0),M2(0))]\equiv E\{E[Y(1,M1(0),M2(0))|\mathbf{X}]\},$$

which is equal to $E\{E[Y(1,M1(0),M2(0))|Z=1,\mathbf{X}]\}$ because the random treatment assignment satisfies Assumption (i). Under assumption (iv), the concurrent mediators $M1(0)$ and $M2(0)$ are independent of each other given z and $\mathbf{x}$. Additionally, under assumptions (ii) and (iii), the above is equal to

$$\int_{\mathbf{x}}\int_{m1}\int_{m2}\int_{y} y \times f(Y(1,m1,m2)=y|Z=1,M1(0)=m1,M2(0)=m2,\mathbf{X}=\mathbf{x})$$

$$\times \mathrm{pr}(M1(0)=m1|Z=1,\mathbf{X}=\mathbf{x}) \times \mathrm{pr}(M2(0)=m2|Z=1,\mathbf{X}=\mathbf{x}) \times h(\mathbf{X}=\mathbf{x})dydm2dm1d\mathbf{x}$$

$$=\int_{\mathbf{x}}\int_{m1}\int_{m2}\int_{y} y \times f(Y(1,m1,m2)=y|Z=1,M1(1)=m1,M2(1)=m2,\mathbf{X}=\mathbf{x})$$

$$\times \mathrm{pr}(M1(0)=m1|Z=1,\mathbf{X}=\mathbf{x}) \times \mathrm{pr}(M2(0)=m2|Z=1,\mathbf{X}=\mathbf{x}) \times h(\mathbf{X}=\mathbf{x})dydm2dm1d\mathbf{x}$$

$$= \int_{\mathbf{x}} \int_{m1} \int_{m2} \int_y y \times f(Y(1,m1,m2)=y|Z=1,M1(1)=m1,M2(1)=m2,\mathbf{X}=\mathbf{x}) \times \mathrm{pr}(M1(1)=m1|Z=1,\mathbf{X}=\mathbf{x})$$

$$\times \mathrm{pr}(M2(1)=m2|Z=1,\mathbf{X}=\mathbf{x}) \times \frac{\mathrm{pr}(M1(0)=m1|Z=0,\mathbf{X}=\mathbf{x}) \times \mathrm{pr}(M2(0)=m2|Z=0,\mathbf{X}=\mathbf{x})}{\mathrm{pr}(M1(1)=m1|Z=1,\mathbf{X}=\mathbf{x}) \times \mathrm{pr}(M2(1)=m2|Z=1,\mathbf{X}=\mathbf{x})}$$

$$\times h(\mathbf{X}=\mathbf{x})dydm2dm1d\mathbf{x}$$

$$= E(Y^*|Z=1)$$

where $Y^* = WY$ and

$$W = \frac{\mathrm{pr}(M1(0)=m1|Z=0,\mathbf{X}=\mathbf{x}) \times \mathrm{pr}(M2(0)=m2|Z=0,\mathbf{X}=\mathbf{x})}{\mathrm{pr}(M1(1)=m1|Z=1,\mathbf{X}=\mathbf{x}) \times \mathrm{pr}(M2(1)=m2|Z=1,\mathbf{X}=\mathbf{x})} = \frac{\theta_{M1(0)=m1}}{\theta_{M1(1)=m1}} \times \frac{\theta_{M2(0)=m2}}{\theta_{M2(1)=m2}}.$$

(c) Estimating $E[Y(1, M1(0), M2(1))]$,

$$E[Y(1,M1(0),M2(1))] \equiv E\{E[Y(1,M1(0),M2(1))|\mathbf{X}]\},$$

which is equal to $E\{E[Y(1,M1(0),M2(1))|Z=1,\mathbf{X}]\}$ with the random treatment assignment that satisfies Assumption (i). Under assumptions (ii), (iii), and (iv), $M1(1)$ and $M2(0)$ are independent of each other; and both are independent of the potential outcomes given z and $\mathbf{x}$. Hence, the above is equal to

$$\int_{\mathbf{x}} \int_{m1} \int_{m2} \int_y y \times f(Y(1,m1,m2)=y|Z=1,M1(0)=m1,M2(1)=m2,\mathbf{X}=\mathbf{x})$$

$$\times \mathrm{pr}(M1(0)=m1|Z=1,\mathbf{X}=\mathbf{x}) \times pr(M2(1)=m2|Z=1,\mathbf{X}=\mathbf{x}) \times h(\mathbf{X}=\mathbf{x})dydm2dm1d\mathbf{x}$$

$$= \int_{\mathbf{x}} \int_{m1} \int_{m2} \int_y y \times [(f(Y(1,m1,m2)=y,M2(1)=m2|Z=1,M1(0)=m1,\mathbf{X}=\mathbf{x}))/(\mathrm{pr}(M2(1)=m2|Z=1,\mathbf{X}=\mathbf{x}))]$$

$$\times \mathrm{pr}(M1(0)=m1|Z=1,\mathbf{X}=\mathbf{x}) \times \mathrm{pr}(M2(1)=m2|Z=1,\mathbf{X}=\mathbf{x}) \times h(\mathbf{X}=\mathbf{x})dydm2dm1d\mathbf{x}$$

$$= \int_{\mathbf{x}} \int_{m1} \int_{m2} \int_y y \times [(f(Y(1,m1,m2)=y,M2(1)=m2|Z=1,M1(1)=m1,\mathbf{X}=\mathbf{x}))/(\mathrm{pr}(M2(1)=m2|Z=1,\mathbf{X}=\mathbf{x}))]$$

$$\times \mathrm{pr}(M1(0)=m1|Z=0,\mathbf{X}=\mathbf{x}) \times \mathrm{pr}(M2(1)=m2|Z=1,\mathbf{X}=\mathbf{x}) \times h(\mathbf{X}=\mathbf{x})dydm2dm1d\mathbf{x}$$

$$= \int_{\mathbf{x}} \int_{m1} \int_{m2} \int_y y \times f(Y(1,m1,m2)=y|Z=1,M1(1)=m1,M2(1)=m2,\mathbf{X}=\mathbf{x})$$

$$\times \mathrm{pr}(M1(0)=m1|Z=0,\mathbf{X}=\mathbf{x}) \times \mathrm{pr}(M2(1)=m2|Z=1,\mathbf{X}=\mathbf{x}) \times h(\mathbf{X}=\mathbf{x})dydm2dm1d\mathbf{x}$$

$$= \int_{\mathbf{x}} \int_{m1} \int_{m2} \int_y y \times f(Y(1,m1,m2)=y|Z=1,M1(1)=m1,M2(1)=m2,\mathbf{X}=\mathbf{x}) \times \mathrm{pr}(M1(1)=m1|Z=1,\mathbf{X}=\mathbf{x})$$

$$\times \mathrm{pr}(M2(1)=m2|Z=1,\mathbf{X}=\mathbf{x}) \times \frac{\mathrm{pr}(M1(0)=m1|Z=0,\mathbf{X}=\mathbf{x})}{\mathrm{pr}(M1(1)=m1|Z=1,\mathbf{X}=\mathbf{x})} \times h(\mathbf{X}=\mathbf{x})dydm2dm1d\mathbf{x}$$

$$= E(Y^*|Z=1)$$

where $Y^* = WY$ and $W = \dfrac{\mathrm{pr}(M1(0)=m1|Z=0,\mathbf{X}=\mathbf{x})}{\mathrm{pr}(M1(1)=m1|Z=1,\mathbf{X}=\mathbf{x})} = \dfrac{\theta_{M1(0)=m1}}{\theta_{M1(1)=m1}}.$

Appendix 13.B: Derivation of RMPW for estimating population average counterfactual outcomes of consecutive mediators

(a) Estimating $E[Y(1, M1(0), M2(0, M1(0)))]$,

$$E[Y(1,M1(0),M2(0,M1(0)))]\equiv E\{E[Y(1,M1(0),M2(0,M1(0)))|\mathbf{X}]\},$$

which is equal to $E\{E[Y(1,M1(0),M2(0,M1(0)))|Z=1,\mathbf{X}]\}$ because the random treatment assignment satisfies Assumption (i). Note that $M2(0,M1(0))=M2(0)$. The above is equal to

$$\int_{\mathbf{x}}\int_{m1}\int_{m2}\int_{y} y \times f(Y(1,m1,m2)=y|Z=1,M1(0)=m1,M2(0,M1(0))=m2,\mathbf{X}=\mathbf{x})$$

$$\times \mathrm{pr}(M2(0,M1(0))=m2|Z=1,M1(0)=m1,\mathbf{X}=\mathbf{x}) \times \mathrm{pr}(M1(0)=m1|Z=1,\mathbf{X}=\mathbf{x})$$

$$\times h(\mathbf{X}=\mathbf{x})dydm2dm1d\mathbf{x}$$

$$=\int_{\mathbf{x}}\int_{m1}\int_{m2}\int_{y} y \times [(f(Y(1,m1,m2)=y,M2(0)=m2|Z=1,M1(0)=m1,\mathbf{X}=\mathbf{x}))/$$

$$(\mathrm{pr}(M2(0)=m2|Z=1,M1(0)=m1,\mathbf{X}=\mathbf{x}))]$$

$$\times \mathrm{pr}(M2(0)=m2|Z=1,M1(0)=m1,\mathbf{X}=\mathbf{x}) \times \mathrm{pr}\,(M1(0)=m1|Z=1,\mathbf{X}=\mathbf{x})$$

$$\times h(\mathbf{X}=\mathbf{x})dydm2dm1d\mathbf{x},$$

which, under Assumptions (i), (ii), and (iii*), is equal to

$$=\int_{\mathbf{x}}\int_{m1}\int_{m2}\int_{y} y \times [(f(Y(1,m1,m2)=y,M2(0)=m2|Z=1,M1(1)=m1,\mathbf{X}=\mathbf{x}))/$$

$$(\mathrm{pr}(M2(0)=m2|Z=1,M1(1)=m1,\mathbf{X}=\mathbf{x}))]$$

$$\times \mathrm{pr}(M2(0)=m2|Z=0,M1(0)=m1,\mathbf{X}=\mathbf{x}) \times \mathrm{pr}(M1(0)=m1|Z=0,\mathbf{X}=\mathbf{x})$$

$$\times h(\mathbf{X}=\mathbf{x})dydm2dm1d\mathbf{x}$$

$$=\int_{\mathbf{x}}\int_{m1}\int_{m2}\int_{y} y \times f(Y(1,m1,m2)=y|Z=1,M1(1)=m1,M2(0)=m2,\mathbf{X}=\mathbf{x})$$

$$\times pr(M2(0)=m2|Z=0,M1(0)=m1,\mathbf{X}=\mathbf{x})\times pr(M1(0)=m1|Z=0,\mathbf{X}=\mathbf{x})$$

$$\times h(\mathbf{X}=\mathbf{x})dydm2dm1d\mathbf{x}$$

$$=\int_{\mathbf{x}}\int_{m1}\int_{m2}\int_{y}y\times f(Y(1,m1,m2)=y|Z=1,M1(1)=m1,M2(1)=m2,\mathbf{X}=\mathbf{x})$$

$$\times pr(M2(1)=m2|Z=1,M1(1)=m1,\mathbf{X}=\mathbf{x})\times pr(M1(1)=m1|Z=1,\mathbf{X}=\mathbf{x})$$

$$\times\frac{pr(M1(0)=m1|Z=0,\mathbf{X}=\mathbf{x})\times pr(M2(0)=m2|Z=0,M1(0)=m1,\mathbf{X}=\mathbf{x})}{pr(M1(1)=m1|Z=1,\mathbf{X}=\mathbf{x})\times pr(M2(1)=m2|Z=1,M1(1)=m1,\mathbf{X}=\mathbf{x})}\times h(\mathbf{X}=\mathbf{x})dydm2dm1d\mathbf{x}$$

$$=E(Y^*|Z=1) \text{ where } Y^*=WY \text{ and}$$

$$W=\frac{pr(M1(0)=m1|Z=0,\mathbf{X}=\mathbf{x})\times pr(M2(0)=m2|Z=0,M1(0)=m1,\mathbf{X}=\mathbf{x})}{pr(M1(1)=m1|Z=1,\mathbf{X}=\mathbf{x})\times pr(M2(1)=m2|Z=1,M1(1)=m1,\mathbf{X}=\mathbf{x})}=\frac{\theta_{M1(0)=m1}}{\theta_{M1(1)=m1}}\times\frac{\theta_{M2(0,m1)=m2}}{\theta_{M2(1,m1)=m2}}.$$

(b) Estimating $E[Y(1, M1(1), M2(0, M1(0)))]$,

$$E[Y(1,M1(1),M2(0,M1(0)))]\equiv E\{E[Y(1,M1(1),M2(0,M1(0)))|\mathbf{X}]\},$$

which is equal to $E\{E[Y(1,M1(1),M2(0,M1(0)))|Z=1,\mathbf{X}]\}$ because the random treatment assignment satisfies Assumption (i). Note that $M2(0,M1(0))=M2(0)$. Under Assumptions (i), (ii), and (iii*), the above is equal to

$$=\int_{\mathbf{x}}\int_{m1}\int_{m2}\int_{y}y\times f(Y(1,m1,m2)=y|Z=1,M1(1)=m1,M2(0)=m2,\mathbf{X}=\mathbf{x})$$

$$\times pr(M2(0)=m2|Z=1,M1(1)=m1,\mathbf{X}=\mathbf{x})\times pr(M1(1)=m1|Z=1,\mathbf{X}=\mathbf{x})\times h(\mathbf{X}=\mathbf{x})dydm2dm1d\mathbf{x}$$

$$=\int_{\mathbf{x}}\int_{m1}\int_{m2}\int_{y}y\times f(Y(1,m1,m2)=y|Z=1,M1(1)=m1,M2(1)=m2,\mathbf{X}=\mathbf{x})$$

$$\times[(pr(M2(0)=m2,M1(1)=m1|Z=0,\mathbf{X}=\mathbf{x}))/(pr(M1(1)=m1|Z=0,\mathbf{X}=\mathbf{x}))]\times pr(M1(1)=m1|Z=1,\mathbf{X}=\mathbf{x})$$

$$\times h(\mathbf{X}=\mathbf{x})dydm2dm1d\mathbf{x}$$

$$=\int_{\mathbf{x}}\int_{m1}\int_{m2}\int_{y}y\times f(Y(1,m1,m2)=y|Z=1,M1(1)=m1,M2(1)=m2,\mathbf{X}=\mathbf{x})$$

$$\times[pr(M2(0)=m2,M1(1)=m1|Z=1,\mathbf{X}=\mathbf{x})/(pr(M1(1)=m1|Z=1,\mathbf{X}=\mathbf{x}))]\times pr(M1(1)=m1|Z=1,\mathbf{X}=\mathbf{x})$$

$$\times h(\mathbf{X}=\mathbf{x})dydm2dm1d\mathbf{x}$$

$$=\int_{\mathbf{x}}\int_{m1}\int_{m2}\int_{y}y\times f(Y(1,m1,m2)=y|Z=1,M1(1)=m1,M2(1)=m2,\mathbf{X}=\mathbf{x})$$

$$\times pr(M2(1)=m2|Z=1,M1(1)=m1,\mathbf{X}=\mathbf{x})\times pr(M1(1)=m1|Z=1,\mathbf{X}=\mathbf{x})$$

$$\times\frac{pr(M2(0)=m2|Z=0,M1(0)=m1,\mathbf{X}=\mathbf{x})}{pr(M2(1)=m2|Z=1,M1(1)=m1,\mathbf{X}=\mathbf{x})}\times h(\mathbf{X}=\mathbf{x})dydm2dm1d\mathbf{x}$$

$$=E(Y^*|Z=1)$$

where $Y^* = WY$ and

$$W = \frac{\mathrm{pr}(M2(0) = m2|Z = 0, M1(0) = m1, \mathbf{X} = \mathbf{x})}{\mathrm{pr}(M2(1) = m2|Z = 1, M1(1) = m1, \mathbf{X} = \mathbf{x})} = \frac{\theta_{M2(0,m1) = m2}}{\theta_{M2(1,m1) = m2}}.$$

(c) Estimating $E[Y(1, M1(0), M2(1, M1(1)))]$,

$$E[Y(1, M1(0), M2(1, M1(1)))] \equiv E\{E[Y(1, M1(0), M2(1, M1(1)))|\mathbf{X}]\},$$

which is equal to $E\{E[Y(1, M1(0), M2(1, M1(1)))|Z = 1, \mathbf{X}]\}$ because the random treatment assignment satisfies assumption (i). Note that $M2(1, M1(1)) = M2(1)$. Under assumptions (i), (ii), and (iii*), the above is equal to

$$\int_{\mathbf{x}}\int_{m1}\int_{m2}\int_{y} y \times f(Y(1, m1, m2) = y|Z = 1, M1(0) = m1, M2(1) = m2, \mathbf{X} = \mathbf{x})$$

$$\times \mathrm{pr}(M2(1) = m2|Z = 1, M1(0) = m1, \mathbf{X} = \mathbf{x}) \times \mathrm{pr}(M1(0) = m1|Z = 1, \mathbf{X} = \mathbf{x})$$
$$\times h(\mathbf{X} = \mathbf{x}) dy dm2 dm1 d\mathbf{x}$$

$$= \int_{\mathbf{x}}\int_{m1}\int_{m2}\int_{y} y \times [(f(Y(1, m1, m2) = y, M2(1) = m2|Z = 1, M1(0) = m1, \mathbf{X} = \mathbf{x}))/$$

$$(\mathrm{pr}(M2(1) = m2|Z = 1, M1(0) = m1, \mathbf{X} = \mathbf{x}))]$$
$$\times \mathrm{pr}(M2(1) = m2|Z = 1, M1(0) = m1, \mathbf{X} = \mathbf{x}) \times \mathrm{pr}(M1(0) = m1|Z = 1, \mathbf{X} = \mathbf{x})$$
$$\times h(\mathbf{X} = \mathbf{x}) dy dm2 dm1 d\mathbf{x}$$

$$= \int_{\mathbf{x}}\int_{m1}\int_{m2}\int_{y} y \times [(f(Y(1, m1, m2) = y, M2(1) = m2|Z = 1, M1(1) = m1, \mathbf{X} = \mathbf{x}))/$$

$$(\mathrm{pr}(M2(1) = m2|Z = 1, M1(1) = m1, \mathbf{X} = \mathbf{x}))]$$
$$\times \mathrm{pr}(M2(1) = m2|Z = 1, M1(1) = m1, \mathbf{X} = \mathbf{x}) \times \mathrm{pr}(M1(0) = m1|Z = 0, \mathbf{X} = \mathbf{x})$$
$$\times h(\mathbf{X} = \mathbf{x}) dy dm2 dm1 d\mathbf{x}$$

$$= \int_{\mathbf{x}}\int_{m1}\int_{m2}\int_{y} y \times f(Y(1, m1, m2) = y|Z = 1, M1(1) = m1, M2(1) = m2, \mathbf{X} = \mathbf{x})$$

$$\times \mathrm{pr}(M2(1) = m2|Z = 1, M1(1) = m1, \mathbf{X} = \mathbf{x}) \times \mathrm{pr}(M1(0) = m1|Z = 0, \mathbf{X} = \mathbf{x})$$
$$\times h(\mathbf{X} = \mathbf{x}) dy dm2 dm1 d\mathbf{x}$$

$$= \int_{\mathbf{x}}\int_{m1}\int_{m2}\int_{y} y \times f(Y(1, m1, m2) = y|Z = 1, M1(1) = m1, M2(1) = m2, \mathbf{X} = \mathbf{x})$$

$$\times \mathrm{pr}(M2(1) = m2|Z = 1, M1(1) = m1, \mathbf{X} = \mathbf{x}) \times \mathrm{pr}(M1(1) = m1|Z = 1, \mathbf{X} = \mathbf{x})$$
$$\times \frac{\mathrm{pr}(M1(0) = m1|Z = 0, \mathbf{X} = \mathbf{x})}{\mathrm{pr}(M1(1) = m1|Z = 1, \mathbf{X} = \mathbf{x})} \times h(\mathbf{X} = \mathbf{x}) dy dm2 dm1 d\mathbf{x}$$

$$= E(Y^*|Z = 1)$$

where $Y^* = WY$ and

$$W = \frac{\text{pr}(M1(0) = m1 | Z = 0, \mathbf{X} = \mathbf{x})}{\text{pr}(M1(1) = m1 | Z = 1, \mathbf{X} = \mathbf{x})} = \frac{\theta_{M1(0) = m1}}{\theta_{M1(1) = m1}}.$$

References

Albert, J.M. and Nelson, S. (2011). Generalized causal mediation analysis. *Biometrics*, *67*, 1028–1038.

Bollen, K.A. (1989). *Structural Equations with Latent Variables*. New York, Wiley. Wiley Series in Probability and Mathematical Statistics.

Briggs, N. (2006). Estimation of the standard error and confidence interval of the indirect effect in multiple mediator models. *Dissertation Abstracts International*, *37*, 4755B.

Cole, D.A. and Maxwell, S.E. (2003). Testing mediational models with longitudinal data: questions and tips in the use of structural equation modeling. *Journal of Abnormal Psychology*, *112*(4), 558–577.

Duncan, G.J., Morris, P.A. and Rodrigues, C. (2011). Does money really matter? Estimating impacts of family income on young children's achievement with data from random-assignment experiments. *Developmental Psychology*, *47*(5), 1263–1279.

Frangakis, C.E. and Rubin, D.B. (2002). Principal stratification in causal inference. *Biometrics*, *58*, 21–29.

Gallop, R., Small, D.S., Lin, J.Y., Elliott, M.R., Joffe, M.T. and Have, T.R. (2009). Mediation analysis with principal stratification. *Statistics in Medicine*, *28*, 1108–1130.

Hayes, A.F. (2013). *Introduction to Mediation, Moderation, and Conditional Process Analysis: A Regression-based Approach*, The Guilford Press, New York.

Hong, G., Deutsch, J., and Hill, H. (2011). *Parametric and non-parametric weighting methods for estimating mediation effects: an application to the national evaluation of welfare-to-work strategies*. In JSM Proceedings, Social Statistics Section. Alexandria, VA, American Statistical Association, pp. 3215–3229.

Huber, M. (2014). Identifying causal mechanisms (primarily) based on inverse probability weighting. *Journal of Applied Econometrics*, *29*(6), 920–943.

Imai, K. and Yamamoto, T. (2013). Identification and sensitivity analysis for multiple causal mechanisms: revisiting evidence from framing experiments. *Political Analysis*, *21*(2), 141–171.

James, L.R. and Brett, J.M. (1984). Mediators, moderators, and test for mediation. *Journal of Applied Psychology*, *69*, 307–321.

Jin, H. and Rubin, D.B. (2008). Principal stratification for causal inference with extended partial compliance. *Journal of the American Statistical Association*, *103*, 101–111.

Jin, H. and Rubin, D.B. (2009). Public schools versus private schools: causal inference with partial compliance. *Journal of Educational and Behavioral Statistics*, *34*, 24–45.

Kling, J.R., Liebman, J.B. and Katz, L.F. (2007). Experimental analysis of neighborhood effects. *Econometrica*, *75*(1), 83–119.

MacKinnon, D.P. (2008). *Introduction to Statistical Mediation Analysis*, Erlbaum, Mahwah, NJ.

Morgan-Lopez, A.A. and MacKinnon, D.P. (2006). Demonstration and evaluation of a method for assessing mediated moderation. *Behavior Research Methods*, *38*, 77–87.

Morgan-Lopez, A.A., Castro, F.G., Chassin, L. and MacKinnon, D.P. (2003). A mediated moderation model of cigarette use among Mexican American youth. *Addictive Behaviors*, *28*, 583–589.

Muller, D., Judd, C.M. and Yzerbyt, V.Y. (2005). When moderation is mediated and mediation is moderated. *Journal of Personality and Social Psychology*, *89*, 852–863.

Page, L.C. (2012). Principal stratification as a framework for investigating mediational processes in experimental settings. *Journal of Research on Educational Effectiveness*, 5(3), 215–244.

Preacher, K.J. and Hayes, A.F. (2008). Asymptotic and resampling strategies for assessing and comparing indirect effects in multiple mediator models. *Behavior Research Methods*, 40(3), 879–891.

Preacher, K.J., Rucker, D.D. and Hayes, A.F. (2007). Addressing moderated mediation hypotheses: theory, methods, and prescriptions. *Multivariate Behavioral Research*, 42, 185–227.

Raudenbush, S.W., Reardon, S.F. and Nomi, T. (2012). Statistical analysis for multisite trials using instrumental variables with random coefficients. *Journal of Research on Educational Effectiveness special issue on the statistical approaches to studying mediator effects in education research*, 5(3), 303–332.

Reardon, S.F. and Raudenbush, S.W. (2013). Under what assumptions do site-by-treatment instruments identify average causal effects? *Sociological Methods & Research*, 42(2), 143–163.

Taylor, A.B., MacKinnon, D.P. and Tein, J.-Y. (2008). Tests of the three-path mediated effect. *Organizational Research Methods*, 11, 241–269.

Weiss, M.J., Bloom, H.S. and Brock, T. (2014). Methods for policy analysis. *Journal of Policy Analysis and Management*, 33(3), 778–808.

Williams, J. and MacKinnon, D.P. (2008). Resampling and distribution of the product methods for testing indirect effects in complex models. *Structural Equation Modeling*, 15, 23–51.

Part IV

SPILL-OVER

14

Spill-over of treatment effects: concepts and methods

14.1 Spill-over: A nuisance, a trifle, or a focus?

This section of the book considers a wide range of cases in which a treatment received by some individuals may have an intended or unintended impact on the outcomes of other individuals. In textbooks on experimental designs, spill-overs of information or other resources from an experimental group to a control group are typically considered a nuisance that the experimenter needs to prevent, whenever possible. For example, in order to determine the impact on child growth of an intervention that provides nutrition supplements in a developing country, researchers may assign children at random to the nutrition program or to a control condition. Naturally, caregivers will be inclined to share the nutrition supplements provided for the experimental children with the control children. Shaddish, Cook, and Campbell (2002) used the term "treatment diffusion" to describe the spill-over phenomenon. In some other cases, diffusion could occur in two opposite directions: those who are treated may affect the outcome values of some individuals in the control group; and conversely, those in the control group may affect the outcome values of some treated individuals. An adolescent assigned to a drug prevention program, for example, may abandon drug use only if his friends are assigned to the same program and may continue drug use if his friends are assigned to the control condition instead. In the presence of treatment diffusion, the control group may operate in a way similar to the experimental group or vice versa, which will diminish the experimental–control contrast and will likely result in an underestimation of the treatment effect. Therefore, preventing spill-over across treatment conditions seems crucial for maintaining the construct validity of "the experimental condition" and "the control condition" such that these labels represent the treatment conditions as initially designed. Preventing spill-over is also deemed crucial for ensuring the internal validity of the experimental results such that a difference in the mean outcome between the experimental group and the control group is caused by the treatment assignment only.

Shaddish, Cook, and Campbell (2002) pointed out that the problem of treatment diffusion "is most acute in cases in which experimental and control units are in physical proximity or

Causality in a Social World: Moderation, Mediation and Spill-over, First Edition. Guanglei Hong.
© 2015 John Wiley & Sons, Ltd. Published 2015 by John Wiley & Sons, Ltd.

can communicate" (p. 81). Therefore, they recommended minimizing common influences or isolating participants who have been assigned to different treatment groups. When isolation is impractical, researchers may instead take advantage of the natural clustering of individuals in households, neighborhoods, villages, or schools and opt for assigning clusters at random to treatments. These are called "cluster randomized designs." In the first example earlier, an entire village may be assigned at random to either the nutrition program or the control condition. Treatment diffusion among residents within a village will no longer be an issue when all the residents in the same village receive the same treatment.

Because an individual is a social being in connection with many other individuals, treatment diffusion through spill-overs is sometimes of key scientific interest. It has been found in vaccination studies (Ali *et al.*, 2005; Halloran and Struchiner, 1995) that unvaccinated individuals can benefit when most other people in the local community are vaccinated. The logic is straightforward: the more people are vaccinated, the less exposure an unvaccinated individual will have to the infectious disease of concern. Hence, in a cost–benefit analysis, one may identify the minimal proportion of a population that needs to be vaccinated in order to prevent an epidemic. Similarly, social scientists have been intrigued by the phenomenon of social contagion—that information, attitudes, and behaviors can spread among people—and its potential implications for social change (Burt, 1987; Granovetter, 1973, 1978; Lindzey and Aronsson, 1985). In social psychological research, for example, an intervention program designed to reduce intergroup prejudice and conflict may deliberately choose to concentrate on the "social referents," also called "opinion leaders," in peer groups as opposed to assign individuals at random to be treated. The rationale is that because individuals serving as social referents for their peers are frequently sought after in personally motivated interactions, they play a key role in changing collective norms and behavior in a social network (Paluck, 2011; Paluck and Shephard, 2012). This example shows that treatment resources can be strategically allocated to take advantage of potential spill-over and thereby maximize the treatment benefit.

However, in the causal inference literature, discussions of spill-over effects have received only limited attention until the very recent years (Aronow, 2012; Hong and Raudenbush, 2006, 2013; Hudgens and Halloran, 2008; Manski, 2000, 2013; Rosenbaum, 2007; Sinclair, McConnell, and Green, 2012; Sobel, 2006: VanderWeele, Hong, Jones, and Brown, 2013; Verbitsky-Savitz and Raudenbush, 2012). In defining the causal effect of a treatment relative to a control condition for a given individual, Rubin (1986) emphasized the key role of the stable unit treatment value assumption (SUTVA). SUTVA states that one's potential outcome value when assigned to a given treatment condition will be stable no matter how the treatment is assigned and no matter what treatments are received by other individuals. Treatment diffusion would be a clear violation of SUTVA. Because causal inference becomes seemingly intractable without SUTVA, most researchers have chosen to invoke SUTVA in theoretical and empirical analyses. The imposition of SUTVA would be inconsequential only if the treatment assignments of other individuals have a negligible effect, if any, on an individual's potential outcome values. Yet as discussed earlier, social science theories often suggest otherwise.

This chapter will reveal the consequences for causal inference when SUTVA is applied to situations in which it is clearly implausible. We will then review a modified framework of causal inference that invokes assumptions relatively weaker than SUTVA in defining causal effects. Under the modified framework, one may define, identify, estimate, and test spill-over effects in experimental and quasiexperimental data. The next chapter will further consider spill-overs as a part of the mediation mechanism of an intervention and will examine conditions that may either enhance or reduce the spill-over effect.

14.2   Stable versus unstable potential outcome values: An example from agriculture

Throughout the previous chapters in this book, we have defined a causal effect as a contrast between two potential outcomes, each associated with a specific treatment assignment for an individual. This idea has developed over the past decades in statistics (Holland, 1986; Neyman, 1923, 1935; Rubin, 1974, 1978) and economics (Heckman, 1979; Roy, 1951). For example, if individual unit i is assigned to an experimental condition denoted by $Z_i = 1$, the subsequent potential outcome is denoted by $Y_i(Z_i = 1)$, which has often been simplified as $Y_i(1)$. If the individual unit is assigned to a control condition instead denoted by $Z_i = 0$, the subsequent potential outcome is then denoted by $Y_i(Z_i = 0)$ and can be simplified as $Y_i(0)$. Hence, the treatment effect for individual unit i can be defined simply as $Y_i(1) - Y_i(0)$. In Neyman's (1923) initial formulation of this framework in the context of agricultural experiments, individual unit i is a plot of land on which a single variety of fertilizer z may be applied, which will generate a potential yield $Y_i(z)$. We may let $z = 0, 1$ when two varieties of fertilizers are compared.

Neyman (1935) discussed the inadequacy of the aforementioned formulation by clarifying that the true yield of plot i with fertilizer z is the mean of observable yields over infinite hypothetical replications of the current experiment "without any change of vegetative conditions or of arrangement." He attributed the differences between the observable yields and the true yield of a given plot to technical errors "due solely to the inaccuracy of the experimental technique" (cited in Rubin, 1990, p. 475). This discussion suggests that, while the value of the true yield of plot i with fertilizer z may be considered stable, the value of an observable yield of this plot with fertilizer z is perhaps inevitably unstable because the experimenter may not have perfect control over all the relevant conditions such as weather that may vary randomly across the replications of the experiment.

Another important source of variation in the observable yield of a given plot with a given fertilizer is the variety of fertilizers applied to the neighboring plots. Suppose that plot i receiving a traditional type of fertilizer $Z_i = 0$ is next to plot i' receiving a new type of fertilizer $Z_{i'} = 1$. And suppose that there is no blockage between these two plots that prevents the flow of water carrying the new type of fertilizer into plot i. In this case, the yield of plot i is affected by the new type of fertilizer applied to the neighboring plot as well as by the traditional type of fertilizer applied to this particular plot and should be denoted by $Y_i(Z_i = 0, Z_{i'} = 1)$. If the next plot i' receives the traditional type of fertilizer $Z_{i'} = 0$ instead, the yield of plot i will then be denoted by $Y_i(Z_i = 0, Z_{i'} = 0)$, the value of which will conceivably be different from $Y_i(Z_i = 0, Z_{i'} = 1)$. Because the type of fertilizer applied to plot i remains unchanged, the change in the potential yield of plot i is solely due to the spill-over of the fertilizer from the next plot. Hence, $Y_i(Z_i = 0, Z_{i'} = 1) - Y_i(Z_i = 0, Z_{i'} = 0)$ defines the spill-over effect. Similarly, if plot i receives the new type of fertilizer $Z_i = 1$, its yield may nonetheless be affected by the treatment assignment of neighboring plot i', in which case $Y_i(Z_i = 1, Z_{i'} = 1) - Y_i(Z_i = 1, Z_{i'} = 0)$ defines the spill-over effect. Note that the above two spill-over effects do not have to be equal.

In summary, depending on the treatment assignment for plot i and for neighboring plot i', plot i may have four potential yields: $Y_i(Z_i = 0, Z_{i'} = 0)$, $Y_i(Z_i = 0, Z_{i'} = 1)$, $Y_i(Z_i = 1, Z_{i'} = 0)$, and

$Y_i(Z_i = 1, Z_{i'} = 1)$. Let $Y_i(Z_i = 0, Z_{i'} = 0)$ be the reference condition; we may define the following three treatment effects on the potential yield of plot i:

1. $Y_i(Z_i = 1, Z_{i'} = 0) - Y_i(Z_i = 0, Z_{i'} = 0)$ is the effect of treating the focal plot with the new type of fertilizer, while the neighboring plot is untreated (i.e., is given the traditional type of fertilizer).

2. $Y_i(Z_i = 0, Z_{i'} = 1) - Y_i(Z_i = 0, Z_{i'} = 0)$ is the spill-over effect of treating the neighboring plot on the potential yield of the focal plot when the latter is untreated.

3. $Y_i(Z_i = 1, Z_{i'} = 1) - Y_i(Z_i = 0, Z_{i'} = 0)$ is the effect of simultaneously treating both plots on the potential yield of the focal plot.

This last treatment effect can be represented as a sum of two treatment effects:

$$[Y_i(Z_i = 1, Z_{i'} = 1) - Y_i(Z_i = 1, Z_{i'} = 0)] + [Y_i(Z_i = 1, Z_{i'} = 0) - Y_i(Z_i = 0, Z_{i'} = 0)].$$

The first component is the spill-over effect of treating the neighboring plot on the potential yield of the focal plot when the latter is treated, while the second is the effect of only treating the focal plot as defined in (1). The decomposition is not unique, because treatment effect (3) can also be represented as

$$[Y_i(Z_i = 1, Z_{i'} = 1) - Y_i(Z_i = 0, Z_{i'} = 1)] + [Y_i(Z_i = 0, Z_{i'} = 1) - Y_i(Z_i = 0, Z_{i'} = 0)].$$

Here, the first component is the effect of treating the focal plot when the neighboring plot is treated, while the second is the spill-over effect of treating the neighboring plot as defined in (2) when the focal plot is untreated.

Similarly, we may define the treatment effects for neighboring plot i'. In a hypothetical population of two plots that are next to each other, we may define the population average treatment effect when a neighboring plot is untreated, the population average treatment effect when a neighboring plot is treated, the effect of spill-over from a neighboring plot when a focal plot is untreated, and the spill-over effect when a focal plot is treated.

The previous example from agriculture is perhaps the simplest case of "interference between units." More generally, in an experiment that involves a total of N individual units, the potential outcome value of a given unit could conceivably be affected by the treatments applied to all the N units due to interference. Rubin (1990) used $Y_i(\mathbf{z})$ to represent the potential outcome of individual unit i where $\mathbf{z} = (z_1, \ldots, z_N)$ is a vector containing N elements indicating the treatments assigned to the N units, in which the ith element indicates the treatment assigned to individual unit i. Because every individual unit may be assigned to either the experimental or the control condition, if we allow the potential outcome value of focal unit i to be influenced by the treatment assignment of any other unit in addition to its own, in theory, the focal unit may have as many as 2^N potential outcomes. When the treatment assigned to focal unit i is fixed, this unit may still have as many as 2^{N-1} potential outcomes. A contrast between any two of these 2^{N-1} potential outcomes can be defined as a spill-over effect.

Rubin (1980, 1986, 1990) discussed conditions under which the potential outcome of individual unit i with treatment z can be adequately represented by $Y_i(Z_i = z)$ rather than $Y_i(\mathbf{z})$. In the previous example from agriculture, the experimenter can eliminate "interference between units" by using guard rows between plots. This additional procedure may minimize the difference between $Y_i(Z_i=0,Z_{i'}=1)$ and $Y_i(Z_i=0,Z_{i'}=0)$ as well as the difference between $Y_i(Z_i=1,Z_{i'}=1)$ and $Y_i(Z_i=1,Z_{i'}=0)$ such that it may seem plausible to assume

$$Y_i(Z_i=0,Z_{i'}=1)=Y_i(Z_i=0,Z_{i'}=0)=Y_i(Z_i=0);$$
$$Y_i(Z_i=1,Z_{i'}=1)=Y_i(Z_i=1,Z_{i'}=0)=Y_i(Z_i=1).$$

Under this assumption, the spill-over effects $Y_i(Z_i=0,Z_{i'}=1)-Y_i(Z_i=0,Z_{i'}=0)$ and $Y_i(Z_i=1,Z_{i'}=1)-Y_i(Z_i=1,Z_{i'}=0)$ will both be zero. Hence, the treatment effects as defined in (1) and (3) will become equivalent and can be simplified as $Y_i(Z_i=1)-Y_i(Z_i=0)$. In this hypothetical population of two plots, the only causal parameter of interest is the average effect of treating a focal plot versus not regardless of the treatment assigned to a neighboring plot.

In general, if the potential outcome of one individual unit is unlikely to be affected by the treatments assigned to other units or, in other words, if spill-over is arguably nonexistent, one may assume that $Y_i(\mathbf{z})=Y_i(Z_i=z)$. This assumption, initially invoked by Cox (1958), was highlighted by Rubin (1990, 1986) as a part of SUTVA. Causal inference is greatly simplified under SUTVA.

Rubin (1990) acknowledged that interference between units "can be a major issue when studying medical treatments for infectious diseases (e.g., malaria, AIDS) or educational treatments given to children who interact with each other" (p. 475). Despite these warnings, SUTVA has been routinely invoked in applications of causal inference. According to Brock and Durlauf (2001), neglecting social interactions and social interdependences has also been a long tradition in standard economic models of individual choice. In the following, we discuss implications for causal inference in social science research when SUTVA is unwarranted.

14.3 Consequences for causal inference when spill-over is overlooked

Social interactions and social interdependences have been central in sociological considerations of neighborhood effects on human development (e.g., Coleman, 1988; Sampson, 2012). It is now widely believed that concentration of poverty in urban neighborhoods in the United States devastates residents' well-being and precludes upward mobility above and beyond the impact of individual poverty (Brooks-Gunn, Duncan, and Aber, 1997; Durlauf, 1996; Wilson, 1987). Would moving a poor family to a low-poverty neighborhood improve individual outcomes with regard to safety, health, and economic self-sufficiency? The Moving to Opportunity (MTO) experiment in the mid-1990s promised definitive answers to these causal questions. In five cities around the country, eligible residents in high-poverty neighborhoods were assigned at random to one of three treatment conditions: counseling and housing vouchers to be used for moving to low-poverty neighborhoods, housing vouchers to be used anywhere, and the control condition. It was expected that providing housing vouchers, especially with counseling, would encourage and enable poor families to relocate to low-poverty neighborhoods and would generate benefits when compared with the control condition. Analyzing the MTO data, public policy researchers have estimated the effects of treatment assignment as well as the effects of using housing vouchers on a wide range of outcomes. Yet as Sobel (2006) pointed out, these are estimates of average treatment effects only if SUTVA is valid.

To simplify, here, we combine the first two treatment conditions into one (i.e., being offered housing voucher) as opposed to the control condition (i.e., not being offered housing voucher). Households that volunteered to participate in the experiment are the units of study. Let us suppose that, at a given site, the experimenter assigned a p proportion of the participating households at random to be offered housing vouchers and assigned a $1-p$ proportion of the households to the control condition. Sobel (2006) pointed out that when there is a nonzero spill-over effect on the untreated, the intent-to-treat (ITT) effect reported by MTO researchers actually estimates the difference between the effect of assigning a p proportion of the participants rather than none to be treated for families that received the offer and the effect for families that did not receive the offer, the latter being the pure spill-over effect for the untreated.

In this particular application, the fraction of participants to be treated becomes an important factor for a number of reasons. Sobel (2006) cited evidence suggesting that when some of the participants receiving housing vouchers left the high-poverty neighborhoods and took jobs elsewhere, the jobs originally held by these treated units might become open to the untreated. Offering housing vouchers to a p proportion of the participants rather than none therefore might have a nonzero impact on the employment and income of the untreated. In general, the treatment assignments of other participants could affect the potential outcome of a focal unit who has been assigned to the control group through affecting the rental market, job market, and school composition in the origin neighborhoods. This reasoning is related to the "general equilibrium effect" in the economics literature (Heckman, Lochner, and Taber, 1998). Additional evidence suggests that participants who received housing vouchers and moved into low-poverty neighborhoods tended to be particularly satisfied with safety in their new neighborhoods in contrast with the origin neighborhoods. Through communicating with those who moved, the untreated participants might rate the safety of their neighborhoods lower than they would have when nobody had received housing vouchers.

In the earlier chapters, by invoking SUTVA, we have defined the causal effect of an experimental condition relative to a control condition for individual unit i as $Y_i(1)-Y_i(0)$ and the corresponding population average treatment effect as $E[Y(1)-Y(0)]$, which is equivalent to $E[Y(1)]-E[Y(0)]$. We have interpreted $E[Y(1)]$ as the population average potential outcome when the entire population is treated and $E[Y(0)]$ as the population average potential outcome when the entire population is untreated. An experiment typically assigns a p proportion of the individual units at random to be treated and a $1-p$ proportion to be untreated. Under SUTVA, the mean observed outcome of the treated units in such an experiment has been taken as unbiased for the population average potential outcome when all are treated. Similarly, the mean observed outcome of the untreated units has been taken as unbiased for the population average potential outcome when none is treated. This interpretation, however, becomes seriously misleading in the presence of spill-over across the treatment conditions. This is because the average potential outcome of the treated individuals when everybody is treated is likely different from their average potential outcome when only a p proportion is treated. For example, when housing vouchers are offered only to a small fraction of participants, a poor family that receives and uses the voucher to move will likely find many middle-class neighbors in the receiving community. However, if all the poor families are offered vouchers, the poor family after moving may again be surrounded by low-income neighbors over time. Similarly, according to our earlier reasoning, an untreated individual's potential response when no one receives voucher may be different from the potential response after some other households have moved into better neighborhoods. In conclusion, the mean difference in the observed outcome between the treated and the untreated in the current experiment that assigns a p proportion of the participants to be treated is biased for the population average effect of treating all rather than treating none.

More generally, an individual's potential outcome under a given treatment may depend on not just what proportion of participants is to be treated but also which participants are to be treated, how those participants are related to the focal individual, and who delivers the treatment. In the following, we present a general framework that addresses all these considerations.

14.4   Modified framework of causal inference

As we have shown earlier, in an experiment involving N participants, if the potential outcome of individual i in the treated group could be influenced by the treatment assignments of the other $N-1$ participants, this individual will have as many as 2^{N-1} potential outcomes when treated. Similarly, the individual will have 2^{N-1} potential outcomes when untreated. As N increases, the number of potential outcomes increases exponentially for a single person. Would causal inference still be possible without invoking SUTVA? We review several attempts that have been made in the past decade to modify the causal inference framework. These modifications have led to a rich class of causal questions that can be empirically addressed with appropriate designs and analyses.

14.4.1   Treatment settings

Hong (2004) proposed a framework that attributes the lack of stability of an individual's potential outcome values under a given treatment to the variability in *treatment settings* often due to a series of random events other than the focal individual's treatment assignment. In the agriculture example discussed earlier, the treatment setting for a plot can be characterized primarily by the treatment assignments of the neighboring plots and the weather condition in a certain year. In the MTO experiment, the treatment setting for a participant can be characterized by the treatment assignments of other participants living in the same housing project as well as the particular state of the local rental market and job market. In a third example that evaluates the impact of class size reduction on student achievement (Finn and Achillis, 1990), students and teachers in a school were assigned at random to either a small class or a regular class. The treatment setting for a student in a small class (or in a regular class) is made up of other students and teachers who have been assigned to the same class. Conceivably, if a few students displaying disruptive behaviors and a teacher inexperienced with class management happen to be assigned to this class, the focal student may achieve lower than he would otherwise at the end of the treatment. Finally, even in a cluster randomized trial, interference is sometimes inevitable between clusters that are spatially adjacent. In a study of "community policing" in Chicago (Verbitsky-Savitz and Raudenbush, 2004, 2012), police districts were assigned to receive either community policing or regular policing. Because effective community policing in one neighborhood may encourage offenders to operate elsewhere, a neighborhood assigned to the control condition may suffer when nearby neighborhoods receive community policing. The treatment setting for residents in a focal neighborhood therefore is defined by the treatments assigned to the surrounding neighborhoods.

Hong (2004) defined a "treatment setting," in the context of social science research, as a specific local environment consisting of one or more agents and a set of participants along with the treatment assignments of those agents and participants. Hong and Raudenbush (2013) further elaborated this rationale by emphasizing two pervasive features of social interventions. First, the interventions are typically delivered by human agents—therapists, teachers, caseworkers, or police officers—who tend to be heterogeneous in beliefs, training, and experience. Rather than viewing all individual participants as autonomous "agents" as in agent-based modeling (Axelrod, 1997), the notion of "agents" is much narrower here and refers only to those who are designated the role of delivering and implementing a social intervention. Second, participants are often clustered in organizations or communities such as schools and neighborhoods. A treatment setting is often determined at least in part by unique characteristics of the organizational or community context. Changing the treatment assignments of agents or peers or changing the local context may modify a participant's potential outcome value even if that participant's treatment assignment remains constant.

According to this reasoning, under a given treatment, a participant may have as many potential outcome values as the number of possible treatment settings. This generalization relaxes SUTVA and enables one to define person-specific causal effects as:

a. Comparisons between potential outcomes associated with alternative treatment assignments for an individual participant under a given treatment setting

b. Comparisons between potential outcomes associated with alternative treatment settings when a focal participant's treatment assignment is fixed

In addition, one may investigate whether the treatment effect depends on the treatment setting or, in other words, whether characteristics of the treatment setting moderate the effect of treatment assignment for an individual participant.

A general causal framework that incorporates heterogeneous agents and social interactions among participants can be formalized as follows: suppose that an experiment involves N participants and J agents. Following Hong (2004), Hong and Raudenbush (2006), and Verbitsky-Savitz and Raudenbush (2012), we describe the potential outcome of individual participant i as a function of the individual's own treatment assignment z_i, the treatment assignments of other participants represented by an $N-1$ row vector $\mathbf{z}_{-i} = (z_1 \ldots z_{i-1} \ z_{i+1} \ldots z_N)$, as well as the treatment assignments of the agents represented by a row vector with J elements $\mathbf{a} = (a_1 \ldots a_J)$. As before, let $z_i = 1$ if individual participant i is assigned to the experimental condition and 0 otherwise; similarly, let $a_j = 1$ if agent j is assigned to the experimental condition and 0 otherwise. In the case that individuals are clustered geographically or organizationally, an individual's potential outcome may additionally depend on the particular state of the local context at a given time and the contextual characteristics experienced by other individuals (e.g., weather condition as in an agricultural study or local rental market as in the MTO study). Suppose that there are K clusters of individual participants. We may represent distinct contextual characteristics of different clusters in a row vector of K elements $\mathbf{c} = (c_1 \ldots c_K)$. For example, we may let $c_k = 1$ if the local rental market in cluster k is favorable to low-income tenants and 0 otherwise. Hence, individual i's potential outcome, in its most general form, is as follows:

$$Y_i(z_i, \mathbf{z}_{-i}, \mathbf{a}, \mathbf{c}). \tag{14.1}$$

It may seem reasonable to argue that once the treatment setting is given, the value of an individual participant's potential outcome associated with a given treatment will become stable. Hence, the individual-specific treatment effect can be defined as

$$Y_i(1, \mathbf{z}_{-i}, \mathbf{a}, \mathbf{c}) - Y_i(0, \mathbf{z}_{-i}, \mathbf{a}, \mathbf{c}). \tag{14.2}$$

14.4.2 Simplified characterization of treatment settings

To simplify the characterization of treatment settings, one may consider the following questions:

1. Would the treatment assignments of the $N-1$ peers in the community have equal impacts on individual participant i's potential outcome value?

2. Or rather, would the spill-over effect come mainly from peers closest to individual participant i?

3. Would the agents other than the one who delivers the treatment to individual i have an impact on the individual's potential outcome?

4. Would the potential outcome of individual i in cluster k be influenced by the treatment assignments in and the contextual characteristics of other clusters?

The first question is especially relevant when the researcher theorizes, as in Sobel (2006), that how a focal individual responds to a given treatment may depend on the proportion of participants to be treated. For example, when individual i is assigned to the control group, one may reason that assigning 40% of the participants to be treated may influence this focal individual differently than assigning none of the participants to be treated. Suppose that all the other participants can be rank ordered by their physical or relational proximity to the focal individual. We then ask whether assigning the first 40% of the participants to be treated would have the same impact for the focal individual as assigning the last 40% of the participants to be treated. If the answer is yes, then the $N-1$ peers may be viewed as *exchangeable* for the causal question at hand. One may summarize the treatment assignments of the $N-1$ peers in a unidimensional function:

$$f(\mathbf{z}_{-i}) = P_{-i} = \frac{\sum_{i'=1}^{N-1} Z_{i'}}{N-1} \tag{14.3}$$

for $i' \neq i$, representing the proportion of peers assigned to the experimental condition.

In some applications, the answer to question (1) might be negative while that to question (2) might be positive. This would be true when social contagion occurs primarily among members of a cohesive group (Festinger, Schachter, and Back, 1950). Suppose that individual i belongs to a network involving h peers $(0 < h < N-1)$, one may reduce $\mathbf{z}_{-i}$ to a row vector with h elements only. Moreover, if these h peers do not have equal impacts on individual i, one may assign a relatively high weight (e.g., $V_{ii'} = 1$) to a strong tie and a relatively low weight (e.g., $V_{ii'} = 0.5$) to a weak tie in summarizing the impact of the h peers. Here, $V_{ii'}$ quantifies individual i's evaluation of the social significance of his or her relationship with peer i'.

Suppose that individual i has strong ties with two peers, one of them treated and the other untreated; at the same time, the focal individual has a weak tie with a third peer who is treated. The treatment assignments of the peers may be summarized as

$$f(\mathbf{z}_{-i}) = P_{-i} = \frac{\sum_{i'=1}^{h} V_{ii'} Z_{i'}}{\sum_{i'=1}^{h} V_{ii'}}, \tag{14.4}$$

which will be equal to $(1+0+0.5)/(1+1+0.5) = 0.6$ in the previous example. The same representation applies when social contagion occurs among individuals holding structurally equivalent positions who are likely in competition. Individuals who are structurally equivalent tend to use one another as a social frame of reference to evaluate one's own relative adequacy, even if direct personal ties do not exist between them. For example, Burt (1987) found evidence that physicians are eager to adopt a medical innovation upon observing (perhaps through mass media or word of mouth) other physicians' adoption of the same innovation. In this case, $V_{ii'}$ quantifies individual i's evaluation of the equivalence of the position held by individual i' to his or her own in a social structure.

Question (3) asks whether individual i's potential outcome could be influenced by the agents other than one's own. The answer would be no especially if the researcher could rule out spill-over between the agents. Suppose that the treatment to be evaluated is a new therapy. The researcher may argue that, for patient i who is assigned to the experimental condition and receives the therapy from therapist j, this patient's potential outcome does not depend on whether other therapists are also trained to deliver the new therapy. In this case, patient i's potential outcome will be a function of a_j rather than $\mathbf{a}$. However, the earlier assumption may be implausible if, in fact, agents can influence one another either by having direct communications or by virtue of holding structurally equivalent positions. One may argue that other therapists may have an indirect impact on patient i's potential outcome partly through influencing the practice of therapist j who directly works with the focal patient. One may also argue that the indirect impact of other therapists may be transmitted partly through influencing the behaviors of other patients to whom the focal patient is connected. Chapter 15 will further modify the causal inference framework to accommodate spill-over as a potential mediation mechanism and will provide application examples.

In considering question (4), the researcher needs to determine whether there could be spill-over between clusters such as households, schools, neighborhoods, and firms. In some applications, it may seem reasonable to assume that individuals do not change cluster memberships during the study and that communications between individuals from different clusters are sporadic and inconsequential for a focal individual's potential outcomes. When this is the case, the researcher may argue that interference between individuals is possible within a cluster but not between clusters (Hong and Raudenbush, 2006; Hudgens and Halloran, 2008). The potential outcome of individual i in cluster k will then depend on the peer treatment assignment within the cluster but not on the treatment assignments in other clusters. Moreover, the individual's potential outcome will reflect the contextual characteristics of his or her own cluster and can be represented as a function of c_k rather than $\mathbf{c}$.

If the researcher answers *yes* to question (1) or (2) and *no* to questions (3) and (4), the researcher can simplify the potential outcome of individual i in cluster k who encounters agent j as a function of the individual's own treatment assignment, a summary index of peers' treatment assignments in the same cluster, the treatment assignment of the agent for the focal individual, and the contextual characteristics of the cluster that the individual belongs to

$$Y_{ijk}(z_{ik}, P_{-ik}, j). \tag{14.5}$$

14.4.3 Causal effects of individual treatment assignment and of peer treatment assignment

SUTVA stipulates that a change in treatment setting should have no impact on an individual's potential outcomes. When this assumption fails, causal inference is nonetheless possible by modeling the potential outcomes as functions of treatment settings. We then define *the causal effect of individual treatment assignment* while holding characteristics of the treatment settings fixed, such as fixing peer treatment assignment p, fixing agent assignment j, and fixing cluster membership k:

$$Y(1, p, j, k) - Y(0, p, j, k). \tag{14.6}$$

The population average causal effect of individual treatment assignment when peer treatment assignment is fixed at p is an average over all the individuals who share the same agent j and the same cluster membership k and is averaged over all the agents and over all the clusters. The population average causal effect is

$$E_c(E_a\{E[Y(1, p, j, k) - Y(0, p, j, k)]\}). \tag{14.7}$$

Moreover, when we hold an individual's treatment assignment fixed at z, we can define *the causal effect of peer treatment assignment*. This is the spill-over effect from one's peers as a contrast between peer treatment assignments p and p'. For example, Hong and Raudenbush (2006) defined the average causal effect of retaining a high proportion of low-achieving students (p) as opposed to retaining a low proportion of such students (p') in a school on the academic learning of student i in school k who is not at risk of retention (i.e., $z_{ik} = 0$). The researchers used school membership to capture the impact of agents and characteristics of the local school context. They defined the average causal effect of peer treatment assignment for the population of students not at risk of retention as an average over all such students in a school and over all the schools:

$$E\{E[Y_{ik}(0, p) - Y_{ik}(0, p')|k]\}. \tag{14.8}$$

Following Halloran and Struchiner (1991, 1995), Hudgens and Halloran (2008) called the causal effect of individual treatment assignment "the individual direct causal effect" or "direct effect" and called the causal effect of peer treatment assignment "the indirect causal effect." We choose not to use the terms "direct effect" and "indirect effect" here to represent the causal effect of individual treatment assignment and the spill-over effect of peer treatment assignments, respectively, because they have particular usage in causal mediation research. While characterizing peer spill-over in a simplified epidemiologic setting, Hudgens and Halloran (2008) did not consider the treatment assignments of agents. Nor did they acknowledge the salience of preexisting social networks within a cluster and of other contextual characteristics that may influence an individual's potential outcome under a given treatment.

14.5   Identification: Challenges and solutions

As clarified in Chapter 2, in causal inference, a population average treatment effect defined in terms of potential outcomes is identifiable if the counterfactual quantities can be related to observable data under certain assumptions. Typically, in an experiment that assigns individual units at random to different treatment conditions, the average observed outcome of the untreated identifies the average counterfactual outcome associated with the control condition for the treated; and the average observed outcome of the treated identifies the average counterfactual outcome associated with the experimental condition for the untreated.

Section 14.3 of the current chapter has revealed the difficulties in identifying the population average treatment effect, despite the randomization of treatment assignment, when there is a possible spill-over between individual participants. In the MTO study, according to Sobel (2006), the mean difference in the observed outcome between the treated and the untreated does not identify the population average treatment effect. Rather, it is equal to the difference between the effect of assigning a p proportion of the participants rather than none to be treated for the treated units and the spill-over effect for the untreated units. The mean contrast between the treated and the untreated identifies the population average treatment effect only when the spill-over effect is zero.

We take on this challenge in the current section. Under the modified causal framework that allows for a nonzero spill-over effect, we propose a range of research designs and clarify the minimal assumptions required for identifying the average treatment effect. We then further discuss research designs that may be employed for identifying the effect of spill-over and assess the accompanying identification assumptions.

14.5.1   Hypothetical experiments for identifying average treatment effects in the presence of social interactions

In an experiment in which the treatment setting is given and can be viewed as universal for all the participants regardless of one's actual treatment assignment, the population average treatment effect under this given treatment setting can be identified. For example, suppose that 50% of all the residents in a densely populated community of population size N are to be vaccinated in a randomized study. If the disease to be prevented would spread in the air, arguably, vaccinating one's acquaintances is no more important than vaccinating any stranger that one may encounter in a public space. The treatment setting characterized by vaccinating $N/2$ residents for an individual who is not vaccinated can be viewed the same as the treatment setting that vaccinates $(N/2)-1$ other residents for an individual who is vaccinated. Because vaccination is a standard procedure for a trained public health practitioner, it becomes reasonable to ignore agent effects (i.e., who delivers the vaccine). Hence, vaccinating 50% of the residents, denoted by $p=0.5$, defines a treatment setting that can be viewed identical for all participants in the study. The *population average causal effect of individual treatment assignment* in this particular treatment setting is defined as

$$E[Y(1,p=0.5)-Y(0,p=0.5)]. \qquad (14.9)$$

The expectation is taken over all the N individuals in the population (i.e., $i = 1 \ldots, N$) under this particular treatment setting.

The causal effect defined in (14.9) is identifiable under the following two assumptions:

1. *Ignorable individual treatment assignment.* An individual's treatment assignment is independent of the potential outcomes.

2. *Exchangeable peer treatment assignments.* Different treatment assignments for other individuals in the population can be viewed exchangeable as long as they all result in a p proportion of the population receiving the treatment.

Under these two assumptions, we have that

$$E[Y(z,p=0.5)] = E[Y|Z=z, P=0.5] \tag{14.10}$$

for $z = 0,1$. Here $P = 0.5$ refers to a subpopulation of individuals who actually experience a treatment setting in which 50% of the residents are vaccinated. The randomization procedure satisfies assumption (1). Hence, whether an individual is assigned to receive vaccination is independent of the individual's health outcome when 50% of the population is vaccinated. Under assumption (2), the individual's health outcome when vaccinated (or unvaccinated) is stable regardless of which 50% of the population is vaccinated. Therefore, the population average potential outcome of receiving vaccination when 50% of the population is vaccinated, denoted earlier by $E[Y(1,p=0.5)]$, is identified by the mean observed outcome of the 50% of the population randomized to be vaccinated $E[Y|Z=1, P=0.5]$. The population average potential outcome of not receiving vaccination in this setting $E[Y(0,p=0.5)]$ can be similarly identified by the mean observed outcome of the 50% of the population randomized to the control condition $E[Y|Z=0, P=0.5]$. The causal parameter therefore is identified by the difference between the mean observable outcome of the vaccinated and that of the unvaccinated:

$$E[Y(1,p=0.5) - Y(0,p=0.5)] = E[Y|Z=1, P=0.5] - E[Y|Z=0, P=0.5].$$

The experiment as described earlier, however, cannot identify the population average treatment effect in an alternative treatment setting in which the proportion of the population randomized to be vaccinated is different from $p = 0.5$.

Assumption (2) is called "stratified interference" by Hudgens and Halloran (2008). It states that an individual's potential outcome associated with a given treatment—for example, not receiving vaccination—can be assumed stable when $N \times p$ individuals in the population are vaccinated, regardless of which particular $N \times p$ individuals receive vaccination. Assumption (2) may become implausible, however, if a disease is spread only through intimate contacts such that the benefit of vaccinating one's intimate contacts, such as one's family members, is expected to dominate the benefit of vaccinating strangers for one's potential outcome.

When this is the case, the researcher may become interested in evaluating the causal effect of individual treatment assignment for individual i in household k when all other members of the individual's household are vaccinated. One may ask whether vaccinating all the other household members provides sufficient protection for the focal individual who is not vaccinated. The treatment assignment for the other household members, denoted by $p_{-ik} = 1$, characterizes the treatment setting for the focal individual. The population average causal effect of individual treatment assignment in this particular treatment setting is

$$E[Y(1, p_{-ik} = 1) - Y(0, p_{-ik} = 1)]. \tag{14.11}$$

Here, the expectation is to be taken over all the individuals in a household and over all the households in the population.

The identification of the aforementioned population average causal effect still requires assumption (1) "ignorable individual treatment assignment." However, we now replace assumption (2) "exchangeable peer treatment assignments" with the following assumption that was highlighted by Hong and Raudenbush (2006):

3. *No interference between clusters.* There are no social interactions between individuals from different clusters such as communities or households that would affect an individual's potential outcome values.

Under assumption (3), the disease of concern does not spread from one household to another. Assumption (3) is violated if individuals enter intimate relationships with those from other families or even change their family membership through marriage after treatment assignment, which may open routes to between-household interference.

To satisfy assumption (1), the experimenter may randomly select an individual from each household and then assign these study participants at random to either the experimental condition or the control condition. Each of the study participants in the experimental group will receive vaccination along with the entire household; yet none of the study participants in the control group will receive vaccination although the rest of the family will be vaccinated. The treatment setting is the same for all the study participants regardless of their individual treatment assignment. This design makes it possible to identify the population average effect of individual vaccination when the rest of one's household is vaccinated. The population average effect of individual vaccination is then identified as the difference between the average observed outcome of the study participants in the experimental group $E[Y|Z = 1, P_{-ik} = 1]$ and that of the study participants in the control group $E[Y|Z = 0, P_{-ik} = 1]$:

$$E[Y(1, p_{-ik} = 1) - Y(0, p_{-ik} = 1)] = E[Y|Z = 1, P_{-ik} = 1] - E[Y|Z = 0, P_{-ik} = 1].$$

The researcher may also ask whether an unvaccinated individual receives no protection when other members of the household are not vaccinated. The treatment setting for focal individual i in household k in this case is denoted by $p_{-ik} = 0$. The population average causal effect of individual treatment assignment in this treatment setting is

$$E[Y(1, p_{-ik} = 0) - Y(0, p_{-ik} = 0)]. \tag{14.12}$$

One may conduct an experiment in which a study participant will be vaccinated if assigned at random to the experimental condition and will not be vaccinated if assigned to the control condition, yet none of the family members will be vaccinated regardless of the treatment assignment of the study participant. This alternative design satisfies assumption (1). When assumption (3) also holds, the research can identify the population average effect of individual vaccination when the rest of one's family is not vaccinated:

$$E[Y(1,p_{-ik}=0)-Y(0,p_{-ik}=0)]=E[Y|Z=1,P_{-ik}=0]-E[Y|Z=0,P_{-ik}=0].$$

In another example examining the impact of a community policing intervention on reducing neighborhood crime rate, one must take into account the fact that each neighborhood is adjacent to a number of other neighborhoods in a big city with a total of N neighborhoods. This is because crime often "spills" over the border shared by two neighborhoods. Let Z_i and $Z_{i'}$ denote the treatment assignments for neighborhood i and neighborhood i', respectively, each taking value 1 if receiving community policing and 0 otherwise. The treatment setting for neighborhood i can be characterized by the faction of adjacent neighborhoods that receive the intervention:

$$P_{-i}=\frac{\sum_{i'=1}^{N-1}V_{ii'}Z_{i'}}{\sum_{i'=1}^{N-1}V_{ii'}},$$

where $V_{ii'}$ indicates whether neighborhood i' is adjacent to focal neighborhood i. One may then define the average effect of treating versus not treating a focal neighborhood under each distinct treatment setting represented by a unique value of P_{-i} (Verbitsky-Savitz and Raudenbush, 2012).

The identification is possible under assumption (1) "ignorable individual treatment assignment" and assumption (2) "exchangeable peer treatment assignments." To satisfy assumption (1), the experimenter may simply assign neighborhoods at random to either the intervention program or a control condition. Instead of assuming "no interference between clusters," however, the problem at hand requires a different assumption. Because the adjacent neighborhoods serve the function of spatially segregating a focal neighborhood from all other neighborhoods in the city, it may seem reasonable to assume that the spill-over effect comes only from one's adjacent neighborhoods. One may therefore invoke the following assumption:

4. *No interference between disjoint units.*

In the current case, assumption (4) suggests that the potential crime rate in a focal neighborhood under a given treatment is unaffected by the treatment assignments of other neighborhoods that do not share borders with the focal neighborhood. An analogy can be found in causal inference studies involving social network data. If a social scientific theory indicates that social contagion spreads beyond one's direct ties in a network, assumption (4) may be modified to allow for interference between, for example, individual units with second-degree connections (Adamic and Adar, 2003) or holding structurally equivalent positions (Burt, 1987).

Finally, the community policing intervention requires skillful implementation by police officers who are the major agents in this study. Hence, it is essential to minimize the agent effects through intensive training as well as through randomization of police officers to different treatment conditions and to different neighborhoods such that it becomes plausible to assume the following:

5. *Exchangeable agent assignments to treatments and to clients.*

If the agents nonetheless show important differences in their skill levels that may have an impact on the potential outcomes, the treatment setting for a focal neighborhood will need to be characterized by the skill level of the agents assigned to this neighborhood as well as by the treatment assignments of the adjacent neighborhoods.

14.5.2   Hypothetical experiments for identifying the impact of social interactions

To identify the spill-over effect of vaccinating a relatively large proportion as opposed to a relatively small proportion of residents in a community requires a study that involves multiple communities. Suppose that the researcher is interested in contrasting a treatment assignment strategy that vaccinates 80% of the residents with one that vaccinates only 40% of the residents. As suggested by Hong and Raudenbush (2006) and by Hudgens and Halloran (2008), a two-stage randomization procedure will minimize the identification assumptions. In the first stage, the researcher will assign the communities at random either to the strategy that vaccinates 80% of the residents, denoted by $p=0.8$, or to the one that vaccinates 40% of the residents, denoted by $p=0.4$. In the second stage, residents in a community assigned to $p=0.8$ will be assigned at random for vaccination with probability 0.8, while residents in a community assigned to $p=0.4$ will be vaccinated at random with probability 0.4. The average spill-over effect of vaccinating 80% rather than 40% of the community residents for unvaccinated individuals is defined as

$$E[Y(0,p=0.8)-Y(0,p=0.4)]. \tag{14.13}$$

This causal effect can be identified, under a series of assumptions explicated in the following, by the difference between the mean observable outcome of unvaccinated individuals in the communities assigned to $p=0.8$ and that of unvaccinated individuals in the communities assigned to $p=0.4$:

$$E[Y(0,p=0.8)-Y(0,p=0.4)]=E[Y|Z=0,P=0.8]-E[Y|Z=0,P=0.4].$$

This average spill-over effect on the unvaccinated is identified under the following assumptions:

1a. *Ignorable cluster treatment assignment.* The cluster-level treatment assignment is independent of individual potential outcomes.

1b. *Ignorable individual treatment assignment within a cluster.* The individual-level treatment assignment within a cluster is independent of individual potential outcomes.

2. *Exchangeable peer treatment assignment,* which now applies to peer treatment assignments within a cluster.

3. *No interference between clusters.* That is, there are no social interactions between individuals from different clusters that would affect an individual's potential outcome values; and additionally, individuals do not change cluster membership after the cluster-level or the individual-level treatment assignment.

If a disease spreads primarily among one's immediate family members, the researcher is perhaps interested in the average spill-over effect for unvaccinated individuals between two alternative treatment settings. In the experimental setting, the rest of the family members are all vaccinated, as denoted by $p_{-ik} = 1$ earlier, while in the control setting, which is denoted by $p_{-ik} = 0$, none of the family members are vaccinated. The population average spill-over effect for the unvaccinated is defined as

$$E[Y(0, p_{-ik} = 1) - Y(0, p_{-ik} = 0)]. \tag{14.14}$$

One may further hypothesize that individual vaccination is more essential for the individual's health outcome when none of the family members are vaccinated than when the rest of one's family are all vaccinated. Focusing on the moderating role of family vaccination, this theoretical hypothesis involves comparing the causal effect of vaccinating a focal individual between the experimental setting and the control setting, that is,

$$E[Y(1, p_{-ik} = 1) - Y(0, p_{-ik} = 1)] - E[Y(1, p_{-ik} = 0) - Y(0, p_{-ik} = 0)]. \tag{14.15}$$

The identification of these causal effects requires assumptions (1a), (1b), and (3) but does not require assumption (2) "exchangeable peer treatment assignments." These examples reveal the fact that depending on the scientific characterization of the treatment setting, assumption (2) may not be necessary for identification.

In a study of neighborhood crime, however, spill-over may occur between any two neighborhoods that are geographically connected rather than clustered. One may define the average spill-over effect of completely or partially treating the adjacent neighborhoods as opposed to leaving the adjacent neighborhoods untreated for the crime rate in a focal neighborhood. Without clustering, a hypothetical experiment will assign neighborhoods and agents (i.e., police officers) at random to treatments. Hence, we will invoke assumptions (1) "ignorable individual treatment assignment" and (5) "exchangeable agent assignments to treatments and to clients" rather than assumptions (1a) and (1b). As discussed in the previous section, additional assumptions required for identification are (2) "exchangeable peer treatment assignments" and (4) "no interference between disjoint units" but not (3) "no interference between clusters."

The importance of the skill level of the agent is often noteworthy in social interventions. For example, in the Tennessee class size study, students and teachers were assigned at random within each participating elementary school to either a small class or a regular class. The primary interest of the study was to determine the average impact of class size reduction. This particular design satisfies assumptions (1b) "ignorable individual treatment assignment within a cluster" and (5) "exchangeable agent assignments to treatments and to clients." The identification of the class size effect also requires assumption (3) "no interference between clusters" but does not need most other assumptions. The experiment has enabled researchers to identify teacher effects on student outcomes as well. Taking advantage of the multisite randomized design, some researchers have detected a teacher effect on student achievement especially in math and a larger effect in schools serving students from a low socioeconomic background than in schools serving more advantaged students (Nye, Konstantopoulos, and Hedges, 2004). Here, we may view the socioeconomic composition of students in a school as a theoretically important contextual characteristic. This example shows that the agent effect may depend on the local context. Others have reported a cumulative effect of the teachers over the elementary school years on sixth-grade achievement (Konstantopoulos and Chung, 2011) and a positive effect of being taught by a relatively experienced teacher in kindergarten on later earnings (Chetty *et al.*, 2011).

The ignorability assumptions (1), (1a), (1b), and (5) will require modification if analysts are working with quasiexperimental rather than experimental data. This will be illustrated in the application example in the next section.

14.5.3   Application to an evaluation of kindergarten retention

Hong and Raudenbush (2006) explicitly defined and identified spill-over effects in their evaluation of the effect of retaining a child in kindergarten rather than promoting the child to first grade. Past studies of this type have implicitly assumed that each child has a single potential outcome if retained and a single potential outcome if promoted. Hong and Raudenbush (2006) reasoned that a child's potential outcome values may depend not only on whether the child is retained but also on who attends the same school and how many of his or her schoolmates are retained. This is because the treatment assignments of other students will determine, during the treatment year, where a focal child stands relative to other students in class, how the teacher may design instruction, and how the students may interact with each other. They characterized peer treatment assignment, therefore, as the fraction of students in the same grade who are retained. According to these researchers, children who are considered for kindergarten retention typically display academic or behavioral difficulties in the previous year. If a school decides to retain a relatively high proportion of students, those who are retained in kindergarten may suffer from frequent disruptions in class and may have to compete with a relatively large number of retained peers for instructional assistance. Moreover, they reasoned that the outcomes of students who are making excellent progress and therefore have no risk of being retained may nonetheless be influenced by their peers' retention. This is because retaining a relatively high proportion of students in kindergarten may improve the learning environment for promoted first graders by making their classrooms more homogeneous in ability and less disrupted and thereby increasing instructional efficiency.

In this application, a student's potential outcome is represented, in general, as a function of the treatment assignment for the focal student z_i, the treatment assignments of other students in the population denoted by $\mathbf{z}_{-i}$, and the school assignments of all the students in the population denoted by $\mathbf{s} = (s_1 \cdots s_N)$ where the possible values for the N elements in this vector are school identification numbers $j = 1, \ldots, J$. Here, the school assignment for each student summarizes information about agents and school characteristics. Hong and Raudenbush (2006) invoked assumption (3) "no interference between schools" and thereby restricted the treatment setting to the school that a child actually attended. They argued that a student's potential outcomes primarily depend on the identities and treatment assignments of that student's schoolmates, whereas the identities and treatment assignments of students attending other schools are assumed to be uninformative of the focal student's outcomes. This assumption seems plausible given the social and geographic isolation of schools, though one could certainly challenge this assumption. The treatment setting for student i in school j therefore is characterized by the student's school membership and the treatment assignment of the schoolmates. The results are to be generalized to the population of existing schools rather than to a hypothetical world in which children are assigned at random to schools.

To further simplify, these researchers focused on two distinct treatment settings: $I_{(P_j > \varphi)} = 1$ if the retention rate in school j is above the national average φ; and $I_{(P_j > \varphi)} = 0$ otherwise. This characterization of treatment settings necessitates the invocation of assumption (2) "exchangeable peer treatment assignments" within a school. In other words, different treatment assignments of one's schoolmates who are at risk of being retained are assumed to have the same impact on one's potential outcome associated with a given individual-level treatment, as long as all such peer treatment assignments lead to a "high" retention rate. The same can be said of different treatment assignments of one's at-risk school mates in schools that have a "low" retention rate. This assumption seems plausible if the spill-over effect of peer treatment assignments operates primarily through classroom instruction as a teacher typically tailors instruction to the composition of students in a class.

Under assumptions (2) and (3), the potential outcome of student i in school j is represented as $Y_{ij}(z_{ij}, I_{(P_j > \varphi)} = 1)$ if the school has a "high" retention rate and as $Y_{ij}(z_{ij}, I_{(P_j > \varphi)} = 0)$ if the school has a "low" retention rate. Furthermore, let q_1 denote a child's probability of being retained under a high retention rate, and let q_0 denote the child's probability of being retained under a low retention rate. For a subpopulation of children who are certainly at risk of being retained regardless of the retention rate, for whom $q_1 \geq q_0 > 0$, the average effect of retention as opposed to promotion in a low-retention school is

$$E[Y(1,0) - Y(0,0) | q_1 \geq q_0 > 0]. \tag{14.16}$$

The average retention effect for such students in a high-retention school is

$$E[Y(1,1) - Y(0,1) | q_1 \geq q_0 > 0]. \tag{14.17}$$

A second subpopulation of children is at risk of retention only in a high-retention school but not in a low-retention school, for whom $q_1 > 0, q_0 = 0$. These are students at *partial* risk of retention because they are highly unlikely to be retained under a low retention rate. The average retention effect for these students in a high-retention school is

$$E[Y(1,1) - Y(0,1) | q_1 > 0, q_0 = 0]. \tag{14.18}$$

Finally, a third subpopulation of children are at *no* risk of being retained regardless of the retention rate, for whom $q_1 = q_0 = 0$. The average causal effect of adopting a high retention rate rather than a low rate by the school for these children is

$$E[Y(0,1) - Y(0,0) | q_1 = q_0 = 0]. \tag{14.19}$$

Hong and Raudenbush (2006) envisioned a two-stage hypothetical experiment. For simplicity, suppose that schools do not differ in their retention rate in the past. At the first stage, the experimenter would assign intact schools at random to a high retention rate or a low rate. The second stage is a multisite randomized trial in which students at risk of being retained in each school would be assigned at random to retention according to the fixed retention rate determined by the school-level treatment assignment. This is similar to a standard split-plot design in agriculture. Here, schools are "whole plots" assigned to a high or low rate, and students are "subplots" to be retained or promoted. Under such a design, the school-level treatment assignment and the individual-level treatment assignment within each school are both ignorable, which satisfies assumptions (1a) and (1b) key to the identification of the aforementioned causal effects.

When the aforementioned identification assumptions hold, we have that, for $z = 0, 1$,

$$E\left[Y\left(z, I_{(P_j > \varphi)} = 1\right) \middle| q_1, q_0\right] = E\left[Y | Z = z, I_{(P_j > \varphi)} = 1, q_1, q_0\right];$$

$$E\left[Y\left(z, I_{(P_j > \varphi)} = 0\right) \middle| q_1, q_0\right] = E\left[Y | Z = z, I_{(P_j > \varphi)} = 0, q_1, q_0\right].$$

That is, for students in the same subpopulation defined by the individual-level probability of being retained under a high rate q_1 and that under a low rate q_0, the average potential outcome of retention under a high rate is identified by the average observable outcome of such students who are actually retained in schools that have been assigned to a high rate; the average potential outcome of promotion under a high rate is identified similarly. Following the same logic, the subpopulation average potential outcome of retention and that of promotion under a low rate are identified by the average observable outcome of retainees and that of promoted students, respectively, in schools with a low rate.

Hong and Raudenbush (2006) analyzed the Early Childhood Longitudinal Study–Kindergarten (ECLS-K) cohort data to address the aforementioned causal questions. The sample includes 471 kindergarten retainees and 10,255 promoted students in 1,080 schools that allowed for kindergarten retention. In this study, the schools were not randomized to a high retention rate or a low rate, nor were the students assigned at random to be retained or promoted. In fact, whether a child would be retained is a highly selective decision that depends on the child's pretreatment characteristics. Similarly, a school may select its retention rate for particular reasons. Nonetheless, the ECLS-K data contain many theoretically important covariates predicting school retention rate or child retention status. The researchers tentatively assumed that identification assumptions (1a) and (1b) hold within levels of the observed covariates $\mathbf{X}$, that is, the assignment of schools to a high retention rate or a low rate and the assignment of at-risk students to retention or promotion are independent of the potential outcomes for schools and students sharing the same pretreatment characteristics $\mathbf{X} = \mathbf{x}$. They assessed the consequences of having omitted confounders through sensitivity analysis. The analytic strategies are described in the next section.

14.6   Analytic strategies for experimental and quasiexperimental data

Depending on the data structure and the nature of spill-over, the analysis may require special techniques for clustered data, spatial data, or network data. The need for removing selection bias in quasiexperimental data poses additional challenges. Continuing the application example of kindergarten retention, we will provide a brief overview of causal effect estimation in the context of a hypothetical experiment. We will then review the propensity score stratification-based analyses of the quasiexperimental data as implemented by Hong and Raudenbush (2006). Finally, we will discuss how to apply the marginal mean weighting through stratification (MMWS) method to the problem at hand.

14.6.1   Estimation with experimental data

In the evaluation of kindergarten retention, analyses of data from a hypothetical two-stage randomized experiment must take into account the clustering of students within schools.

The mean difference in the observed outcome between retainees and at-risk students randomized to be promoted in a low-retention school, averaged over all the low-retention schools, estimates the retention effect under a low rate as defined in (14.16) for students always at risk. Similarly, one may estimate the retention effect under a high rate for students at partial risk or always at risk of retention. The estimated difference between (14.17) and (14.16) for students always at risk of retention regardless of the school-level retention rate, if statistically significant, will indicate that the retention effect depends on the retention rate for such students. The analytic model will include the indicator for individual retention treatment as a student-level predictor and the indicator for retention rate as a school-level predictor. The interaction effect of these two predictors represents the moderating effect of retaining a relatively high fraction rather than a low fraction of students in a school on the impact of retention for those who are always at risk of being retained. Moreover, for students who would never be considered for retention even under a high rate, the mean difference in the observed outcome of such students between the high-retention schools and low-retention schools will estimate the spill-over effect of a high rate relative to a low rate as defined in (14.19), hypothesized to be beneficial for students at no risk.

14.6.2   Propensity score stratification

In analyzing the quasiexperimental data, in order to account for selection bias, Hong and Raudenbush (2006) specified a model predicting a school's propensity of adopting a high retention rate $Q = Q(\mathbf{x}) = \mathrm{pr}(I_{(P_j > \varphi)} = 1 | \mathbf{X} = \mathbf{x})$, a model predicting a student's propensity of being retained under a high retention rate $q_1 = q_1(\mathbf{x}) = \mathrm{pr}(Z = 1 | I_{(P_j > \varphi)} = 1, \mathbf{X} = \mathbf{x})$, and a model predicting a student's propensity of being retained under a low rate $q_0 = q_0(\mathbf{x}) = \mathrm{pr}(Z = 1 | I_{(P_j > \varphi)} = 0, \mathbf{X} = \mathbf{x})$. The researchers used the predicted student propensity of retention under a high rate and that under a low rate to empirically identify students who are always at risk, those at partial risk, and those at no risk of retention. After removing students at no risk of retention under a low rate according to the predicted q_0, they divided the sample of students attending low-retention schools into eight strata on this propensity score. The estimated average within-stratum retention effect, obtained from analyzing a two-level random-effects model with students nested within schools, is more than half a standard deviation of the reading outcome suggesting a detrimental effect of retention. There is statistically significant variation in the retention effect, ranging from about -1.5 to 0.5 standard deviations among 95% of the low-retention schools, suggesting considerable heterogeneity in how students are taught if retained or promoted across these schools. The random variation in the retention effect reflects an important role played by agents—in this case, teachers and principals—that would require further investigation. These researchers conducted a stratified analysis of the data from the high-retention schools on the basis of the predicted q_1. The results parallel those for the low-retention schools. The overall average and the between-school variation of the retention effect under a high retention rate are similar to those under a low rate.

Students always at risk of retention are those who could be retained even in schools with a low retention rate. The researchers did not attempt to estimate the impact of a school's retention rate for these students because the sample size of such students was too small to support precise inference. Therefore, the question of whether the retention effect for students always at risk of retention depends on the retention rate remains unanswered. To estimate the effect of a high rate as opposed to a low rate for students at no risk of retention, the researchers stratified schools on the predicted school-level propensity Q. The estimated average within-stratum effect of retention rate, despite showing a positive sign for the point estimate favoring those attending low-retention schools, cannot be statistically distinguished from zero.

14.6.3   MMWS

The MMWS method provides an alternative to propensity score-based matching or stratification for bias removal. This method transforms the quasiexperimental data through nonparametric weighting with the goal of approximating a two-stage experimental design described in Section 14.5.3. See Chapter 4 for a detailed explanation of the theoretical rationale as well as the general analytic procedure of the MMWS method. In particular, in constructing the weight, the researcher must be explicit about the population relevant for each causal effect. We have defined the causal effects of interest in (14.16)–(14.19) for three different subpopulations of students—those who are always at risk of retention regardless of the school-level retention rate, those who are at partial risk of retention under a high rate only, and those who are at no risk of retention. All three subpopulations of students attend schools that allow for kindergarten retention. Therefore, we do not generalize the causal inference to schools that have banned kindergarten retention.

To reduce pretreatment differences between high-retention schools and low-retention schools, the ECLS-K sample of schools that allowed for kindergarten retention is stratified on the basis of the estimated school propensity of adopting a high rate denoted by Q as described in the previous section. We then compute and assign the marginal mean weight to low-retention schools and high-retention schools within each stratum such that schools of similar pretreatment characteristics will have equal representation in the two groups after weighting. In stratum $S_Q = s$, the weight for a high-retention school is $\mathrm{pr}(I_{(P_j > \varphi)} = 1)/\mathrm{pr}(I_{(P_j > \varphi)} = 1 | S_Q = s)$; and that for a low-retention school is $\mathrm{pr}(I_{(P_j > \varphi)} = 0)/\mathrm{pr}(I_{(P_j > \varphi)} = 0 | S_Q = s)$. In each case, the numerator is the proportion of high-retention schools (or low-retention schools) in the entire sample, while the denominator is the proportion of such schools in the given stratum. Under the identification assumption that school assignment to a high rate versus a low rate becomes ignorable given the observed pretreatment school characteristics, in the weighted data, the group of high-retention schools and the group of low-retention schools are each expected to display the same observed pretreatment composition as that of the entire sample. Once the researcher has identified students who are at no risk of retention, a weighted analysis comparing the mean outcome of such students in high-retention schools and their counterparts in low-retention schools generates an estimate of (14.19), that is, the spill-over effect of retaining a high fraction as opposed to a low fraction of the schoolmates for students at no risk.

Within a low-retention school, some students are selected to be retained, while their peers similarly at risk may be selected for promotion. We can use MMWS to reduce pretreatment differences between the retained and the promoted students who are always at risk of retention under a low rate, After stratifying such students in low-retention schools on the basis of their estimated propensity of retention denoted by q_0, we compute a weight at the student level that is equal to $\mathrm{pr}(Z = 1 | I_{(P_j > \varphi)} = 0)/\mathrm{pr}(Z = 1 | I_{(P_j > \varphi)} = 0, S_{q_0} = s)$ for a retained student in stratum $S_{q_0} = s$ and is equal to $\mathrm{pr}(Z = 0 | I_{(P_j > \varphi)} = 0)/\mathrm{pr}(Z = 0 | I_{(P_j > \varphi)} = 0, S_{q_0} = s)$ for a promoted student in the same stratum who is similarly at risk of retention. In each case, the numerator is the proportion of students always at risk of retention who are actually retained (or promoted) in low-retention schools, while the denominator is the proportion of such students in the given stratum. The weighting equates the observed pretreatment composition of the retained group and the group of students who were similarly at risk yet received promotion in low-retention schools. By applying this student-level weight along with the school-level weight in a two-level analysis of the observed outcome of students always at risk of retention, we may estimate the retention effect under a low rate for this subpopulation of students as defined in (14.16). The result is generalizable to a population of schools that allows for kindergarten retention.

In parallel, we may stratify high-retention school students at risk of being retained under a high rate on the basis of the estimated student propensity score q_1. The weight is equal to $pr(Z=1|I_{(P_j>\varphi)}=1)/pr(Z=1|I_{(P_j>\varphi)}=1, S_{q_1}=s)$ for a retained student in stratum $S_{q_1}=s$ and is equal to $pr(Z=0|I_{(P_j>\varphi)}=1)/pr(Z=0|I_{(P_j>\varphi)}=1, S_{q_1}=s)$ for a promoted student in the same stratum who is similarly at risk of retention under a high rate. Again applying both the student-level weight and the school-level weight to at-risk students attending high-retention schools, we estimate the retention effect under a high rate that is generalizable to the population of schools that allows for kindergarten retention. The MMWS analysis with the data from high-retention schools, if restricted to the sample of students always at risk of retention regardless of the school-level retention rate, will generate an estimate of the retention effect under a high rate for students always at risk as defined in (14.17). Restricting the analysis to the sample of students at partial risk of retention under a high rate only, we will obtain an estimate of the retention effect under a high rate for students at partial risk as defined in (14.18).

14.7   Summary

Spill-over through social interactions is a phenomenon of great scientific interest. Overlooking nonnegligible spill-over effects can be consequential because it may lead to misinterpretations of the results of causal analyses. A modified framework for causal inference views an individual's potential outcome as a function of the treatment setting as well as of the individual's treatment assignment. A treatment setting is defined as a specific local environment consisting of one or more agents and a set of participants along with the treatment assignments of those agents and participants (Hong, 2004). This chapter illustrates different characterizations of treatment settings on the basis of different scientific theories with regard to primary sources of spill-over. The causal effect of individual treatment assignment is well-defined and can be identified when the treatment setting is given. Moreover, the effect of treating versus not treating an individual may depend on the treatment setting, in which case the latter is an important moderator of the former. Examples from research on vaccination, residential mobility, neighborhood crime, and grade retention have further suggested that when experimental manipulation of the treatment setting is feasible, a change in the treatment setting may generate a spill-over effect for untreated individuals and possibly for treated individuals as well.

Past research on peer effects has been plagued by self-selection of social affiliation (Evans, Oates, and Schwab, 1992; Gaviria and Raphael, 2001; Sacerdote, 2011). In general, individuals are not randomized to schools, neighborhoods, or peer networks. The modified framework takes the composition of a social organization or a social network as it is and defines a treatment setting by the treatment assignments of other units within the same organization or network. In other words, the experimenter can change the characteristics of a treatment setting for a focal individual without changing the membership composition of the organization or the network that the individual is affiliated with. The results of causal inference are to be generalized to the population of existing organizations or networks as well as to the population of individuals.

We have shown that the assumptions required for identifying causal effects depend on how the treatment setting is characterized. Randomized experiments can be designed accordingly to facilitate identification. For example, while randomization of individual treatment assignment is always of great value, randomization of clusters to different treatment settings is relevant if spill-over occurs within clusters. Randomization of agents to treatments and to clients becomes important if agents can have an impact on the potential outcomes. Without clustering, the assumption of "no interference between clusters" does not apply. Yet a different assumption such as "no interference between disjoint units" may be invoked to represent the limit on spill-over. Finally, when the treatment settings to be contrasted are either treating all of one's peers or treating none of the peers, there is no need to assume "exchangeable peer treatment assignments." Although spill-over tends to occur sporadically, a contrast between "treating all" versus "treating none" maximizes the effect size and hence may provide the best occasion for testing theories about social interactions. In conclusion, researchers interested in spill-over effects must carefully consider the social context under study, operationalize theories of spill-over in a meaningful way, and select identification assumptions and analytic strategies applicable to the problem at hand.

References

Adamic, L.A. and Adar, E. (2003). Friends and neighbors on the web. *Social Networks*, *25*(3), 211–230.

Ali, M., Emch, M., von Seidlein, L. *et al* (2005). Herd immunity conferred by killed oral cholera vaccines in Bangladesh: a reanalysis. *Lancet*, *366*, 44–49.

Aronow, P.M. (2012). A general method for detecting interference between units in randomized experiments. *Sociological Methods & Research*, *41*(1), 3–16.

Axelord, R. (1997). *The Complexity of Cooperation: Agent-based Models of Competition and Collaboration*, Princeton University Press, Princeton.

Brock, W.A. and Durlauf, S.N. (2001). Interactions-based models, in *Handbook of Econometrics*, vol. 5 (eds J. Heckman and E. Leamer), Elsevier, Amsterdam, pp. 3297–3380.

Brooks-Gunn, J., Duncan, G.J. and Aber, J.L. (eds) (1997). *Neighborhood Poverty: Context and Consequences for Children*, vol. 1, Russell Sage Foundation, New York.

Burt, R.S. (1987). Social contagion and innovation: cohesion versus structural equivalence. *American Journal of Sociology*, *92*(6), 1287–1335.

Chetty, R., Friedman, J.N., Hilger, N. *et al* (2011). How does your kindergarten classroom affect your earnings? Evidence from Project STAR. *The Quarterly Journal of Economics*, *126*(4), 1593–1660.

Coleman, J. (1988). Social capital in the creation of human capital. *American Journal of Sociology*, *94*, S95–S12.

Cox, D. (1958). *Planning of Experiments*, John Wiley & Sons, New York.

Dabrowska, D.M., Speed, T.P. and Neyman, J.with cooperation of K. Iwaskiewicz and St. Kolodziejczyk (1935). Statistical problems in agricultural experimentation (with discussion). *Supplemental Journal of the Royal Statistical Society, Series B*, *2*, 107–180.

Durlauf, S. (1996). A theory of persistent income inequality. *Journal of Economic Growth*, *1*, 75–93.

Evans, W.N., Oates, W.E. and Schwab, R.M. (1992). Measuring peer group effects: a study of teenage behavior. *Journal of Political Economy*, 100(5), 966–991.

Festinger, L., Schachter, S. and Back, K.W. (1950). *Social Pressures in Informal Groups*, Stanford University Press, Palo Alto, CA.

Finn, J.D. and Achilles, C.M. (1990). Answers and questions about class size: a statewide experiment. *American Educational Research Journal*, 27(3), 557–577.

Gaviria, A. and Raphael, S. (2001). School-based peer effects and juvenile behavior. *Review of Economics and Statistics*, 83(2), 257–268.

Granovetter, M. (1973). The strength of weak ties. *American Journal of Sociology*, 78(6), 1360–1380.

Granovetter, M. (1978). Threshold models of collective behavior. *American Journal of Sociology*, 83(6), 1420–1443.

Halloran, M.E. and Struchiner, C.J. (1991). Study designs for dependent happenings. *Epidemiology*, 2, 331–338.

Halloran, M.E. and Struchiner, C.J. (1995). Causal inference in infectious diseases. *Epidemiology*, 6, 142–151.

Heckman, J. (1979). Sample selection bias as a specification error. *Econometrika*, 47, 153–161.

Heckman, J., Lochner, L. and Taber, C. (1998). General equilibrium treatment effects: a study of tuition policy. *American Economic Review*, 88, 381–386.

Holland, P.W. (1986). Statistics and causal inference. *Journal of the American Statistical Association*, 81, 945–960.

Hong, G. (2004). *Causal inference for multi-level observational data with application to kindergarten retention*. Ph.D. dissertation. Department of Educational Studies, University of Michigan, Ann Arbor, MI.

Hong, G. and Raudenbush, S.W. (2006). Evaluating kindergarten retention policy: a case study of causal inference for multi-level observational data. *Journal of the American Statistical Association*, 101(475), 901–910.

Hong, G. and Raudenbush, S.W. (2013). Heterogeneous agents, social interactions, and causal inference, in *Handbook of Causal Analysis for Social Research* (ed S.L. Morgan), Springer, New York, pp. 331–352.

Hudgens, M.G. and Halloran, M.E. (2008). Toward causal inference with interference. *Journal of the American Statistical Association*, 103(482), 832–842.

Konstantopoulos, S. and Chung, V. (2011). The persistence of teacher effects in elementary grades. *American Educational Research Journal*, 48(2), 361–386.

Lindzey, G. and Aronsson, E. (1985). *Handbook of Social Psychology: Group Psychology and the Phenomena of Interaction*, 3rd edn, Lawrence Erlbaum, Mahwah, NJ.

Manski, C. F. (2000). Economic analysis of social interactions (Working Paper No. 7580). Retrieved from National Bureau of Economic Research website: http://www.nber.org/papers/w7580. Accessed December 9, 2014.

Manski, C.F. (2013). Identification of treatment response with social interactions. *The Econometrics Journal*, 1(16), S1–S23.

Neyman, J. (1923). On the application of probability theory to agricultural experiments. Essay on principles. Section 9. *Statistical Science*, 5(4), 465–472. Translated by Dabrowska, D. M. and Speed, T. P.

Neyman, J. and with cooperation of K. Iwaskiewicz and St. Kolodziejczyk (1935). Statistical problems in agricultural experimentation (with discussion). *Supplemental Journal of the Royal Statistical Society, Series B*, 2, 107–180.

Nye, B., Konstantopoulos, S. and Hedges, L.V. (2004). How large are teacher effects? *Educational Evaluation and Policy Analysis*, 26(3), 237–257.

Paluck, E.L. (2011). Peer pressure against prejudice: a high school field experiment examining social network change. *Journal of Experimental Social Psychology*, *47*, 350–358.

Paluck, E.L. and Shepherd, H. (2012). The salience of social referents: a field experiment on collective norms and harassment behavior in a school social network. *Journal of Personality and Social Psychology*, *103*, 899–915.

Rosenbaum, P.R. (2007). Interference between units in randomized experiments. *Journal of the American Statistical Association*, *102*, 191–200.

Roy, A.D. (1951). Some thoughts on the distribution of earnings. *Oxford Economic Papers (New Series)*, *3*, 135–146.

Rubin, D.B. (1974). Estimating causal effects of treatments in randomized and nonrandomized studies. *Journal of Educational Psychology*, *66*, 688–701.

Rubin, D.B. (1978). Bayesian inference for causal effects: the role of randomization. *The Annals of Statistics*, *6*, 34–58.

Rubin, D.B. (1980). Comment on "Randomizatoin analysis of experimental data: The Fisher randomization test" by D. Basu. *Journal of the American Statistical Association*, *75*, 591–593.

Rubin, D.B. (1986). Comment: which ifs have causal answers. *Journal of the American Statistical Association*, *81*, 961–962.

Rubin, D.B. (1990). Neyman (1923) and causal inference in experiments and observational studies. *Statistical Sciences*, *5*, 472–480.

Sacerdote, B. (2011). Peer effects in education: how might they work, how big are they and how much do we know thus far?, in *Handbook of the Economics of Education*, Vol. 3, Number 3 (eds E. Hanushek, S. Machin and L. Woessmann), Elsevier, Amsterdam, The Netherlands, pp. 249–277.

Sampson, R.J. (2012). *Great American City: Chicago and the Enduring Neighborhood Effect*, University of Chicago Press, Chicago, IL.

Shadish, W.R., Cook, T.D. and Cambell, D.T. (2002). *Experimental and Quasi-Experimental Designs for Generalized Causal Inference*, Houghton Mifflin, Boston, MA.

Sinclair, B., McConnell, M. and Green, D.P. (2012). Detecting spillover effects: design and analysis of multilevel experiments. *American Journal of Political Science*, *56*(4), 1055–1069.

Sobel, M.E. (2006). What do randomized studies of housing mobility demonstrate? Causal inference in the face of interference. *Journal of the American Statistical Association*, *101*, 1398–1407.

VanderWeele, T.J., Hong, G., Jones, S.M. and Brown, J.L. (2013). Mediation and spillover effects in group-randomized trials: a case study of the 4R's educational intervention. *Journal of the American Statistical Association*, *108*(502), 469–482.

Verbitsky-Savitz, N. and Raudenbush, S.W. (2004). Causal inference in spatial settings. Proceedings of the American Statistical Association. Social Statistics Section, 2369–2374.

Verbitsky-Savitz, N. and Raudenbush, S.W. (2012). Causal inference under interference in spatial settings: a case study evaluating community policing program in Chicago. *Epidemiologic Methods*, *1*(1), 107–130.

Wilson, W.J. (1987). *The Truly Disadvantaged: The Inner City, the Underclass, and Public Policy*, University of Chicago Press, Chicago, IL.

15

Mediation through spill-over

In this chapter, we consider spill-over as an intermediate process that may constitute a part of the mediation mechanism through which a treatment exerts a causal impact on an outcome. As discussed in the previous chapter, researchers have often adopted cluster randomized trials to minimize treatment diffusion between treated and untreated units under two crucial assumptions. One is that individuals do not change their cluster membership in response to the treatment assignment; and the second is that there are no social interactions between individuals from different clusters that would affect the outcome of interest. Yet even when these assumptions hold, an individual's potential outcome under a cluster-level treatment may nonetheless depend on how other individuals in the same cluster respond to the treatment at the intermediate stage. In the following are two examples.

Banning smoking in office buildings is an intervention that applies to all those who work in the same building. One of the main goals of this intervention is to limit the impact of secondhand smoke on nonsmokers. Therefore, the intervention is expected to benefit individual health through changing not only one's own smoking behavior but also the smoking behavior of one's colleagues. The health consequence for the focal individual will be of concern if the person does not comply with the new rule. However, even if the focal individual is a nonsmoker or chooses to quit smoking in response to the intervention, the individual's health outcome remains uncertain if other people in the same office building do not comply with the smoking ban.

In a second example, assuming that an adolescent's social network is typically confined to those attending the same school, researchers may assign all students in a school who are at risk of juvenile delinquency to a prevention program that offers intensive mentoring to each participant. Hence, schools are clusters to be randomized in an evaluation of the program. After an adolescent and all his at-risk peers have been assigned to the program, the adolescent's antisocial behavior may nonetheless depend on whether the peers actually participate in the program and comply with the requirements set by the program.

In these two hypothetical examples as well as in many other program evaluations, individual compliance with treatment assignment has typically been viewed as a key mediator through which the treatment exerts an impact on the individual's outcome. Here, we highlight the compliance behaviors of other individuals as an additional mediator of the treatment effect

Causality in a Social World: Moderation, Mediation and Spill-over, First Edition. Guanglei Hong.
© 2015 John Wiley & Sons, Ltd. Published 2015 by John Wiley & Sons, Ltd.

on a focal individual's outcome if social interactions are inevitable. Researchers sometimes choose clusters rather than individuals as experimental units with the anticipation that the desired changes in targeted social behaviors will be reinforced through social interactions among individuals in the same cluster. Therefore, social interactions between individuals under the same treatment may become an important mechanism through which a cluster-level intervention exerts its impact on individual outcomes.

At the intermediate stage of a cluster randomized trial, spill-over may occur between individuals in the same cluster who have experienced the treatment program differently due to between-agent differences in treatment implementation. Agents are responsible for delivering a given treatment to the clients. For example, mentors may vary in how they communicate with their mentees; teachers may enact a new curriculum differently; and similarly, therapists may use discretion in treating their patients. We will illustrate this with an application study in which a school-wide program was designed to improve students' social–emotional well-being through improving the quality of teacher–student interactions and student–student interactions within each class. After nearly a year of implementation, the quality of these interactions improved to different degrees across the classrooms in the same school. Because students additionally interacted in the hallways, in the cafeteria, and on the playground, researchers reasoned that the program could exert its impact through not only improving one's own class quality but also the quality of other classes in the same school and perhaps the quality of campus life in general (VanderWeele *et al.*, 2013).

Finally, if an intervention involves regrouping individuals for treatment delivery, a treatment-induced change in peer composition and the subsequent group dynamics may mediate the treatment effects on individual outcomes. An examination of a policy requiring all ninth graders to take algebra revealed a change in peer composition as a potential mediator. When algebra was not required, low-achieving students typically took remedial math with other low-achieving peers. Following the change of course requirement, many low-achieving students sat in the algebra classes with high-achieving students instead. Researchers found that while the exposure to the algebra content was potentially beneficial, the benefit was offset by the change in class composition possibly due to unfavorable comparisons experienced by the low-achieving students or due to a lack of necessary differentiation in instructional pace to accommodate low-achieving students' needs for extra scaffolding (Hong and Nomi, 2012).

Mediation through spill-over is especially likely in a multisite trial in which individuals at the same experimental site are assigned to different treatment conditions and may subsequently have different experiences at the intermediate stage. Social interactions between individuals may allow a focal individual's potential outcome to be affected by other individual's intermediate experiences. The Moving to Opportunity (MTO) experiment, reviewed in Chapter 14, provided such an example. Many poor families that were offered housing vouchers and moved from high-poverty neighborhoods to low-poverty neighborhoods often maintained contacts with their previous neighbors who were assigned to the control condition and therefore stayed in the high-poverty neighborhoods. As Sobel (2006) argued, through these social communications, individuals in the control group might develop an acute awareness of their disadvantaged situation in the high-poverty neighborhoods and might subsequently increase their desire for change. Hence, it seems likely that the information flow from the treated to the untreated mediates the treatment effect on the outcome of the untreated.

In Section 15.1, we define the treatment effect mediated through spill-over in a single cluster in cluster randomized trials in which individuals in the same cluster are all assigned to the same treatment. We define each causal effect in terms of potential outcomes. Identification and estimation of the spill-over effect are discussed in Section 15.2. Section 15.3 defines the treatment effects mediated through spill-over when individuals at the same site are assigned to different treatments in multisite randomized trials. Section 15.4 shows the corresponding identification and estimation results. In each case, we discuss an ideal randomized experiment that generates data satisfying arguably the weakest identification assumptions. We then explicate the assumptions required for identification when the mediators are not randomized as is typically the case in most existing social experiments. The estimation procedure is focused on obtaining an unbiased sample estimate of each of the population average potential outcomes. Section 15.5 reveals the consequences for causal inference when mediation through spill-over is overlooked. Section 15.6 provides an application example involving quasiexperimental data.

15.1   Definition of mediated effects through spill-over in a cluster randomized trial

15.1.1   Notation

In order to examine causal mechanisms, an individual's potential outcomes are each defined as a function of the treatment and the mediator of interest (Angrist, Imbens, and Rubin, 1996; Pearl, 2001; Robins and Greenland, 1992). We start by considering the case in which all at-risk students in a school have been assigned to mentoring services with the aim of improving their behavioral outcomes. The cluster-level treatment assignment $z_j = 1$ therefore applies to all at-risk students in school j. Yet the compliance behavior, measured by actual participation in the program, may vary among individual students. We use $M_{ij}(z_j = 1)$, simplified to be $M_{ij}(1)$, to denote the participation behavior of student i in school j under the experimental condition. Let $M_{ij}(1)$ take value 1 if the student actually participates in the program and 0 otherwise. In parallel, we use $M_{ij}(z_j = 0)$, simplified to be $M_{ij}(0)$, to denote the student's participation under the control condition. For simplicity, let us suppose that if a school has been assigned to the control condition, none of the at-risk students at the school will have access to mentoring services and, therefore, $M_{ij}(0) = 0$ under the control condition. As we reasoned earlier, the behavioral improvement of student i in school j that has been assigned to the experimental condition may depend not only on whether this particular student participates in the program but also on whether other at-risk students in the same school participate. With n_j individuals in cluster j, the compliance behaviors of individuals other than focal individual i in this cluster are denoted by an $n_j - 1$ row vector:

$$\mathbf{M}_{-ij}(1) = \left(M_{1j}(1) \quad \cdots \quad M_{i-1,j}(1)\, M_{i+1,j}(1) \quad \cdots \quad M_{n_j,j}(1) \right).$$

The elements in this vector are either 0 or 1 indicating whether each individual student participates in the program. Hence, the potential outcome of individual i under the experimental condition can be denoted as $Y_{ij}\left(1, M_{ij}(1), \mathbf{M}_{-ij}(1)\right)$. The individual's potential outcome under the control condition is denoted as $Y_{ij}\left(0, M_{ij}(0), \mathbf{M}_{-ij}(0)\right)$ where $\mathbf{M}_{-ij}(0)$ is an $n_j - 1$ row vector of 0's in a simplified scenario in which mentoring is unavailable under the control condition. The definitions of causal effects described in the subsequent sections can be easily extended to the general case in which mentoring is available to some students even if the school has been assigned to the control condition.

15.1.2   Treatment effect mediated by a focal individual's compliance

Individual compliance with treatment assignment has been a mediator of central interest in past literature. Here, we focus on at-risk student i in experimental school j for whom we have that $M_{ij}(0) = 0$ and that $M_{ij}(1) = 1$. That is, the student would not participate in mentoring if the school is assigned to the control condition and would participate if the school is assigned to the experimental condition. In theory, the treatment effect on the student's outcome is at least partially mediated by the student's compliance. This indirect effect on one's outcome transmitted through the treatment-induced change in one's own mediator value is defined as

$$Y_{ij}\big(1, M_{ij}(1), \mathbf{M}_{-ij}(1)\big) - Y_{ij}\big(1, M_{ij}(0), \mathbf{M}_{-ij}(1)\big), \tag{15.1}$$

which, in this student's case, is $Y_{ij}\big(1, 1, \mathbf{M}_{-ij}(1)\big) - Y_{ij}\big(1, 0, \mathbf{M}_{-ij}(1)\big)$. The indirect effect will be zero if the student, for some reason, fails to participate despite the fact that the school is assigned to the experimental condition. In such a case, we will have that $M_{ij}(1) = M_{ij}(0) = 0$ and, therefore, $Y_{ij}\big(1, 0, \mathbf{M}_{-ij}(1)\big) - Y_{ij}\big(1, 0, \mathbf{M}_{-ij}(1)\big) = 0$.

15.1.3   Treatment effect mediated by peers' compliance through spill-over

Even if a focal student fails to participate in the mentoring program, the school-level program may indirectly affect this student's behavioral outcome through changing the mediator values of the students' at-risk peers. One may hypothesize that if most of the peers participate in the mentoring program, the focal student will likely benefit due to a reduction of risky behaviors among the peers. This indirect effect of the treatment on a focal individual's outcome transmitted through treatment-induced changes in peers' mediator values is defined by

$$Y_{ij}\big(1, M_{ij}(0), \mathbf{M}_{-ij}(1)\big) - Y_{ij}\big(1, M_{ij}(0), \mathbf{M}_{-ij}(0)\big). \tag{15.2}$$

$\mathbf{M}_{-ij}(0)$ and $\mathbf{M}_{-ij}(1)$ are unequal if at least one of the at-risk peers participates in the mentoring program under the experimental condition. Because $M(0) = 0$ for all individuals under the control condition, the indirect effect of the treatment due to the spill-over of peers' compliance can be rewritten as $Y_{ij}\big(1, 0, \mathbf{M}_{-ij}(1)\big) - Y_{ij}(1, 0, \mathbf{0})$.

When compliance is the mediator of interest, the *total indirect effect* of the treatment on an individual's outcome is a sum of the first indirect effect transmitted through the individual's own compliance defined in (15.1) and a second indirect effect transmitted through peers' compliance defined in (15.2). That is,

$$\begin{aligned}
&Y_{ij}\big(1, M_{ij}(1), \mathbf{M}_{-ij}(1)\big) - Y_{ij}\big(1, M_{ij}(0), \mathbf{M}_{-ij}(0)\big) \\
&= Y_{ij}\big(1, M_{ij}(1), \mathbf{M}_{-ij}(1)\big) - Y_{ij}\big(1, M_{ij}(0), \mathbf{M}_{-ij}(1)\big) \\
&\quad + Y_{ij}\big(1, M_{ij}(0), \mathbf{M}_{-ij}(1)\big) - Y_{ij}\big(1, M_{ij}(0), \mathbf{M}_{-ij}(0)\big).
\end{aligned} \tag{15.3}$$

Table 15.1   Decomposition of the total treatment effect in a cluster randomized trial

Causal effect	Definition
Total effect	$Y_{ij}\left(1,M_{ij}(1),\mathbf{M}_{-ij}(1)\right)-Y_{ij}\left(0,M_{ij}(0),\mathbf{M}_{-ij}(0)\right)$
Indirect effect mediated by a focal individual's compliance	$=\left[Y_{ij}\left(1,M_{ij}(1),\mathbf{M}_{-ij}(1)\right)-Y_{ij}\left(1,M_{ij}(0),\mathbf{M}_{-ij}(1)\right)\right]$
Indirect effect mediated by peers' compliance	$+\left[Y_{ij}\left(1,M_{ij}(0),\mathbf{M}_{-ij}(1)\right)-Y_{ij}\left(1,M_{ij}(0),\mathbf{M}_{-ij}(0)\right)\right]$
Direct effect	$+\left[Y_{ij}\left(1,M_{ij}(0),\mathbf{M}_{-ij}(0)\right)-Y_{ij}\left(0,M_{ij}(0),\mathbf{M}_{-ij}(0)\right)\right]$

15.1.4   Decomposition of the total treatment effect

The direct effect of the treatment is defined as the effect of the school-level program on at-risk students' behavioral outcome when none of these students actually participate:

$$Y_{ij}\left(1,M_{ij}(0),\mathbf{M}_{-ij}(0)\right)-Y_{ij}\left(0,M_{ij}(0),\mathbf{M}_{-ij}(0)\right). \qquad (15.4)$$

The direct effect would likely be positive if assigning a school to the experimental condition heightens the educators' awareness of at-risk students' needs and improves their communications with these students. The direct effect and the total indirect effect sum up to the total effect of the treatment:

$$Y_{ij}\left(1,M_{ij}(1),\mathbf{M}_{-ij}(1)\right)-Y_{ij}\left(0,M_{ij}(0),\mathbf{M}_{-ij}(0)\right). \qquad (15.5)$$

The decomposition is summarized in Table 15.1.

The population average causal effects are defined accordingly. Each population average, denoted by $E[\cdot]$ representing the expected value of a random variable, is taken over all the individuals in a cluster and over all the clusters. To simplify the notation, henceforth, we omit the subscripts except for making a distinction between symbols M and $\mathbf{M}_-$. The former is a shorthand for M_{ij}, while the latter represents $\mathbf{M}_{-ij}$. The population average indirect effect of the treatment transmitted through a focal individual's compliance is $E[Y(1,M(1),\mathbf{M}_-(1))-Y(1,M(0),\mathbf{M}_-(1))]$; the population average indirect effect transmitted through peers' compliance is $E[Y(1,M(0),\mathbf{M}_-(1))-Y(1,M(0),\mathbf{M}_-(0))]$; and the population average direct effect of the treatment is $E[Y(1,M(0),\mathbf{M}_-(0))-Y(0,M(0),\mathbf{M}_-(0))]$.

15.2   Identification and estimation of the spill-over effect in a cluster randomized design

15.2.1   Identification in an ideal experiment

We have defined, in (15.1), the indirect effect of the treatment mediated by a focal individual's compliance and, in (15.2), the indirect effect mediated by peers' compliance. The corresponding population average causal effects are identifiable if they can be equated with observable quantities under certain assumptions. Chapter 13 discussed the case of two concurrent mediators each representing a distinct aspect of an individual's intermediate experience with the treatment. In contrast, here the second mediator represents the intermediate experiences of peers who are in the same cluster with the focal individual. An ideal experiment for identifying the spill-over effect in a cluster randomized trial, which is described in the following text, is therefore different from the ideal experiment for studying two concurrent mediators described in Chapter 13.

In the example of student mentoring, identification of the population average causal effects requires that the following four population average potential outcomes be equated with observable quantities: $E[Y(1,M(1),\mathbf{M}_-(1))]$, $E[Y(0,M(0),\mathbf{M}_-(0))]$, $E[Y(1,M(0),\mathbf{M}_-(1))]$, and $E[Y(1,M(0),\mathbf{M}_-(0))]$. The first two are the population average potential outcomes under the experimental condition and the control condition, respectively; the third is the population average counterfactual outcome under the experimental condition when a focal student fails to participate; and the fourth is the population average counterfactual outcome under the experimental condition when none of the students participate. The identification becomes possible under the following assumptions:

1. There are no social interactions between individuals from different clusters. Therefore, an at-risk student's potential outcome cannot be affected by the treatments assigned to other schools or by the participation of at-risk students in other schools.

2. A school's treatment assignment is independent of the students' potential outcomes and potential mediators.

3. A student's participation in mentoring in an experimental school is independent of the student's potential outcomes.

4. The peers' participation in mentoring in an experimental school is independent of a focal student's potential outcomes.

An ideal experiment may involve three steps. In the first step, the experimenter assigns schools at random to either a school-level mentoring program or the control condition, which will satisfy assumption (2). Step 2 is to assign trained mentors at random to the experimental schools. In Step 3, at-risk students within an experimental school are assigned at random to the mentors. It is conceivable that mentors may not display the same level of effectiveness in engaging the at-risk students. Suppose that an at-risk student may fail to participate only if the assigned mentor is ineffective. The proportion of at-risk students in a school actually participating in mentoring therefore depends on the proportion of mentors assigned to the school who are effective in service delivery. These proportions are likely uneven across the experimental schools during the second step randomization. There is a random chance that no effective mentor is assigned to a particular experimental school. The randomization in Steps 2 and 3 helps satisfy assumptions (3) and (4).

Because the first step randomization ensures that a school's treatment assignment is independent of the students' potential outcomes, the population average potential outcome under the experimental condition can be identified by the average observable outcome of at-risk students attending schools that have been assigned to the experimental condition. That is,

$$E[Y(1,M(1),\mathbf{M}_-(1))] = E[Y(1,M(1),\mathbf{M}_-(1))|Z=1] = E(Y|Z=1)$$

Similarly, the population average potential outcome under the control condition can be identified by the average observable outcome of at-risk students attending schools that have been assigned to the control condition. That is,

$$E[Y(0,M(0),\mathbf{M}_-(0))] = E[Y(0,M(0),\mathbf{M}_-(0))|Z=0] = E(Y|Z=0).$$

Next, we show how to identify the population average counterfactual outcome of having one's school assigned to the experimental condition yet none of the at-risk students in the school participate in mentoring $E[Y(1,M(0),\mathbf{M}_-(0))]$. Due to the randomizations in Step 2 and Step 3, in an experimental school, a focal student's participation in mentoring denoted by M_{ij} and the peers' participation denoted by $\mathbf{M}_{-ij}$ are both independent of the potential outcomes. Additionally, suppose that the peers' participation is independent of a focal student's participation. Therefore, $E[Y(1,M(0),\mathbf{M}_-(0))]$ can be identified by the average observable outcome of at-risk students attending experimental schools in which no one participates in mentoring:

$$E[Y(1,M(0),\mathbf{M}_-(0))]$$
$$= E[Y(1,M(0),\mathbf{M}_-(0))|Z=1,M=0,\mathbf{M}_-=0]$$
$$= E(Y|Z=1,M=0,\mathbf{M}_-=0).$$

Finally, the population average counterfactual outcome under the experimental condition when a focal individual fails to participate, $E[Y(1,M(0),\mathbf{M}_-(1))]$, is an average over the distribution of peers' participation under the experimental condition $\mathbf{M}_-(1)$. We use $\mathbf{m}$ to denote the observable participation of one's peers in an experimental school. The average observable outcome of students in experimental schools who fail to participate in mentoring, while their peers' mediator values are represented by vector $\mathbf{m}$, is denoted by $E(Y|Z=1,M=0,\mathbf{M}_-=\mathbf{m})$. Under the same identification assumptions listed earlier, this observable quantity is unbiased for $E[Y(1,M(0)=0,\mathbf{M}_-(1)=\mathbf{m})]$, the average potential outcome associated with the experimental condition when a focal individual fails to participate and when the peers adopt mediator values $\mathbf{m}$. Its contribution to the identification of $E[Y(1,M(0),\mathbf{M}_-(1))]$ is proportional to the fraction of students in experimental schools whose peers' actual mediator values are $\mathbf{m}$, the latter being denoted by $\mathrm{pr}(\mathbf{M}_-=\mathbf{m}|Z=1)$:

$$E[Y(1,M(0),\mathbf{M}_-(1))]$$
$$= \sum_{\mathbf{m}} E[Y(1,M(0)=0,\mathbf{M}_-(1)=\mathbf{m})|Z=1,M=0,\mathbf{M}_-=\mathbf{m}] \times \mathrm{pr}(\mathbf{M}_-=\mathbf{m}|Z=1)$$
$$= \sum_{\mathbf{m}} E(Y|Z=1,M=0,\mathbf{M}_-=\mathbf{m}) \times \mathrm{pr}(\mathbf{M}_-=\mathbf{m}|Z=1).$$

Chapter 14 discussed a number of strategies for summarizing peers' treatment assignments in a theoretically meaningful index. These strategies can be adapted to the current problem for summarizing peers' participation. For example, we may use $f(\mathbf{M}_-(1))=q$ for $0 \leq q \leq 1$ to represent the proportion of one's at-risk peers who participate in mentoring in an experimental school under the theoretical assumption that the impacts of the at-risk peers on a focal individual's potential outcome are exchangeable. We will have the following identification result:

$$E[Y(1,M(0),\mathbf{M}_-(1))] = \sum_q E(Y|Z=1,M=0,f(\mathbf{M}_-(1))=q) \times \text{pr}(f(\mathbf{M}_-(1))=q).$$

15.2.2   Identification when the mediators are not randomized

In most cluster randomized experiments that have been administered in the past, clusters are assigned at random to the experimental condition or the control condition; yet individuals are not assigned at random to different mediator values. We illustrate this with an evaluation of a school-wide intervention. The Reading, Writing, Respect, and Resolution (4Rs) program is aimed at promoting not only literacy development but also intergroup understanding and conflict resolution among students in urban elementary schools. The intervention theory focuses on the mediating role of changes in class quality induced by the 4Rs intervention in promoting child academic learning and social–emotional well-being. Class quality encompasses instructional support, emotional support, and organizational climate. The intervention, however, is designed for an entire school to mobilize educators' collective efforts and to utilize reinforcement among students in the same building so that positive social interactions will be maximized and negative interactions will be minimized. In particular, one may hypothesize that a student's depressive symptoms may be reduced not only due to an enhanced environment within one's own classroom but also due to an enhanced environment in the school as a whole. This is because students interact not only within a classroom but also in the hallways, in the cafeteria, on the playground, in school buses, and even in the surrounding neighborhoods.

A cluster randomized trial conducted in New York City assigned 18 elementary schools in matched pairs at random to either the 4Rs program or the control condition (Brown *et al.*, 2010; Jones, Brown, and Aber, 2011). A sample from the first year of the study includes 82 third-grade classrooms and 942 students. The intervention was found to be beneficial in reducing students' depressive symptoms and also in improving classroom quality. In a reanalysis of the experimental data, VanderWeele *et al.* (2013) evaluated the effect of the 4Rs intervention on student depressive symptoms associated with an improvement in the quality of a student's own classroom, the treatment effect associated with an improvement in the quality of other classrooms, and the treatment effect transmitted through potential pathways other than classroom quality.

For student i in class j in school k, $Y_{ijk}(1,M_{jk}(1),\mathbf{M}_{-jk}(1))$ denotes the student's potential outcome if the school is assigned to the 4Rs intervention, and $Y_{ijk}(0,M_{jk}(0),\mathbf{M}_{-jk}(0))$ denotes the student's potential outcome if the school is assigned to the control condition. Here, $M_{jk}(1)$ and $\mathbf{M}_{-jk}(1)$ denote the quality of the student's classroom and the quality of other classrooms in the same school, respectively, under the 4Rs intervention; $M_{jk}(0)$ and $\mathbf{M}_{-jk}(0)$ denote the corresponding mediators under the control condition. According to prior research and the current data, the overall measure of classroom quality shows a threshold effect on child outcomes. In other words, only after exceeding the threshold does an improvement in class quality show an impact on child behavior. The researchers therefore pinpointed a cutoff score and dichotomized the measure so that $M_{jk}=1$ if class j in school k displays a relatively high quality and 0 otherwise. The quality of other classrooms in the same school was averaged and then dichotomized as well so that, under treatment z for $z=0,1$, we let $G(\mathbf{M}_{-jk}(z))=G_{jk}(z)$

take value 1 if, for class j in school k, the average quality of the other classrooms is relatively high and $G_{jk}(z) = 0$ otherwise. Unlike the earlier example of student mentoring in which students in control schools do not have access to mentors, in the 4Rs study, some classrooms in the control schools nonetheless displayed a relatively high quality. Therefore, $M_{jk}(0)$ and $G_{jk}(0)$ are not necessarily zero.

The total indirect effect of the treatment mediated by classroom quality has two components: the first is the indirect effect mediated by a treatment-induced change in the quality of one's own classroom; and the second is the indirect effect mediated by a treatment-induced change in the quality of other classrooms:

$$E[Y(1,M(1),G(1)) - Y(1,M(0),G(0))]$$
$$= E[Y(1,M(1),G(0)) - Y(1,M(0),G(0))] + E[Y(1,M(1),G(1)) - Y(1,M(1),G(0))]. \quad (15.6)$$

The direct effect of the treatment that is not transmitted through a change in classroom quality is defined as $E[Y(1,M(0),G(0)) - Y(0,M(0),G(0))]$.

VanderWeele $et\ al.$ (2013) discussed the conditions under which these causal effects can be identified. Corresponding to identification assumptions (1) and (2) listed in Section 15.2.1, they assumed that there is no interference between schools and that the school-level treatment assignment is independent of potential mediators and potential outcomes. However, because students were not assigned at random to classrooms of different quality, the M–Y relationship and the G–Y relationship are potentially confounded by selection factors. This is because classroom quality arises from a combination of the particular teacher and the particular students in that class. For example, in both the experimental and control schools, the classrooms taught by teachers who reported higher confidence in behavior management prior to the treatment tended to display higher quality at the end of the treatment year. Other selection factors are possibly associated with baseline student characteristics, classroom characteristics, and additional teacher characteristics. To simplify the notation, here we use $\mathbf{X}$ to represent a collection of these pretreatment measures. In order to identify the mediated effects, the researchers invoked the following assumptions that are considerably stronger than assumptions (3) and (4):

3* For those whose pretreatment characteristics $\mathbf{X}$ take on the same particular values $\mathbf{x}$, the quality of one's own classroom is independent of all the potential outcomes.

4* For those whose pretreatment characteristics $\mathbf{X}$ take on the same particular values $\mathbf{x}$, the quality of the classrooms other than one's own is independent of all the potential outcomes.

5 For those whose pretreatment characteristics $\mathbf{X}$ take on the same particular values $\mathbf{x}$, the quality of one's own classroom is independent of the quality of other classrooms across treatment conditions.

Assumptions (3*) and (4*) are different from assumptions (3) and (4), respectively, in that the independences between mediator values and potential outcomes are now assumed within levels of $\mathbf{X} = \mathbf{x}$. Moreover, because the mediator values may vary under the control condition as well as under the experimental condition in the current study, the independences are assumed not only within a treatment but also across treatment conditions. That is, under assumptions (3^*) and (4^*), $M_{jk}(0)$ and $G_{jk}(0)$ are assumed to be independent of $Y_{ijk}(1, M_{jk}(1), G_{jk}(1))$ for those sharing the same pretreatment characteristics. In other words, for control school students who share the same baseline conditions, whether one would experience a high-quality environment in one's own classroom and whether one would attend schools in which other classrooms would be high quality under the control condition are unrelated to the degree to which one would display depressive symptoms under the 4Rs program. In addition, the conditional independence between $M_{jk}(1)$ and $G_{jk}(0)$ stated in assumption (5) becomes necessary for partitioning the total indirect effect. Under assumption (5), whether a student would experience a high-quality environment in one's own classroom under the 4Rs program is unrelated to whether the student would have high-quality (or low-quality) neighboring classrooms under the control condition.

When the aforementioned assumptions hold, the data can be viewed as if they were generated from a sequential randomized experiment in which the initial school-level treatment assignment is followed by a block randomized design within each school. In such an experiment, students of the same characteristics are assigned at random to classes; and teachers of the same characteristics are assigned at random to classes as well. The researchers discussed the plausibility of these assumptions in the 4Rs context (VanderWeele et al., 2013).

15.2.3   Estimation of mediated effects through spill-over

Analyzing the 4Rs data, VanderWeele et al. (2013) examined the average effects of the treatment on the outcome under the following four conditions: (1) when the quality of one's own classroom and the quality of other classrooms are fixed at a low level, (2) are fixed at a high level, or (3) (4) are fixed at a low level for one and a high level for the other. They also assessed the average effects of raising the quality of one's own classroom from a low level to a high level either under the control condition or under the 4Rs intervention when the quality of other classrooms is fixed at either a low level or a high level. Finally, the average effects of raising the quality of other classrooms were investigated similarly. The researchers regressed the outcome on dummy indicators each representing a possible combination of the quality of one's own class and the quality of other classes in a school. In an attempt to meet the ignorability assumptions, they made linear covariance adjustment for a number of pretreatment covariates including classroom composition and teacher self-reported confidence in class management at the baseline. This is a three-level model with students nested within classes that in turn nested within schools. Interestingly, being assigned to the 4Rs program rather than the control condition showed a statistically significant benefit when the quality of one's own class was low, while the quality of other classes in school was high. Under these conditions, the 4Rs program seemed to have the potential of reducing child depressive symptoms by about 60% of a standard deviation through means other than improving classroom quality. This treatment effect may be mediated by conflict resolution or reduction in bullying among students in the hallways or on the playground.

However, these researchers did not estimate the indirect effect mediated by a treatment-induced change in the quality of one's own classroom and the indirect effect mediated by a treatment-induced change in the quality of other classrooms as defined in (15.6). Nor did they estimate the direct effect of the treatment. The treatment decomposition involves the estimation of the four population average potential outcomes described earlier: $E[Y(1, M(1), G(1))]$, $E[Y(1, M(1), G(0))]$, $E[Y(1, M(0), G(0))]$, and $E[Y(0, M(0), G(0))]$. In the following, we discuss the estimation of each of these under the identification assumptions explained in the previous section.

The population average potential outcome under the 4Rs intervention $E[Y(1, M(1), G(1))]$ and that under the control condition $E[Y(0, M(0), G(0))]$, each defined for a population of students, can be estimated by the average observed outcome of students attending the 4Rs schools and the average observed outcome of students attending the control schools, respectively. Each average is taken over n_{jk} students enrolled in class j in school k, over n_k classes in school k, and over K schools. A sample estimator of the population average potential outcome under the 4Rs intervention is

$$\frac{\sum_{k=1}^{K} \sum_{j=1}^{n_k} \sum_{i=1}^{n_{jk}} Z_k Y_{ijk}}{\sum_{k=1}^{K} \sum_{j=1}^{n_k} \sum_{i=1}^{n_{jk}} Z_k};$$

and a sample estimator of the population average potential outcome under the control condition is

$$\frac{\sum_{k=1}^{K} \sum_{j=1}^{n_k} \sum_{i=1}^{n_{jk}} (1-Z_k) Y_{ijk}}{\sum_{k=1}^{K} \sum_{j=1}^{n_k} \sum_{i=1}^{n_{jk}} (1-Z_k)}.$$

These estimators are unbiased under assumptions (1) and (2).

We apply the ratio-of-mediator-probability weighting (RMPW) strategy (Hong, 2010a; Hong, Deutsch, and Hill, 2011, in press) to the estimation of the two population average counterfactual outcomes. Chapter 11 has presented the rationale for RMPW. In estimating the population average potential outcome under the 4Rs intervention when the quality of one's own class and that of other classes counterfactually remain at the same level as they would be under the control condition, we use weighting to transform the distribution of one's own class quality and the distribution of other class quality such that they resemble the corresponding distributions under the control condition. A sample estimator of $E[Y(1, M(0), G(0))]$ is

$$\frac{\sum_{k=1}^{K} \sum_{j=1}^{n_k} \sum_{i=1}^{n_{jk}} Z_k W_{jk} Y_{ijk}}{\sum_{k=1}^{K} \sum_{j=1}^{n_k} \sum_{i=1}^{n_{jk}} Z_k W_{jk}},$$

where, for students attending class j in school k with baseline characteristics $\mathbf{X}_{jk} = \mathbf{x}$ and for whom the school has been assigned to the 4Rs program, the quality level of their classroom is m, the quality level of other classrooms at the same school is g, and the weight is

$$W_{jk} = \frac{\text{pr}\left(M_{jk} = m, G_{jk} = g | Z_k = 0, \mathbf{X}_{jk} = \mathbf{x}\right)}{\text{pr}\left(M_{jk} = m, G_{jk} = g | Z_k = 1, \mathbf{X}_{jk} = \mathbf{x}\right)}.$$

The estimation is unbiased under assumptions (1), (2), (3*), and (4*).

Finally, to estimate the population average potential outcome under the 4Rs intervention when only the quality of other classes counterfactually remains at the same level as they would be under the control condition $E[Y(1, M(1), G(0))]$, we apply the following weight to students attending class j in school k that has been assigned to the 4Rs program and for whom the quality level of other classrooms at the same school is g:

$$W_{jk} = \frac{\text{pr}\left(G_{jk} = g | Z_k = 0, \mathbf{X}_{jk} = \mathbf{x}\right)}{\text{pr}\left(G_{jk} = g | Z_k = 1, \mathbf{X}_{jk} = \mathbf{x}\right)}.$$

Unbiased estimation requires identification assumptions (1), (2), (3^*), (4^*), and (5). Similar weights have been derived in Chapter 13 for estimating treatment effects transmitted through two concurrent mediators. As shown in Chapter 13, the causal effects of interest can be estimated as mean contrasts through a weighted analysis.

15.3 Definition of mediated effects through spill-over in a multisite trial

In a multisite randomized trial, treatments are randomly assigned within each site. Taking into consideration social interactions among individuals at the same site, one may reason that an individual's potential mediator is a function not only of the individual's own treatment assignment but also of the treatment assignments of the individual's peers. Moreover, the individual's potential outcome is a function of not only the individual's own treatment assignment and mediator values but also the peers' treatment assignments and mediator values. We use the MTO experiment to illustrate the case.

15.3.1 Notation

For simplicity, we consider a single experimental site j in which the household of individual i living in poverty is assigned at random to either an experimental condition that offers a housing voucher, denoted by $z_{ij} = 1$, or a control condition denoted by $z_{ij} = 0$. If the site has n_j households participating in the study, then $\mathbf{z}_j$ is a vector of n_j elements denoting the treatment assignment of all the participating households at this site. There are a wide range of outcomes of interest in the MTO experiment including self-reported perception of neighborhood safety. MTO researchers observe the outcome of a randomly sampled grown-up individual from each household.

Let the mediator for a treated individual take value 1 if the household actually uses the housing voucher and moves from a high-poverty neighborhood to a low-poverty neighborhood and 0 otherwise. We assume that social interactions among individuals from different experimental sites are essentially nonexistent given that the sites are scattered around the country. Therefore, an individual's potential mediator will unlikely be affected by the treatment assignment of households in other sites. We will highlight this assumption again in the discussion of identification. Whether a household actually moves, however, may depend not only on its treatment status but also on the treatment assignment of other households at the same site. As discussed in Chapter 14, for example, a household may choose to move along with its close neighbors who have also received housing vouchers and may choose to stay if the neighbors have been assigned to the control condition. Therefore, the potential mediator for a focal individual's household is denoted by

$$M_{ij}\left(\mathbf{z}_j\right) \equiv M_{ij}\left(z_{ij}, \mathbf{z}_{-ij}\right).$$

Here, $\mathbf{z}_{-ij}$ is an $n_j - 1$ row vector indicating the treatment status of other households at the site. In theory, a household assigned to the control condition does not receive a voucher and therefore cannot use a voucher to move to a low-poverty neighborhood regardless of the treatment assignment of other households at the same site. For simplicity, we let

$$M_{ij}\left(0,\mathbf{z}_{-ij}\right) = M_{ij}(0) = 0$$

for all i, j in the current example.

How one perceives the neighborhood safety will then depend on whether the household is offered a housing voucher z_{ij}, whether the household uses the housing voucher to move to a low-poverty neighborhood $M_{ij}(z_j)$, whether other households are offered housing vouchers $\mathbf{z}_{-ij}$, and whether they actually use housing vouchers to move $\mathbf{M}_{-ij}\left(z_j\right)$. The potential outcome of individual i in site j is then denoted by

$$Y_{ij}\left(\mathbf{z}_j, \mathbf{M}_j\left(\mathbf{z}_j\right)\right) \equiv Y_{ij}\left(z_{ij}, \mathbf{z}_{-ij}, M_{ij}\left(\mathbf{z}_j\right), \mathbf{M}_{-ij}\left(\mathbf{z}_j\right)\right).$$

Here, $\mathbf{M}_j(\mathbf{z}_j)$ is an n_j row vector indicating the moving status of all households in site j as a result of the treatment assignments of all these households.

Assuming that the impacts of other households at the same site are exchangeable for focal individual i, an assumption that we will revisit in a later section, we may summarize the treatment assignments of all other households in site j in a scalar function representing the proportion of other households in site j that are offered vouchers. This is approximately equal to the proportion of households in site j that are offered vouchers when n_j is sufficiently large:

$$f\left(\mathbf{z}_{-ij}\right) \approx f\left(\mathbf{z}_j\right) = P_j = \frac{\sum_{i=1}^{n_j} Z_{ij}}{n_j}.$$

Let $P_j = p$ when a p proportion of the participating households are offered vouchers. Hence, $M_{ij}(p)$ denotes whether the household of individual i in site j, if offered a voucher, uses the voucher to move when a p proportion of the participating households at the site are offered vouchers. We use $\mathbf{M}_j(p)$ to denote the compliance of all participating households at the site under the same treatment assignment. In contrast, $P_j = 0$ indicates that none of the participating households in site j are offered vouchers, in which case we will have that

$$\mathbf{M}_j(0) = \mathbf{0},$$

that is, none of the participating households use vouchers to move simply because they are not offered vouchers.

If the household of individual i in site j is offered a voucher when a p proportion of other households at the same site are also offered vouchers, using this simplified notation, we represent the potential outcome of the focal individual as $Y_{ij}\left(1, p, M_{ij}(p), \mathbf{M}_{-ij}(p)\right)$. If the focal individual's household is not offered voucher, however, when a p proportion of other households at the same site are offered vouchers, the potential outcome of the focal individual becomes $Y_{ij}\left(0, p, M_{ij}(0), \mathbf{M}_{-ij}(p)\right)$ where $M_{ij}(0) = 0$. Finally, if none of the households at the site are offered a voucher, the focal individual's potential outcome is $Y_{ij}\left(0, 0, M_{ij}(0), \mathbf{M}_{-ij}(0)\right)$, which will be equal to $Y_{ij}(0, 0, 0, \mathbf{0})$.

15.3.2   Treatment effect mediated by a focal individual's compliance

We first define the indirect effect of the treatment assignment for individual i in site j on this individual's outcome transmitted through this focal individual's compliance behavior. This is the difference between two potential outcomes of the individual when a $p > 0$ proportion of the participating households in site j including the focal individual's household receive vouchers and when the mediator value of the focal individual's household changes from $M_{ij}(0) = 0$ to $M_{ij}(p)$. That is,

$$Y_{ij}\left(1, p, M_{ij}(p), \mathbf{M}_{-ij}(p)\right) - Y_{ij}\left(1, p, 0, \mathbf{M}_{-ij}(p)\right). \tag{15.7}$$

Suppose that a focal individual's household will choose to move if offered a voucher. Moving to a low-poverty neighborhood will likely change the focal individual's perception of neighborhood safety. The indirect effect of the treatment on the outcome transmitted through the compliance behavior of the focal individual's household is defined by $Y_{ij}\left(1, p, 1, \mathbf{M}_{-ij}(p)\right) - Y_{ij}\left(1, p, 0, \mathbf{M}_{-ij}(p)\right)$ when $z_{ij} = 1$. If, however, the focal individual's household will not move despite that it is offered a voucher, this indirect effect will become zero.

The indirect effect is also zero if the focal individual's household does not receive a voucher when a p proportion of other participating households in site j are offered vouchers. This is because, when not offered a voucher, the focal individual's household cannot use a voucher to move, in which case the indirect effect is simply

$$Y_{ij}\left(0, p, 0, \mathbf{M}_{-ij}(p)\right) - Y_{ij}\left(0, p, 0, \mathbf{M}_{-ij}(p)\right) = 0.$$

15.3.3   Treatment effect mediated by peers' compliance through spill-over

In the case that a focal individual's household is not offered a voucher or if the household does not move even though it is offered a voucher, as we have discussed earlier, the focal individual's subjective perception of neighborhood safety can nonetheless be influenced by some of the neighbors who use vouchers to move and report back to the focal individual their perception of the stark contrast in crime rate between the origin high-poverty neighborhood and the receiving low-poverty neighborhood. Hence, the second indirect effect of the treatment assignments in site j on the outcome of individual i is transmitted through the neighbors' compliance. This is defined as the difference between two potential outcomes of the individual when the neighbors' mediator values change from $\mathbf{M}_{-ij}(0) = 0$ to $\mathbf{M}_{-ij}(p)$. When a p proportion of the households at the site including the focal individual's household are offered vouchers yet the focal individual's household fails to use the voucher to move, the indirect effect is

$$Y_{ij}\left(1, p, 0, \mathbf{M}_{-ij}(p)\right) - Y_{ij}\left(1, p, 0, \mathbf{M}_{-ij}(0)\right). \tag{15.8}$$

If, instead, the focal individual's household is not offered a voucher yet a p proportion of other households at the site are offered vouchers, the indirect effect becomes

$$Y_{ij}\left(0, p, 0, \mathbf{M}_{-ij}(p)\right) - Y_{ij}\left(0, p, 0, \mathbf{M}_{-ij}(0)\right).$$

The indirect effect transmitted through peers' compliance will be zero if none of the neighbors who receive vouchers actually move, that is, if $\mathbf{M}_{-ij}(p) = \mathbf{M}_{-ij}(0) = \mathbf{0}$.

The total indirect effect of the treatment on the focal individual's outcome is therefore the sum of the treatment effect mediated by the focal individual's compliance behavior as defined in (15.7) and the treatment effect mediated by neighbors' compliance through spill-over as defined in (15.8). When the focal individual's household is offered a voucher, the total indirect effect is

$$
\begin{aligned}
Y_{ij}\big(1,p,M_{ij}(p),\mathbf{M}_{-ij}(p)\big) &- Y_{ij}\big(1,p,0,\mathbf{M}_{-ij}(0)\big) \\
= \big[Y_{ij}\big(1,p,M_{ij}(p),\mathbf{M}_{-ij}(p)\big) &- Y_{ij}\big(1,p,0,\mathbf{M}_{-ij}(p)\big)\big] \qquad (15.9)\\
+ \big[Y_{ij}\big(1,p,0,\mathbf{M}_{-ij}(p)\big) &- Y_{ij}\big(1,p,0,\mathbf{M}_{-ij}(0)\big)\big].
\end{aligned}
$$

If the focal individual's household is not offered a voucher, the total indirect effect is simply equal to the treatment effect mediated by peers' compliance:

$$
\begin{aligned}
Y_{ij}\big(0,p,M_{ij}(p),\mathbf{M}_{-ij}(p)\big) &- Y_{ij}\big(0,p,0,\mathbf{M}_{-ij}(0)\big) \\
= Y_{ij}\big(0,p,0,\mathbf{M}_{-ij}(p)\big) &- Y_{ij}\big(0,p,0,\mathbf{M}_{-ij}(0)\big).
\end{aligned}
$$

15.3.4   Direct effect of individual treatment assignment on the outcome

The direct effect of the individual treatment assignment in site j on individual i's outcome is the difference between two potential outcomes of the individual when vouchers are offered to a p proportion of the participating households in site j including the focal individual's household as opposed to excluding the focal individual's household. Perhaps counterfactually, none of the households use vouchers to move as would have been the case had they all been assigned to the control condition. Receiving the voucher may nonetheless change the focal individual's perception of neighborhood safety if the individual subsequently gathers information about some low-poverty neighborhoods that the household might potentially move into. This direct effect is represented as follows:

$$
Y_{ij}\big(1,p,0,\mathbf{M}_{-ij}(0)\big) - Y_{ij}\big(0,p,0,\mathbf{M}_{-ij}(0)\big). \qquad (15.10)
$$

15.3.5   Direct effect of peer treatment assignment on the outcome

The direct effect of the peer treatment assignment in site j on individual i's outcome is the difference between two potential outcomes of the individual when vouchers are offered to a p proportion of the participating households excluding the focal individual's household as opposed to offering vouchers to none of the households at this site. Again, perhaps counter-factually, none of the households use vouchers to move as would have been the case had they all been assigned to the control condition. The focal individual may nonetheless change his perception of neighborhood safety when informed by his neighbors who have visited low-poverty neighborhoods. This second direct effect is represented as follows:

$$
Y_{ij}\big(0,p,0,\mathbf{M}_{-ij}(0)\big) - Y_{ij}\big(0,0,0,\mathbf{M}_{-ij}(0)\big). \qquad (15.11)
$$

15.3.6   Decomposition of the total treatment effect

The direct effect and the total indirect effect for individual i sum up to the total effect of the treatment assignments in site j:

Table 15.2   Decomposition of the total treatment effect in a multisite randomized trial

Causal effect	Definition
Total effect	$Y_{ij}\big(1,p,M_{ij}(p),\mathbf{M}_{-ij}(p)\big)-Y_{ij}\big(0,p,0,\mathbf{M}_{-ij}(0)\big)$
Indirect effect mediated by a focal individual's compliance	$=\big[Y_{ij}\big(1,p,M_{ij}(p),\mathbf{M}_{-ij}(p)\big)-Y_{ij}\big(1,p,0,\mathbf{M}_{-ij}(p)\big)\big]$
Indirect effect mediated by peers' compliance	$+\big[Y_{ij}\big(1,p,M_{ij}(0),\mathbf{M}_{-ij}(p)\big)-Y_{ij}\big(1,p,0,\mathbf{M}_{-ij}(0)\big)\big]$
Direct effect of individual treatment assignment	$+\big[Y_{ij}\big(1,p,M_{ij}(0),\mathbf{M}_{-ij}(0)\big)-Y_{ij}\big(0,p,0,\mathbf{M}_{-ij}(0)\big)\big]$
Direct effect of peer treatment assignment	$+\big[Y_{ij}\big(0,p,M_{ij}(0),\mathbf{M}_{-ij}(0)\big)-Y_{ij}\big(0,0,0,\mathbf{M}_{-ij}(0)\big)\big]$

$$Y_{ij}\big(1,p,M_{ij}(p),\mathbf{M}_{-ij}(p)\big)-Y_{ij}\big(0,0,0,\mathbf{M}_{-ij}(0)\big). \tag{15.12}$$

The total effect is defined as the difference between an individual's potential outcome when a p proportion of the participating households at the site including the focal individual's household receive vouchers and the individual's potential outcome when none of the households receive vouchers. The decomposition is summarized in Table 15.2.

The population average causal effects can be defined accordingly. Henceforth, we omit the subscripts again to simplify the notation. The *population average indirect effect of the treatment transmitted through a focal individual's compliance* is

$$E[Y(1,p,M(p),\mathbf{M}_-(p))-Y(1,p,0,\mathbf{M}_-(p))].$$

The *population average indirect effect of the treatment transmitted through peers' compliance* is

$$E[Y(1,p,0,\mathbf{M}_-(p))-Y(1,p,0,\mathbf{M}_-(0))].$$

The *population average direct effect of individual treatment assignment* is

$$E[Y(1,p,0,\mathbf{M}_-(0))-Y(0,p,0,\mathbf{M}_-(0))].$$

And finally, the *population average direct effect of peer treatment assignment* is

$$E[Y(0,p,0,\mathbf{M}_-(0))-Y(0,0,0,\mathbf{M}_-(0))].$$

The expectations are taken over all the individuals at a site and over all the sites.

15.4   Identification and estimation of spill-over effects in a multisite trial

The causal effects defined earlier involve five population average potential outcomes. The first of these, denoted by $E[Y(1,p,M(p),\mathbf{M}_-(p))]$, is the population average potential outcome when a p proportion of the participating households at a site including a focal individual's household are offered vouchers. The second one, denoted by $E[Y(1,p,0,\mathbf{M}_-(p))]$, is the population average potential outcome when a p proportion of the participating households at a site including a focal individual's household are offered vouchers yet the focal individual's household, perhaps counterfactually, fails to use the voucher to move. The third one, denoted by $E[Y(1,p,0,\mathbf{M}_-(0))]$, is the population average potential outcome when a p proportion of the participating households at a site including a focal individual's household are offered vouchers yet, perhaps counterfactually, none of them use the vouchers to move. The fourth one, denoted by $E[Y(0,p,0,\mathbf{M}_-(0))]$, is the population average potential outcome when a focal individual's household is not offered a voucher while a p proportion of other participating households at the site are offered vouchers yet, perhaps counterfactually, none of them use the vouchers to move. The last one is the population average potential outcome when none of the participating households at a site are offered vouchers, denoted by $E[Y(0,0,0,\mathbf{M}_-(0))]$.

15.4.1   Identification in an ideal experiment

The aforementioned causal effects can be identified under a series of assumptions as follows:

1. *No interference between sites.* We assume that individuals from different experimental sites do not interact.

2. *Treatment assignments are independent of potential mediators and potential outcomes.* This includes three assumptions:
    2a. The site-specific probability of treatment assignment is independent of an individual's potential mediators and potential outcomes.
    2b. Within a site, an individual's treatment assignment is independent of the individual's potential mediators and potential outcomes.
    2c. Within a site, the peers' treatment assignments are independent of a focal individual's potential mediators and potential outcomes.

3. *Mediator value assignments are independent of potential outcomes.* This includes two assumptions:
    3a. Within a site, an individual's mediator value assignment is independent of the individual's potential outcomes.
    3b. Within a site, the peers' mediator value assignments are independent of a focal individual's potential outcomes.

4. *Exchangeable peer treatment assignments.* Different treatment assignments for peers at the same site can be viewed exchangeable as long as they all result in a p proportion of the individual units at the site receiving the treatment.

These assumptions are easy to satisfy in an ideal experiment that involves a four-step randomization. In Step 1, experimental sites are assigned at random to different probabilities of treatment assignment. To keep it simple, we have considered a contrast between two site-specific probabilities of treatment assignment: $P=p$ versus $P=0$. Here P denotes a random variable taking values between 0 and 1 while p denotes a specific nonzero proportion. In the MTO context, this is a contrast between offering vouchers to a p proportion of the participating households versus to none of the households. In Step 2, individuals at each site are assigned at random to the experimental condition or the control condition according to the site-specific probability of treatment assignment denoted by p or 0 in this simplified case. Hence, for individual i at site j, if a p proportion of the households at this site are to receive vouchers, we will have that $\mathrm{pr}(Z_{ij}=1|P_j=p)=p$, that is, the individual's household will have a p probability of receiving a voucher; if, instead, the site has been assigned to have a zero proportion of the households receiving vouchers, we will have that $\mathrm{pr}(Z_{ij}=1|P_j=0)=0$, and therefore, the individual's household will not be offered a voucher. The same site-specific probability applies to all other participating households at the same site. In Step 3, every experimental site that has been assigned to $P=p$ is then linked at random to a number of receiving low-poverty neighborhoods. However, the amount of real estate properties available for renting in the receiving low-poverty neighborhoods may vary across the sites. We use $R_j=r$ to denote the availability of rental units for the participating households in site j. It is conceivable that, in an extreme case, the receiving neighborhoods for a site may not have rental availability during the experiment, which is represented by $R=0$. For simplicity, let $R=1$ if the receiving neighborhoods provide abundant rental opportunities. In Step 4, households that have been offered vouchers at a site with a site-specific probability of treatment assignment $P=p$ are then assigned at random to different mediator values. That is, these households are assigned at random to either having an opportunity of using vouchers to move or not having such an opportunity, depending on the rental availability in the receiving low-poverty neighborhoods.

Under this ideal experimental design, the population average potential outcome of assigning a p proportion of households including a focal individual's household to receive vouchers at each site, $E[Y(1,p,M(p),\mathbf{M}_-(p))]$, can be identified. This is equal to the average observable outcome of the individuals at the sites in which a p proportion of the households including these individuals' households receive vouchers at each site:

$$E[Y(1,p,M(p),\mathbf{M}_-(p))]$$
$$= E[Y(1,p,M(p),\mathbf{M}_-(p))|Z=1,P=p] \qquad (15.13)$$
$$= E[Y|Z=1,P=p].$$

The first equation holds under identification assumptions (1), (2), and (4).

Under the same assumptions, the population average potential outcome of not treating anybody at a site, $E[Y(0,0,0,\mathbf{M}_-(0))]$, can be identified by the average observable outcome of the individuals at the sites in which a zero proportion of the participating households are treated:

$$E[Y(0,0,0,\mathbf{M}_-(0))]$$
$$= E[Y(0,0,0,\mathbf{M}_-(0))|Z=0,P=0] \qquad (15.14)$$
$$= E[Y|Z=0,P=0].$$

The population average potential outcome when a p proportion of the participating households at a site including a focal individual's household are offered vouchers yet the focal individual's household, perhaps counterfactually, fails to use the voucher to move, $E[Y(1,p,$ $0,\mathbf{M}_-(p))]$, can be identified under assumptions (1), (2), (3), and (4). This is equal to the average observable outcome of the individuals who do not move after receiving vouchers at the sites where a p proportion of the participating households are offered vouchers:

$$
\begin{aligned}
E[Y(1,p,0,\mathbf{M}_-(p))] \\
= E[Y(1,p,0,\mathbf{M}_-(p))|Z=1,M=0,P=p] \\
= [Y|Z=1,M=0,P=p].
\end{aligned}
\tag{15.15}
$$

The population average potential outcome when a focal individual's household and a p proportion of other participating households at the site are offered vouchers yet, perhaps counterfactually, none of them use the vouchers to move $E[Y(1,p,0,\mathbf{M}_-(0))]$ is identifiable under the same assumptions. This is equal to the average observable outcome of the individuals who do not move after receiving vouchers at the sites where a p proportion of the participating households are offered vouchers yet none of these households use the vouchers to move (i.e., $R=0$). At a site where a p proportion of the participating households are offered vouchers and some of the households actually use vouchers to move (i.e., $R>0$), even if a focal individual's household receives a voucher yet fails to use the voucher to move, this focal individual's observable outcome will not contribute to the identification of this population average potential outcome:

$$
\begin{aligned}
E[Y(1,p,0,\mathbf{M}_-(0))] \\
= E[Y(1,p,0,\mathbf{M}_-(0))|Z=1,R=0,P=p] \\
= [Y|Z=1,R=0,P=p].
\end{aligned}
\tag{15.16}
$$

Finally, the population average potential outcome when a focal individual's household is not offered a voucher while a p proportion of other participating households at the site are offered vouchers yet, perhaps counterfactually, none of them use the vouchers to move $E[Y(1,p,0,\mathbf{M}_-(0))]$ can be identified under the same set of assumptions. This is equal to the average observable outcome of the individuals who do not receive vouchers at the sites where a p proportion of the participating households are offered vouchers yet none of these households use the vouchers to move:

$$
\begin{aligned}
E[Y(0,p,0,\mathbf{M}_-(0))] \\
= E[Y(0,p,0,\mathbf{M}_-(0))|Z=0,R=0,P=p] \\
= [Y|Z=0,R=0,P=p].
\end{aligned}
\tag{15.17}
$$

15.4.2   Identification when the mediators are not randomized

In the MTO study, about half of the households receiving vouchers actually moved to low-poverty neighborhoods. Apparently, a household's decision to use the voucher to move depended not only on the rental availability in the receiving neighborhood but also on the unique circumstances of the household. For example, it was reported that a household was more likely to move if the residents perceived their neighborhood to be very unsafe and their apartment conditions to be very unsatisfactory and if they did not have teenage children at the baseline (Kling, Liebman, and Katz, 2007). Let $\mathbf{X}$ be a vector of baseline covariates that may predict the likelihood that a household will use the voucher to move. A household's propensity to move to a low-poverty neighborhood when receiving a voucher at a site where a p proportion of other participating households are offered vouchers can be represented by

$$pr(M = 1|Z = 1, P = p, R = r, \mathbf{X} = \mathbf{x}).$$

Note that the propensity is zero if $P = 0$, if $Z = 0$, or if $R = 0$.

Because the MTO sites are cities that are geographically separate, assumption (1) "no interference between sites" seems highly plausible. Assumption (2) "independence of treatment assignments" is warranted by the experimental design. However, due to the self-selection of mediator values, assumption (3) "independence of mediator value assignment" needs to be modified as follows:

3* Mediator value assignments are independent of potential outcomes for individuals with the same pretreatment characteristics. This includes three assumptions:

3a* Within a site and among individuals with the same pretreatment characteristics, an individual's mediator value assignment is independent of the individual's potential outcomes.

3b* Within a site and among individuals with the same pretreatment characteristics, peers' mediator value assignments are independent of a focal individual's potential outcomes.

3c* Within a site and among individuals with the same pretreatment characteristics, an individual's mediator value assignment is independent of the peers' mediator value assignments within and across treatment conditions.

In other words, it is assumed that, among individual households with the same pretreatment characteristics at a site, whether a household actually moves is as if a result of randomization that assigns the household to either having a definite possibility of moving or not having such a possibility.

15.4.3   Estimation of mediated effects through spill-over

15.4.3.1   Estimating $E[Y(1, p, M(p), M_-(p))]$ and $E[Y(0, 0, 0, M_-(0))]$

Under assumptions (1) and (2), the population average potential outcome of offering vouchers to a p proportion of households $E[Y(1, p, M(p), \mathbf{M}_-(p))]$ and that of offering vouchers to none of the households at a site $E[Y(0, 0, 0, \mathbf{M}_-(0))]$ can be identified as shown in (15.13) and (15.14), respectively. Let $I_j(P = p)$ be an indicator that takes value 1 if a p proportion of the households at site j are assigned to receive vouchers and 0 otherwise. Analyzing the experimental data, one may estimate $E[Y(1, p, M(p), \mathbf{M}_-(p))]$ with the following:

$$\frac{\sum_{j=1}^{J} \sum_{i=1}^{n_j} I_j(P = p) Z_{ij} Y_{ij}}{\sum_{j=1}^{J} \sum_{i=1}^{n_j} I_j(P = p) Z_{ij}}.$$

Here, the numerator is the sum of the observed outcome of individuals whose households are offered vouchers at the sites where a p proportion of the households are offered vouchers. The denominator is the total number of such individuals. Let $I_j(P=0)$ be another indicator that takes value 1 if none of the households at site j receive vouchers and 0 otherwise. One may estimate $E[Y(0,0,0,\mathbf{M}_-(0))]$ with the following:

$$\frac{\sum_{j=1}^{J}\sum_{i=1}^{n_j} I_j(P=0)Y_{ij}}{\sum_{j=1}^{J} I_j(P=0)n_j}.$$

15.4.3.2   Estimating $E[Y(1,p,0,\mathbf{M}_-(p))]$

Next, we estimate the population average potential outcome $E[Y(1,p,0,\mathbf{M}_-(p))]$ under a site-specific probability of treatment assignment $P=p$ for individuals whose households are offered vouchers yet, perhaps counterfactually, do not use the vouchers to move. Due to self-selection in a households' decision to use a voucher to move, the average potential outcome of those who are offered vouchers and do not use the vouchers to move may not be equal to the average of the same potential outcome for those who actually use the vouchers to move. When assumption (3*) holds, however, we can use weighting to transform the distribution of the observed outcome of those who do not move after being offered vouchers at each site with site-specific probability of treatment assignment $P=p$. The estimator with weighting is

$$\frac{\sum_{j=1}^{J}\sum_{i=1}^{n_j} I_j(P=p)Z_{ij}\left(1-M_{ij}\right)W_{ij}Y_{ij}}{\sum_{j=1}^{J}\sum_{i=1}^{n_j} I_j(P=p)Z_{ij}\left(1-M_{ij}\right)W_{ij}},$$

where, for individual i at site j whose household is offered a voucher yet does not move, the weight is

$$W_{ij} = \frac{\text{pr}\left(M_{ij}=0|Z_{ij}=1,P_j=p\right)}{\text{pr}\left(M_{ij}=0|Z_{ij}=1,P_j=p,\mathbf{X}_{ij}=\mathbf{x}\right)}.$$

Here, the denominator is the household's propensity of not moving after being offered a voucher given its pretreatment characteristics; the numerator is the proportion of households being offered vouchers at the site that do not move. Appendix 15.1 derives this weight. The weight can be obtained nonparametrically by stratifying the sample of households receiving vouchers at a site on the estimated propensity score. For individual i at site j in stratum s whose household is offered a voucher yet does not move, the weight is the ratio of the proportion of households receiving vouchers that do not move to the proportion of such households within the stratum:

$$W_{ij} = \frac{\text{pr}\left(M_{ij}=0|Z_{ij}=1,P_j=p\right)}{\text{pr}\left(M_{ij}=0|Z_{ij}=1,P_j=p,S_{ij}=s\right)}.$$

This is the marginal mean weighting through stratification (MMWS) strategy (Hong, 2010b, 2012) introduced in Chapters 4 and 5.

15.4.3.3 Estimating $E[Y(1, p, 0, M\_(0))]$

The population average potential outcome when a p proportion of the participating households at a site are offered vouchers and, perhaps counterfactually, none of them use the voucher to move can be estimated in correspondence with the identification result in (15.16). The estimator is the average observed outcome of individuals whose households are offered vouchers at the sites where the site-specific probability of treatment assignment is $P = p$ yet the receiving low-poverty neighborhoods have no rental availability. Let $I_j(R = 0)$ be the indicator that takes value one if no rental units are available to the households at site j and zero otherwise. Below is the sample estimator of this population average potential outcome:

$$\frac{\sum_{j=1}^{J}\sum_{i=1}^{n_j}I_j(P=p)I_j(R=0)Z_{ij}Y_{ij}}{\sum_{j=1}^{J}\sum_{i=1}^{n_j}I_j(P=p)I_j(R=0)Z_{ij}}.$$

15.4.3.4 Estimating $E[Y(0, p, 0, M\_(0))]$

Finally, we estimate the population average potential outcome when a focal individual's household is not offered a voucher while a p proportion of other households at the site are offered vouchers yet, perhaps counterfactually, none of them use the voucher to move. Applying the identification result in (15.17), we can estimate this with the average observed outcome of individuals whose households are not offered vouchers at the sites where the site-specific probability of treatment assignment is $P = p$ yet the receiving low-poverty neighborhoods have no rental availability:

$$\frac{\sum_{j=1}^{J}\sum_{i=1}^{n_j}I_j(P=p)I_j(R=0)\left(1-Z_{ij}\right)Y_{ij}}{\sum_{j=1}^{J}\sum_{i=1}^{n_j}I_j(P=p)I_j(R=0)\left(1-Z_{ij}\right)}.$$

We estimate the causal effects of interest, each defined as a contrast between two population average potential outcomes as shown in Section 15.3, by computing the difference between the corresponding sample statistics while taking into account the nesting of households within experimental sites.

15.5   Consequences of omitting spill-over effects in causal mediation analyses

Past mediation analyses typically invoke the stable unit treatment value assumption (SUTVA) (Rubin, 1986). Under this assumption, the potential mediator of individual i at site j depends only on the individual's treatment assignment z_{ij} and does not depend on the treatment assignments of other individuals at the site denoted by $\mathbf{z}_{-ij}$. That is, $M_{ij}\left(z_{ij}, \mathbf{z}_{-ij}\right) = M_{ij}\left(z_{ij}\right)$. Similarly, the potential outcome of the individual depends only on z_{ij} and $M_{ij}(z_{ij})$ but not on $\mathbf{z}_{-ij}$. Nor does it depend on the mediator values of other individuals at the site denoted by $\mathbf{M}_{-ij}(\mathbf{z}_j)$. Therefore, $Y_{ij}\left(z_{ij}, \mathbf{z}_{-ij}, M_{ij}\left(z_{ij}, \mathbf{z}_{-ij}\right), \mathbf{M}_{-ij}\left(\mathbf{z}_j\right)\right)$ can be simplified as $Y_{ij}(z_{ij}, M_{ij}(z_{ij}))$.

15.5.1   Biased inference in a cluster randomized trial

As summarized in Table 15.1, the total effect of a cluster-level treatment assignment can be decomposed into an indirect effect mediated by a focal individual's compliance, an indirect effect mediated by peers' compliance, and a direct effect of the treatment assignment. Under SUTVA, the population average treatment effect $E[Y(1,M(1),\mathbf{M}_-(1)) - Y(0,M(0),\mathbf{M}_-(0))]$ is simplified to be $E[Y(1,M(1)) - Y(0,M(0))]$; the population average indirect effect mediated by a focal individual's compliance $E[Y(1,M(1),\mathbf{M}_-(1)) - Y(1,M(0),\mathbf{M}_-(1))]$ is simplified to be $E[Y(1,M(1)) - Y(1,M(0))]$; the population average indirect effect mediated by peers' compliance $E[Y(1,M(0),\mathbf{M}_-(1)) - Y(1,M(0),\mathbf{M}_-(0))]$ is simplified to be $E[Y(1,M(0)) - Y(1,M(0))]$, which becomes zero; and the population average direct effect $E[Y(1,M(0),\mathbf{M}_-(0)) - Y(0,M(0),\mathbf{M}_-(0))]$ is simplified to be $E[Y(1,M(0)) - Y(0,M(0))]$. This is the conceptual framework for causal mediation analysis that has dominated the causal inference literature (Pearl, 2001; Robins and Greenland, 1992) and has been adopted in Chapters 9–13 of this book. VanderWeele *et al.* (2013) have shown, in the context of the 4Rs study, that when data analysts fail to consider a possibly nonzero indirect effect mediated by peers' compliance, this indirect effect will be misinterpreted as a part of the direct effect, which will lead to an inaccurate interpretation of the direct effect.

15.5.2   Biased inference in a multisite randomized trial

Table 15.2 displays the decomposition of the total treatment effect in a multisite trial that contrasts treating a $p > 0$ proportion of the participants with treating none at a site. Under SUTVA, the population average total effect $E[Y(1,p,M(p),\mathbf{M}_-(p)) - Y(0,0,M(0),\mathbf{M}_-(0))]$ becomes $E[Y(1,M(1)) - Y(0,M(0))]$. The population average indirect effect mediated by a focal individual's compliance $E[Y(1,p,M(p),\mathbf{M}_-(p)) - Y(1,p,M(0),\mathbf{M}_-(p))]$ is simplified to be $E[Y(1,M(1)) - Y(1,M(0))]$. The population average indirect effect mediated by peers' compliance $E[Y(1,p,M(0),\mathbf{M}_-(p)) - Y(1,p,M(0),\mathbf{M}_-(0))]$ is simplified to be $E[Y(1,M(0)) - Y(1,M(0))] = 0$. The population average direct effect of individual treatment assignment $E[Y(1,p,M(0),\mathbf{M}_-(0)) - Y(0,p,M(0),\mathbf{M}_-(0))]$ is simplified to be $E[Y(1,M(0)) - Y(0, M(0))]$. And finally, the population average direct effect of peer treatment assignment $E[Y(0,p,M(0),\mathbf{M}_-(0)) - Y(0,0,M(0),\mathbf{M}_-(0))]$ is simplified to be $E[Y(0,M(0)) - Y(0, M(0))] = 0$. These simplifications make no distinction, for example, between treating a $0 < p < 1$ proportion of the participants rather than none and treating all the participants (i.e., $p = 1$) rather than none. In the following, we discuss the consequences of omitting peer spill-over effects in these cases.

When SUTVA is invoked, the population average treatment effect is generally defined as $E[Y(1,M(1)) - Y(0,M(0))]$. This is equal to $E[Y(1,M(1))] - E[Y(0,M(0))]$, typically interpreted as the mean difference between the potential outcome when the entire population is treated and the potential outcome when the entire population is untreated, and hence is usually considered equal to

$$E[Y(1,1,M(1),\mathbf{M}_-(1))-Y(0,0,M(0),\mathbf{M}_-(0))]. \qquad (15.18)$$

However, if an experiment assigns a $0<p<1$ proportion of the participants to be treated at every site, following Sobel (2006), we can see that the intent-to-treat (ITT) effect that researchers typically report on the basis of the experimental data is an estimate of

$$E[Y(1,p,M(p),\mathbf{M}_-(p))|Z=1]-E[Y(0,p,M(0),\mathbf{M}_-(p))|Z=0]. \qquad (15.19)$$

In the presence of spill-over, this ITT effect is biased for the population average causal effect of treating all versus none defined in (15.18). According to the derivation in Appendix 15.2, the bias is

$$\begin{aligned} &E[Y(1,p,M(p),\mathbf{M}_-(p))-Y(1,1,M(1),\mathbf{M}_-(1))|Z=1]\\ &+E[Y(0,0,M(0),\mathbf{M}_-(0))-Y(0,p,M(0),\mathbf{M}_-(p))|Z=0]. \end{aligned} \qquad (15.20)$$

The first term of the bias is the causal effect of treating a $0<p<1$ proportion of the participants rather than treating all the participants at each site for the treated participants in the current experiment; the second term of the bias is the causal effect of treating none of the participants rather than a $0<p<1$ proportion of them at each site for the untreated participants in the current experiment.

The first term of the bias can be further decomposed into a direct effect of peer treatment assignment and an indirect effect of treating a $0<p<1$ proportion rather than all the participants for the treated units:

$$\begin{aligned} &E[Y(1,p,M(p),\mathbf{M}_-(p))-Y(1,1,M(1),\mathbf{M}_-(1))|Z=1]\\ &=E[Y(1,p,M(p),\mathbf{M}_-(p))-Y(1,1,M(p),\mathbf{M}_-(p))|Z=1]\\ &+E[Y(1,1,M(p),\mathbf{M}_-(p))-Y(1,1,M(1),\mathbf{M}_-(1))|Z=1]. \end{aligned} \qquad (15.21)$$

Similarly, the second term of the bias can be further decomposed into a direct effect of peer treatment assignment and an indirect effect of treating none of the participants rather than a $0<p<1$ proportion of them for the untreated units:

$$\begin{aligned} &E[Y(0,0,M(0),\mathbf{M}_-(0))-Y(0,p,M(0),\mathbf{M}_-(p))|Z=0]\\ &=E[Y(0,0,M(0),\mathbf{M}_-(0))-Y(0,p,M(0),\mathbf{M}_-(0))|Z=0]\\ &+E[Y(0,p,M(0),\mathbf{M}_-(0))-Y(0,p,M(0),\mathbf{M}_-(p))|Z=0]. \end{aligned} \qquad (15.22)$$

We now evaluate the first bias shown in (15.21) in the MTO example. Suppose that housing vouchers are offered additionally to the households initially assigned to the control condition such that 100% of the participants at a site are offered vouchers. Also, suppose that those initially assigned to the control condition are the only ones who are unable to use the vouchers to move due to the constraint in the rental market in the receiving low-poverty neighborhoods. In this hypothetical scenario, the compliance of all the households will remain unchanged, and hence, the potential outcome of a treated individual is denoted by $Y(1,1,M(p),\mathbf{M}_-(p))$. One may reason that increasing the proportion of participants receiving vouchers without increasing the proportion of participants actually using vouchers to move may have no impact on the treated individuals' posttreatment perception of safety in their neighborhoods. Therefore, the average direct effect of peer treatment assignment for the treated $E[Y(1,p,M(p),\mathbf{M}_-(p)) - Y(1,1,M(p),\mathbf{M}_-(p))|Z=1]$ may be zero.

We now suppose that there is no rental constraint in the receiving low-poverty neighborhoods. When the offering of vouchers is extended to those initially assigned to the control condition, if many of them also use the vouchers to move, the additional influx of poor households into the receiving low-poverty neighborhoods will likely further change the demographic composition and the crime rate in those neighborhoods. For the treated individuals who have moved, the perceived safety in their new neighborhoods may decline accordingly. For this reason, the average indirect effect of treating a $0<p<1$ proportion rather than all the participants for the treated $E[Y(1,1,M(p),\mathbf{M}_-(p)) - Y(1,1,M(1),\mathbf{M}_-(1))|Z=1]$ is likely positive.

In evaluating the second bias shown in (15.22), one may reason that offering vouchers to none rather than a $0<p<1$ proportion of the participants is unlikely to change the untreated individuals' perception of the high-poverty neighborhood that they live in if none of the treated actually use vouchers to move. Therefore, the average direct effect of peer treatment assignment for the untreated $E[Y(0,0,M(0),\mathbf{M}_-(0)) - Y(0,p,M(0),\mathbf{M}_-(0))|Z=0]$ is arguably zero. However, when a $0<p<1$ proportion of the participants are offered vouchers, if many of them actually move to low-poverty neighborhoods, their continuing communications with the untreated individuals back in the high-poverty neighborhoods may change the untreated individuals' perceptions of neighborhood safety. The untreated individuals may view the high-poverty neighborhoods less favorably if their treated neighbors move than if those neighbors do not move. Therefore, the average indirect effect of treating none of the peers rather than a p proportion of them for the untreated $E[Y(0,p,M(0),\mathbf{M}_-(0)) - Y(0,p,M(0),\mathbf{M}_-(p))|Z=0]$ may be positive.

In summary, in the presence of peer spill-over as a part of the mediation process, a positive indirect effect for the treated and a positive indirect effect for the untreated will add up to a positive bias when MTO researchers mistakenly regard the ITT effect as the population average causal effect of treating all versus treating none.

As discussed earlier, when SUTVA is invoked, the population average indirect effect mediated by peers' compliance $E[Y(1,p,M(0),\mathbf{M}_-(p)) - Y(1,p,M(0),\mathbf{M}_-(0))]$ is assumed to be zero; so is the population average direct effect of peer treatment assignment $E[Y(0,p,0,\mathbf{M}_-(0)) - Y(0,0,0,\mathbf{M}_-(0))]$. One may argue that changing the proportion of households receiving vouchers may generate no impact if none of the households actually move. However, if some of the peers actually use the vouchers to move, one may instead argue that their communications with a focal individual may change his or her perception of the current neighborhood even if the individual's household is offered a voucher yet does not move. A nonzero indirect effect mediated by peers' compliance, if ignored, could potentially distort the identification of the indirect effect mediated by a focal individual's compliance and of the direct effect of individual treatment assignment.

15.5.3   Biased inference of the local average treatment effect

When compliance is of central interest, researchers often additionally invoke the exclusion restriction, which assumes that the effect of treatment assignment on an individual's outcome must be via the treatment effect on the individual's compliance (Angrist, Imbens, and Rubin, 1996). Hence, the individual's potential outcome is a function of $M_{ij}(z_{ij})$ only. Under SUTVA and the exclusion restriction, potential outcome $Y_{ij}\left(z_{ij}, p_j, M_{ij}\left(p_j\right), \mathbf{M}_{-ij}\left(p_j\right)\right)$ can be simplified as $Y_{ij}(M_{ij}(z_{ij}))$. These two assumptions, along with several others explicated by Angrist and colleagues (1996), enable researchers to employ the instrumental variable (IV) method for estimating the average treatment effect for the compliers, which has also been named "the local average treatment effect." In the MTO example, the treatment assignment may arguably have no impact on an individual's outcome if none of the households at a site receive a voucher or if none of the households receiving vouchers actually move to a low-poverty neighborhood. However, as long as some households actually use vouchers to move, the indirect effect mediated by peers' compliance is likely nonzero, which will then violate the exclusion restriction and may invalidate some of the empirical results that are based on the IV strategy.

15.6   Quasiexperimental application

At the beginning of this chapter, we introduced the case in which individuals are regrouped for treatment delivery if assigned to the experimental condition. In theory, a treatment-induced change in peer composition may shape a focal individual's intermediate experience with the treatment and may eventually exert an impact on the individual's outcome. Hong and Nomi (2012) examined the role of spill-over between peers in the group context as part of the causal mediation mechanism in a quasiexperimental study. Here, we briefly review this study, focusing on the analytic strategy employed by these researchers.

The Chicago Public Schools (CPS) introduced a policy in 1997 that required all students to take algebra by the end of ninth grade. Prior to the introduction of the "algebra-for-all" policy, low-achieving students typically took remedial math with other low-achieving peers. As schools increased algebra enrollment after 1997, they often created mixed-ability algebra classes by enrolling low-achieving students in the same algebra classes with high-achieving students. Hong and Nomi (2012) reasoned that class peer ability represents the amount of math knowledge and skills collectively brought by students in a class. Ability composition of a class may influence the instructional content, pace, participation structure, peer interactions, and evaluation, which may subsequently influence a student's relative standing and math learning. Specifically, for low-achieving students, experiencing a rise in peer ability may heighten peer competition, increase anxiety for failure, trigger frustration and alienation, and lower one's self-esteem due to unfavorable social comparisons. These negative experiences may largely offset the potential benefits of gaining exposure to advanced academic discourse involving high-achieving peers.

On the basis of the earlier reasoning, rather than assuming that the curricular policy change would affect low-achieving students' math learning only through changing their math course taking, Hong and Nomi (2012) investigated a second pathway through which the policy may affect these students' math learning by changing the classroom setting. They placed a particular emphasis on the policy-induced change in peer ability composition within a math class. Specifically, student i in school k might be assigned to algebra class j and experience peer ability denoted by $C_{ijk}(1)$ if attending the ninth grade after the policy was introduced and might be assigned to remedial math class j' and experience peer ability $C_{ij'k}(0)$ if attending the ninth grade before the policy was introduced. When a student's school membership is given, class peer ability is an immediate result of the policy and therefore may serve as a mediator of the policy effect. The researchers focused on decomposing the total policy effect for low-achieving students, denoted by $E[Y(1,C(1))-Y(0,C(0))]$, into an indirect effect mediated by class peer composition change and a direct effect, the latter being primarily attributable to the change from taking remedial math to taking algebra. The average indirect effect is denoted by $E[Y(1,C(1))-Y(1,C(0))]$; and the average direct effect is denoted by $E[Y(1,C(0))-Y(0,C(0))]$.

Hong and Nomi (2012) analyzed data from all 59 CPS neighborhood high schools. Among them, 14 schools offered algebra to all ninth graders even prior to 1997. They selected one prepolicy cohort and one postpolicy cohort of first-time ninth graders and, by specifying prediction models, empirically identified 997 prepolicy students and 541 postpolicy students who would likely experience a rise in class peer ability as well as a change in course taking due to the policy. These students displayed low incoming skills relative to other students in the same cohort before entering high school. About 42% of these low-achieving students enrolled in algebra prepolicy; all of them enrolled in algebra postpolicy. The average peer ability within a class was measured by the class median math incoming ability. After accounting for the system-wide improvement in ninth graders' incoming skills from the prepolicy to the post-policy year, the researchers found the class peer ability for the postpolicy low-achieving students to be a full standard deviation higher than that for their prepolicy counterparts.

Because the data are quasiexperimental, there are a number of potential threats to valid causal inference. The first is that between-cohort differences in observed and unobserved pre-treatment student characteristics and school characteristics may confound the estimation of the total policy effect on math learning. The researchers used MMWS (Hong, 2010b, 2012), explained at length in Chapters 4 and 5, to adjust for between-cohort differences in the math outcome associated with the observed pretreatment covariates. The weight was computed nonparametrically on the basis of the individual propensity of policy exposure. Yet additional confounding may be associated with unobserved covariates reflecting the impacts of historical events including concurrent CPS policies. The researchers reasoned that the historical con-founding may affect all students including those attending the 14 comparison schools that offered algebra to all ninth graders in the prepolicy year and therefore were unaffected by the policy. For the students in these 14 schools, any between-cohort difference in the average math outcome was to be attributed to historical factors concurrent to the algebra-for-all policy. Modifying the conventional difference-in-differences (DID) strategy, they developed a prognostic score-based DID strategy to reduce historical confounding. This strategy identi-fied subgroups of students in the schools affected by the policy and those in the comparison schools unaffected by the policy who would display similar prepolicy math outcomes and similar postpolicy outcomes had they all attended the comparison schools. After estimating the average historical confounding effect for each homogeneous subgroup of students, the researchers obtained an adjusted postpolicy math outcome. Combining MMWS and the prognostic score-based DID strategy, they were able to estimate the population average poten-tial outcomes $E[Y(1, C(1))]$ and $E[Y(0, C(0))]$ for low-achieving prepolicy students under the assumption that there is no additional confounding of the policy effect. The estimated total policy effect was 0.23 (SE = 1.15, $t = 0.20$). The effect size was only about 2% of a standard deviation of the outcome.

To estimate the population average counterfactual outcome $E[Y(1, C(0))]$ is even more challenging. This is the math outcome that low-achieving students would have displayed in the postpolicy year if the school had, perhaps counterfactually, decided to continue the same practice of sorting students to math classes by ability as they did prior to the policy. Such a practice would lead to a lack of change in class peer ability despite the curricular policy change. However, those who experienced different class peer ability under each policy tended to be systematically different in their pretreatment characteristics. The researchers used RMPW to transform the distribution of class peer ability in the postpolicy cohort to resemble its distribution in the prepolicy cohort within levels of the pretreatment characteristics. Let $c = 0, 1, 2, 3, 4$ denote the lowest, lower, medium, higher, and highest class ability levels within each cohort. The weight assigned to each postpolicy student who actually experienced class peer ability level c is

$$W = \frac{\text{pr}(C(0) = c | Z = 0, \mathbf{X}, \bar{\mathbf{X}}, \mathbf{L}(1), \mathbf{L}(0))}{\text{pr}(C(1) = c | Z = 1, \mathbf{X}, \bar{\mathbf{X}}, \mathbf{L}(1), \mathbf{L}(0))}.$$

Here, the denominator is a student's propensity of experiencing class peer ability level c in the postpolicy year; the numerator is the student's propensity of experiencing class peer ability level c had the student attended the same school in the prepolicy year instead. Predictors of each propensity score included pretreatment student covariates $\mathbf{X}$, pretreatment school covariates $\bar{\mathbf{X}}$ (e.g., demographic and SES composition and school mean and standard deviation of math incoming skills), and posttreatment school covariates that were measured twice—once in the prepolicy year $\mathbf{L}(0)$ and once in the postpolicy year $\mathbf{L}(1)$ (e.g., algebra enrollment rate, advanced math course enrollment rate, and within-school variability in ninth-grade math class size). The posttreatment school characteristics are potential intermediate outcomes of the policy and can be adjusted in the RMPW analysis when both $\mathbf{L}(1)$ and $\mathbf{L}(0)$ are observed for each school. The propensity score models also included a school-specific random intercept and random slopes for student incoming skills, allowing the selection mechanism to vary across schools.

Applying the product of MMWS and RMPW to the DID-adjusted outcome data, Hong and Nomi (2012) estimated the direct effect of the policy and the indirect effect mediated by the policy-induced change in class peer ability. The estimated average direct effect is 2.70 (SE $= 1.20$, $t = 2.24$, $p < 0.05$); and the estimated average indirect effect is -2.33 (SE $= 0.88$, $t = -2.63$, $p < 0.01$). Although the policy did not raise math achievement on average, the above results indicated that the policy effect was partly mediated by class peer ability change. The evidence for a negative indirect effect is consistent with the theoretical hypothesis that, for low-achieving students, a rise in class peer ability might put them at a disadvantage due to unfavorable social comparisons or because instruction pitched to the middle of the class ability distribution was perhaps beyond their grasp. The positive direct effect of the policy, primarily due to the replacement of remedial math with algebra, indicates that exposing low-achieving students to algebra would have improved their math learning as intended had their class peer ability remained unchanged. These results are unbiased only if there is no omitted confounding of the treatment–mediator relationship, of the treatment–outcome relationship, and of the mediator–outcome relationship. The researchers investigated the consequences associated with some of the potential confounders.

This and other application examples have shown that, in empirical investigations of causal mediation through peer spill-over, we can again apply the weighting strategies described earlier in this book. Specifically, under the assumption that the treatment assignment is ignorable given the observed pretreatment covariates in quasiexperimental data, MMWS is useful for removing pretreatment differences between treatment groups. Under the additional assumption that the mediator value assignment is ignorable under each treatment condition given the observed pretreatment covariates, RMPW becomes handy for decomposing the total treatment effect into direct and indirect effects with adjustment for mediator value selection. Hong and Nomi (2012) took advantage of some unique features of the education accountability data to increase the plausibility of the ignorability assumptions. In particular, repeated observations of schools before and after the policy initiation made it possible to adjust for observed posttreatment covariates; data from a comparison group of schools unaffected by the policy enabled the researchers to assess the amount of historical confounding associated with concurrent events; and finally, repeated observations of each student's achievement trajectory prior to ninth grade provided a projection of ninth-grade math achievement in the absence of the treatment.

15.7   Summary

As we have shown in this last unit of the book, theoretical considerations of spill-over call for a major alteration of the individual-based conceptual framework of causal inference. We have defined the spill-over effects associated with peer treatment assignment, agent treatment assignment, peer compliance, and agent compliance. Because peer treatment assignment and agent treatment assignment may shape the treatment setting for a focal individual, they can be viewed as moderators of the effect of individual treatment assignment. Chapter 14 has presented a number of examples revealing how individual treatment assignment and peer/agent treatment assignment may jointly affect an individual's outcome. Peer compliance and agent compliance, on the other hand, may provide pathways through which peer and agent treatment assignments indirectly affect an individual's outcome even when the individual is untreated or is noncompliant. Chapter 15 has focused on the spill-over effect of peer/agent compliance by addressing it as a mediation problem. Researchers may further investigate how individual compliance and peer compliance operate as concurrent mediators. One may assess, for example, the extent to which the effect of individual treatment assignment mediated by individual compliance depends on peer compliance. Future research may also examine agent compliance as an initial mediator that may influence the subsequent mediators including individual compliance and peer compliance. Analytic tools for studying moderation and mediation problems, including MMWS and RMPW, may be extended to these new applications. Future advancements along these lines of research have the potential of greatly enrich social scientific understanding of how social interventions may improve individual and societal well-being.

Appendix 15.1: Derivation of the weight for estimating the population average counterfactual outcome $E[Y(1, p, 0, M\_(p))]$

In Section 15.4.3, we discussed that, in a site in which a p proportion of participating households are offered vouchers, the potential outcome of being offered a voucher and not moving is observed for individuals who are offered vouchers and do not move but cannot be observed for those who actually use vouchers to move. Because these two subgroups tend to be different in pretreatment characteristics X, we may apply a weighting adjustment for the pretreatment differences between the two groups. We derive the weight as follows:

$$E[Y(1,p,0,\mathbf{M}_-(p))]$$
$$= E[Y(1,p,0,\mathbf{M}_-(p))|Z=1,P=p]$$
$$= \int_{\mathbf{x}} E[Y(1,p,0,\mathbf{M}_-(p))|Z=1,P=p,\mathbf{X}=\mathbf{x}]pr(\mathbf{X}=\mathbf{x}|Z=1,P=p)d\mathbf{x}.$$

Under assumption (3^*) that the mediator value assignment M of an individual household is independent of the potential outcome within levels of pretreatment characteristics $\mathbf{X}$ when the treatment assignment is given, the above is equal to

$$\int_{\mathbf{x}} E[Y(1,p,0,\mathbf{M}_-(p))|Z=1,M=0,P=p,\mathbf{X}=\mathbf{x}]pr(\mathbf{X}=\mathbf{x}|Z=1,P=p)d\mathbf{x}$$

$$= \int_{\mathbf{x}} E[WY|Z=1,M=0,P=p,\mathbf{X}=\mathbf{x}] \times pr(\mathbf{X}=\mathbf{x}|Z=1,M=0,P=p)d\mathbf{x}$$

$$= E(WY|Z=1,M=0,P=p).$$

The weight is

$$W = \frac{pr(\mathbf{X}=\mathbf{x}|Z=1,P=p)}{pr(\mathbf{X}=\mathbf{x}|Z=1,M=0,P=p)}.$$

Applying Bayes theorem to the denominator, we have that

$$W = pr(\mathbf{X}=\mathbf{x}|Z=1,P=p)$$
$$\times \frac{pr(M=0|Z=1,P=p)}{pr(M=0|Z=1,P=p,\mathbf{X}=\mathbf{x}) \times pr(\mathbf{X}=\mathbf{x}|Z=1,P=p)}$$
$$= \frac{pr(M=0|Z=1,P=p)}{pr(M=0|Z=1,P=p,\mathbf{X}=\mathbf{x})}.$$

Appendix 15.2: Derivation of bias in the ITT effect due to the omission of spill-over effects

In a multisite randomized trial, the ITT effect is biased for the total treatment effect due to spill-over. The bias is the difference between (15.19) and (15.18):

$$\{E[Y(1,p,M(p),\mathbf{M}_-(p))|Z=1]-E[Y(0,p,M(0),\mathbf{M}_-(p))|Z=0]\}$$
$$-\{E[Y(1,1,M(1),\mathbf{M}_-(1))-Y(0,0,M(0),\mathbf{M}_-(0))]\}.$$

Due to randomization at each site, individual treatment assignment is independent of the potential outcomes. We therefore have that

$$E[Y(1,1,M(1),\mathbf{M}_-(1))] = E[Y(1,1,M(1),\mathbf{M}_-(1))|Z=1]$$

and that

$$E[Y(0,0,M(0),\mathbf{M}_-(0))] = E[Y(0,0,M(0),\mathbf{M}_-(0))|Z=0].$$

The above bias is then equal to

$$E[Y(1,p,M(p),\mathbf{M}_-(p))|Z=1] - E[Y(0,p,M(0),\mathbf{M}_-(p))|Z=0]$$
$$-E[Y(1,1,M(1),\mathbf{M}_-(1))|Z=1] + E[Y(0,0,M(0),\mathbf{M}_-(0))|Z=0]$$
$$= E[Y(1,p,M(p),\mathbf{M}_-(p)) - Y(1,1,M(1),\mathbf{M}_-(1))|Z=1]$$
$$+ E[Y(0,0,M(0),\mathbf{M}_-(0)) - Y(0,p,M(0),\mathbf{M}_-(p))|Z=0].$$

References

Angrist, J.D., Imbens, G.W. and Rubin, D.B. (1996). Identification of causal effects using instrumental variables. *Journal of the American Statistical Association, 91*(434), 444–472.

Brown, J.L., Jones, S.M., LaRusso, M. and Aber, J.L. (2010). Improving classroom quality: teacher influences and experimental impacts of the 4Rs program. *Journal of Educational Psychology, 102*, 153–169.

Hong, G. (2010a). *Ratio of mediator probability weighting for estimating natural direct and indirect effects.* JSM Proceedings, Biometrics Section. Alexandria, VA, American Statistical Association, pp. 2401–2415.

Hong, G. (2010b). Marginal mean weighting through stratification: adjustment for selection bias in multilevel data. *Journal of Educational and Behavioral Statistics, 35*(5), 499–531.

Hong, G. (2012). Marginal mean weighting through stratification: a generalized method for evaluating multi-valued and multiple treatments with non-experimental data. *Psychological Methods, 17*(1), 44–60.

Hong, G. and Nomi, T. (2012). Weighting methods for assessing policy effects mediated by peer change. *Journal of Research on Educational Effectiveness special issue on the statistical approaches to studying mediator effects in education research, 5*(3), 261–289.

Hong, G., Deutsch, J., and Hill, H. (2011). Parametric and non-parametric weighting methods for estimating mediation effects: an application to the national evaluation of welfare-to-work strategies. JSM Proceedings, Social Statistics Section. Alexandria, VA, American Statistical Association, pp. 3215–3229.

Hong, G., Deutsch, J., & Hill, H. D. (in press). Ratio-of-mediator-probability weighting for causal mediation analysis in the presence of treatment-by-mediator interaction. *Journal of Educational and Behavioral Statistics.*

Jones, S.M., Brown, J.L. and Aber, J.L. (2011). The longitudinal impact of a universal school-based social-emotional and literacy intervention: an experiment in translational developmental research. *Child Development, 82*, 533–554.

Kling, J.R., Liebman, J.B. and Katz, L.F. (2007). Experimental analysis of neighborhood effects. *Econometrica, 75*(1), 83–119.

Pearl, J. (2001). *Direct and indirect effects. Proceedings of the Seventeenth Conference on Uncertainty in Artificial Intelligence.* San Francisco, CA, Morgan Kaufmann, pp. 411–420.

Robins, J.M. and Greenland, S. (1992). Identifiability and exchangeability for direct and indirect effects. *Epidemiology, 3*(2), 143–155.

Rubin, D.B. (1986). Comment: which ifs have causal answers. *Journal of the American Statistica Association, 81,* 961–962.

Sobel, M.E. (2006). What do randomized studies of housing mobility demonstrate? Causal inference in the face of interference. *Journal of the American Statistical Association, 101,* 1398—1407.

VanderWeele, T., Hong, G., Jones, S. and Brown, J. (2013). Mediation and spill-over effects in group-randomized trials: a case study of the 4R's educational intervention. *Journal of the American Statistical Association, 108*(502), 469–482.

Index

Printed and bound by CPI Group (UK) Ltd, Croydon, CR0 4YY

27/10/2024

14580219-0004